# Turbulence Modeling
# for CFD

### Second Edition

**by**

**David C. Wilcox**

**Turbulence Modeling for CFD**

First Printing:     July, 1998
Second Printing:    March, 2000
Third Printing:     December, 2002

DCW Industries, Inc.
5354 Palm Drive, La Cañada, California 91011
818/790-3844  (FAX) 818/952-1272
e-mail: dcwilcox@ix.netcom.com
World Wide Web: http://dcwindustries.com

This book was prepared with LaTeX as implemented by Personal TeX, Inc. of Mill Valley, California. It was printed and bound in the United States of America by KNI, Inc., Anaheim, California.

**Library of Congress Cataloging in Publication Data**

Wilcox, David C.
  Turbulence Modeling for CFD / David C. Wilcox—2nd ed.
  Includes bibliography, index and Compact Disk.
  1. Turbulence–Mathematical Models.
  2. Fluid Dynamics–Mathematical Models.
Catalog Card Number 2002096210

ISBN     1-928729-10-X

*Dedicated to my Wife*

BARBARA

*my Children*

KINLEY and BOB

*and my Dad*

# About the Author

**Dr. David C. Wilcox**, was born in Wilmington, Delaware. He did his under-graduate studies from 1963 to 1966 at the Massachusetts Institute of Technology, graduating with a Bachelor of Science degree in Aeronautics and Astronautics. From 1966 to 1967, he was employed by the McDonnell Douglas Aircraft Division in Long Beach, California, and began his professional career under the guidance of the late A. M. O. Smith. His experience with McDonnell Douglas focused on subsonic and transonic flow calculations. From 1967 to 1970, he attended the California Institute of Technology, graduating with a Ph.D. in Aeronautics. In 1970 he joined TRW Systems, Inc. in Redondo Beach, California, where he performed studies of both high- and low-speed fluid-mechanical and heat-transfer problems, such as turbulent hypersonic flow and thermal radiation from a flame. From 1972 to 1973, he was a staff scientist for Applied Theory, Inc., in Los Angeles, California. He participated in many research efforts involving numerical computation and analysis of fluid flows such as separated turbulent flow, transitional flow and hypersonic plume-body interaction.

In 1973, he founded DCW Industries, Inc., a La Cañada, California firm engaged in engineering research, software development and publishing, for which he is currently the President. He has taught several fluid mechanics and applied mathematics courses at the University of Southern California and at the University of California, Los Angeles.

Dr. Wilcox has numerous publications on turbulence modeling, computational fluid dynamics, boundary-layer separation, boundary-layer transition, thermal radiation, and rapidly rotating fluids. His book publications include an under-graduate text entitled *Basic Fluid Mechanics* and a graduate-level text entitled *Perturbation Methods in the Computer Age*. He is an Associate Fellow of the American Institute of Aeronautics and Astronautics (AIAA) and has served as an Associate Editor for the AIAA Journal.

# Contents

# Notation

This section includes the most commonly used notation in this book. In order to avoid departing too much from conventions normally used in literature on turbulence modeling and general fluid mechanics, a few symbols denote more than one quantity.

## English Symbols

| Symbol | Definition |
|---|---|
| $a$ | Speed of sound; strain rate |
| $a_{ijkl}$ | Tensor in rapid-pressure-strain term |
| $A_n, B_n, C_n, D_n$ | Coefficients in tridiagonal matrix equation |
| $A^+$ | Van Driest damping coefficient |
| $A_{ij}$ | Slow pressure-strain tensor |
| $b_{ij}$ | Dimensionless Reynolds-stress anisotropy tensor |
| $c_f$ | Skin friction based on edge velocity, $\tau_w/(\frac{1}{2}\rho U_e^2)$ |
| $c_{f_\infty}$ | Skin friction based on freestream velocity, $\tau_w/(\frac{1}{2}\rho U_\infty^2)$ |
| $c_p, c_v$ | Specific heat at constant pressure, volume |
| $C$ | Additive constant in the law of the wall |
| $C_\kappa$ | Kolmogorov constant |
| $C_p$ | Pressure coefficient |
| $C_s$ | Smagorinsky constant |
| $C_{ij}$ | LES cross-term stress tensor |
| $C_{ijk}$ | Turbulent transport tensor |
| $d$ | Distance from closest surface |
| $D$ | Drag per unit body width; diameter |
| $D_{ij}$ | The tensor $\tau_{im}\partial U_m/\partial x_j + \tau_{jm}\partial U_m/\partial x_i$ |
| $e$ | Specific internal energy |
| $E$ | Total energy |
| $E(\kappa)$ | Energy spectral density |
| $E(\eta)$ | Dimensionless self-similar dissipation rate |

| | |
|---|---|
| $E_h$ | Discretization error |
| $f(x;r)$ | Longitudinal correlation function |
| $f_\beta$ | Vortex-stretching function |
| $f_{\beta^*}$ | Cross-diffusion function |
| $\mathbf{f}, \mathbf{f}_v$ | Turbulence-flux vectors |
| $F(\eta)$ | Dimensionless self-similar streamfunction |
| $F_{Kleb}(y;\delta)$ | Klebanoff intermittency function |
| $\mathbf{F}, \mathbf{F}_v$ | Mean-flow flux vectors |
| $G$ | Amplitude factor in von Neumann stability analysis |
| $G(\mathbf{x} - \boldsymbol{\xi})$ | LES filter |
| $h$ | Specific enthalpy |
| $H$ | Total enthalpy; channel height; shape factor, $\delta^*/\theta$ |
| $\mathcal{H}(x)$ | Heaviside step function |
| $\mathbf{i}, \mathbf{j}, \mathbf{k}$ | Unit vectors in $x, y, z$ directions |
| $I$ | Unit (identity) matrix |
| $II, III$ | Stress-tensor invariants |
| $j$ | Two-dimensional ($j = 0$), axisymmetric ($j = 1$) index |
| $J$ | Specific momentum flux (flux per unit mass) |
| $k$ | Kinetic energy of turbulent fluctuations per unit mass |
| $k_g$ | Geometric progression ratio |
| $k_s$ | Surface roughness height |
| $K$ | Distortion parameter |
| $K(\eta)$ | Dimensionless self-similar turbulence kinetic energy |
| $Kn$ | Knudsen number |
| $\ell$ | Turbulence length scale; characteristic eddy size |
| $\ell_{mfp}$ | Mean free path |
| $\ell_{mix}$ | Mixing length |
| $\ell_\mu$ | von Kármán length scale |
| $L$ | Characteristic length scale |
| $L_a$ | Reattachment length |
| $L_{ij}$ | Leonard-stress tensor |
| $m$ | Molecular mass; round/radial jet index |
| $M$ | Mach number |
| $M_{ijkl}$ | Tensor in rapid-pressure-strain term |
| $M_c$ | Convective Mach number |
| $M_t$ | Turbulence Mach number, $\sqrt{2k}/a$ |
| $M_\tau$ | Turbulence Mach number, $u_\tau/a_w$ |
| $n$ | Normal distance; number density |
| $N(\eta)$ | Dimensionless self-similar eddy viscosity |
| $N_{CFL}$ | CFL number |
| $\mathcal{N}(u_i)$ | Navier-Stokes operator |
| $p$ | Instantaneous static pressure |

| | |
|---|---|
| $p_{ij}$ | Instantaneous momentum-flux tensor |
| $P$ | Mean static pressure; turbulence-energy production, $\frac{1}{2}P_{ii}$ |
| $P_{ij}$ | Production tensor, $\tau_{im}\partial U_j/\partial x_m + \tau_{jm}\partial U_i/\partial x_m$ |
| $P_k, P_\omega, P_\epsilon$ | Net production per unit dissipation of $k$, $\omega$, $\epsilon$ |
| $Pr_L, Pr_T$ | Laminar, turbulent Prandtl number |
| $P^+$ | Dimensionless pressure-gradient parameter, $(\nu_w/\rho u_\tau^3)dP/dx$ |
| $q_j$ | Heat-flux vector |
| $q_w$ | Surface heat flux |
| $q_{L_j}, q_{T_j}$ | Laminar, turbulent mean heat-flux vector |
| $Q_{ij}$ | LES stress tensor, $C_{ij} + R_{ij}$ |
| $\mathbf{q}, \mathbf{Q}$ | Dependent-variable vectors |
| $r, \theta, x$ | Cylindrical polar coordinates |
| $R$ | Pipe radius; channel half height; perfect-gas constant |
| $R_{ij}$ | SGS Reynolds stress tensor |
| $R_{ij}(\mathbf{x},t;\mathbf{r})$ | Two-point velocity correlation tensor |
| $\mathcal{R}$ | Radius of curvature |
| $\mathcal{R}_E(\mathbf{x};t')$ | Eulerian time-correlation coefficient |
| $\mathcal{R}_{ij}(\mathbf{x}, t; t')$ | Autocorrelation tensor |
| $R^+$ | Sublayer scaled radius or half height, $u_\tau R/\nu$ |
| $Re_L$ | Reynolds number based on length $L$ |
| $Re_T$ | Turbulence Reynolds number, $k^{1/2}\ell/\nu$ |
| $Re_\tau$ | Sublayer scaled radius or half height, $R^+$ |
| $Ri_T$ | Turbulence Richardson number |
| $R_y$ | Near-wall turbulence Reynolds number, $k^{1/2}y/\nu$ |
| $s$ | Arc length |
| $s_{ij}$ | Instantaneous strain-rate tensor |
| $\mathbf{s}, \mathbf{S}$ | Source-term vectors |
| $S$ | Source term; shear rate |
| $S_{ij}$ | Mean strain-rate tensor |
| $\overset{\circ}{S}_{ij}$ | Oldroyd derivative of $S_{ij}$ |
| $S_e, S_k, S_u, S_w$ | Source terms in a similarity solution |
| $S_B$ | Dimensionless surface mass injection function |
| $S_R$ | Dimensionless surface roughness function |
| $t$ | Time |
| $t_{ij}$ | Instantaneous viscous stress tensor |
| $T$ | Temperature; characteristic time scale |
| $T'$ | Freestream turbulence intensity |
| $u, v, w$ | Instantaneous velocity components in $x$, $y$, $z$ directions |
| $u_i$ | Instantaneous velocity in tensor notation |
| $\mathbf{u}$ | Instantaneous velocity in vector notation |
| $u', v', w'$ | Fluctuating velocity components in $x$, $y$, $z$ directions |
| $u'_i$ | Fluctuating velocity in tensor notation |

| | |
|---|---|
| $\mathbf{u}'$ | Fluctuating velocity in vector notation |
| $\hat{u}, \hat{v}, \hat{w}$ | Relative turbulence intensity, $\sqrt{u'^2}/U_e$, $\sqrt{v'^2}/U_e$, $\sqrt{w'^2}/U_e$ |
| $\tilde{u}, \tilde{v}, \tilde{w}$ | Favre-averaged velocity components in $x$, $y$, $z$ directions |
| $\tilde{u}_i$ | Favre-averaged velocity in tensor notation |
| $\tilde{\mathbf{u}}$ | Favre-averaged velocity in vector notation |
| $u'', v'', w''$ | Favre fluctuating velocity components in $x$, $y$, $z$ directions |
| $u''_i$ | Favre fluctuating velocity in tensor notation |
| $\mathbf{u}''$ | Favre fluctuating velocity; fluctuating molecular velocity |
| $u_{rms}, v_{rms}$ | RMS fluctuating velocity components in $x$, $y$ directions |
| $\overline{u'_i u'_j}$ | Temporal average of fluctuating velocities |
| $u_\tau$ | Friction velocity, $\sqrt{\tau_w/\rho_w}$ |
| $\hat{\mathbf{u}}$ | Velocity perturbation vector |
| $u^*$ | Van Driest scaled velocity |
| $U, V, W$ | Mean velocity components in $x$, $y$, $z$ directions |
| $\mathbf{U}$ | Mean velocity in vector notation |
| $U_e$ | Shear-layer edge velocity |
| $U_i$ | Mean velocity in tensor notation |
| $U_\infty$ | Freestream velocity |
| $U^+$ | Dimensionless, sublayer-scaled, velocity, $U/u_\tau$ |
| $U_m$ | Maximum or centerline velocity |
| $\mathcal{U}(\eta)$ | Dimensionless self-similar streamwise velocity |
| $v_{mix}$ | Mixing velocity |
| $v_{th}$ | Thermal velocity |
| $v_w$ | Surface injection velocity |
| $\mathcal{V}(\eta)$ | Dimensionless self-similar normal velocity |
| $W(\eta)$ | Dimensionless self-similar specific dissipation rate |
| $x, y, z$ | Rectangular Cartesian coordinates |
| $x_i$ | Position vector in tensor notation |
| $\mathbf{x}$ | Position vector in vector notation |
| $y^+$ | Dimensionless, sublayer-scaled, distance, $u_\tau y/\nu$ |
| $y_2^+$ | $y^+$ at first grid point above surface |
| $y_m$ | Inner/outer layer matching point |

# Greek Symbols

| Symbol | Definition |
|---|---|
| $\alpha_T, \sigma_T, \omega_T$ | Defect-layer similarity parameters |
| $\beta_r$ | Bradshaw's constant |
| $\beta_T$ | Equilibrium parameter, $(\delta^*/\tau_w)dP/dx$ |
| $\gamma$ | Specific-heat ratio, $c_p/c_v$ |
| $\delta$ | Boundary-layer or shear-layer thickness |

| | |
|---|---|
| $\delta_{vi}$ | Viscous-interface layer thickness |
| $\delta'$ | Free shear layer spreading rate |
| $\delta^*$ | Displacement thickness, $\int_0^\delta \left(1 - \frac{\rho}{\rho_e}\frac{U}{U_e}\right) dy$ |
| $\delta_v^*$ | Velocity thickness, $\int_0^\delta \left(1 - \frac{U}{U_e}\right) dy$ |
| $\delta_x$ | Finite-difference matrix operator |
| $\delta_{ij}$ | Kronecker delta |
| $\Delta$ | LES filter width |
| $\Delta(x)$ | Clauser thickness, $U_e \delta^*/u_\tau$ |
| $\Delta\mathbf{q}, \Delta\mathbf{Q}$ | Incremental change in $\mathbf{q}$, $\mathbf{Q}$ |
| $\Delta x, \Delta y$ | Incremental change in $x$, $y$ |
| $\Delta t$ | Timestep |
| $\epsilon$ | Dissipation per unit mass |
| $\epsilon_d$ | Dilatation dissipation |
| $\epsilon_s$ | Solenoidal dissipation |
| $\epsilon_{ij}$ | Dissipation tensor |
| $\epsilon_{ijk}$ | Permutation tensor |
| $\zeta$ | Second viscosity coefficient |
| $\eta$ | Kolmogorov length scale; similarity variable |
| $\theta$ | Momentum thickness, $\int_0^\delta \frac{\rho}{\rho_e}\frac{U}{U_e}\left(1 - \frac{U}{U_e}\right) dy$ |
| $\kappa$ | Kármán constant; wavenumber; thermal conductivity |
| $\kappa_v$ | Effective Kármán constant for flows with mass injection |
| $\kappa_\epsilon, \kappa_\omega$ | Effective Kármán constants for compressible flows |
| $\lambda$ | Taylor microscale; RNG $k$-$\epsilon$ model parameter |
| $\lambda_{max}$ | Largest eigenvalue |
| $\mu$ | Molecular viscosity |
| $\mu_T$ | Eddy viscosity |
| $\mu_{T_i}$ | Inner-layer eddy viscosity |
| $\mu_{T_o}$ | Outer-layer eddy viscosity |
| $\nu$ | Kinematic molecular viscosity, $\mu/\rho$ |
| $\nu_T$ | Kinematic eddy viscosity, $\mu_T/\rho$ |
| $\nu_{T_i}$ | Inner-layer kinematic eddy viscosity |
| $\nu_{T_o}$ | Outer-layer kinematic eddy viscosity |
| $\xi$ | Dimensionless streamwise distance |
| $\Pi$ | Coles' wake-strength parameter |
| $\Pi_{ij}$ | Pressure-strain correlation tensor |
| $\Pi_{ij}^{(w)}$ | Wall-reflection (pressure-echo) term |
| $\rho$ | Mass density |
| $\sigma(x)$ | Nonequilibrium parameter |
| $\tau$ | Kolmogorov time scale; turbulence dissipation time |
| $\tau_E$ | Micro-time scale |

| | |
|---|---|
| $\tau_{ij}$ | Specific Reynolds stress tensor, $-\overline{u_i' u_j'}$ |
| $\tau_{turnover}$ | Eddy turnover time |
| $\tau_{xy}$ | Specific Reynolds shear stress, $-\overline{u'v'}$ |
| $\tau_{xx}, \tau_{yy}, \tau_{zz}$ | Specific normal Reynolds stresses, $-\overline{u'^2}, -\overline{v'^2}, -\overline{w'^2}$ |
| $\tau_w$ | Surface shear stress |
| $\upsilon$ | Kolmogorov velocity scale |
| $\chi_k$ | Dimensionless cross-diffusion parameter |
| $\chi_p, \chi_\omega$ | Dimensionless vortex-stretching parameter |
| $\psi$ | Streamfunction |
| $\psi_k, \psi_\epsilon, \psi_\omega$ | Parabolic marching scheme coefficients |
| $\omega$ | Specific dissipation rate; vorticity vector magnitude |
| $\Omega_{ij}$ | Mean-rotation tensor |

# Other

| Symbol | Definition |
|---|---|
| $\partial \mathbf{f}/\partial \mathbf{q}$ | Turbulence flux-Jacobian matrix |
| $\partial \mathbf{F}/\partial \mathbf{Q}$ | Mean-flow flux-Jacobian matrix |
| $\partial \mathbf{s}/\partial \mathbf{q}$ | Source-Jacobian matrix |

# Subscripts

| Symbol | Definition |
|---|---|
| $aw$ | Adiabatic wall |
| $DNS$ | Direct Numerical Simulation |
| $e$ | Boundary-layer-edge value |
| $eq$ | Equilibrium value |
| $LES$ | Large Eddy Simulation |
| $o$ | Centerline value |
| $sep$ | Separation |
| $t$ | Transition |
| $v$ | Viscous |
| $w$ | Wall (surface) value |
| $\infty$ | Freestream value |

# Superscripts

| Symbol | Definition |
|---|---|
| $+$ | Sublayer-scaled value |

# Preface

The second edition of this book has been motivated by the overwhelmingly favorable response to the first edition. *Turbulence Modeling for CFD* has been adopted for course use in more than 30 universities and the author has presented a short course based on the book an average of three times a year since it was first published in 1993. Its popularity has exceeded all expectations, and this author is especially thankful to the turbulence-research and CFD communities. To express those thanks in a tangible manner, I have opted to create a second edition rather than a third printing of the first edition. While new developments in the field have come far less frequently during the five years since the book first appeared than in some previous periods, a few important advances have been made. Every attempt has been made to integrate new developments into the second edition.

All chapters and appendices have undergone improvement and expansion. Most notably, the introduction includes an expanded discussion of the physics of turbulence, making the book a more self-contained study aid. Chapter 2 now includes a discussion of two-point statistics as they pertain to engineering turbulence-model development. Chapter 3 subjects algebraic and 1/2-equation models to an increased number of baseline applications, which are repeated for other models in subsequent chapters. Chapter 4 includes an expanded discussion of recent one-equation models, and presents a new version of the $k$-$\omega$ model that yields close agreement with measurements for both boundary layers with pressure gradient (for which the $k$-$\epsilon$ model is very inaccurate) and for classical free shear flows (for which the $k$-$\epsilon$ model is marginally accurate). The improved $k$-$\omega$ model should provide a source for fresh research ideas. New insight into compressible turbulence gleaned from basic analyses is included to bring Chapter 5 up to date in this rapidly advancing field. The extensive efforts of Speziale and various co-authors in devising nonlinear stress/strain-rate relations are included in Chapter 6. The chapter also introduces a new stress-transport (second-order closure) model that uses the $\omega$ equation. The discussion of DNS and LES in Chapter 8 has been expanded. Finally, to enhance the book's utility in the classroom, the number of homework problems has increased by 50%.

The second edition of *Turbulence Modeling for CFD* includes a few changes in notation that have been made to bring the text into closer harmony with standard conventions. Two of the changes pertain to the classical law of the wall and law of the wake. The additive constant in the law of the wall has been changed from $B$ to $C$. Also, Coles' wake-strength parameter is now denoted by $\Pi$ as opposed to $\tilde{\Pi}$ in the first edition — the tilde has been dropped. By far the most significant change pertains to the Reynolds-stress tensor. The second edition defines $\tau_{ij} \equiv -\overline{u'_i u'_j}$, so that $\tau_{ij}$ is the *specific* Reynolds-stress tensor with dimensions length$^2$/time$^2$. In the first edition, $\tau_{ij}$ was the true Reynolds-stress tensor, $-\overline{\rho u'_i u'_j}$.

This book has been developed from the author's lecture notes used in presenting a graduate course on turbulence modeling at the University of Southern California. While several computational fluid dynamics (CFD) texts include some information about turbulence modeling, very few texts dealing exclusively with turbulence modeling have been written. As a consequence, turbulence modeling is regarded by many CFD researchers as "black magic," lacking in rigor and physical foundation. This book has been written to show that turbulence modeling can be done in a systematic and physically sound manner. This is not to say all turbulence modeling has been done in such a manner, for indeed many ill-conceived and ill-fated turbulence models have appeared in engineering journals. However, with judicious use of relatively simple mathematical tools, systematic construction of a well-founded turbulence model is not only possible but can be an exciting and challenging research project.

Thus, the primary goal of this book is to provide a systematic approach to developing a set of constitutive equations suitable for computation of turbulent flows. The engineer who feels no existing turbulence model is suitable for his or her needs and wishes to modify an existing model or to devise a new model will benefit from this feature of the text. A methodology is presented in Chapters 3 and 4 for devising and testing such equations. The methodology is illustrated in great detail for two-equation turbulence models. However, it is by no means limited to such models and is used again in Chapter 6 for a full stress-transport model, but with less detail.

A secondary goal of this book is to provide a rational way for deciding how complex a model is required for a given problem. The engineer who wishes to select an existing model that is sufficient for his or her needs will benefit most from this feature of the text. Chapter 3 begins with the simplest turbulence models and subsequent chapters chart a course leading to some of the most complex models that have been applied to a nontrivial turbulent-flow problem. Two things are done at each level of complexity. First, the range of applicability of the model is estimated. Second, many of the models are applied to the same flows, including comparisons with measurements, to illustrate how accuracy changes with complexity.

The methodology makes extensive use of tensor analysis, similarity solutions, singular-perturbation methods, and numerical procedures. The text assumes the user has limited prior knowledge of these mathematical concepts and provides what is needed in the main text or in the Appendices. For example, Appendix A introduces Cartesian tensor analysis to facilitate manipulation of the Navier-Stokes equation, which is done extensively in Chapter 2. Chapter 3 shows, in detail, the way a similarity solution is generated. Similarity solutions are then obtained for the turbulent mixing layer, jet and far wake. Appendix B presents elements of singular-perturbation theory. Chapters 4, 5 and 6 use these tools to dissect model-predicted features of the turbulent boundary layer.

No book on turbulence-model equations is complete without a discussion of numerical-solution methods. Anyone who has ever tried to obtain a numerical solution to a set of turbulence-transport equations can attest to this. Often, standard numerical procedures just won't work and alternative methods must be found to obtain accurate converged solutions. Chapter 7 focuses on numerical methods and elucidates some of the commonly encountered problems such as stiffness, sharp turbulent-nonturbulent interfaces, and difficulties attending turbulence related time scales.

The concluding chapter presents a brief overview of new horizons, including direct numerical simulation (DNS), large-eddy simulation (LES) and the interesting mathematical theory of chaos.

Because turbulence modeling is a key ingredient in CFD work, the text would be incomplete without companion software implementing numerical solutions to standard turbulence-model equations. Appendices C and D describe several computer programs that are included on the diskette accompanying the book. The programs all have a similar structure and can be easily modified to include new turbulence models.

The material presented in this book is appropriate for a one-semester, first or second year graduate course, or as a reference text for a CFD course. Successful study of this material requires an understanding of viscous-flow and boundary-layer theory. Some degree of proficiency in solving partial differential equations is also needed. A working knowledge of computer programming, preferably in FORTRAN, the most common programming language in engineering, will help the reader gain maximum benefit from the companion software described in the Appendices.

A friend of many years, Dr. P. Bradshaw, reviewed the entire manuscript of both the first and second editions as I wrote them, and taught me a lot through numerous discussions, comments and suggestions that greatly improved the final draft. Dr. Bradshaw also assisted in preparation of key material in Chapters 5 and 8, adding physical insight and state-of-the-art information. Another long-time friend, Dr. C. G. Speziale, reviewed the manuscripts of both editions and offered a great deal of physical insight in the process.

I extend my thanks to Dr. L. G. Redekopp of USC for encouraging and supporting development of the course for which this book was originally intended. Dr. D. D. Knight helped me understand why I had to write this book, reviewed the entire first-edition manuscript and assisted in its preparation. One of my best graduate students, Patrick Yee, was very thorough in reviewing the second edition, including the solutions manual. My favorite Caltech mathematics teacher, Dr. D. S. Cohen, made sure I omitted the dot over every $\iota$ and crossed every $z$ in Appendix B. Drs. F. R. Menter, C. C. Horstman and P. R. Spalart were kind enough to provide results of several of their computations in digital form. Thanks are also due for the support and help of several other friends and colleagues, most notably Drs. P. J. Roache, J. A. Domaradzki, R. M. C. So. Also, Raymond Jones provided a critical test of the new $k$-$\omega$ model by applying it to the backward-facing step and Steven Ciesla did a thorough review of the software appendices, both on very short notice.

I thank the nine students who were the first to take the course that this book was written for. Their patience was especially noteworthy, particularly in regard to typographical errors in the homework problems! That outstanding group of young engineers is D. Foley, R. T. Holbrook, N. Kale, T.-S. Leu, H. Lin, T. Magee, S. Tadepalli, P. Taniguchi and D. Hammond.

Finally, I owe a lifelong debt to my loving wife Barbara for tolerating the hectic pace first in college and then in the business world. Without her, this book would not have been possible.

*David C. Wilcox*

---

**NOTE:** We have taken great pains to assure the accuracy of this manuscript. However, if you find errors, please report them to DCW Industries' Home Page on the Worldwide Web at **http://dcwindustries.com**. As long as we maintain a WWW page, we will provide an updated list of known typographical errors.

# Chapter 1

# Introduction

## 1.1 Definition of an Ideal Turbulence Model

Turbulence modeling is one of three key elements in Computational Fluid Dynamics (CFD). Very precise mathematical theories have evolved for the other two key elements, viz., grid generation and algorithm development. By its nature — in creating a mathematical model that approximates the physical behavior of turbulent flows — far less precision has been achieved in turbulence modeling. This is not really a surprising event since our objective has been to approximate an extremely complicated phenomenon.

The field is, to some extent, a throwback to the days of Prandtl, Taylor, von Kármán and all the many other clever engineers who spent a good portion of their time devising engineering approximations and models describing complicated physical flows. Simplicity combined with physical insight seems to have been a common denominator of the work of these great men. Using their work as a gauge, **an ideal model should introduce the minimum amount of complexity while capturing the essence of the relevant physics**. This description of an ideal model serves as the keystone of this text.

## 1.2 How Complex Must a Turbulence Model Be?

Aside from any physical considerations, turbulence is inherently three dimensional and time dependent. Thus, an enormous amount of information is required to completely describe a turbulent flow. Fortunately, we usually require something less than a complete time history over all spatial coordinates for every flow property. Thus, for a given turbulent-flow application, we must pose the following question. Given a set of initial and/or boundary conditions, how do we

1

predict the relevant properties of the flow? What properties of a given flow are relevant is generally dictated by the application. For the simplest applications, we may require only the skin-friction and heat-transfer coefficients. More esoteric applications may require detailed knowledge of energy spectra, turbulence fluctuation magnitudes and scales.

Certainly, we should expect the complexity of the mathematics required for a given application to increase as the amount of required flowfield detail increases. On the one hand, if all we require is skin friction for an attached flow, a simple mixing-length model (Chapter 3) may suffice. Such models are well developed and can be implemented with very little specialized knowledge. On the other hand, if we desire a complete time history of every aspect of a turbulent flow, only a solution to the complete Navier-Stokes equation will suffice. Such a solution requires an extremely accurate numerical solver and may require use of subtle transform techniques, not to mention vast computer resources. Most engineering problems fall somewhere between these two extremes.

**Thus, once the question of how much detail we need is answered, the level of complexity of the model follows, qualitatively speaking.**[1] In the spirit of Prandtl, Taylor and von Kármán, the conscientious engineer will strive to use as conceptually simple an approach as possible to achieve his ends. Overkill is often accompanied by unexpected difficulties that, in CFD applications, almost always manifest themselves as numerical difficulties!

## 1.3  Comments on the Physics of Turbulence

Before plunging into the mathematics of turbulence, it is worthwhile to first discuss physical aspects of the phenomenon. The following discussion is not intended as a complete description of this complex topic. Rather, we focus upon a few features of interest in engineering applications, and in construction of a mathematical model. For a more-complete introduction, refer to basic texts on the physics of turbulence such as those by Hinze (1975), Tennekes and Lumley (1983), Landahl and Mollo-Christensen (1992) or Libby (1996).

### 1.3.1  Importance of Turbulence in Practical Situations

For "small enough" scales and "low enough" velocities, in the sense that the Reynolds number is not too large, the equations of motion for a viscous fluid have well-behaved, steady solutions. Such flows are controlled by viscous diffusion of vorticity and momentum. The motion is termed laminar and can be observed experimentally and in nature.

---

[1]This is not a foolproof criterion, however. For example, a complicated model may be required to predict even the simplest properties of a very complex flow.

Figure 1.1: *Examples of turbulent motion. Upper left: a cumulus cloud; Upper right: flow in the wake of a cylinder; Bottom: flow in the wake of a bullet [Bottom photograph courtesy of Corrsin and Kistler (1954)].*

At larger Reynolds numbers, the viscous stresses are overcome by the fluid's inertia, and the laminar motion becomes unstable. Rapid velocity and pressure fluctuations appear and the motion becomes inherently three dimensional and unsteady. When this occurs, we describe the motion as being turbulent. In the cases of fully-developed Couette flow and pipe flow, for example, laminar flow is assured only if the Reynolds number based on maximum velocity and channel height or pipe radius is less than 1500 and 2300, respectively.

Virtually all flows of practical engineering interest are turbulent. Flow past vehicles such as rockets, airplanes, ships and automobiles, for example, is always turbulent. Turbulence dominates in geophysical applications such as river currents, the planetary boundary layer and the motion of clouds (Figure 1.1). Turbulence even plays a role at the breakfast table, greatly enhancing the rate at which sugar and cream mix in a cup of coffee!

Turbulent flows always occur when the Reynolds number is large. For slightly viscous fluids such as water and air, "large" Reynolds number corresponds to anything stronger than a small breeze or a puff of wind. Thus, to analyze fluid motion for general applications, we must deal with turbulence. Although vigorous research has been conducted to help discover the mysteries of turbulence, it has been called the major unsolved problem of classical physics! In the following subsections, we will explore some of the most important aspects of turbulence.

## 1.3.2   General Properties of Turbulence

- **Basic Definition.** In 1937, von Kármán defined turbulence in a presentation at the Twenty-Fifth Wilbur Wright Memorial Lecture entitled "Turbulence." He quoted G. I. Taylor as follows [see von Kármán (1937)]:

    *"Turbulence is an irregular motion which in general makes its appearance in fluids, gaseous or liquid, when they flow past solid surfaces or even when neighboring streams of the same fluid flow past or over one another."*

As the understanding of turbulence has progressed, researchers have found the term "irregular motion" to be too imprecise. Simply stated, an irregular motion is one that is typically aperiodic and that cannot be described as a straightforward function of time and space coordinates. An irregular motion might also depend strongly and sensitively upon initial conditions. The problem with the Taylor-von Kármán definition of turbulence lies in the fact that there are nonturbulent flows that can be described as irregular.

Turbulent motion is indeed irregular in the sense that it can be described by the laws of probability. Even though instantaneous properties in a turbulent flow are extremely sensitive to initial conditions, statistical averages of the instantaneous properties are not. To provide a sharper definition of turbulence, Hinze (1975) offers the following revised definition:

    *"Turbulent fluid motion is an irregular condition of flow in which the various quantities show a random variation with time and space coordinates, so that statistically distinct average values can be discerned."*

To complete the definition of turbulence, Bradshaw [cf. Cebeci and Smith (1974)] adds the statement that *turbulence has a wide range of scales*. Time and length scales of turbulence are represented by frequencies and wavelengths that are revealed by a Fourier analysis of a turbulent-flow time history.

The irregular nature of turbulence stands in contrast to laminar motion, so called historically because the fluid was imagined to flow in smooth laminae, or layers. In describing turbulence, many researchers refer to **eddying motion**, which is a local swirling motion where the vorticity can often be very intense. **Turbulent eddies** of a wide range of sizes appear and give rise to vigorous mixing and effective turbulent stresses (a consequence of the "mixing" of momentum) that can be enormous compared to laminar values.

- **Instability and Nonlinearity.** Analysis of solutions to the Navier-Stokes equation, or more typically to its boundary-layer form, shows that turbulence develops as an instability of laminar flow. To analyze the stability of laminar flows, classical methods begin by linearizing the equations of motion. Although linear theories achieve some degree of success in predicting the onset of instabilities that ultimately lead to turbulence, the inherent nonlinearity of the Navier-Stokes equation precludes a complete analytical description of the actual transition process, let alone the fully-turbulent state. For a real (i.e., viscous) fluid, mathematically speaking, the instabilities result mainly[2] from interaction between the Navier-Stokes equation's nonlinear inertial terms and viscous terms. The interaction is very complex because it is rotational, fully three dimensional and time dependent.

  As an overview, the nonlinearity of the Navier-Stokes equation leads to interactions between fluctuations of differing wavelengths and directions. As discussed below, the wavelengths of the motion usually extend all the way from a maximum comparable to the width of the flow to a minimum fixed by viscous dissipation of energy. The main physical process that spreads the motion over a wide range of wavelengths is vortex stretching. The turbulence gains energy if the vortex elements are primarily oriented in a direction in which the mean velocity gradients can stretch them. Most importantly, wavelengths that are not too small compared to the mean-flow width interact most strongly with the mean flow. **Consequently, the larger-scale turbulent motion carries most of the energy and is mainly responsible for the enhanced diffusivity and attending stresses.** In turn, the larger eddies randomly stretch the vortex elements that comprise the smaller eddies, cascading energy to them. Energy is dissipated by viscosity in the shortest wavelengths, although the *rate* of dissipation of energy is set by the long-wavelength motion at the start of the cascade. The shortest wavelengths simply adjust accordingly.

- **Statistical Aspects.** The time-dependent nature of turbulence also contributes to its intractability. The additional complexity goes beyond the

---

[2]Inviscid instabilities, such as the Kelvin-Helmholtz instability, also play a role.

introduction of an additional dimension. Turbulence is characterized by random fluctuations thus mandating the use of statistical methods to analyze it. On the one hand, this aspect is not really a problem from the engineer's viewpoint. Even if we had a complete time history of a turbulent flow, we would usually integrate the flow properties of interest over time to extract **time averages**, or **mean values**. On the other hand, as we will see in Chapter 2, time-averaging operations lead to terms in the equations of motion that cannot be determined a priori.

- **Turbulence is a Continuum Phenomenon.** In principle, we know that the time-dependent, three-dimensional continuity and Navier-Stokes equations contain all of the physics of a given turbulent flow. That this is true follows from the fact that turbulence is a continuum phenomenon. As noted by Tennekes and Lumley (1983),

    *"Even the smallest scales occurring in a turbulent flow are ordinarily far larger than any molecular length scale."*

    Nevertheless, the smallest scales of turbulence are still extremely small (we will see just how small in the next subsection). They are generally many orders of magnitude smaller than the largest scales of turbulence, the latter often being of the same order of magnitude as the dimension of the object about which the fluid is flowing. Furthermore, the ratio of smallest to largest scales decreases rapidly as the Reynolds number increases. To make an accurate numerical simulation (i.e., a fully time-dependent three-dimensional solution) of a turbulent flow, all physically relevant scales must be resolved.

    While more and more progress is being made with such simulations, computers of the 1990's have insufficient memory and speed to solve any turbulent-flow problem of practical interest. To underscore the magnitude of the problem, Speziale (1985) notes that a numerical simulation of turbulent pipe flow at a Reynolds number of 500,000 would require a computer 10 million times faster than a Cray Y/MP. This is true because, as discussed in Chapter 8, the number of numerical operations in such a computation is proportional to $Re^{9/4}$, where $Re$ is a characteristic Reynolds number. However, the results are very useful in developing and testing approximate methods.

- **Vortex Stretching.** The strongly rotational nature of turbulence goes hand-in-hand with its three dimensionality. The vorticity in a turbulent flow is itself three dimensional so that vortex lines in the flow are nonparallel. The resulting vigorous stretching of vortex lines maintains the ever-present fluctuating vorticity in a turbulent flow. Vortex stretching is

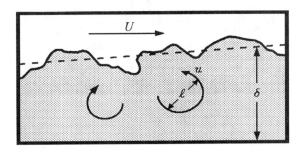

Figure 1.2: *Schematic of large eddies in a turbulent boundary layer. The flow above the boundary layer has a steady velocity U ; the eddies move at randomly-fluctuating velocities of the order of a tenth of U. The largest eddy size (ℓ) is comparable to the boundary-layer thickness (δ). The interface and the flow above the boundary is quite sharp [Corrsin and Kistler (1954)].*

absent in two-dimensional flows so that turbulence must be three dimensional. This inherent three dimensionality means there are no satisfactory two-dimensional approximations for determining fine details of turbulent flows. This is true even when the average motion is two dimensional. The induced velocity field attending these skewed vortex lines further increases three dimensionality and, at all but very low Reynolds numbers, the vorticity is drawn out into a tangle of thin tubes or sheets. Therefore, **most of the vorticity in a turbulent flow resides in the smallest eddies**.

- **Turbulence Scales and the Cascade.** Turbulence consists of a continuous spectrum of scales ranging from largest to smallest, as opposed to a discrete set of scales. In order to visualize a turbulent flow with a spectrum of scales we often cast the discussion in terms of eddies. As noted above, a turbulent eddy can be thought of as a local swirling motion whose characteristic dimension is the local turbulence scale (Figure 1.2). Alternatively, from a more mathematical point of view, we sometimes speak in terms of wavelengths. When we think in terms of wavelength, we imagine we have done a Fourier analysis of the fluctuating flow properties.

We observe that eddies overlap in space, large ones carrying smaller ones. Turbulence features a **cascade process** whereby, as the turbulence decays, its kinetic energy transfers from larger eddies to smaller eddies. Ultimately, the smallest eddies dissipate into heat through the action of molecular viscosity. Thus, we observe that, like any viscous flow, **turbulent flows are always dissipative**.

Figure 1.3: *Laser-induced fluorescence image of an incompressible turbulent boundary layer. Flow is from left to right and has been visualized with disodium fluorescein dye in water. Reynolds number based on momentum thickness is 700. [From C. Delo — Used with permission.]*

- **Large Eddies and Turbulent Mixing.** An especially striking feature of a turbulent flow is the way large eddies migrate across the flow, carrying smaller-scale disturbances with them. The arrival of these large eddies near the interface between the turbulent region and nonturbulent fluid distorts the interface into a highly convoluted shape (Figures 1.2 and 1.3). In addition to migrating across the flow, they have a lifetime so long that they persist for distances as much as 30 times the width of the flow [Bradshaw (1972)]. **Hence, the state of a turbulent flow at a given position depends upon upstream history and cannot be uniquely specified in terms of the local strain-rate tensor as in laminar flow.**

- **Enhanced Diffusivity.** Perhaps the most important feature of turbulence from an engineering point of view is its enhanced diffusivity. Turbulent diffusion greatly enhances the transfer of mass, momentum and energy. Apparent stresses in turbulent flows are often several orders of magnitude larger than in corresponding laminar flows.

In summary, turbulence is dominated by the large, energy-bearing, eddies. The large eddies are primarily responsible for the enhanced diffusivity and stresses observed in turbulent flows. Because large eddies persist for long distances, the diffusivity and stresses are dependent upon flow history, and cannot necessarily be expressed as functions of local flow properties. Also, while the small eddies ultimately dissipate turbulence energy through viscous action, the rate at which they dissipate is controlled by the rate at which they receive energy from the largest eddies. These observations must play an important role in the formulation of any rational turbulence model. As we progress through the following chapters, we will introduce more specific details of turbulence properties for common flows on an as-needed basis.

### 1.3.3 The Smallest Scales of Turbulence

As stated in the preceding subsection, we regard turbulence as a continuum phenomenon because the smallest scales of turbulence are much larger than any molecular length scale. We can estimate the magnitude of the smallest scales by appealing to dimensional analysis, and thereby confirm this claim. Of course, to establish the relevant dimensional quantities, we must first consider the physics of turbulence at very small length scales.

We begin by noting that the cascade process present in all turbulent flows involves a transfer of **turbulence kinetic energy** (per unit mass), $k$, from larger eddies to smaller eddies. Dissipation of kinetic energy to heat through the action of molecular viscosity occurs at the scale of the smallest eddies. Because small-scale motion tends to occur on a short time scale, we can reasonably assume that such motion is independent of the relatively slow dynamics of the large eddies and of the mean flow. Hence, the smaller eddies should be in a state where the rate of receiving energy from the larger eddies is very nearly equal to the rate at which the smallest eddies dissipate the energy to heat. This is one of the premises of Kolmogorov's (1941) **universal equilibrium theory**. Hence, the motion at the smallest scales should depend only upon: (a) the rate at which the larger eddies supply energy, $\epsilon = -dk/dt$, and (b) the kinematic viscosity, $\nu$.

Having established $\epsilon$ (whose dimensions are length$^2$/time$^3$) and $\nu$ (whose dimensions are length$^2$/time) as the appropriate dimensional quantities, it is a simple matter to form the following length ($\eta$), time ($\tau$) and velocity ($v$) scales.

$$\eta \equiv \left(\nu^3/\epsilon\right)^{1/4}, \qquad \tau \equiv \left(\nu/\epsilon\right)^{1/2}, \qquad v \equiv \left(\nu\epsilon\right)^{1/4} \qquad (1.1)$$

These are the **Kolmogorov scales** of length, time and velocity.

To appreciate how small the Kolmogorov length scale is, for example, estimates based on properties of typical turbulent boundary layers indicate the following. For an automobile moving at 65 mph, the Kolmogorov length scale near the driver's window is about $\eta \approx 1.8 \cdot 10^{-4}$ inch. Also, on a day when the temperature is 68° F, the mean free path of air, i.e., the average distance traveled by a molecule between collisions, is $\ell_{mfp} \approx 2.5 \cdot 10^{-6}$ inch. Therefore,

$$\frac{\eta}{\ell_{mfp}} \approx 72 \qquad (1.2)$$

so that the Kolmogorov length is indeed much larger than the mean free path of air, which, in turn, is typically 10 times the molecular diameter.

### 1.3.4 Spectral Representation and the Kolmogorov -5/3 Law

To provide further insight into the description of turbulence presented above, it is worthwhile to cast the discussion in a bit more quantitative form. Since turbulence contains a continuous spectrum of scales, it is often convenient to do our analysis in terms of the **spectral distribution** of energy. In general, a spectral representation is a Fourier decomposition into wavenumbers, $\kappa$, or, equivalently, wavelengths, $\lambda = 2\pi/\kappa$. While this text, by design, makes only modest use of Fourier-transform methods, there are a few interesting observations we can make now without considering all of the complexities involved in the mathematics of Fourier transforms. In the present context, we think of the reciprocal of $\kappa$ as the eddy size.

If $E(\kappa)d\kappa$ is the turbulence kinetic energy contained between wavenumbers $\kappa$ and $\kappa + d\kappa$, we can say

$$k = \int_0^\infty E(\kappa)\,d\kappa \tag{1.3}$$

Recall that $k$ is the kinetic energy per unit mass of the fluctuating turbulent velocity. Correspondingly, the **energy spectral density** or **energy spectrum function**, $E(\kappa)$, is related to the Fourier transform of $k$.

Observing that turbulence is so strongly driven by the large eddies, we expect $E(\kappa)$ to be a function of a length characteristic of the larger eddies, $\ell$, and the mean strain rate, $S$, which feeds the turbulence through direct interaction of the mean flow and the large eddies. Additionally, since turbulence is always dissipative, we expect $E(\kappa)$ to depend upon $\nu$ and $\epsilon$. By definition, it also must depend upon $\kappa$. For high Reynolds number turbulence, dimensional analysis suggests, and measurements confirm, that $k$ can be expressed in terms of $\epsilon$ and $\ell$ according to [Taylor (1935)]

$$\epsilon \sim \frac{k^{3/2}}{\ell} \quad \Longrightarrow \quad k \sim (\epsilon\ell)^{2/3} \tag{1.4}$$

Although we have not yet quantified the length scale $\ell$, it is the primary length scale most turbulence models are based on. In our discussion of two-point correlations in Chapter 2, an alternative to the spectral representation of turbulence, we will find that one measure of $\ell$ is known as the **integral length scale**. In most turbulence-modeling analysis, we assume there is a wide separation of scales, which means we implicitly assume $\ell$ is very large compared to the Kolmogorov length scale, viz.,

$$\ell \gg \eta \tag{1.5}$$

Substituting the estimate of $\epsilon$ from Equation (1.4) into the Kolmogorov length scale, we find

$$\frac{\ell}{\eta} = \frac{\ell}{(\nu^3/\epsilon)^{1/4}} \sim \frac{\ell\left(k^{3/2}/\ell\right)^{1/4}}{\nu^{3/4}} \sim Re_T^{3/4} \quad \text{where} \quad Re_T \equiv \frac{k^{1/2}\ell}{\nu} \tag{1.6}$$

The quantity $Re_T$ is the **turbulence Reynolds number**. It is based on the velocity characteristic of the turbulent motions as represented by the square root of $k$, the turbulence length scale, $\ell$, and the kinematic viscosity of the fluid, $\nu$. Thus, the condition $\ell \gg \eta$ holds provided we have high Reynolds number turbulence in the sense that

$$Re_T \gg 1 \qquad (1.7)$$

The existence of a wide separation of scales is a central assumption Kolmogorov made as part of his universal equilibrium theory. That is, he hypothesized that for very large Reynolds number, there is a range of eddy sizes between the largest and smallest for which the cascade process is independent of the statistics of the energy-containing eddies (so that $S$ and $\ell$ can be ignored) and of the direct effects of molecular viscosity (so that $\nu$ can be ignored). The idea is that a range of wavenumbers exists in which the energy transferred by inertial effects dominates, wherefore $E(\kappa)$ depends only upon $\epsilon$ and $\kappa$. On dimensional grounds, he thus concluded that

$$E(\kappa) = C_K \epsilon^{2/3} \kappa^{-5/3}, \qquad \frac{1}{\ell} \ll \kappa \ll \frac{1}{\eta} \qquad (1.8)$$

where $C_K$ is the **Kolmogorov constant**. Because inertial transfer of energy dominates, Kolmogorov identified this range of wavenumbers as the **inertial subrange**. The existence of the inertial subrange has been verified by many experiments and numerical simulations, although many years passed before definitive data were available to confirm its existence. Figure 1.4 shows a typical energy spectrum for a turbulent flow.

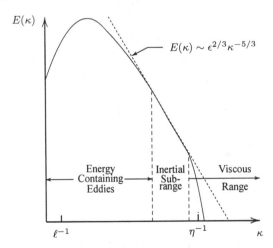

Figure 1.4: *Energy spectrum for a turbulent flow — log-log scales.*

While Equation (1.8) is indeed consistent with measurements, it is not the only form that can be deduced from dimensional analysis. Unfortunately, this is one of the shortcomings of dimensional analysis, i.e., the results we obtain are rarely unique. For example, lacking Kolmogorov's physical intuition, some researchers would retain $\nu$ as a dimensional quantity upon which $E(\kappa)$ depends as well as $\epsilon$ and $\kappa$. Then, a perfectly valid alternative to Equation (1.8) is

$$E(\kappa) = \epsilon^{1/4}\nu^{5/4}f(\kappa\eta), \qquad \eta = (\nu^3/\epsilon)^{1/4} \qquad (1.9)$$

where $f(\kappa\eta)$ is an undetermined function. This form reveals nothing regarding the variation of $E(\kappa)$ with $\kappa$, which is a straightforward illustration of how dimensional analysis, although helpful, is insufficient to deduce physical laws.

Afzal and Narasimha (1976) use the more-powerful concepts from perturbation theory (Appendix B) to remove this ambiguity and determine the asymptotic variation of the function $f$ in the inertial subrange. In their analysis, they assume that for small scales, corresponding to large wavenumbers, the energy spectrum function is given by Equation (1.9). This represents the *inner solution*.

Afzal and Narasimha also assume that viscous effects are unimportant for the largest eddies, and that the energy spectrum function is given, if the only relevant scales are $k$ and $\ell$, by

$$E(\kappa) = k\ell g(\kappa\ell) \qquad (1.10)$$

where $k$ is the turbulence kinetic energy, $\ell$ is the large-eddy length scale discussed above, and $g(\kappa\ell)$ is a second undetermined function. Although we omit the details here for the sake of brevity, we can exclude explicit dependence of $E(\kappa)$ on strain rate, $S$, since it is proportional to $k^{1/2}/\ell$ for high Reynolds number boundary layers. This represents the *outer solution*.

Finally, they *match* the two solutions, which means they insist that the inner and outer solutions are identical when $\kappa\eta$ is small and $\kappa\ell$ is large, i.e.,

$$\epsilon^{1/4}\nu^{5/4}f(\kappa\eta) = k\ell g(\kappa\ell) \qquad \text{for} \qquad \kappa\eta \ll 1 \text{ and } \kappa\ell \gg 1 \qquad (1.11)$$

In words, this matching operation assumes that

> *"Between the viscous and the energetic scales in any turbulent flow exists an overlap domain over which the solutions [characterizing] the flow in the two corresponding limits must match as Reynolds number tends to infinity."*

Note that the qualification regarding Reynolds number means the Reynolds number must be large enough to permit a wide separation of scales, i.e., so that $\ell \gg \eta$. To complete the matching operation, Afzal and Narasimha proceed as follows. In the spirit of singular-perturbation theory, the matching operation presumes that the *functional forms* of the inner and outer solutions are the same in the overlap region. This is a much stronger condition than requiring the two solutions to have the same value at a given point. Hence, if their functional forms are the same, so are their first derivatives. Thus, differentiating both sides of Equation (1.11) with respect to $\kappa$, we have

$$\eta \epsilon^{1/4} \nu^{5/4} f'(\kappa \eta) = k \ell^2 g'(\kappa \ell) \qquad \text{for} \qquad \kappa \eta \ll 1 \text{ and } \kappa \ell \gg 1 \qquad (1.12)$$

Then, noting that the Kolmogorov length scale is $\eta = \nu^{3/4} \epsilon^{-1/4}$ while Equation (1.4) tells us $k = \epsilon^{2/3} \ell^{2/3}$, we can rewrite Equation (1.12) as

$$\nu^2 f'(\kappa \eta) = \epsilon^{2/3} \ell^{8/3} g'(\kappa \ell) \qquad \text{for} \qquad \kappa \eta \ll 1 \text{ and } \kappa \ell \gg 1 \qquad (1.13)$$

Finally, multiplying through by $\kappa^{8/3} \epsilon^{-2/3}$ and using the fact that $\nu^2 \epsilon^{-2/3} = \eta^{8/3}$, we arrive at the following equation.

$$(\kappa \eta)^{8/3} f'(\kappa \eta) = (\kappa \ell)^{8/3} g'(\kappa \ell) \qquad \text{for} \qquad \kappa \eta \ll 1 \text{ and } \kappa \ell \gg 1 \qquad (1.14)$$

If there is a wide separation of scales, we can regard $\kappa \eta$ and $\kappa \ell$ as separate independent variables. Thus, Equation (1.14) says that a function of one independent variable, $\kappa \eta$, is equal to a function of a different independent variable, $\kappa \ell$. This can be true only if both functions tend to a constant value in the indicated limits. Thus, in the Afzal-Narasimha overlap domain, which is the inertial subrange,

$$(\kappa \eta)^{8/3} f'(\kappa \eta) = \text{constant} \quad \Longrightarrow \quad f(\kappa \eta) = C_K (\kappa \eta)^{-5/3} \qquad (1.15)$$

where $C_K$ is a constant. Combining Equations (1.9) and (1.15), we again arrive at the Kolmogorov inertial-subrange relation, viz.,

$$E(\kappa) = C_K \epsilon^{2/3} \kappa^{-5/3} \qquad (1.16)$$

which is identical to Equation (1.8).

Although the Kolmogorov $-5/3$ law is of minimal use in conventional turbulence models, it is of central importance in work on Direct Numerical Simulation (DNS) and Large Eddy Simulation (LES), which we discuss in Chapter 8. The Kolmogorov $-5/3$ law is so well established that, as noted by Rogallo and Moin (1984), theoretical or numerical predictions are regarded with skepticism if they fail to reproduce it. Its standing is as important as the law of the wall, which we discuss in the next subsection.

## 1.3.5 The Law of the Wall

The **law of the wall** is one of the most famous empirically-determined relationships in turbulent flows near solid boundaries. Measurements show that, for both internal and external flows, the streamwise velocity in the flow near the wall varies logarithmically with distance from the surface. This behavior is known as the law of the wall. In this section, we use both dimensional analysis and matching arguments to infer this logarithmic variation.

Observation of high Reynolds number turbulent boundary layers reveals a useful, approximate description of the near-surface turbulence statistics. We find that effects of the fluid's inertia and the pressure gradient are small near the surface. Consequently, the statistics of the flow near the surface in a turbulent boundary layer are established by two primary mechanisms. The first is the rate at which momentum is transferred to the surface, per unit area per unit time, which is equal to the local shear stress, $\tau$. The second is molecular diffusion of momentum, which plays an important role very close to the surface. Observations also indicate that the details of the eddies farther from the surface are of little importance to the near-wall flow statistics.

The validity of this approximate description improves with decreasing $y/\delta$, where $\delta$ is the boundary-layer thickness. This is true because the ratio of typical eddy size far from the surface to eddy size close to the surface increases as $y/\delta$ decreases. In other words, since $\delta$ increases with Reynolds number, we find a wide separation of scales at high Reynolds numbers. The astute reader will note interesting parallels between this description of the turbulent boundary layer and the general description of turbulence presented in Subsection 1.3.2. Note, however, that the analogy is mathematical rather than physical. This analogy is discussed, for example, by Mellor (1972) and by Afzal and Narasimha (1976).

Although $\tau$ varies near the surface, the variation with distance from the surface, $y$, is fairly slow. Hence, for the dimensional-analysis arguments to follow, we can use the surface shear stress, $\tau_w$, in place of the local shear stress. Also, we denote the molecular viscosity of the fluid by $\mu$. Since turbulence behaves the same in gases as in liquids, it is reasonable to begin with $\tau_w/\rho$ and kinematic viscosity, $\nu = \mu/\rho$, as our primary dimensional quantities, effectively eliminating fluid density, $\rho$, as a primary dimensional quantity.

Following Landau and Lifshitz (1966), since the dimensions of the quantity $\tau_w/\rho$ are length$^2$/time$^2$, while those of $\nu$ are length$^2$/time, clearly we can derive a velocity scale, $u_\tau$, defined by

$$u_\tau \equiv \sqrt{\frac{\tau_w}{\rho}} \tag{1.17}$$

and a length scale, $\nu/u_\tau$. The quantity $u_\tau$ is known as the **friction velocity**, and is a velocity scale representative of velocities close to a solid boundary. If we now postulate that the mean velocity gradient, $\partial U/\partial y$, can be correlated as a function of $u_\tau$, $\nu/u_\tau$ and $y$, dimensional analysis yields

$$\frac{\partial U}{\partial y} = \frac{u_\tau}{y} F\left(u_\tau y/\nu\right) \tag{1.18}$$

where $F(u_\tau y/\nu)$ is presumed to be a universal function. Examination of experimental data for a wide range of boundary layers [see, for example, Coles and Hirst (1969)], indicates that, as a good leading-order approximation,

$$F(u_\tau y/\nu) \rightarrow \frac{1}{\kappa} \quad \text{as} \quad u_\tau y/\nu \rightarrow \infty \tag{1.19}$$

where $\kappa$ is **Kármán's constant**. The function $F(u_\tau y/\nu)$ approaching a constant value is consistent with the notion that viscous effects cease to matter far from the surface, i.e., if it varies with $u_\tau y/\nu$ it would thus depend upon $\nu$. Integrating over $y$, we arrive at the famous law of the wall, viz.,

$$\frac{U}{u_\tau} = \frac{1}{\kappa} \ell n \frac{u_\tau y}{\nu} + C \tag{1.20}$$

where $C$ is a dimensionless integration constant. Correlation of measurements indicate $C \approx 5.0$ for smooth surfaces and $\kappa \approx 0.41$ for smooth and rough surfaces [see Kline et al. (1969)].

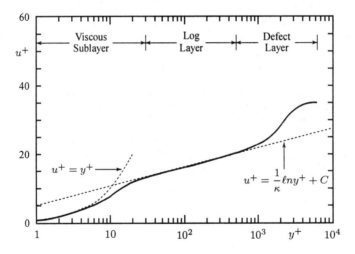

Figure 1.5: *Typical velocity profile for a turbulent boundary layer.*

Figure 1.5 shows a typical velocity profile for a turbulent boundary layer. The graph displays the dimensionless velocity, $u^+$, and distance, $y^+$, defined as:

$$u^+ \equiv \frac{U}{u_\tau} \quad \text{and} \quad y^+ \equiv \frac{u_\tau y}{\nu} \tag{1.21}$$

The velocity profile matches the law of the wall for values of $y^+$ in excess of about 30. As Reynolds number increases, the maximum value of $y^+$ at which the law of the wall closely matches the actual velocity increases.

Observe that three distinct regions are discernible, viz., the **viscous sublayer**, the **log layer** and the **defect layer**. By definition, the log layer is the portion of the boundary layer where the sublayer and defect layer merge and the law of the wall accurately represents the velocity. It is not a distinct layer. Rather, it is an overlap region between the inner and outer parts of the boundary layer. As we will see in the following discussion, originally presented by Millikan (1938), it is an overlap domain similar to that of the Afzal-Narasimha analysis of the preceding subsection.

Assuming the velocity in the viscous sublayer should depend only upon $u_\tau$, $\nu$ and $y$, we expect to have a relationship of the form

$$U = u_\tau f\left(y^+\right) \tag{1.22}$$

where $f(y^+)$ is a dimensionless function. This general functional form is often referred to as the **law of the wall**, and Equation (1.20) is simply a more explicit form. By contrast, in the defect layer, numerous experimenters including Darcy, von Kármán and Clauser found that velocity data correlate reasonably well with the so-called **velocity-defect law** or **Clauser defect law**:

$$U = U_e - u_\tau g(\eta), \qquad \eta \equiv \frac{y}{\Delta} \tag{1.23}$$

where $U_e$ is the velocity at the boundary-layer edge and $g(\eta)$ is another dimensionless function. The quantity $\Delta$ is a thickness characteristic of the outer portion of the boundary layer.

Hence, we have an inner length scale $\nu/u_\tau$ and an outer length scale $\Delta$. Millikan's postulate is that if a wide separation of scales exists in the sense that

$$\frac{\nu}{u_\tau} \ll \Delta \tag{1.24}$$

then an overlap domain exists such that

$$u_\tau f\left(y^+\right) = U_e - u_\tau g(\eta) \qquad \text{for} \qquad y^+ \gg 1 \text{ and } \eta \ll 1 \tag{1.25}$$

We can complete the matching without explicit knowledge of the functions $f$ and $g$ by differentiating Equation (1.25) with respect to $y$. Hence,

$$\frac{u_\tau^2}{\nu} f'\left(y^+\right) = -\frac{u_\tau}{\Delta} g'(\eta) \qquad \text{for} \qquad y^+ \gg 1 \text{ and } \eta \ll 1 \tag{1.26}$$

Then, multiplying through by $y/u_\tau$, we find

$$y^+ f'\left(y^+\right) = -\eta g'(\eta) \qquad \text{for} \qquad y^+ \gg 1 \text{ and } \eta \ll 1 \tag{1.27}$$

Thus, since a wide separation of scales means we can regard $y^+$ and $\eta$ as independent variables, clearly the only way a function of $y^+$ can be equal to a function of $\eta$ is for both to be equal to a constant. Therefore,

$$y^+ f'(y^+) = \text{constant} = \frac{1}{\kappa} \implies f(y^+) = \frac{1}{\kappa}\ell n y^+ + C \qquad (1.28)$$

which, when combined with Equation (1.22), yields Equation (1.20).

As noted earlier, the value of $C$ for a perfectly-smooth surface is $C \approx 5.0$. For surfaces with roughness elements of average height $k_s$, the law of the wall still holds, although $C$ is a function of $k_s$. Figure 1.6 illustrates how $C$ varies as a function of the dimensionless roughness height given by

$$k_s^+ \equiv \frac{u_\tau k_s}{\nu} \qquad (1.29)$$

As shown, as $k_s$ increases, the value of $C$ decreases. For large roughness height, measurements of Nikuradse [Schlichting (1979)] show that

$$C \to 8.5 - \frac{1}{\kappa}\ell n k_s^+, \qquad k_s^+ \gg 1 \qquad (1.30)$$

Substituting this value of $C$ into the law of the wall as represented in Equation (1.20) yields:

$$\frac{U}{u_\tau} = \frac{1}{\kappa}\ell n\left(\frac{y}{k_s}\right) + 8.5 \qquad \text{(completely-rough wall)} \qquad (1.31)$$

The absence of viscosity in this equation is consistent with the notion that the surface "shear stress" is due to pressure drag on the roughness elements.

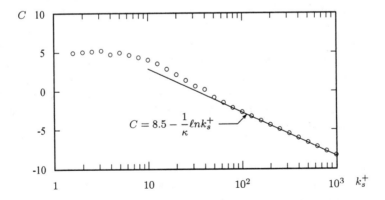

Figure 1.6: *Constant in the law of the wall, C, as a function of surface roughness;* o *based on measurements of Nikuradse [Schlichting (1979)].*

The defect layer lies between the log layer and the edge of the boundary layer. The velocity asymptotes to the law of the wall as $y/\delta \to 0$, and makes a noticeable departure from logarithmic behavior approaching the freestream. Again, from correlation of measurements, the velocity behaves as

$$U^+ = \frac{1}{\kappa}\ell n y^+ + C + \frac{2\Pi}{\kappa}\sin^2\left(\frac{\pi}{2}\frac{y}{\delta}\right) \tag{1.32}$$

where $\Pi$ is **Coles' wake-strength parameter** and $\delta$ is boundary-layer thickness. It varies with pressure gradient, and for constant pressure, correlation of measurements suggests $\Pi \approx 0.6$. Equation (1.32) is often referred to as the composite law of the wall and **law of the wake** profile.

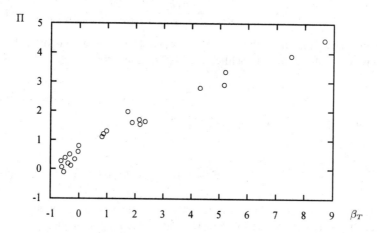

Figure 1.7: *Coles' wake-strength parameter,* $\Pi$, *as a function of pressure gradient;* ∘ *based on data tabulated by Coles and Hirst (1969).*

Figure 1.7 shows how $\Pi$ varies with pressure gradient for the so-called **equilibrium turbulent boundary layer**. As demonstrated by Clauser (1956) experimentally and justified with perturbation methods by others analytically [see, for example, Kevorkian and Cole (1981), Van Dyke (1975) or Wilcox (1995a)], the velocity in the defect layer varies in a self-similar manner provided the **equilibrium parameter** defined by

$$\beta_T \equiv \frac{\delta^*}{\tau_w}\frac{dP}{dx} \tag{1.33}$$

is constant. The quantities $\delta^*$ and $P$ in Equation (1.33) are displacement thickness and mean pressure, respectively. As demonstrated by Wilcox (1993b), even when $\beta_T$ is not constant, if it is not changing too rapidly, the value for $\Pi$ is close to the value shown in Figure 1.7.

### 1.3.6    Power Laws

Often, as an approximation, turbulent boundary-layer profiles are represented by a **power-law** relationship. That is, we sometimes say

$$\frac{U}{U_e} = \left(\frac{y}{\delta}\right)^{1/n} \tag{1.34}$$

where $n$ is typically an integer between 6 and 8. A value of $n = 7$, first suggested by Prandtl [Schlichting (1979)], yields a good approximation at high Reynolds number for the flat-plate boundary layer. Figure 1.8 compares a 1/7 power-law profile with measured values. As shown, the agreement between measured values for a plate-length Reynolds number of $Re_x = 1.09 \cdot 10^7$ and the approximate profile is surprisingly good with differences everywhere less than 3%.

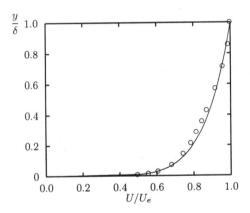

Figure 1.8: *Power-law velocity profile;* —— $U/U_e = (y/\delta)^{1/7}$; $\circ$ *Wieghardt data at* $Re_x = 1.09 \cdot 10^7$ *[Coles and Hirst (1969)].*

Recently, Barenblatt and others [see, for example, Barenblatt (1991), George, Knecht and Castillo (1992), Barenblatt (1993) and Barenblatt, Chorin and Prostokishin (1997)] have challenged the validity of the law of the wall. Their contention is that a power-law variation of the velocity in the inner layer better correlates pipe-flow measurements and represents a more realistic description of the turbulence in a boundary layer.

The critical assumption that Barenblatt et al. challenge is the existence of a wide separation of scales, i.e., large $\delta/(\nu/u_\tau)$. They maintain that the turbulence in the overlap region is Reynolds-number dependent. If this is true, the law of the wall and defect-law Equations (1.22) and (1.23), respectively, must be replaced by

$$U = u_\tau \tilde{f}\left(y^+, Re\right) \quad \text{and} \quad U = U_e - u_o\tilde{g}(\eta, Re) \tag{1.35}$$

where $Re$ is an appropriate Reynolds number, $\tilde{f}$ and $\tilde{g}$ are universal functions, and $u_o$ is a velocity scale that is not necessarily equal to $u_\tau$. Equivalently, the Barenblatt et al. hypothesis replaces Equation (1.18) by

$$\frac{\partial U}{\partial y} = \frac{u_\tau}{y} \Phi\left(y^+, Re\right) \qquad (1.36)$$

where the universal function $\Phi(y^+, Re)$ appears in place of $F(y^+)$.

In the Millikan argument, the assumption of a wide separation of scales implies that the boundary layer possesses self-similar solutions both in the defect layer and the sublayer, in the sense that a similarity variable, e.g., $y^+ = u_\tau y/\nu$ and $\eta = y/\Delta$, exists in each region. The assumption that we can regard $y^+$ and $\eta$ as distinct independent variables in the overlap region is described as a condition of **complete similarity**. By contrast, the Barenblatt hypothesis corresponds to **incomplete similarity**. Barenblatt (1979) discusses the distinction between complete and incomplete similarity in detail.

Under the assumption of incomplete similarity, there is no a priori reason for the function $\Phi(y^+, Re)$ to approach a constant value in the limit $y^+ \rightarrow \infty$, even when $Re \rightarrow \infty$. Rather, Barenblatt et al. argue that for large $y^+$,

$$\Phi\left(y^+, Re\right) = A\left(y^+\right)^\alpha \qquad (1.37)$$

where the coefficient $A$ and the exponent $\alpha$ are presumed to be functions of Reynolds number. In the nomenclature of Barenblatt, Chorin and Prostokishin (1997), they assume "incomplete similarity in the parameter $[y^+]$ and no similarity in the parameter $Re$." Combining Equations (1.36) and (1.37) yields

$$\frac{\partial U^+}{\partial y^+} = A\left(y^+\right)^{\alpha-1} \quad \Longrightarrow \quad U^+ = \frac{A}{\alpha}\left(y^+\right)^\alpha \qquad (1.38)$$

Based primarily on experimental data for pipe flow gathered by Nikuradse in the 1930's [Schlichting (1979)], Barenblatt, Chorin and Prostokishin conclude that

$$A = 0.577\ell n Re + 2.50 \quad \text{and} \quad \alpha = \frac{1.5}{\ell n Re} \qquad (1.39)$$

where $Re$ is Reynolds number based on average velocity and pipe diameter.

To test the Barenblatt et al. alternative to the law of the wall, Zagarola, Perry and Smits (1997) have performed an analysis based on more recent experiments by Zagarola (1996). The advantage of these data lies in the much wider range of Reynolds numbers considered, especially large values, relative to those considered by Nikuradse. They conclude that the classical law of the wall provides closer correlation with measurements than the power law given by combining Equations (1.38) and (1.39), although they recommend a somewhat larger value for $\kappa$ of 0.44.

To remove the possibility that the 60-year-old data of Nikuradse provide a poor correlation of $A$ and $\alpha$, Zagarola, Perry and Smits determine their values from the Zagarola data, concluding that

$$A = 0.7053 \ell n Re + 0.3055 \quad \text{and} \quad \alpha = \frac{1.085}{\ell n Re} + \frac{6.535}{(\ell n Re)^2} \quad (1.40)$$

Even with these presumably more-accurate values, the logarithmic law of the wall still provides closer correlation with measurements than the power-law form.

This is not the end of the controversy however, as Barenblatt, Chorin and Prostokishin (1997) attempt, with a questionable argument, to demonstrate that at high Reynolds number, the Zagarola experiments have significant surface roughness. Zagarola, Perry and Smits (1997) reject this possibility in stating that "the pipe surface was shown to be smooth." Given the fragile nature of dimensional analysis and perturbation methods combined with measurement uncertainties, this controversy may persist for some time to come.

# 1.4 A Brief History of Turbulence Modeling

The primary emphasis in this book is upon the time-averaged Navier-Stokes equation. The origin of this approach dates back to the end of the nineteenth century when Reynolds (1895) published results of his research on turbulence. His pioneering work proved to have such profound importance for all future developments that we refer to the standard time-averaging process as one type of Reynolds averaging.

The earliest attempts at developing a mathematical description of turbulent stresses sought to mimic the molecular gradient-diffusion process. In this spirit, Boussinesq (1877) introduced the concept of a so-called eddy viscosity. As with Reynolds, Boussinesq has been immortalized in turbulence literature. The Boussinesq eddy-viscosity approximation is so widely known that few authors find a need to reference his original paper.

Neither Reynolds nor Boussinesq attempted a solution of the Reynolds-averaged Navier-Stokes equation in any systematic manner. Much of the physics of viscous flows was a mystery in the nineteenth century, and further progress awaited Prandtl's discovery of the boundary layer in 1904. Focusing upon turbulent flows, Prandtl (1925) introduced the mixing length (an analog of the mean-free path of a gas) and a straightforward prescription for computing the eddy viscosity in terms of the mixing length. The mixing-length hypothesis, closely related to the eddy-viscosity concept, formed the basis of virtually all turbulence-modeling research for the next twenty years. Important early contributions were made by several researchers, most notably by von Kármán (1930). In modern terminology, we refer to a model based on the mixing-length hypothesis as an

**algebraic model** or a **zero-equation model of turbulence.** By definition, an **n-equation model** signifies a model that requires solution of **n** additional differential transport equations in addition to those expressing conservation of mass, momentum and energy for the mean flow.

To improve the ability to predict properties of turbulent flows and to develop a more realistic mathematical description of the turbulent stresses, Prandtl (1945) postulated a model in which the eddy viscosity depends upon the kinetic energy of the turbulent fluctuations, $k$. He proposed a modeled partial-differential equation approximating the exact equation for $k$. This improvement, on a conceptual level, takes account of the fact that the turbulent stresses, and thus the eddy viscosity, are affected by where the flow has been, i.e., upon flow history. Thus was born the concept of the so-called **one-equation model of turbulence.**

While having an eddy viscosity that depends upon flow history provides a more physically realistic model, the need to specify a turbulence length scale remains. That is, on dimensional grounds, viscosity has dimensions of velocity times length. Since the length scale can be thought of as a characteristic eddy size and since such scales are different for each flow, turbulence models that do not provide a length scale are **incomplete.** That is, we must know something about the flow, other than initial and boundary conditions, in advance in order to obtain a solution. Incomplete models are not without merit and, in fact, have proven to be of great value in many engineering applications.

To elaborate a bit further, an incomplete model generally defines a turbulence length scale in a prescribed manner from the mean flow, e.g., the displacement thickness, $\delta^*$, for an attached boundary layer. However, a different length scale in this example would be needed when the boundary layer separates since $\delta^*$ may be negative. Yet another length might be needed for free shear flows, etc. In essence, incomplete models usually define quantities that may vary more simply or more slowly than the Reynolds stresses (e.g., eddy viscosity and mixing length). Presumably such quantities are easier to correlate than the actual stresses.

A particularly desirable type of turbulence model would be one that can be applied to a given turbulent flow by prescribing at most the appropriate boundary and/or initial conditions. Ideally, no advance knowledge of any property of the turbulence should be required to obtain a solution. We define such a model as being **complete.** Note that our definition implies nothing regarding the accuracy or universality of the model, only that it can be used to determine a flow with no prior knowledge of any flow details.

Kolmogorov (1942) introduced the first **complete** model of turbulence. In addition to having a modeled equation for $k$, he introduced a second parameter $\omega$ that he referred to as "the rate of dissipation of energy in unit volume and time." The reciprocal of $\omega$ serves as a turbulence time scale, while $k^{1/2}/\omega$ serves as the analog of the mixing length and $k\omega$ is the analog of the dissipation rate, $\epsilon$. In this model, known as a $k$-$\omega$ model, $\omega$ satisfies a differential equation somewhat similar to the equation for $k$. The model is thus termed a **two-equation model of turbulence**. While this model offered great promise, it went with virtually no applications for the next quarter century because of the unavailability of computers to solve its nonlinear differential equations.

Chou (1945) and Rotta (1951) laid the foundation for turbulence models that obviate use of the Boussinesq approximation. Rotta devised a plausible model for the differential equation governing evolution of the tensor that represents the turbulent stresses, i.e., the Reynolds-stress tensor. Such models are most appropriately described as **stress-transport models**. Many authors refer to this approach as **second-order closure** or **second-moment closure**. The primary conceptual advantage of a stress-transport model is the natural manner in which nonlocal and history effects are incorporated. Although quantitative accuracy often remains difficult to achieve, such models automatically accommodate complicating effects such as streamline curvature, rigid-body rotation, and body forces. This stands in contrast to eddy-viscosity models that account for these effects only if empirical terms are added. For a three-dimensional flow, a stress-transport model introduces seven equations, one for the turbulence (length or equivalent) scale and six for the components of the Reynolds-stress tensor. As with Kolmogorov's $k$-$\omega$ model, stress-transport models awaited adequate computer resources.

Thus, by the early 1950's, four main categories of turbulence models had evolved, viz.,

1. Algebraic (Zero-Equation) Models

2. One-Equation Models

3. Two-Equation Models

4. Stress-Transport Models

With the coming of the age of computers since the 1960's, further development of all four classes of turbulence models has occurred. The following overview lists a few of the most important modern developments for each of the four classes.

**Algebraic Models.** Van Driest (1956) devised a viscous damping correction for the mixing-length model that is included in virtually all algebraic models in use today. Cebeci and Smith (1974) refined the eddy-viscosity/mixing-length model to a point that it can be used with great confidence for most attached boundary layers. To remove some of the difficulties in defining the turbulence length scale from the shear-layer thickness, Baldwin and Lomax (1978) proposed an alternative algebraic model that enjoyed widespread use for many years.

**One-Equation Models.** Of the four types of turbulence models described above, the one-equation model has enjoyed the least popularity and success. Perhaps the most successful early model of this type was formulated by Bradshaw,

Ferriss and Atwell (1967). In the 1968 Stanford Conference on Computation of
Turbulent Boundary Layers [Coles and Hirst (1969)] the best turbulence models
of the day were tested against the best experimental data of the day. In this author's opinion, of all the models used, the Bradshaw-Ferriss-Atwell model most
faithfully reproduced measured flow properties. There has been some renewed
interest in one-equation models based on a postulated equation for eddy viscosity [c.f. Baldwin and Barth (1990), Goldberg (1991) and Spalart and Allmaras
(1992)]. This work has been motivated primarily by the ease with which such
model equations can be solved numerically, relative to two-equation models and
stress-transport models. Of these recent one-equation models, that of Spalart and
Allmaras appears to be the most accurate for practical turbulent-flow applications.

**Two-Equation Models.** While Kolmogorov's $k$-$\omega$ model was the first of this
type, it remained in obscurity until the coming of the computer. By far the most
extensive work on two-equation models has been done by Launder and Spalding
(1972) and a continuing succession of students and colleagues. Launder's $k$-$\epsilon$
model is as well known as the mixing-length model and is the most widely used
two-equation model. Even the model's demonstrable inadequacy for flows with
adverse pressure gradient [c.f. Rodi and Scheuerer (1986), Wilcox (1988a, 1993b)
and Henkes (1998a)] has done little to discourage its widespread use. With no
prior knowledge of Kolmogorov's work, Saffman (1970) formulated a $k$-$\omega$ model
that enjoys advantages over the $k$-$\epsilon$ model, especially for integrating through the
viscous sublayer and for predicting effects of adverse pressure gradient. Wilcox
and Alber (1972), Saffman and Wilcox (1974), Wilcox and Traci (1976), Wilcox
and Rubesin (1980), and Wilcox (1988a), for example, have pursued further
development and application of $k$-$\omega$ models. This text, in Chapter 4, introduces
a new version of the $k$-$\omega$ model, a significant improvement over that described
in the first edition. As pointed out by Lakshminarayana (1986), $k$-$\omega$ models are
the second most widely used type of two-equation turbulence model.

**Stress-Transport Models.** By the 1970's, sufficient computer resources
became available to permit serious development of this class of model. The
most noteworthy efforts were those of Donaldson [Donaldson and Rosenbaum
(1968)], Daly and Harlow (1970) and Launder, Reece and Rodi (1975). The
latter has become the baseline stress-transport model: more recent contributions
by Lumley (1978), Speziale (1985, 1987a, 1991) and Reynolds (1987) have added
mathematical rigor to the closure process. However, because of the large number
of equations and complexity involved in stress-transport models, they have thus
far found their way into a relatively small number of applications compared to
algebraic and two-equation models.

This book investigates all four classes of turbulence models. The primary
emphasis is upon examining the underlying physical foundation and upon developing the mathematical tools for analyzing and testing the models. **The text**

**is not intended to be a catalog of all turbulence models.** Rather, the text approaches each class of models in a generic sense. Detailed information is provided for models that have stood the test of time; additionally, references are given for most models.

As a concluding comment, turbulence models have been created that fall beyond the bounds of the four categories cited above. This is true because model developers have tried unconventional approaches in an attempt to remove deficiencies of existing models of the four basic classes. Given the erratic track record of most turbulence models, new ideas are always welcome. While a chapter could be added to discuss some of the less conventional approaches, it would be inconsistent with one of the themes of this book. Specifically, the text focuses on concepts that have stood the test of time by gaining wide acceptance and/or demonstrating their accuracy and usefulness. Thus, no such chapter is included.

# Problems

**1.1** To appreciate why laminar flow is of minimal importance in many engineering applications, compute the percent of the vehicle over which laminar flow exists for the following situations. In each case, consider the boundary layer on a flat portion of the vehicle and assume transition occurs at a (very high) Reynolds number of $Re_{x_t} = 5 \cdot 10^5$.

(a) A 20-foot automobile moving at 25 mph ($\nu = 1.62 \cdot 10^{-4}$ ft$^2$/sec).

(b) A 20-foot automobile moving at 65 mph $\nu = 1.62 \cdot 10^{-4}$ ft$^2$/sec).

(c) A small aircraft with an average wing chord length of 8 feet moving at 150 mph $\nu = 1.67 \cdot 10^{-4}$ ft$^2$/sec).

(d) A *Boeing 747* with an average wing chord length of 30 feet moving at 570 mph ($\nu = 4.27 \cdot 10^{-4}$ ft$^2$/sec).

**1.2** To appreciate why laminar flow is of minimal importance in many engineering applications, compute the percent of the vehicle over which laminar flow exists for the following situations. In each case, consider the boundary layer on a flat portion of the vehicle and assume transition occurs at a (very high) Reynolds number of $Re_{x_t} = 10^6$. Note that 1 knot = 0.514 m/sec.

(a) A 10-meter sailboat moving at 3 knots ($\nu = 1.00 \cdot 10^{-6}$ m$^2$/sec).

(b) A 10-meter sailboat moving at 7.5 knots ($\nu = 1.00 \cdot 10^{-6}$ m$^2$/sec).

(c) A 25-meter yacht moving at 12 knots ($\nu = 0.80 \cdot 10^{-6}$ m$^2$/sec).

(d) A 100-meter tanker moving at 15 knots ($\nu = 1.50 \cdot 10^{-6}$ m$^2$/sec).

**1.3** Using dimensional analysis, deduce the Kolmogorov length, time and velocity scales defined in Equation (1.1).

**1.4** As noted in Subsection 1.3.3, for an automobile moving at 65 mph, the Kolmogorov length scale near the driver's window is $\eta \approx 1.8 \cdot 10^{-4}$ inch. If $\nu = 1.62 \cdot 10^{-4}$ ft$^2$/sec, what are the Kolmogorov time and velocity scales? Repeat the computations for a point farther from the surface where $\eta = 0.018$ inch.

**1.5** Using dimensional analysis, deduce the Kolmogorov $-5/3$ law, Equation (1.8), beginning with the assumption that the energy spectral density, $E(\kappa)$, depends only upon wavenumber, $\kappa$, and dissipation rate, $\epsilon$.

**1.6** The viscous sublayer of a turbulent boundary layer extends from the surface up to $y^+ \approx 30$. To appreciate how thin this layer is, consider the boundary layer on the side of your freshly washed and waxed (and therefore smooth) automobile. When you are moving at $U = 65$ mph, the skin friction coefficient, $c_f$, just below your rear-view mirror is 0.0028. Noting that $U/u_\tau = \sqrt{2/c_f}$, estimate the sublayer thickness and compare to the diameter of the head of a pin, $d_{pin} = 0.05$ inch. Assume $\nu = 1.62 \cdot 10^{-4}$ ft$^2$/sec.

**1.7** The viscous sublayer of a turbulent boundary layer extends from the surface up to $y^+ \approx 30$. To appreciate how thin this layer is, consider the boundary layer on the hull of a large tanker moving at speed $U$. Assuming the boundary layer has negligible pressure gradient over most of the hull, you can assume

$$\delta \approx 0.37 x Re_x^{-1/5} \quad \text{and} \quad c_f \approx 0.0576 Re_x^{-1/5}$$

(a) Noting that $U/u_\tau = \sqrt{2/c_f}$, verify that the sublayer thickness, $\delta_{sl} = 30\nu/u_\tau$, is given by

$$\delta_{sl} \approx \frac{478}{Re_x^{7/10}} \delta$$

(b) Compute $\delta_{sl}$ at points on the hull where $Re_x = 2.8 \cdot 10^7$ and $\delta = 2.5$ in, and where $Re_x = 5.0 \cdot 10^8$ and $\delta = 25$ in. Express your answer in terms of $h_\delta/\delta_{sl}$, to the nearest integer, where $h_\delta = 1/10$ inch is the height of the symbol $\delta_{sl}$ on this page.

**1.8** Combining Equations (1.25) and (1.28), verify that the function $g(\eta)$ must be

$$g(\eta) = A - \frac{1}{\kappa} \ell n \eta$$

where $A$ is a function of $U_e$, $u_\tau$, $\Delta$, $\nu$, $\kappa$ and $C$. To have a wide separation of scales, $A$ must be a constant, i.e., it must be independent of Reynolds number. Noting that $U_e/u_\tau = \sqrt{2/c_f}$ and using Clauser's thickness, $\Delta = U_e \delta^*/u_\tau$, where $\delta^*$ is displacement thickness, determine the skin friction, $c_f$, as a function of $A$, $C$ and $Re_{\delta^*} = U_e \delta^*/\nu$.

**1.9** According to Equation (1.32), at the boundary-layer edge we have

$$\frac{U_e}{u_\tau} = \frac{1}{\kappa} \ell n \frac{u_\tau \delta}{\nu} + C + \frac{2\Pi}{\kappa}$$

We would like to determine how skin friction, $c_f = 2u_\tau^2/U_e^2$, is affected by changes in the constant $C$.

(a) Assuming only $u_\tau$ varies with $C$, verify that

$$\frac{1}{c_f} \frac{dc_f}{dC} = -\frac{2\kappa\sqrt{c_f/2}}{\kappa + \sqrt{c_f/2}}$$

(b) Based on the result of Part (a) and assuming $c_f = 0.003$, how much of a change in $C$ is required to give a 5% change in $c_f$? Be sure to include a sign in your result.

**1.10** We would like to determine the values of Reynolds number, $Re$, for which the Barenblatt exponent, $\alpha$, is 1/6, 1/7 and 1/8. Compare the values inferred by using the Barenblatt correlation, Equation (1.39), and the Zagarola correlation, Equation (1.40).

# Chapter 2

# The Closure Problem

Because turbulence consists of random fluctuations of the various flow properties, we use a statistical approach. Our purposes are best served by the procedure introduced by Reynolds (1895) in which all quantities are expressed as the sum of mean and fluctuating parts. We then form the mean of the continuity and Navier-Stokes equations term by term. As we will see in this chapter, the nonlinearity of the Navier-Stokes equation leads to the appearance of momentum fluxes that act as apparent stresses throughout the flow. These momentum fluxes are unknown a priori. We then derive equations for these stresses, which include additional unknown quantities. This illustrates the issue of closure, i.e., establishing a sufficient number of equations for all of the unknowns. The chapter concludes with a discussion of turbulence scales and more-advanced statistical concepts.

To illustrate the nature of turbulence statistics, it is instructive to observe how the velocity field behaves for a turbulent flow. Figure 2.1 shows measured velocity profiles, $u(y)$, for a flat-plate boundary layer. Plotted with a series

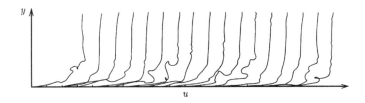

Figure 2.1: *Instantaneous boundary-layer velocity profiles at the same distance from the leading edge of a flat plate at 17 different instants. The profiles are shown with a series of staggered origins. [From Cebeci and Smith (1974) — Copyright © Academic Press 1974 — Used with permission.]*

of staggered origins, all 17 profiles correspond to the same distance from the plate leading edge, and have been measured at several different times using the hydrogen-bubble technique. While the experimental method is a bit crude, e.g., the profiles appear incorrectly multivalued[1] in a few locations, the measured velocity profiles correctly show that the velocity profile changes shape rather dramatically from one instant to the next.

Figure 2.2(a) displays all of the velocity profiles, only this time with a common origin. Clearly, there is a large scatter in the value of the velocity at each distance $y$ from the surface. Figure 2.2(b) shows a standard mean velocity profile for a boundary layer at the same Reynolds number. Comparison of the profiles in (a) and (b) clearly illustrates that the turbulent fluctuations in the velocity cannot be regarded as a small perturbation relative to the mean value. In the following sections, we explore the classical statistical methods used to analyze this inherently complex behavior.

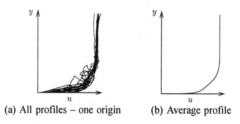

(a) All profiles – one origin          (b) Average profile

Figure 2.2: *Instantaneous and average boundary-layer velocity profiles at the same distance from the leading edge of a flat plate. [From Cebeci and Smith (1974) — Copyright © Academic Press 1974 — Used with permission.]*

## 2.1   Reynolds Averaging

We begin with the averaging concepts introduced by Reynolds (1895). In general, Reynolds averaging assumes a variety of forms involving either an integral or a summation. The three forms most pertinent in turbulence-model research are the **time average**, the **spatial average** and the **ensemble average**: the general term used to describe these averaging processes is "mean."

**Time averaging** is appropriate for **stationary turbulence**, i.e., a turbulent flow that, on the average, does not vary with time, such as flow in a pipe driven

---

[1]The hydrogen-bubble technique cannot isolate the velocity component parallel to the surface, so that the profiles include effects of vertical motion as well, and the apparently multivalued profiles are really a kind of velocity-vector plot. This also illustrates that fluctuating velocities are large in all directions.

by a constant-speed blower. For such a flow, we express an instantaneous flow variable as $f(\mathbf{x}, t)$. Its time average, $F_T(\mathbf{x})$, is defined by

$$F_T(\mathbf{x}) = \lim_{T \to \infty} \frac{1}{T} \int_t^{t+T} f(\mathbf{x}, t)\, dt \tag{2.1}$$

The velocity profile depicted in Figure 2.2(b), for example, was obtained using time averaging for accurate measurements of a similar boundary layer. The applicability of Reynolds averaging (of whatever kind) implicitly depends upon this steadiness of mean values. Time averaging is the most commonly used form of Reynolds averaging because most turbulent flows of interest in engineering are stationary. There are important exceptions, of course, such as the motion of the atmosphere.

**Spatial averaging** can be used for **homogeneous** turbulence, which is a turbulent flow that, on the average, is uniform in all directions. We average over all spatial coordinates by doing a volume integral. Calling the average $F_V$, we have

$$F_V(t) = \lim_{V \to \infty} \frac{1}{V} \iiint_V f(\mathbf{x}, t)\, dV \tag{2.2}$$

**Ensemble averaging** is the most general type of Reynolds averaging suitable for, e.g., flows that decay in time. As an idealized example, in terms of measurements from $N$ identical experiments (with initial and boundary conditions that differ by random infinitesimal perturbations) where $f(\mathbf{x}, t) = f_n(\mathbf{x}, t)$ in the $n^{th}$ experiment, the average is $F_E$, defined by

$$F_E(\mathbf{x}, t) = \lim_{N \to \infty} \frac{1}{N} \sum_{n=1}^N f_n(\mathbf{x}, t) \tag{2.3}$$

For turbulence that is both stationary and homogeneous, we may assume that these three averages are all equal. This assumption is known as the **ergodic hypothesis**.

From this point on, we will consider only time averaging. There is no loss of generality however as virtually all of our results are valid for other kinds of Reynolds averaging. Consider a stationary turbulent flow so that Equation (2.1) holds. For such a flow, we express the instantaneous velocity, $u_i(\mathbf{x}, t)$, as the sum of a mean, $U_i(\mathbf{x})$, and a fluctuating part, $u_i'(\mathbf{x}, t)$, so that

$$u_i(\mathbf{x}, t) = U_i(\mathbf{x}) + u_i'(\mathbf{x}, t) \tag{2.4}$$

**NOTE:** *By convention, throughout this text the instantaneous variable is denoted by a lower-case symbol, the mean is denoted by the corresponding upper-case symbol and the fluctuating part is the lower-case symbol with a prime.*

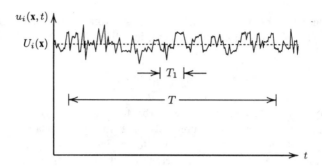

Figure 2.3: *Time averaging for stationary turbulence. Although obscured by the scale of the graph, the instantaneous velocity, $u_i(\mathbf{x}, t)$, has continuous derivatives of all order.*

As in Equation (2.1), the quantity $U_i(\mathbf{x})$ is the time-averaged, or mean, velocity defined by

$$U_i(\mathbf{x}) = \lim_{T \to \infty} \frac{1}{T} \int_t^{t+T} u_i(\mathbf{x}, t)\, dt \tag{2.5}$$

The time average of the mean velocity is again the same time-averaged value, i.e.,

$$\overline{U_i}(\mathbf{x}) = \lim_{T \to \infty} \frac{1}{T} \int_t^{t+T} U_i(\mathbf{x})\, dt = U_i(\mathbf{x}) \tag{2.6}$$

where an overbar is shorthand for the time average. The time average of the fluctuating part of the velocity is zero. That is, using Equation (2.6),

$$\overline{u_i'} = \lim_{T \to \infty} \frac{1}{T} \int_t^{t+T} [u_i(\mathbf{x}, t) - U_i(\mathbf{x})]\, dt = U_i(\mathbf{x}) - \overline{U_i}(\mathbf{x}) = 0 \tag{2.7}$$

While Equation (2.5) is mathematically well defined, we can never truly realize infinite $T$ in any physical flow. This is not a serious problem in practice. In forming our time average, as illustrated in Figure 2.3, we just select a time $T$ that is very long relative to the maximum period of the velocity fluctuations, $T_1$, which we don't need to define precisely. In other words, rather than formally taking the limit $T \to \infty$, we do the indicated integration in Equation (2.5) with $T \gg T_1$. As an example, for flow at 10 m/sec in a 5 cm diameter pipe, an integration time of 20 seconds would probably be adequate. In this time the flow moves 4,000 pipe diameters.

There are some flows for which the mean flow contains very slow variations with time that are not turbulent in nature. For instance, we might impose a slowly varying periodic pressure gradient in a duct or we might wish to compute flow over a helicopter blade or flow through an automobile muffler. Clearly,

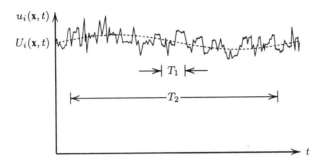

Figure 2.4: *Time averaging for nonstationary turbulence. Although obscured by the scale of the graph, the instantaneous velocity, $u_i(\mathbf{x}, t)$, has continuous derivatives of all order.*

Equations (2.4) and (2.5) must be modified to accommodate such applications. The simplest, but a bit arbitrary, method is to replace Equations (2.4) and (2.5) with

$$u_i(\mathbf{x}, t) = U_i(\mathbf{x}, t) + u_i'(\mathbf{x}, t) \tag{2.8}$$

and

$$U_i(\mathbf{x}, t) = \frac{1}{T} \int_t^{t+T} u_i(\mathbf{x}, t)\, dt, \qquad T_1 \ll T \ll T_2 \tag{2.9}$$

where $T_2$ is the time scale characteristic of the slow variations in the flow that we do not wish to regard as belonging to the turbulence. Figure 2.4 illustrates these concepts.

A word of caution is in order regarding Equation (2.9). We are implicitly assuming that time scales $T_1$ and $T_2$ exist that differ by several orders of magnitude. Very few unsteady flows of engineering interest satisfy this condition. We cannot use Equations (2.8) and (2.9) for such flows because there is no distinct boundary between our imposed unsteadiness and turbulent fluctuations. For such flows, the mean [as defined in Equation (2.9)] and fluctuating components are correlated, i.e., the time average of their product is non-vanishing. In meteorology, for example, this is known as the **spectral gap problem**. If the flow is periodic, **Phase Averaging** (see problems section) can be used; otherwise, full ensemble averaging is required. Phase averaging is a type of ensemble averaging with phase angle replacing time. For a rigorous approach, an alternative method such as Large Eddy Simulation (Chapter 8) will be required.

Clearly our time averaging process, involving integrals over time, commutes with spatial differentiation. Thus, for any scalar $p$ and vector $u_i$,

$$\overline{p_{,i}} = P_{,i} \quad \text{and} \quad \overline{u_{i,j}} = U_{i,j} \tag{2.10}$$

Because we are dealing with definite integrals, time averaging is a linear operation. Thus if $c_1$ and $c_2$ are constants while $a$ and $b$ denote any two flow properties with mean values $A$ and $B$, respectively, then

$$\overline{c_1 a + c_2 b} = c_1 A + c_2 B \qquad (2.11)$$

The time average of an unsteady term like $\partial u_i / \partial t$ is obviously zero for stationary turbulence. For nonstationary turbulence, we must look a little closer. We know that

$$\frac{1}{T} \int_t^{t+T} \frac{\partial}{\partial t}(U_i + u_i') \, dt = \frac{U_i(\mathbf{x}, t+T) - U_i(\mathbf{x}, t)}{T} + \frac{u_i'(\mathbf{x}, t+T) - u_i'(\mathbf{x}, t)}{T}$$

$$(2.12)$$

The second term on the right-hand side of Equation (2.12) can be neglected provided $|u_i'|$ is small relative to $|U_i|$. Since we are assuming $T$ is very small relative to the time scale of the mean flow, i.e. that $T \ll T_2$, the first term is the value corresponding to the limit $T \to 0$, i.e., $\partial U_i / \partial t$. Hence,

$$\overline{\frac{\partial u_i}{\partial t}} \approx \frac{\partial U_i}{\partial t} \qquad (2.13)$$

The approximation that $|u_i'| \ll |U_i|$ is always questionable, especially for free shear flows and for flows very close to a solid boundary. This is one of the inherent complications of turbulence, namely that the fluctuations cannot be assumed to be small relative to the mean values.

Using time averaging in this manner is nevertheless useful for analysis, especially for time-marching numerical methods implemented for solving steady-flow problems. Because Equation (2.13) depends on the doubtful approximation that $|u_i'| \ll |U_i|$ while fluctuations are often in excess of 10% of the mean, a degree of caution must be exercised when such methods are used for time-varying flows.

## 2.2   Correlations

Thus far we have considered averages of linear quantities. When we average the product of two properties, say $\phi$ and $\psi$, we have the following:

$$\overline{\phi\psi} = \overline{(\Phi + \phi')(\Psi + \psi')} = \overline{\Phi\Psi + \Phi\psi' + \Psi\phi' + \phi'\psi'} = \Phi\Psi + \overline{\phi'\psi'} \qquad (2.14)$$

where we take advantage of the fact that the product of a mean quantity and a fluctuating quantity has zero mean because the mean of the latter is zero. There is no a priori reason for the mean of the product of two fluctuating quantities to vanish. Thus, Equation (2.14) tells us that the mean value of a product, $\overline{\phi\psi}$, differs from the product of the mean values, $\Phi\Psi$. The quantities $\phi'$ and $\psi'$ are said to be **correlated** if $\overline{\phi'\psi'} \neq 0$. They are **uncorrelated** if $\overline{\phi'\psi'} = 0$.

Similarly, for a triple product, we find

$$\overline{\phi\psi\xi} = \Phi\Psi\Xi + \overline{\phi'\psi'}\Xi + \overline{\psi'\xi'}\Phi + \overline{\phi'\xi'}\Psi + \overline{\phi'\psi'\xi'} \tag{2.15}$$

Again, terms linear in $\phi'$, $\psi'$ or $\xi'$ have zero mean. As with terms quadratic in fluctuating quantities, there is no a priori reason for the cubic term, $\overline{\phi'\psi'\xi'}$, to vanish.

## 2.3 Reynolds-Averaged Equations

For simplicity we confine our attention to incompressible, constant-property flow. Effects of compressibility will be addressed in Chapter 5. The equations for conservation of mass and momentum are

$$\frac{\partial u_i}{\partial x_i} = 0 \tag{2.16}$$

$$\rho\frac{\partial u_i}{\partial t} + \rho u_j\frac{\partial u_i}{\partial x_j} = -\frac{\partial p}{\partial x_i} + \frac{\partial t_{ji}}{\partial x_j} \tag{2.17}$$

The vectors $u_i$ and $x_i$ are velocity and position, $t$ is time, $p$ is pressure, $\rho$ is density and $t_{ij}$ is the viscous stress tensor defined by

$$t_{ij} = 2\mu s_{ij} \tag{2.18}$$

where $\mu$ is molecular viscosity and $s_{ij}$ is the strain-rate tensor,

$$s_{ij} = \frac{1}{2}\left(\frac{\partial u_i}{\partial x_j} + \frac{\partial u_j}{\partial x_i}\right) \tag{2.19}$$

Note that $s_{ji} = s_{ij}$, so that $t_{ji} = t_{ij}$ for simple viscous fluids (but not for some anisotropic liquids).

To simplify the time-averaging process, we rewrite the convective term in "conservation" form, i.e.,

$$u_j\frac{\partial u_i}{\partial x_j} = \frac{\partial}{\partial x_j}(u_j u_i) - u_i\frac{\partial u_j}{\partial x_j} = \frac{\partial}{\partial x_j}(u_j u_i) \tag{2.20}$$

where we take advantage of mass conservation [Equation (2.16)] in order to drop $u_i\partial u_j/\partial x_j$. Combining Equations (2.17) through (2.20) yields the Navier-Stokes equation in conservation form.

$$\rho\frac{\partial u_i}{\partial t} + \rho\frac{\partial}{\partial x_j}(u_j u_i) = -\frac{\partial p}{\partial x_i} + \frac{\partial}{\partial x_j}(2\mu s_{ji}) \tag{2.21}$$

Time (ensemble) averaging Equations (2.16) and (2.21) yields the **Reynolds averaged equations of motion in conservation form**, viz.,

$$\frac{\partial U_i}{\partial x_i} = 0 \qquad (2.22)$$

$$\rho \frac{\partial U_i}{\partial t} + \rho \frac{\partial}{\partial x_j}\left(U_j U_i + \overline{u_j' u_i'}\right) = -\frac{\partial P}{\partial x_i} + \frac{\partial}{\partial x_j}(2\mu S_{ji}) \qquad (2.23)$$

The time-averaged mass-conservation Equation (2.22) is identical to the instantaneous Equation (2.16) with the mean velocity replacing the instantaneous velocity. Subtracting Equation (2.22) from Equation (2.16) shows that the fluctuating velocity, $u_i'$, also has zero divergence. Aside from replacement of instantaneous variables by mean values, the only difference between the time-averaged and instantaneous momentum equations is the appearance of the correlation $\overline{u_i' u_j'}$. This is a time-averaged rate of momentum transfer due to the turbulence.

**Herein lies the fundamental problem of turbulence.** *In order to compute all mean-flow properties of the turbulent flow under consideration, we need a prescription for computing $\overline{u_i' u_j'}$.*

Equation (2.23) can be written in its most recognizable form by using Equation (2.20) in reverse. The resulting equation is

$$\rho \frac{\partial U_i}{\partial t} + \rho U_j \frac{\partial U_i}{\partial x_j} = -\frac{\partial P}{\partial x_i} + \frac{\partial}{\partial x_j}\left(2\mu S_{ji} - \rho\overline{u_j' u_i'}\right) \qquad (2.24)$$

Equation (2.24) is usually referred to as the **Reynolds-averaged Navier-Stokes equation (RANS)**. The quantity $-\rho\overline{u_i' u_j'}$ is known as the **Reynolds-stress tensor** and we denote it by $\rho\tau_{ij}$, so that $\tau_{ij}$ is the **specific Reynolds stress tensor** given by

$$\tau_{ij} = -\overline{u_i' u_j'} \qquad (2.25)$$

By inspection, $\tau_{ij} = \tau_{ji}$ so that this is a symmetric tensor, and thus has six independent components. Hence, we have produced six unknown quantities as a result of Reynolds averaging. Unfortunately, we have gained no additional equations. Now, for general three-dimensional flows, we have four unknown mean-flow properties, viz., pressure and the three velocity components. Along with the six Reynolds-stress components, we thus have ten unknowns. Our equations are mass conservation [Equation (2.22)] and the three components of Equation (2.24) for a grand total of four. This means our system is not yet **closed**. To close the system, we must find enough equations to solve for our unknowns.

## 2.4   The Reynolds-Stress Equation

In quest of additional equations, we can take moments of the Navier-Stokes equation. That is, we multiply the Navier-Stokes equation by a fluctuating property and time average the product. Using this procedure, we can derive a differential equation for the Reynolds-stress tensor. To illustrate the process, we introduce some special notation. Let $\mathcal{N}(u_i)$ denote the "Navier-Stokes operator," viz.,

$$\mathcal{N}(u_i) = \rho\frac{\partial u_i}{\partial t} + \rho u_k\frac{\partial u_i}{\partial x_k} + \frac{\partial p}{\partial x_i} - \mu\frac{\partial^2 u_i}{\partial x_k \partial x_k} \tag{2.26}$$

The viscous term has been simplified by noting from mass conservation (for incompressible flow) that $s_{ki,k} = u_{i,kk}$. The Navier-Stokes equation can be written symbolically as

$$\mathcal{N}(u_i) = 0 \tag{2.27}$$

In order to derive an equation for the Reynolds stress tensor, we form the following time average.

$$\overline{u_i'\mathcal{N}(u_j) + u_j'\mathcal{N}(u_i)} = 0 \tag{2.28}$$

Note that, consistent with the symmetry of the Reynolds stress tensor, the resulting equation is also symmetric in $i$ and $j$. For the sake of clarity, we proceed term by term. Also, for economy of space, we use tensor notation for derivatives throughout the time averaging process. Non-obvious results in the following equations usually involve the continuity equation ($\partial u_i/\partial x_i = \partial u_i'/\partial x_i = 0$) in various ways. First, we consider the **unsteady term**.

$$\begin{aligned}
\overline{u_i'(\rho u_j)_{,t} + u_j'(\rho u_i)_{,t}} &= \overline{\rho u_i'(U_j + u_j')_{,t}} + \overline{\rho u_j'(U_i + u_i')_{,t}} \\
&= \overline{\rho u_i' U_{j,t}} + \overline{\rho u_i' u_{j,t}'} + \overline{\rho u_j' U_{i,t}} + \overline{\rho u_j' u_{i,t}'} \\
&= \overline{\rho u_i' u_{j,t}'} + \overline{\rho u_j' u_{i,t}'} \\
&= \rho\overline{(u_i' u_j')_{,t}} \\
&= -\rho\frac{\partial \tau_{ij}}{\partial t} \tag{2.29}
\end{aligned}$$

Turning to the **convective term**, we have

$$\begin{aligned}
\overline{\rho u_i' u_k u_{j,k} + \rho u_j' u_k u_{i,k}} &= \overline{\rho u_i'(U_k + u_k')(U_j + u_j')_{,k}} \\
&+ \overline{\rho u_j'(U_k + u_k')(U_i + u_i')_{,k}} \\
&= \overline{\rho u_i' U_k u_{j,k}'} + \overline{\rho u_i' u_k'(U_j + u_j')_{,k}} \\
&+ \overline{\rho u_j' U_k u_{i,k}'} + \overline{\rho u_j' u_k'(U_i + u_i')_{,k}} \\
&= \rho U_k\overline{(u_i' u_j')_{,k}} + \overline{\rho u_i' u_k'}U_{j,k} \\
&+ \overline{\rho u_j' u_k'}U_{i,k} + \overline{\rho u_k'(u_i' u_j')_{,k}} \\
&= -\rho U_k\frac{\partial \tau_{ij}}{\partial x_k} - \rho\tau_{ik}\frac{\partial U_j}{\partial x_k} - \rho\tau_{jk}\frac{\partial U_i}{\partial x_k} \\
&+ \rho\frac{\partial}{\partial x_k}(\overline{u_i' u_j' u_k'}) \tag{2.30}
\end{aligned}$$

In order to arrive at the final line of Equation (2.30), we use the fact that $\partial u_k'/\partial x_k = 0$. The **pressure gradient** term is straightforward.

$$
\begin{aligned}
\overline{u_i' p_{,j} + u_j' p_{,i}} &= \overline{u_i'(P + p')_{,j}} + \overline{u_j'(P + p')_{,i}} \\
&= \overline{u_i' p_{,j}'} + \overline{u_j' p_{,i}'} \\
&= \overline{u_i' \frac{\partial p'}{\partial x_j}} + \overline{u_j' \frac{\partial p'}{\partial x_i}}
\end{aligned}
\tag{2.31}
$$

Finally, the **viscous term** yields

$$
\begin{aligned}
\overline{\mu(u_i' u_{j,kk} + u_j' u_{i,kk})} &= \overline{\mu u_i'(U_j + u_j')_{,kk}} + \overline{\mu u_j'(U_i + u_i')_{,kk}} \\
&= \overline{\mu u_i' u_{j,kk}'} + \overline{\mu u_j' u_{i,kk}'} \\
&= \overline{\mu (u_i' u_{j,k}')_{,k}} + \overline{\mu (u_j' u_{i,k}')_{,k}} - 2\overline{\mu u_{i,k}' u_{j,k}'} \\
&= \overline{\mu (u_i' u_j')_{,kk}} - 2\overline{\mu u_{i,k}' u_{j,k}'} \\
&= -\mu \frac{\partial^2 \tau_{ij}}{\partial x_k \partial x_k} - 2\mu \overline{\frac{\partial u_i'}{\partial x_k} \frac{\partial u_j'}{\partial x_k}}
\end{aligned}
\tag{2.32}
$$

Collecting terms, we arrive at the equation for the Reynolds stress tensor.

$$
\begin{aligned}
\frac{\partial \tau_{ij}}{\partial t} + U_k \frac{\partial \tau_{ij}}{\partial x_k} = {}&- \tau_{ik} \frac{\partial U_j}{\partial x_k} - \tau_{jk} \frac{\partial U_i}{\partial x_k} + 2\nu \overline{\frac{\partial u_i'}{\partial x_k} \frac{\partial u_j'}{\partial x_k}} + \overline{\frac{u_i'}{\rho} \frac{\partial p'}{\partial x_j}} + \overline{\frac{u_j'}{\rho} \frac{\partial p'}{\partial x_i}} \\
&+ \frac{\partial}{\partial x_k} \left[ \nu \frac{\partial \tau_{ij}}{\partial x_k} + \overline{u_i' u_j' u_k'} \right]
\end{aligned}
\tag{2.33}
$$

We have gained six new equations, one for each independent component of the Reynolds-stress tensor. However, we have also generated 22 new unknowns! Specifically, accounting for all symmetries, we have the following.

$$
\overline{u_i' u_j' u_k'} \;\rightarrow\; 10 \text{ unknowns}
$$

$$
2\nu \overline{\frac{\partial u_i'}{\partial x_k} \frac{\partial u_j'}{\partial x_k}} \;\rightarrow\; 6 \text{ unknowns}
$$

$$
\overline{\frac{u_i'}{\rho} \frac{\partial p'}{\partial x_j}} + \overline{\frac{u_j'}{\rho} \frac{\partial p'}{\partial x_i}} \;\rightarrow\; 6 \text{ unknowns}
$$

With a little rearrangement of terms, we can cast the **Reynolds-stress equation** in suitably compact form, viz.,

$$
\frac{\partial \tau_{ij}}{\partial t} + U_k \frac{\partial \tau_{ij}}{\partial x_k} = -\tau_{ik} \frac{\partial U_j}{\partial x_k} - \tau_{jk} \frac{\partial U_i}{\partial x_k} + \epsilon_{ij} - \Pi_{ij} + \frac{\partial}{\partial x_k} \left[ \nu \frac{\partial \tau_{ij}}{\partial x_k} + C_{ijk} \right] \tag{2.34}
$$

where

$$\Pi_{ij} = \overline{\frac{p'}{\rho} \left( \frac{\partial u_i'}{\partial x_j} + \frac{\partial u_j'}{\partial x_i} \right)} \tag{2.35}$$

$$\epsilon_{ij} = 2\nu \overline{\frac{\partial u_i'}{\partial x_k} \frac{\partial u_j'}{\partial x_k}} \tag{2.36}$$

$$\rho C_{ijk} = \rho \overline{u_i' u_j' u_k'} + \overline{p' u_i'} \delta_{jk} + \overline{p' u_j'} \delta_{ik} \tag{2.37}$$

This exercise illustrates the closure problem of turbulence. Because of the nonlinearity of the Navier-Stokes equation, as we take higher and higher moments, we generate additional unknowns at each level. At no point will this procedure balance our unknowns/equations ledger. On physical grounds, this is not a particularly surprising situation. After all, such operations are strictly mathematical in nature, and introduce no additional physical principles. *In essence, Reynolds averaging is a brutal simplification that loses much of the information contained in the Navier-Stokes equation.* The function of turbulence modeling is to devise approximations for the unknown correlations in terms of flow properties that are known so that a sufficient number of equations exists. In making such approximations, we close the system.

## 2.5  The Scales of Turbulence

In Chapter 1, we introduced the **Kolmogorov length, velocity and time scales**, which are characteristic of the smallest eddies. We also discussed the **integral length scale**, $\ell$, which is representative of the energy-bearing eddies. While these are some of the most useful scales for describing turbulence, there are others that are commonly used. The purpose of this section is to further quantify the most commonly used turbulence scales, and briefly introduce the concept of **two-point correlations**.

### 2.5.1  Turbulence Intensity

The Kolmogorov scales, defined in Equation (1.1), provide an estimate of the length, velocity and time scales for the smallest eddies in a turbulent flow. The integral length scale, whose definition has been deferred to this chapter, is a characteristic size of the energy-bearing eddies. Another important measure of any turbulent flow is how intense the turbulent fluctuations are. We quantify this in terms of the specific normal Reynolds stress components, $\overline{u'^2}$, $\overline{v'^2}$ and $\overline{w'^2}$. These three normal Reynolds stresses can also be regarded as the kinetic energy

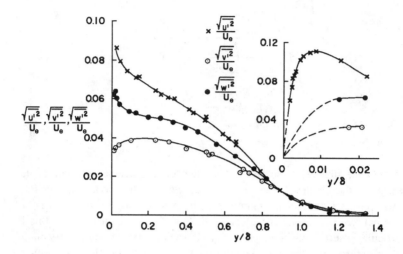

Figure 2.5: *Turbulence intensities for a flat-plate boundary layer of thickness $\delta$. The inset shows values very close to the surface [From Klebanoff (1955)].*

per unit mass of the fluctuating velocity field in the three coordinate directions. These Reynolds stresses are often normalized relative to the freestream mean-flow velocity, $U_e$, according to

$$\hat{u} \equiv \frac{\sqrt{\overline{u'^2}}}{U_e}, \quad \hat{v} \equiv \frac{\sqrt{\overline{v'^2}}}{U_e}, \quad \hat{w} \equiv \frac{\sqrt{\overline{w'^2}}}{U_e} \qquad (2.38)$$

The quantities $\hat{u}$, $\hat{v}$ and $\hat{w}$ are known as the **relative intensities** in the $x$, $y$ and $z$ directions, respectively.

Figure 2.5 displays the relative intensities for an incompressible flat-plate boundary layer. As shown, the three intensities have different values throughout most of the boundary layer. This is true because the turbulence is **anisotropic**, i.e., the normal stress components are unequal. As a rough but useful approximation for a flat-plate boundary layer, we find that

$$\overline{u'^2} : \overline{v'^2} : \overline{w'^2} = 4 : 2 : 3 \qquad (2.39)$$

These ratios are of course not constant through the layer; also, they are by no means universal for boundary layers, being strongly influenced by pressure gradient and compressibility.

Note that the streamwise intensity, $\hat{u} = \sqrt{\overline{u'^2}}/U_e$, exceeds 0.10, or 10%, very close to the surface. This is consistent with the instantaneous velocity profiles shown in Figure 2.2, and further reinforces the claim that the turbulent fluctuations cannot be adequately treated as a small perturbation about the mean.

If we sum the three normal Reynolds stresses and multiply by $\frac{1}{2}$, we have the **turbulence kinetic energy**, which we denote by the symbol $k$. Thus, by definition,

$$k \equiv \frac{1}{2}(\overline{u'^2} + \overline{v'^2} + \overline{w'^2}) = \frac{1}{2}\overline{u_i' u_i'} \tag{2.40}$$

This is the kinetic energy of the turbulent fluctuations per unit mass, and is the same as the quantity defined in Equation (1.3).

As a concluding comment, many turbulence models in current use cannot distinguish the individual normal Reynolds stresses. Rather, only $k$ is provided from the turbulence model. When this is true, we often specify relative turbulence intensity by assuming the fluctuations are more-or-less **isotropic**, i.e., that $\overline{u'^2} \approx \overline{v'^2} \approx \overline{w'^2}$. We then define

$$T' \equiv 100\sqrt{\frac{2}{3}\frac{k}{U_e^2}} \tag{2.41}$$

which gives the relative intensity in percent.

## 2.5.2  Two-Point Correlation Tensors and Related Scales

All of the discussion in this chapter thus far has dealt with **single-point correlations**. That is, we have been dealing with correlations of turbulent fluctuations at a single point in the flowfield. However, as discussed at the end of Subsection 1.3.2, turbulent eddies are large (and long lived). Consequently, it cannot in general be described entirely in terms of local flow properties. Townsend (1976) states this succinctly as follows.

> *Unlike the molecular motion of gases, the motion at any point in a turbulent flow affects the motion at other distant points through the pressure field, and an adequate description cannot be obtained by considering only mean values associated with single fluid particles. This might be put by saying that turbulent motion is less random and more [organized] than molecular motion, and that to describe the [organization] of the flow requires mean values of the functions of the flow variables for two or more particles at two or more positions.*

In this subsection, we introduce the notion of **two-point correlations**, and introduce related time and length scales characteristic of turbulent motion.

There are two types of two-point correlations commonly used in experimental and theoretical turbulence studies. One involves a separation in time while the other is based on a spatial separation. The two are related by Taylor's (1935) hypothesis, which assumes temporal and spatial separations are related by

$$\frac{\partial}{\partial t} = -U\frac{\partial}{\partial x} \tag{2.42}$$

This implies the turbulent fluctuations are convected along at the mean-flow speed, $U$. The Taylor hypothesis is valid provided the turbulent fluctuations are sufficiently weak to avoid inducing significant alterations in the rate at which they are convected. This relationship permits inferring more-relevant two-point space-correlation information from easier-to-measure one-point time-correlation data.

Considering first correlation of velocities at one point and two different times, we define the **autocorrelation tensor**, viz.,

$$\mathcal{R}_{ij}(\mathbf{x}, t; t') = \overline{u'_i(\mathbf{x}, t) u'_j(\mathbf{x}, t + t')} \tag{2.43}$$

That is, we time average the fluctuating quantities at the same point in space but at different times. To see the connection to single-point statistics, note that the turbulence kinetic energy is half the trace of $\mathcal{R}_{ij}$ with $t' = 0$, i.e.,

$$k(\mathbf{x}, t) = \frac{1}{2}\mathcal{R}_{ii}(\mathbf{x}, t; 0) \tag{2.44}$$

A useful time scale characteristic of the energy-bearing eddies can be obtained by integrating $\mathcal{R}_{ii}$ over all possible values of $t'$. Thus, we arrive at the **integral time scale**,

$$\tau(\mathbf{x}, t) = \int_0^\infty \frac{\mathcal{R}_{ii}(\mathbf{x}, t; t')}{2k(\mathbf{x}, t)} dt' \tag{2.45}$$

In defining $\tau(\mathbf{x}, t)$, we have normalized $\mathcal{R}_{ii}$ relative to $k$. For experimental work involving stationary turbulence, i.e., turbulence for which mean values are independent of time, we commonly work with the single streamwise component $\mathcal{R}_{11}(\mathbf{x}; t') = \overline{u'(\mathbf{x}, t) u'(\mathbf{x}, t + t')}$. Normalizing with respect to $\overline{u'^2}$, we arrive at the **Eulerian time-correlation coefficient** defined by

$$\mathcal{R}_E(\mathbf{x}; t') \equiv \frac{\overline{u'(\mathbf{x}, t) u'(\mathbf{x}, t + t')}}{\overline{u'^2}(\mathbf{x})} \tag{2.46}$$

By definition, $\mathcal{R}_E = 1$ when $t' = 0$. For large values of $t'$, we expect the fluctuations to be uncorrelated so that $\mathcal{R}_E \to 0$ as $|t'| \to \infty$. Finally, shifting the time origin shows that $\mathcal{R}_{11}(\mathbf{x}; t') = \mathcal{R}_{11}(\mathbf{x}; -t')$, so that $\mathcal{R}_E$ is an even function of $t'$. Figure 2.6 shows a typical Eulerian time-correlation coefficient.

We can determine another time scale by noting the shape of the Eulerian time-correlation coefficient for small time displacement, $t'$. This is determined, of course, mainly by the small dissipating eddies. That is, expanding in Taylor series about $t' = 0$, we have

$$\mathcal{R}_E(\mathbf{x}; t') = 1 + \frac{t'^2}{2}\left(\frac{\partial^2 \mathcal{R}_E}{\partial t'^2}\right)_{t'=0} + \cdots \approx 1 - \left(\frac{t'}{\tau_E}\right)^2 \tag{2.47}$$

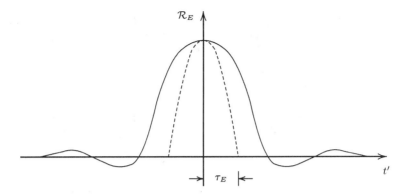

Figure 2.6: *A typical Eulerian time-correlation coefficient (——) with its osculating parabola (- - -).*

where we define the **micro-time scale**, $\tau_E$, as

$$\tau_E \equiv \sqrt{\frac{-2}{(\partial^2 \mathcal{R}_E / \partial t'^2)_{t'=0}}} \tag{2.48}$$

Figure 2.6 shows how Equation (2.47) relates geometrically to the exact time-correlation coefficient. Truncating beyond the term quadratic in $t'$ yields a parabola known as the **osculating parabola**. Its curvature matches that of the exact curve at $t' = 0$. The intercept with the horizontal axis is at $t = \tau_E$.

Turning now to the two-point space correlation, we consider two points in the flow, say $\mathbf{x}$ and $\mathbf{x}+\mathbf{r}$, and do our time average. The **two-point velocity correlation tensor** is defined as

$$R_{ij}(\mathbf{x}, t; \mathbf{r}) = \overline{u_i'(\mathbf{x}, t) u_j'(\mathbf{x} + \mathbf{r}, t)} \tag{2.49}$$

Here, the vector $\mathbf{r}$ is the displacement vector between the two points in the flow. As with the autocorrelation tensor, the turbulence kinetic energy is simply one half the trace of $R_{ij}$ with zero displacement. viz.,

$$k(\mathbf{x}, t) = \frac{1}{2} R_{ii}(\mathbf{x}, t; \mathbf{0}) \tag{2.50}$$

Normalizing $R_{ij}$ with respect to $k$, the **integral length scale**, $\ell$, is defined as the integral of $R_{ii}$ over all displacements, $r = |\mathbf{r}|$, so that

$$\ell(\mathbf{x}, t) \equiv \frac{3}{16} \int_0^\infty \frac{R_{ii}(\mathbf{x}, t; r)}{k(\mathbf{x}, t)} dr \tag{2.51}$$

where $3/16$ is a scaling factor.

In an entirely analogous manner to our analysis of two-point time correlations, we can determine a length scale corresponding to the smallest eddies. We work with the **longitudinal correlation function** for stationary turbulence defined by

$$f(x;r) \equiv \frac{R_{11}(x;r)}{\overline{u'^2}(x)} \tag{2.52}$$

Constructing the osculating parabola for $f(x)$, we find the **Taylor microscale** given by

$$\lambda \equiv \sqrt{\frac{-2}{(\partial^2 f/\partial x^2)_{x=0}}} \tag{2.53}$$

Taylor's hypothesis tells us the micro-time scale, $\tau_E$, is related to $\lambda$ by

$$\lambda = U\tau_E \tag{2.54}$$

As a final comment, when the turbulence is homogeneous and isotropic, the analysis of Taylor (1935) shows that the turbulence kinetic energy decays according to

$$\frac{dk}{dt} = -\frac{10\nu k}{\lambda^2} \tag{2.55}$$

To see how $\lambda$ relates to the Kolmogorov length, we note that for homogeneous, isotropic turbulence, the rate of decay of $k$ is simply the dissipation rate, $\epsilon$, so that

$$\frac{dk}{dt} = -\epsilon \quad \Longrightarrow \quad \epsilon = \frac{10\nu k}{\lambda^2} \tag{2.56}$$

Provided the Reynolds number is not too small, Taylor argues that $\epsilon \sim k^{3/2}/\ell$. We can sharpen the estimate by appealing to measurements that indicate

$$\epsilon \approx 0.09\frac{k^{3/2}}{\ell} \quad \Longrightarrow \quad k \approx \left(\frac{\epsilon\ell}{0.09}\right)^{2/3} \tag{2.57}$$

Then, using the definition of the Kolmogorov length, $\eta = (\nu^3/\epsilon)^{1/4}$, [see Equation (1.1)] combining Equations (2.56) and (2.57) yields

$$\frac{\lambda}{\eta} \approx 7\left(\frac{\ell}{\eta}\right)^{1/3} \tag{2.58}$$

Since $\ell/\eta$ must be at least $10^3$ to have a well-defined inertial subrange, the Taylor microscale will be at least 70 times the Kolmogorov length. It will typically lie within the inertial subrange, but well above the range of the very smallest eddies.

Such a hybrid parameter is of questionable value in turbulence modeling research, which, for the sake of simplicity, attempts to separate the physics of the large eddies from that of the small eddies. Recall, for example, how the assumption of a "wide separation of scales" is used to deduce the Kolmogorov $-5/3$ law (Subsection 1.3.4) and the logarithmic law of the wall (Subsection 1.3.5). To understand why the Taylor microscale is a hybrid length scale, observe that we can use Equation (2.56) to solve for $\lambda$, viz.,

$$\lambda = \sqrt{\frac{10\nu k}{\epsilon}} \qquad (2.59)$$

Hence, this length scale involves a quantity characteristic of the large, energy-bearing eddies, $k$, as well as quantities characteristic of the small, dissipating eddies, $\nu$ and $\epsilon$. Because the Taylor microscale is generally too small to characterize large eddies and too large to characterize small eddies, it has generally been ignored in most turbulence-modeling research. The same comments apply to the micro-time scale, $\tau_e$.

# Problems

**2.1** Develop the time-averaged form of the equation of state for a perfect gas, $p = \rho R T$, accounting for turbulent fluctuations in the instantaneous pressure, $p$, density, $\rho$, and temperature, $T$.

**2.2** Suppose we have a velocity field that consists of: (i) a slowly varying component $U(t) = U_0 e^{-t/\tau}$ where $U_0$ and $\tau$ are constants and (ii) a rapidly varying component $u' = a U_0 \cos(2\pi t/\epsilon^2 \tau)$ where $a$ and $\epsilon$ are constants with $\epsilon \ll 1$. We want to show that by choosing $T = \epsilon\tau$, the limiting process in Equation (2.9) makes sense.

(a) Compute the exact time average of $u = U + u'$.

(b) Replace T by $\epsilon\tau$ in the slowly varying part of the time average of $u$ and let $t_f = \epsilon^2 \tau$ in the fluctuating part of $u$ to show that

$$\overline{U + u'} = U(t) + O(\epsilon)$$

where $O(\epsilon)$ denotes a quantity that goes to zero linearly with $\epsilon$ as $\epsilon \to 0$.

(c) Repeat Parts (a) and (b) for $du/dt$, and verify that in order for Equation (2.13) to hold, necessarily $a \ll \epsilon$.

**2.3** For an imposed periodic mean flow, a standard way of decomposing flow properties is to write

$$u(\mathbf{x}, t) = U(\mathbf{x}) + \hat{u}(\mathbf{x}, t) + u'(\mathbf{x}, t)$$

where $U(\mathbf{x})$ is the mean-value, $\hat{u}(\mathbf{x}, t)$ is the organized response component due to the imposed organized unsteadiness, and $u'(\mathbf{x}, t)$ is the turbulent fluctuation. $U(\mathbf{x})$ is defined as in Equation (2.5). We also use the **Phase Average** defined by

$$<u(\mathbf{x}, t)> \; \equiv \; \lim_{N \to \infty} \frac{1}{N} \sum_{n=0}^{N-1} u(\mathbf{x}, t + n\tau)$$

where $\tau$ is the period of the imposed excitation. Then, by definition,

$$<u(\mathbf{x}, t)> \; = U(\mathbf{x}) + \hat{u}(\mathbf{x}, t), \quad \overline{<u(\mathbf{x}, t)>} \; = U(\mathbf{x}), \quad <\hat{u}(\mathbf{x}, t)> \; = \hat{u}(\mathbf{x}, t)$$

Verify the following. Do not assume that $\hat{u}$ is sinusoidal.

| | | |
|---|---|---|
| (a) $<U> \; = U$ | (d) $<u'> \; = 0$ | (g) $<\hat{u} u'> \; = 0$ |
| (b) $\bar{\hat{u}} = 0$ | (e) $\overline{\hat{u} v'} = 0$ | (h) $<U v> \; = U <v>$ |
| (c) $\overline{u'} = 0$ | (f) $<\hat{u} v> \; = \hat{u} <v>$ | |

**2.4** Compute the difference between the Reynolds average of a quadruple product $\phi \psi \xi \upsilon$ and the product of the means, $\Phi \Psi \Xi \Upsilon$.

**2.5** For an incompressible flow, we have an imposed freestream velocity given by

$$u(x,t) = U_o(1 - ax) + U_o ax \sin 2\pi ft$$

where $a$ is a constant of dimension 1/length, $U_o$ is a constant reference velocity, and $f$ is frequency. Integrating over one period, compute the average pressure gradient, $dP/dx$, for $f = 0$ and $f \neq 0$ in the freestream where the inviscid Euler equation holds, i.e.,

$$\rho\frac{\partial u}{\partial t} + \rho u\frac{\partial u}{\partial x} = -\frac{\partial p}{\partial x}$$

**2.6** Consider the Reynolds-stress equation as stated in Equation (2.34).

(a) Show how Equation (2.34) follows from Equation (2.33).

(b) Contract Equation (2.34), i.e., set $i = j$ and perform the indicated summation, to derive a differential equation for the kinetic energy of the turbulence per unit mass defined by $k \equiv \frac{1}{2}\overline{u_i'u_i'}$.

**2.7** If we rotate the coordinate system about the $z$ axis by an angle $\theta$, the Reynolds stresses for an incompressible two-dimensional boundary layer transform according to:

$$\tau_{xx}' = \frac{1}{2}(\tau_{xx} + \tau_{yy}) + \frac{1}{2}(\tau_{xx} - \tau_{yy})\cos 2\theta + \tau_{xy}\sin 2\theta$$

$$\tau_{yy}' = \frac{1}{2}(\tau_{xx} + \tau_{yy}) - \frac{1}{2}(\tau_{xx} - \tau_{yy})\cos 2\theta - \tau_{xy}\sin 2\theta$$

$$\tau_{xy}' = \tau_{xy}\cos 2\theta - \frac{1}{2}(\tau_{xx} - \tau_{yy})\sin 2\theta$$

$$\tau_{zz}' = \tau_{zz}$$

Assume the normal Reynolds stresses, $\tau_{xx} = -\overline{u'^2}$, etc. are given by Equation (2.39), and that the Reynolds shear stress is $\tau_{xy} = -\overline{u'v'} \approx \frac{3}{10}k$.

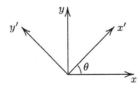

**Problem 2.7**

(a) Determine the angle the *principal axes* make with the $xy$ axes, i.e., the angle that yields $\tau_{xy}' = 0$.

(b) What is the Reynolds-stress tensor, $\tau_{ij}'$, in the *principal axis* system?

**2.8** Verify that, for homogeneous-isotropic turbulence, the ratio of the *micro-time scale*, $\tau_E$, to the Kolmogorov time scale varies linearly with the isotropic turbulence-intensity parameter, $T'$.

# Chapter 3

# Algebraic Models

The simplest of all turbulence models are described as algebraic. These models use the **Boussinesq eddy-viscosity approximation**[1] to compute the Reynolds stress tensor as the product of an eddy viscosity and the mean strain-rate tensor. For computational simplicity, the eddy viscosity, in turn, is often computed in terms of a mixing length that is analogous to the mean free path in a gas. In contrast to the molecular viscosity, which is an intrinsic property of the fluid, the eddy viscosity (and hence the mixing length) depends upon the flow. Because of this, the eddy viscosity and mixing length must be specified in advance, most simply, by an *algebraic* relation between eddy viscosity and length scales of the mean flow. Thus, algebraic models are, by definition, **incomplete** models of turbulence. This is by no means a pejorative term as incomplete models have proven to be useful in many engineering fields.

We begin this chapter by first discussing molecular transport of momentum. Next we introduce Prandtl's mixing-length hypothesis and discuss its physical implications and limitations. The mixing-length model is then applied to free shear flows for which self-similar solutions hold. We discuss two modern algebraic turbulence models that are based on the mixing-length hypothesis, including applications to attached and separated wall-bounded flows. The latter applications illustrate the limit to the algebraic model's range of applicability. An interesting separated-flow replacement for algebraic models, known as the **Half-Equation Model**, improves agreement between computed and measured flow properties. The chapter concludes with a discussion of the range of applicability of algebraic and half-equation models.

---

[1] Throughout this text, we use the terminology Boussinesq **approximation** for consistency with general turbulence literature. Strictly speaking, we could more aptly describe it as the Boussinesq **assumption** since it is not an approximation in any useful sense. By contrast, specific formulas for the eddy viscosity are.

## 3.1  Molecular Transport of Momentum

To understand the motivation for the Boussinesq approximation, it is instructive to discuss momentum transport at the molecular level. However, as a word of caution, **molecules and turbulent eddies are fundamentally different**. They are so different that we will ultimately find, in Section 3.2, that the analogy between turbulent and molecular mixing is false! It is nevertheless fruitful to pursue the analogy to illustrate how important it is to check the premises underlying turbulence closure approximations. At first glance, mimicking the molecular mixing process appears to be a careful exercise in physics. As we will see, the model just cannot stand up under close scrutiny.

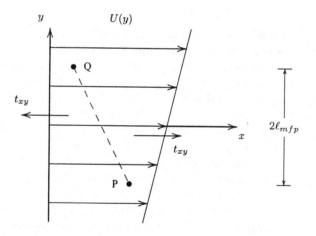

Figure 3.1: *Shear-flow schematic.*

We begin by considering a shear flow in which the velocity is given by

$$\mathbf{U} = U(y)\,\mathbf{i} \tag{3.1}$$

where $\mathbf{i}$ is a unit vector in the $x$ direction. Figure 3.1 depicts such a flow. We consider the flux of momentum across the plane $y = 0$, noting that molecular motion is random in both magnitude and direction. Molecules migrating across $y = 0$ are **typical of where they come from**. That is, molecules moving up bring a momentum deficit and vice versa. This gives rise to a shear stress $t_{xy}$.

At the molecular level, we decompose the velocity according to

$$\mathbf{u} = \mathbf{U} + \mathbf{u}'' \tag{3.2}$$

where $\mathbf{U}$ is the average velocity defined in Equation (3.1) and $\mathbf{u}''$ represents the random molecular motion. The instantaneous flux of any property across $y = 0$ is proportional to the velocity normal to the plane which, for this flow, is simply $v''$. Thus, the instantaneous flux of $x$-directed momentum, $dp_{xy}$, across a differential surface area $dS$ normal to the $y$ direction is

$$dp_{xy} = \rho(U + u'')v'' dS \qquad (3.3)$$

Performing an ensemble average over all molecules, we find

$$dP_{xy} = \overline{\rho u'' v''} dS \qquad (3.4)$$

By definition, the stress acting on $y = 0$ is given by $\sigma_{xy} = dP_{xy}/dS$. It is customary in fluid mechanics to set $\sigma_{ij} = p\delta_{ij} - t_{ij}$, where $t_{ij}$ is the viscous stress tensor. Thus,

$$t_{xy} = -\overline{\rho u'' v''} \qquad (3.5)$$

Equation (3.5) bears a strong resemblance to the Reynolds-stress tensor. This is not a coincidence. As pointed out by Tennekes and Lumley (1983), a stress that is generated as a momentum flux can always be written in this form. The only real difference is that, at the macroscopic level, the turbulent fluctuations, $u'$ and $v'$, appear in place of the random molecular fluctuations, $u''$ and $v''$. **This similarity is the basis of the Boussinesq eddy-viscosity approximation.**

Referring again to Figure 3.1, we can appeal to arguments from the kinetic theory of gases [e.g., Jeans (1962)] to determine $t_{xy}$ in terms of $U(y)$ and the fluid viscosity, $\mu$. First, consider the average number of molecules moving across unit area in the positive $y$ direction. For a perfect gas, molecular velocities follow the Maxwellian distribution so that all directions are equally probable. The average molecular velocity is the thermal velocity, $v_{th}$, which is approximately 4/3 times the speed of sound in air. On average, half of the molecules move in the positive $y$ direction while the other half move downward. The average vertical component of the velocity is $v_{th} \cos \phi$ where $\phi$ is the angle from the vertical. Integrating over a hemispherical shell, the average vertical speed is $v_{th}/2$. Thus, the average number of molecules moving across unit area in the positive $y$ direction is $n v_{th}/4$, where $n$ is the number of molecules per unit volume.

Now consider the transfer of momentum that occurs when molecules starting from point P cross the $y = 0$ plane. As stated earlier, we assume molecules are **typical of where they come from.** On the molecular scale, this is one mean free path away, the mean free path being the average distance a molecule travels between collisions with other molecules. Each molecule starting from a point P below $y = 0$ brings a momentum deficit of $m[U(0) - U(-\ell_{mfp})]$, where $m$ is the molecular mass and $\ell_{mfp}$ is the mean free path. Hence, the momentum flux from below is

$$\Delta P_- = \frac{1}{4}\rho v_{th}[U(0) - U(-\ell_{mfp})] \approx \frac{1}{4}\rho v_{th}\ell_{mfp}\frac{dU}{dy} \qquad (3.6)$$

We have replaced $U(-\ell_{mfp})$ by the first two terms of its Taylor-series expansion in Equation (3.6) and used the fact that $\rho = mn$. Similarly, each molecule moving from a point Q above the plane $y = 0$ brings a momentum surplus of $m[U(\ell_{mfp}) - U(0)]$, and the momentum flux from above is

$$\Delta P_+ = \frac{1}{4}\rho v_{th}[U(\ell_{mfp}) - U(0)] \approx \frac{1}{4}\rho v_{th}\ell_{mfp}\frac{dU}{dy} \tag{3.7}$$

Consequently, the net shearing stress is the sum of $\Delta P_-$ and $\Delta P_+$, wherefore

$$t_{xy} = \Delta P_- + \Delta P_+ \approx \frac{1}{2}\rho v_{th}\ell_{mfp}\frac{dU}{dy} \tag{3.8}$$

Hence, we conclude that the shear stress resulting from molecular transport of momentum in a perfect gas is given by

$$t_{xy} = \mu\frac{dU}{dy} \tag{3.9}$$

where $\mu$ is the molecular viscosity defined by

$$\mu = \frac{1}{2}\rho v_{th}\ell_{mfp} \tag{3.10}$$

The arguments leading to Equations (3.9) and (3.10) are approximate and only roughly represent the true statistical nature of molecular motion. Interestingly, Jeans (1962) indicates that a precise analysis yields $\mu = 0.499\rho v_{th}\ell_{mfp}$, wherefore our approximate analysis is remarkably accurate! However, we have made two implicit assumptions in our analysis that require justification.

**First,** we have truncated the Taylor series appearing in Equations (3.6) and (3.7) at the linear terms. For this approximation to be valid, we must have $\ell_{mfp}|d^2U/dy^2| \ll |dU/dy|$. The quantity $L$ defined by

$$L \equiv \frac{|dU/dy|}{|d^2U/dy^2|} \tag{3.11}$$

is a length scale characteristic of the mean flow. Thus, the linear relation between stress and strain-rate implied by Equation (3.9) is valid provided the Knudsen number, $Kn$, is very small, i.e.,

$$Kn = \ell_{mfp}/L \ll 1 \tag{3.12}$$

For most practical flow conditions,[2] the mean free path is several orders of magnitude smaller than any characteristic length scale of the mean flow. Thus, Equation (3.12) is satisfied for virtually all engineering problems.

---

[2]Two noteworthy exceptions are very-high altitude flight and micron-scale flows such as those encountered in micro-machinery.

**Second**, in computing the rate at which molecules cross $y = 0$, we assumed that $u''$ remained Maxwellian even in the presence of shear. This will be true if molecules experience many collisions on the time scale of the mean flow. Now, the average time between collisions is $\ell_{mfp}/v_{th}$. The characteristic time scale for the mean flow is $|dU/dy|^{-1}$. Thus, we also require that

$$\ell_{mfp} \ll \frac{v_{th}}{|dU/dy|} \tag{3.13}$$

Since $v_{th}$ is of the same order of magnitude as the speed of sound, the right-hand side of Equation (3.13) defines yet another mean-flow length scale. As above, the mean free path is several orders smaller than this length scale for virtually all flows of engineering interest.

## 3.2 The Mixing-Length Hypothesis

Prandtl (1925) put forth the mixing-length hypothesis. He visualized a simplified model for turbulent motion in which fluid particles coalesce into lumps that cling together and move as a unit. He further visualized that in a shear flow such as that depicted in Figure 3.1, the lumps retain their $x$-directed momentum for a distance in the $y$ direction, $\ell_{mix}$, that he called the mixing length. In analogy to the molecular momentum transport process with Prandtl's lump of fluid replacing the molecule and $\ell_{mix}$ replacing $\ell_{mfp}$, we can say that similar to Equation (3.8),

$$\rho \tau_{xy} = \frac{1}{2} \rho v_{mix} \ell_{mix} \frac{dU}{dy} \implies \tau_{xy} = \frac{1}{2} v_{mix} \ell_{mix} \frac{dU}{dy} \tag{3.14}$$

The formulation is not yet complete because the **mixing velocity**, $v_{mix}$, has not been specified. Prandtl further postulated that

$$v_{mix} = \text{constant} \cdot \ell_{mix} \left| \frac{dU}{dy} \right| \tag{3.15}$$

which makes sense on dimensional grounds. Because $\ell_{mix}$ is not a physical property of the fluid, we can always absorb the constant in Equation (3.15) and the factor 1/2 in Equation (3.14) in the mixing length. Thus, in analogy to Equations (3.9) and (3.10), Prandtl's mixing-length hypothesis leads to

$$\tau_{xy} = \nu_T \frac{dU}{dy} \tag{3.16}$$

where $\nu_T$ is the kinematic **eddy viscosity** given by

$$\nu_T = \ell_{mix}^2 \left| \frac{dU}{dy} \right| \tag{3.17}$$

Our formulation still remains incomplete since we have replaced Boussinesq's empirical eddy viscosity, $\nu_T$, with Prandtl's empirical mixing length, $\ell_{mix}$. Prandtl postulated further that for flows near solid boundaries the mixing length is proportional to distance from the surface. This turns out to be a reasonably good approximation over a limited portion of a turbulent wall flow. As we will see in Section 3.3, for free shear flows such as jets, wakes and mixing layers, the mixing length is proportional to the width of the layer, $\delta$. However, each of these flows requires a different coefficient of proportionality between $\ell_{mix}$ and $\delta$. The point is, the mixing length is different for each flow (its ratio to the flow width, for example) and must be known in advance to obtain a solution.

Note that Equation (3.17) can be deduced directly from dimensional analysis. Assuming molecular transport of momentum is unimportant relative to turbulent transport, we expect molecular viscosity has no significance in a dimensional analysis. The only other dimensional parameters available in a shear flow are the assumed mixing length, $\ell_{mix}$, and the velocity gradient, $dU/dy$. (The eddy viscosity cannot depend upon $U$ since that would violate Galilean invariance.) A straightforward dimensional analysis yields Equation (3.17).

Another interesting observation follows from replacing $\tau_{xy}$ by its definition so that

$$-\overline{u'v'} = \ell_{mix}^2 \left| \frac{dU}{dy} \right| \frac{dU}{dy} \tag{3.18}$$

The mixing velocity, $v_{mix}$, must be proportional to an appropriate average of $v'$ such as the RMS value defined by $v_{rms} = (\overline{v'^2})^{1/2}$. Also, Townsend (1976) states that in most turbulent shear flows, measurements indicate

$$\left| -\overline{u'v'} \right| \approx 0.4 u_{rms} v_{rms} \tag{3.19}$$

Consequently, if $v_{rms} \sim v_{mix}$, comparison of Equations (3.15) and (3.18) shows that the mixing-length model implies $v_{rms}$ and $u_{rms}$ are of the same order of magnitude. This is generally true although $u_{rms}$ is usually 25% to 75% larger than $v_{rms}$ for typical shear flows.

At this point, we need to examine the appropriateness of the mixing-length hypothesis in representing the turbulent transport of momentum. Because we have made a direct analogy to the molecular transport process, we have implicitly made the same two basic assumptions we made for molecular transport. Specifically, we have assumed that the Boussinesq approximation holds and that the turbulence is unaltered by the mean shear. Unfortunately, neither condition is rigorously satisfied in practice!

**Concerning the Boussinesq approximation**, its applicability depends upon the Knudsen number being small. Close to a solid boundary, for example, the mixing length is approximately linear with distance from the surface, $y$. Specifically, measurements indicate that $\ell_{mix} \approx 0.41y$. In the same vicinity, the velocity follows the well-known law of the wall [see Subsection 1.3.5], and the velocity gradient varies inversely with $y$. Thus, the length $L$ defined in Equation (3.11) is equal to $y$. Consequently, the Knudsen number is of order one, i.e.,

$$Kn = \ell_{mix}/L \approx 0.41 \qquad (3.20)$$

Hence, the linear stress/strain-rate relation of Equation (3.16) is suspect.

**Concerning the effect of the mean shear on the turbulence,** the assumed lifetime of Prandtl's lumps of fluid is $\ell_{mix}/v_{mix}$. Reference to Equation (3.15) shows that this time is proportional to $|dU/dy|^{-1}$. Hence, the analog to Equation (3.13) is

$$\ell_{mix} \sim \frac{v_{mix}}{|dU/dy|} \qquad (3.21)$$

Because we do not have $\ell_{mix} \ll v_{mix}/|dU/dy|$, Equation (3.21) tells us that the lumps of fluid will undergo changes as they travel from points P and Q toward $y = 0$. This is indeed consistent with the observed nature of turbulent shear flows. Tennekes and Lumley (1983) describe the situation by saying, "the general conclusion must be that turbulence in a shear flow cannot possibly be in a state of equilibrium which is independent of the flow field involved. The turbulence is continually trying to adjust to its environment, without ever succeeding."

Thus, the theoretical foundation of the mixing-length hypothesis is a bit flimsy to say the least. On the one hand, this is a forewarning that a turbulence model built on this foundation is unlikely to possess a very wide range of applicability. On the other hand, as the entire formulation is empirical in its essence, the usefulness of and justification for any of its approximations ultimately lies in how well the model performs in applications, and we defer to the applications of the following sections as its justification.

As a pleasant surprise, we will see that despite its theoretical shortcomings, the mixing-length model does an excellent job of reproducing measurements. It can be easily calibrated for a specific class of flows, and the model's predictions are consistent with measurements provided we don't depart too far from the established data base used to calibrate the mixing length. Eddy-viscosity models based on the mixing length have been fine tuned for many flows since 1925, most notably by Cebeci and Smith (1974). Strictly speaking, the term **equilibrium** is nonsensical in the context of turbulent shear flows since, as noted above, turbulence is continually attempting to adjust to its environment, without ever succeeding. Nevertheless, most turbulence researchers describe certain flows as **equilibrium turbulent flows.** What they actually mean is a relatively simple flow with slowly-varying properties. Most flows of this type can be accurately described by a mixing-length computation. In this spirit, a fitting definition of equilibrium turbulent flow might be a flow that can be accurately described using a mixing-length model!

## 3.3  Application to Free Shear Flows

Our first applications will be to incompressible **free shear flows**. A flow is termed free if it is not bounded by solid surfaces. Figure 3.2 illustrates three different types of free shear flows, viz., the **far wake**, the **mixing layer**, and the **jet**. A wake forms downstream of any object placed in a stream of fluid; we will consider only the two-dimensional wake. A mixing layer occurs between two parallel streams moving at different speeds; for the case shown in the figure, the lower stream is at rest. A jet occurs when fluid is ejected from a nozzle or orifice. We will assume the jet issues into a quiescent fluid, and we will analyze both the (two-dimensional) plane jet and the (axisymmetric) round jet.

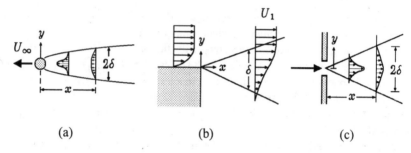

Figure 3.2: *Free shear flows:  (a) far wake;  (b) mixing layer;  (c) jet.*

All three of these flows approach what is known as **self similarity** far enough downstream that details of the geometry and flow conditions near $x = 0$ become unimportant. The velocity component $U(x, y)$, for example, can be expressed as

$$U(x, y) = u_o(x)F(y/\delta(x)) \qquad (3.22)$$

Similar expressions hold for $\tau_{xy}$ and $\nu_T$. This amounts to saying that two velocity profiles located at different $x$ stations have the same shape when plotted in the scaled form $U(x,y)/u_o(x)$ [or $(U_\infty - U(x,y))/u_o(x)$ for the wake] versus $y/\delta(x)$. Flows with this property are also referred to as **self preserving**.

Free shear flows are interesting building-block cases to test a turbulence model on for several reasons. First, there are no solid boundaries so that we avoid the complications boundaries add to the complexity of a turbulent flow. Second, they are mathematically easy to calculate because similarity solutions exist, where the Reynolds-averaged equations of motion can be reduced to ordinary differential equations. This greatly simplifies the task of obtaining a solution. Third, there is a wealth of experimental data available to test model predictions against.

The standard boundary-layer approximations hold for all three of the shear flows considered in this section. Additionally, molecular transport of momentum is negligible compared to turbulent transport. Since all three flows have constant pressure, the equations of motion are (with $j = 0$ for two-dimensional flow and $j = 1$ for axisymmetric flow):

$$\frac{\partial U}{\partial x} + \frac{1}{y^j} \frac{\partial}{\partial y} \left( y^j V \right) = 0 \qquad (3.23)$$

$$U \frac{\partial U}{\partial x} + V \frac{\partial U}{\partial y} = \frac{1}{y^j} \frac{\partial}{\partial y} \left( y^j \tau_{xy} \right) \qquad (3.24)$$

where $y$ is the radial coordinate when $j = 1$. Of course, while the equations are the same for all three flows, boundary conditions are different. The appropriate boundary conditions will be stated when we discuss each flow.

As a historical note, in addition to the mixing-length model, Prandtl also proposed a simpler eddy-viscosity model specifically for free shear flows, viz.,

$$\nu_T = \chi [U_{max}(x) - U_{min}(x)] \delta(x) \qquad (3.25)$$

where $U_{max}$ and $U_{min}$ are the maximum and minimum values of mean velocity in the layer, $\delta$ is the half width of the layer, and $\chi$ is a dimensionless empirical parameter that we refer to as a **closure coefficient**. This model is very convenient for free shear flows because it is a function only of $x$ by construction, and acceptable results can be obtained if $\chi$ is assumed to be constant across the layer. Consequently, laminar-flow solutions can be generalized for turbulent flow with, at most, minor notation changes. We leave application of this model to the problems section. All of the applications in this section will be done using Equations (3.16) and (3.17).

We begin by analyzing the far wake in Subsection 3.3.1. Complete details of the similarity-solution method are given for the benefit of the reader who has not had much experience with this method. The far wake is especially attractive as our first application because a simple closed-form solution can be obtained using the mixing-length model. Then, we proceed to the mixing layer in Subsection 3.3.2. While an analytical solution is possible for the mixing layer, numerical integration of the equations proves to be far simpler. Finally, we study the plane jet and the round jet in Subsection 3.3.3.

### 3.3.1 The Far Wake

Clearly, the flow in the wake of the body indicated in Figure 3.2(a) is symmetric about the $x$ axis. Thus, we solve for $0 \leq y < \infty$. Boundary conditions follow from symmetry on the axis and the requirement that the velocity approach its freestream value far from the body. Hence,

$$U(x, y) \to U_\infty \quad \text{as} \quad y \to \infty \qquad (3.26)$$

$$\frac{\partial U}{\partial y} = 0 \quad \text{at} \quad y = 0 \tag{3.27}$$

The classical approach to this problem is to linearize the momentum equation, an approximation that is strictly valid only in the far wake [Schlichting (1979)]. Thus, we say that

$$\mathbf{U}(x, y) = U_\infty \mathbf{i} - \hat{\mathbf{u}} \tag{3.28}$$

where $|\hat{\mathbf{u}}| \ll U_\infty$. The linearized momentum equation and boundary conditions become

$$U_\infty \frac{\partial \hat{u}}{\partial x} = -\frac{\partial \tau_{xy}}{\partial y} \tag{3.29}$$

$$\hat{u}(x, y) \to 0 \quad \text{as} \quad y \to \infty \tag{3.30}$$

$$\frac{\partial \hat{u}}{\partial y} = 0 \quad \text{at} \quad y = 0 \tag{3.31}$$

There is also an integral constraint that must be satisfied by the solution. If we consider a control volume surrounding the body and extending to infinity, conservation of momentum leads to the following requirement [see Schlichting (1979)],

$$\int_0^\infty \rho U \left(U_\infty - U\right) dy = \frac{1}{2} D \tag{3.32}$$

where $D$ is the drag of the body per unit width.

We use the mixing-length model to specify the Reynolds shear stress, $\tau_{xy}$, which means we write

$$\tau_{xy} = -\ell_{mix}^2 \left|\frac{\partial \hat{u}}{\partial y}\right| \frac{\partial \hat{u}}{\partial y} \tag{3.33}$$

Finally, to close our set of equations, we assume the mixing length is proportional to the half-width of the wake, $\delta(x)$ [see Figure 3.2(a)]. Thus, we say that

$$\ell_{mix} = \alpha \delta(x) \tag{3.34}$$

where $\alpha$ is a closure coefficient. Our fondest hope would be that the same value of $\alpha$ works for all free shear flows, and is independent of $y/\delta$. Unfortunately, this is not the case, which means the mixing-length model must be recalibrated for each type of shear flow.

To obtain the similarity solution to Equations (3.29) through (3.34), we proceed in a series of interrelated steps. The sequence is as follows.

1. Assume the form of the solution.

2. Transform the equations of motion.

3. Transform the boundary conditions and the integral constraint.

4. Determine the conditions required for existence of the similarity solution.

5. Solve the resulting ordinary differential equation subject to the transformed boundary conditions.

In addition to these 5 steps, we will also determine the value of the closure coefficient $\alpha$ in Equation (3.34) by comparison with experimental data.

**Step 1**. We begin by assuming the similarity solution can be written in terms of an as yet unknown velocity scale function, $u_o(x)$, and the wake half width, $\delta(x)$. Thus, we assume that the velocity can be written as

$$\hat{u}(x, y) = u_o(x)F(\eta) \qquad (3.35)$$

where the **similarity variable**, $\eta$, is defined by

$$\eta = y/\delta(x) \qquad (3.36)$$

**Step 2**. In order to transform Equation (3.29), we have to take account of the fact that we are making a formal change of dependent variables. We are transforming from $(x, y)$ space to $(x, \eta)$ space which means that derivatives must be transformed according to the chain rule of calculus. Thus, derivatives transform according to the following rules. Note that a subscript means that differentiation is done holding the subscripted variable constant.

$$
\begin{aligned}
\left(\frac{\partial}{\partial x}\right)_y &= \left(\frac{\partial x}{\partial x}\right)_y \left(\frac{\partial}{\partial x}\right)_\eta + \left(\frac{\partial \eta}{\partial x}\right)_y \left(\frac{\partial}{\partial \eta}\right)_x \\
&= \left(\frac{\partial}{\partial x}\right)_\eta + \left(\frac{\partial \eta}{\partial x}\right)_y \left(\frac{\partial}{\partial \eta}\right)_x \\
&= \left(\frac{\partial}{\partial x}\right)_\eta - \frac{\delta'(x)}{\delta(x)}\eta \left(\frac{\partial}{\partial \eta}\right)_x \qquad (3.37)
\end{aligned}
$$

$$
\begin{aligned}
\left(\frac{\partial}{\partial y}\right)_x &= \left(\frac{\partial x}{\partial y}\right)_x \left(\frac{\partial}{\partial x}\right)_\eta + \left(\frac{\partial \eta}{\partial y}\right)_x \left(\frac{\partial}{\partial \eta}\right)_x \\
&= \left(\frac{\partial \eta}{\partial y}\right)_x \left(\frac{\partial}{\partial \eta}\right)_x \\
&= \frac{1}{\delta(x)} \left(\frac{\partial}{\partial \eta}\right)_x \qquad (3.38)
\end{aligned}
$$

A prime denotes ordinary differentiation so that $\delta'(x) = d\delta/dx$ in Equation (3.37). We now proceed to transform Equation (3.29). For example, the derivatives of $\hat{u}$ are

$$\frac{\partial \hat{u}}{\partial x} = u_o' F(\eta) - \frac{u_o \delta'}{\delta} \eta \frac{dF}{d\eta} \tag{3.39}$$

$$\frac{\partial \hat{u}}{\partial y} = \frac{u_o}{\delta} \frac{dF}{d\eta} \tag{3.40}$$

Proceeding in this manner for all terms in Equation (3.29) and using the mixing-length prescription for the Reynolds stress leads to the transformed momentum equation.

$$\frac{U_\infty \delta u_o'}{u_o^2} F(\eta) - \frac{U_\infty \delta'}{u_o} \eta \frac{dF}{d\eta} = \alpha^2 \frac{d}{d\eta} \left( \left| \frac{dF}{d\eta} \right| \frac{dF}{d\eta} \right) \tag{3.41}$$

**Step 3.** Clearly, $y \to \infty$ corresponds to $\eta \to \infty$ and $y \to 0$ corresponds to $\eta \to 0$. Thus, the boundary conditions in Equations (3.30) and (3.31) transform to

$$F(\eta) \to 0 \quad \text{as} \quad \eta \to \infty \tag{3.42}$$

$$\frac{dF}{d\eta} = 0 \quad \text{at} \quad \eta = 0 \tag{3.43}$$

and the integral constraint becomes

$$\int_0^\infty F(\eta)\, d\eta = \frac{D}{2\rho U_\infty u_o \delta} \tag{3.44}$$

**Step 4.** In seeking a similarity solution, we are attempting to make a separation of variables. The two terms on the left-hand side of Equation (3.41) have coefficients that, in general, vary with $x$. Also, the right-hand side of Equation (3.44) is a function of $x$. **The condition for existence of the similarity solution is that these three coefficients be independent of $x$.** Thus, we require the following three conditions.

$$\frac{U_\infty \delta u_o'}{u_o^2} = a_1, \quad \frac{U_\infty \delta'}{u_o} = a_2, \quad \frac{D}{2\rho U_\infty u_o \delta} = 1 \tag{3.45}$$

The quantities $a_1$ and $a_2$ must, of course, be constant. Note that we could have introduced a third constant in the integral constraint, but it is unnecessary (we, in effect, absorb the third constant in $\delta$). The solution to these three simultaneous equations is simply

$$\delta(x) = \sqrt{\frac{a_2 D x}{\rho U_\infty^2}} \tag{3.46}$$

$$u_o(x) = \frac{1}{2}\sqrt{\frac{D}{a_2 \rho x}} \tag{3.47}$$

$$a_1 = -a_2 \tag{3.48}$$

**Step 5.** Finally, we expect that $F(\eta)$ will have its maximum value on the axis, and then fall monotonically to zero approaching the freestream. If this is true, then $F'(\eta)$ will be negative for all values of $\eta$ and we can replace its absolute value with $-F'(\eta)$. Taking account of Equations (3.45) through (3.48), the momentum equation now simplifies to

$$\alpha^2 \frac{d}{d\eta}\left[(F')^2\right] - a_2(\eta F' + F) = 0 \tag{3.49}$$

The second term is a perfect differential so that Equation (3.49) can be rewritten as

$$\frac{d}{d\eta}\left[\alpha^2(F')^2 - a_2\eta F\right] = 0 \tag{3.50}$$

Integrating once and imposing the symmetry condition at $\eta = 0$ [Equation (3.43)] yields

$$\alpha \frac{dF}{d\eta} = -\sqrt{a_2 \eta F} \tag{3.51}$$

where we observe that $F'(\eta)$ is everywhere less than zero. Integrating once more, we find that the solution for $F(\eta)$ is

$$F(\eta) = C^2 \left[1 - (\eta/\eta_e)^{3/2}\right]^2 \tag{3.52}$$

where $C$ is a constant of integration and $\eta_e$ is given by

$$\eta_e = (3\alpha C/\sqrt{a_2})^{2/3} \tag{3.53}$$

This solution has a peak value at $\eta = 0$ and decreases monotonically to zero as $\eta \to \eta_e$. It then increases without limit for $\eta > \eta_e$. The only way we can satisfy the far field boundary condition [Equation (3.42)] is to use Equation (3.52) for $0 \le \eta \le \eta_e$ and to use the trivial solution, $F(\eta) = 0$, for values of $\eta$ in excess of $\eta_e$.

With no loss of generality, we can set $\eta_e = 1$. To understand this, note that $\eta/\eta_e = y/[\eta_e\delta(x)]$. Hence, by setting $\eta_e = 1$ we simply rescale the $\eta$ coordinate so that $\delta(x)$ is the wake half width as originally planned. Therefore,

$$3\alpha C = \sqrt{a_2} \tag{3.54}$$

Finally, imposing the integral constraint, Equation (3.44), yields an equation for the constant $C$. Performing the integration, we have

$$\int_0^1 C^2 [1 - \eta^{3/2}]^2 \, d\eta = \frac{9}{20} C^2 = 1 \tag{3.55}$$

Therefore,

$$C = \sqrt{20}/3 = 1.491 \tag{3.56}$$

and

$$\alpha = \sqrt{a_2/20} \tag{3.57}$$

If the closure coefficient $\alpha$ were known, our solution would be completely determined at this point with Equation (3.57) specifying $a_2$. This is the nature of an **incomplete** turbulence model. The coefficient $\alpha$ is unknown because the mixing length [Equation (3.34)] is unknown a priori for this flow. To complete the solution, we appeal to experimental data [c.f. Schlichting (1979)], which show that the wake half width grows according to

$$\delta(x) \approx 0.805 \sqrt{\frac{Dx}{\rho U_\infty^2}} \tag{3.58}$$

Comparison of Equations (3.46) and (3.58) shows that the value of $a_2$ is

$$a_2 = 0.648 \tag{3.59}$$

The value of the coefficient $\alpha$ then follows from Equation (3.57), i.e.,

$$\alpha = 0.18 \tag{3.60}$$

Collecting all of this, the final solution for the far wake, according to the mixing-length model is

$$U(x, y) = U_\infty - 1.38 \sqrt{\frac{D}{\rho x}} \left[ 1 - (y/\delta)^{3/2} \right]^2 \tag{3.61}$$

where $\delta(x)$ is given by Equation (3.58). Figure 3.3 compares this profile with data of Fage and Falkner (1932) and the slightly asymmetrical wake data of Weygandt and Mehta (1995). As shown, the mixing-length model, once calibrated, does an excellent job of reproducing measured values. This solution has an interesting feature that we will see in many of our applications. Specifically, we have found a sharp turbulent/nonturbulent interface. This manifests itself in the nonanalytic behavior of the solution at $y/\delta = 1$, i.e., all derivatives of $U$ above $\partial^2 U/\partial y^2$ are discontinuous at $y/\delta = 1$. Measurements confirm existence of such

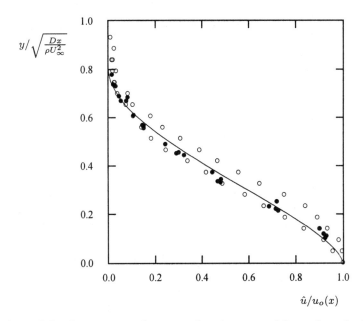

Figure 3.3: *Comparison of computed and measured far-wake velocity profiles:*
—— *Mixing length;* • *Fage and Falkner (1932);* ○ *Weygandt and Mehta (1995).*

interfaces in all turbulent flows. However, the time-averaged interface is continuous to high order, being subjected to a near-Gaussian jitter. Time averaging would thus smooth out the sharpness of the physical interface. Consistent with this smoothing, we should actually expect analytical behavior approaching the freestream. Hence, the mixing-length model is predicting a nonphysical feature.

## 3.3.2   The Mixing Layer

For the mixing layer, we consider two parallel streams with velocities $U_1$ and $U_2$. By convention, the stream with velocity $U_1$ lies above $y = 0$ and $U_1 > U_2$. The boundary conditions are thus

$$U(x, y) \to U_1 \quad \text{as} \quad y \to +\infty \tag{3.62}$$

$$U(x, y) \to U_2 \quad \text{as} \quad y \to -\infty \tag{3.63}$$

The most convenient way to solve this problem is to introduce the stream-function, $\psi$. The velocity components are given in terms of $\psi$ as follows.

$$U = \frac{\partial \psi}{\partial y} \quad \text{and} \quad V = -\frac{\partial \psi}{\partial x} \tag{3.64}$$

Equation (3.23) is automatically satisfied and the momentum equation becomes

$$\frac{\partial\psi}{\partial y}\frac{\partial^2\psi}{\partial x\partial y} - \frac{\partial\psi}{\partial x}\frac{\partial^2\psi}{\partial y^2} = \frac{\partial}{\partial y}\left[\ell_{mix}^2\left|\frac{\partial^2\psi}{\partial y^2}\right|\frac{\partial^2\psi}{\partial y^2}\right] \tag{3.65}$$

The boundary conditions on $\psi$ are

$$\frac{\partial\psi}{\partial y} \to U_1 \quad \text{as} \quad y \to +\infty \tag{3.66}$$

$$\frac{\partial\psi}{\partial y} \to U_2 \quad \text{as} \quad y \to -\infty \tag{3.67}$$

Because the velocity is obtained from the streamfunction by differentiation, $\psi$ involves a constant of integration. For the sake of uniqueness, we can specify an additional boundary condition on $\psi$, although at this point it is unclear where we should impose the extra boundary condition. The choice will become obvious when we set up the similarity solution. As with the far wake, we assume

$$\psi(x, y) = \psi_o(x)F(\eta) \tag{3.68}$$

where the similarity variable, $\eta$, is defined by

$$\eta = y/\delta(x) \tag{3.69}$$

As can be verified by substituting Equations (3.68) and (3.69) into Equation (3.65), a similarity solution exists provided we choose

$$\psi_o(x) = AU_1x \tag{3.70}$$

$$\delta(x) = Ax \tag{3.71}$$

where A is a constant to be determined. Using Equation (3.34) to determine the mixing length, after some algebra Equation (3.65) transforms to

$$\alpha^2\frac{d}{d\eta}\left[(F'')^2\right] + AFF'' = 0 \tag{3.72}$$

Note that we have removed the absolute value sign in Equation (3.65) by noting that we expect a solution with $\partial U/\partial y = \partial^2\psi/\partial y^2 > 0$. As an immediate consequence, we can simplify Equation (3.72). Specifically, expanding the first term leads to the following **linear** equation for the transformed streamfunction, $F(\eta)$.

$$2\alpha^2\frac{d^3F}{d\eta^3} + AF = 0 \tag{3.73}$$

To determine the constant of integration in the streamfunction, note that our assumed form for $\psi$ [Equation (3.68)] is consistent with letting $F(\eta)$ vanish at $\eta = 0$. This is known as the dividing streamline. Thus, our boundary conditions are

$$\frac{dF}{d\eta} \to 1 \quad \text{as} \quad \eta \to +\infty \qquad (3.74)$$

$$\frac{dF}{d\eta} \to U_2/U_1 \quad \text{as} \quad \eta \to -\infty \qquad (3.75)$$

$$F(0) = 0 \qquad (3.76)$$

For simplicity, we consider the limiting case $U_2 = 0$. This problem can be solved in closed form using elementary methods. Unfortunately, the solution is a bit complicated. Furthermore, as with the far-wake solution, the mixing-length model predicts a sharp turbulent/nonturbulent interface and it becomes a rather difficult chore to determine a straightforward relationship between the closure coefficient $\alpha$ and the constant $A$. The easier way to proceed is to solve the equation numerically for various values of $\alpha^2/A$ and compare with measurements to infer the values of $\alpha$ and $A$. Proceeding in this manner (see Program **MIXER** in Appendix C), optimum agreement between computed and measured [Liepmann and Laufer (1947)] velocity profiles occurs if we choose

$$A = 0.247 \quad \text{and} \quad \alpha = 0.071 \quad \text{(Mixing Layer)} \qquad (3.77)$$

This value of $\alpha$ is nearly identical to the value (0.070) quoted by Launder and Spalding (1972). Figure 3.4 compares computed and measured velocity profiles. The traditional definition of spreading rate, $\delta'$, for the mixing layer is the difference between the values of $y/x$ where $(U - U_2)^2/(U_1 - U_2)^2$ is 9/10 and 1/10. The values of $A$ and $\alpha$ have been selected to match the experimentally measured spreading rate, viz.,

$$\delta' = 0.115 \qquad (3.78)$$

While the computed velocity goes to zero more rapidly than measured on the low speed side of the mixing layer, the overall agreement between theory and experiment is remarkably good.

### 3.3.3    The Jet

We now analyze the two-dimensional, or **plane jet**, and the axisymmetric, or **round jet**. Referring to Figure 3.2(c), we assume the jet issues into a stagnant fluid. The jet entrains fluid from the surrounding fluid and grows in width

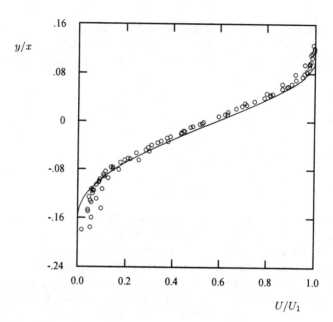

Figure 3.4: *Comparison of computed and measured velocity profiles for a mixing layer:* —— *Mixing length;* ○ *Liepmann and Laufer (1947).*

downstream of the origin. Equations (3.23) and (3.24) govern the motion with $j = 0$ corresponding to the plane jet and $j = 1$ corresponding to the round jet. As with the far wake, we take advantage of the symmetry about the $x$ axis and solve for $0 \leq y < \infty$. The boundary conditions for both the plane and the round jet are

$$U(x, y) \rightarrow 0 \quad \text{as} \quad y \rightarrow \infty \tag{3.79}$$

$$\frac{\partial U}{\partial y} = 0 \quad \text{at} \quad y = 0 \tag{3.80}$$

To insure that the momentum in the jet is conserved, our solution must satisfy the following integral constraint:

$$\pi^j \int_0^\infty U^2 y^j \, dy = \frac{1}{2} J \tag{3.81}$$

where $J$ is the momentum flux per unit mass, or, **specific momentum flux**.

To solve, we introduce the streamfunction, which can be generalized to account for the axisymmetry of the round jet, i.e.,

$$y^j U = \frac{\partial \psi}{\partial y} \quad \text{and} \quad y^j V = -\frac{\partial \psi}{\partial x} \tag{3.82}$$

The momentum equation thus becomes

$$
y^{-j} \frac{\partial \psi}{\partial y} \frac{\partial^2 \psi}{\partial x \partial y} - \frac{\partial \psi}{\partial x} \frac{\partial}{\partial y} \left( y^{-j} \frac{\partial \psi}{\partial y} \right)
$$

$$
= \frac{\partial}{\partial y} \left[ y^j \ell_{mix}^2 \left| \frac{\partial}{\partial y} \left( y^{-j} \frac{\partial \psi}{\partial y} \right) \right| \frac{\partial}{\partial y} \left( y^{-j} \frac{\partial \psi}{\partial y} \right) \right] \quad (3.83)
$$

Assuming a similarity solution of the form given in Equations (3.68) and (3.69), the appropriate forms for $\psi_o(x)$ and $\delta(x)$ are

$$
\psi_o(x) = \sqrt{\frac{J A^{j+1} x^{j+1}}{2 \pi^j}} \quad (3.84)
$$

$$
\delta(x) = A x \quad (3.85)
$$

where $A$ is a constant that will be determined by comparison with experimental data. For the jet, we expect to have $\partial U / \partial y \leq 0$. Using this fact to replace the absolute value in Equation (3.83) with a minus sign, the following ordinary differential equation for the transformed streamfunction, $F(\eta)$, results.

$$
\alpha^2 \eta^j \left[ \frac{d}{d\eta} \left( \frac{F'}{\eta^j} \right) \right]^2 = \frac{j+1}{2} A F \left( \frac{F'}{\eta^j} \right) \quad (3.86)
$$

This equation must be solved subject to the following conditions.

$$
F(0) = 0 \quad (3.87)
$$

$$
\frac{1}{\eta^j} \frac{dF}{d\eta} \to 0 \quad \text{as} \quad y \to \infty \quad (3.88)
$$

$$
\frac{d}{d\eta} \left[ \frac{1}{\eta^j} \frac{dF}{d\eta} \right] \quad \text{as} \quad y \to 0 \quad (3.89)
$$

$$
\int_0^\infty \frac{(F')^2}{\eta^j} \, d\eta = 1 \quad (3.90)
$$

Doing a numerical solution of Equation (3.86) subject to Equations (3.87) through (3.90), and comparing with experiment yields

$$
A = 0.246 \quad \text{and} \quad \alpha = 0.098 \quad \text{(Plane Jet)} \quad (3.91)
$$

$$
A = 0.233 \quad \text{and} \quad \alpha = 0.080 \quad \text{(Round Jet)} \quad (3.92)
$$

The values for the mixing-length coefficient, $\alpha$, are about 8% larger than corresponding values (0.090 and 0.075) quoted by Launder and Spalding (1972), which is within the bounds of experimental error. The values quoted in Equations (3.91) and (3.92) have been obtained using an accurate solver (see Program **JET** in Appendix C). Figures 3.5 and 3.6 compare computed and measured [Bradbury (1965), Heskestad (1965), Wygnanski and Fiedler (1969), Rodi (1975)] velocity profiles for the plane and round jets. Somewhat larger discrepancies between theory and experiment are present for the plane jet than for the round jet.

The traditional definition of spreading rate, $\delta'$, for the jet is the value of $y/x$ where the velocity is half its peak value. Experimental data indicate $\delta'$ is between 0.100 and 0.110 for the plane jet and between 0.086 and 0.095 for the round jet. The mixing-length computational results shown in Figures 3.5 and 3.6 correspond to

$$\delta' = \begin{cases} 0.100 & \text{(Plane Jet)} \\ 0.086 & \text{(Round Jet)} \end{cases} \qquad (3.93)$$

This concludes our application of the mixing-length model to free shear flows. A few final comments will help put this model into proper perspective. We postulated in Equation (3.34) that the mixing length is proportional to the width of the shear layer. Our theory thus has a single **closure coefficient**, $\alpha$, and we have found that it must be changed for each flow. The following values are optimum for the four cases considered.

| | |
|---|---|
| Far Wake | $\alpha = 0.180$ |
| Mixing Layer | $\alpha = 0.071$ |
| Plane Jet | $\alpha = 0.098$ |
| Round Jet | $\alpha = 0.080$ |

While fairly close agreement has been obtained between computed and measured velocity profiles, we have not predicted the all important spreading rate. In fact, we established the value of our closure coefficient by forcing agreement with the measured spreading rate. If we are only interested in far-wake applications or round jets we might use this model with the appropriate closure coefficient for a parametric study in which some flow property might be varied. However, we must proceed with some degree of caution knowing that our formulation lacks in universality.

## 3.4  Modern Variants of the Mixing-Length Model

For free shear flows, we have seen that the mixing length is constant across the layer and proportional to the width of the layer. For flow near a solid boundary, turbulence behaves differently and, not too surprisingly, we must use a different prescription for the mixing length. Prandtl originally postulated that for flows

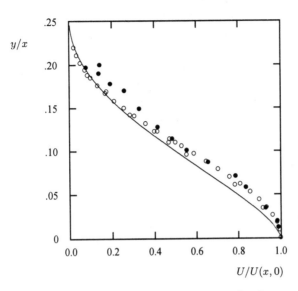

Figure 3.5: *Comparison of computed and measured velocity profiles for the plane jet: —— Mixing length; ○ Bradbury (1965); ● Heskestad (1965).*

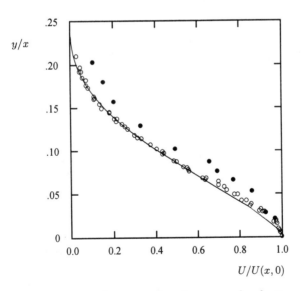

Figure 3.6: *Comparison of computed and measured velocity profiles for the round jet: —— Mixing length; ○ Wygnanski and Fiedler (1969); ● Rodi (1975).*

near solid boundaries the mixing length is proportional to the distance from the surface. As we will demonstrate shortly, this postulate is consistent with the well-known **law of the wall**, which has been observed for a wide range of wall-bounded flows over roughly the nearest 10% of the flow width from the surface (see Subsection 1.3.5).

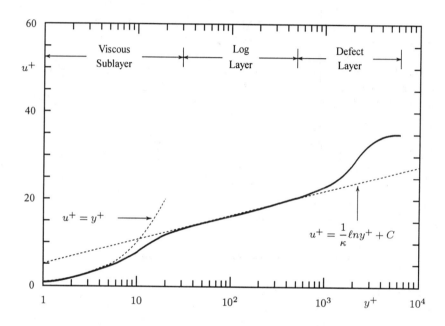

Figure 3.7: *Typical velocity profile for a turbulent boundary layer.*

Figure 3.7 shows a typical velocity profile for a turbulent boundary layer. The quantity $y^+$, defined in Equation (1.21), is dimensionless distance from the surface. As discussed in Subsection 1.3.5, three distinct regions are discernible, viz., the **viscous sublayer** (or "viscous wall region"), the **log layer** and the **defect layer**. By definition, the log layer is the portion of the boundary layer sufficiently close to the surface that inertial terms can be neglected yet sufficiently distant that the molecular, or viscous, stress is negligible compared to the Reynolds stress. This region typically lies between $y^+ = 30$ and $y = 0.1\delta$, where the value of $y^+$ at the upper boundary is dependent upon Reynolds number. Of particular interest to the present discussion, the **law of the wall** holds in the log layer. The viscous sublayer is the region between the surface and the log layer. Close to the surface, the velocity varies approximately linearly with $y^+$, and gradually asymptotes to the law of the wall for large values of $y^+$. The defect layer lies between the log layer and the edge of the boundary layer. The velocity asymptotes to the law of the wall as $y/\delta \to 0$, and makes a noticeable departure from the law of the wall approaching the freestream. Chapter 4 discusses these three layers in great detail.

From a mathematician's point of view, there are actually only two layers, viz., the viscous sublayer and the defect layer, and they overlap. In the parlance of singular-perturbation theory (Appendix B), the defect layer is the region in which the outer expansion is valid, while the viscous sublayer is the region where the inner expansion holds. In performing the classical matching procedure, we envision the existence of an overlap region, in which both the viscous sublayer and defect-layer solutions are valid. In the present context, matching shows that $U$ varies logarithmically with $y$ in the overlap region, which we choose to call the log layer. Strictly speaking, the log layer is not a distinct layer, but rather the asymptotic limit of the inner and outer layers. Nevertheless, we will find the log layer to be useful because of the simplicity of the equations of motion in the layer.

Consider an incompressible, constant-pressure boundary layer. The flow is governed by the standard boundary-layer equations.

$$\frac{\partial U}{\partial x} + \frac{\partial V}{\partial y} = 0 \qquad (3.94)$$

$$U\frac{\partial U}{\partial x} + V\frac{\partial U}{\partial y} = \frac{\partial}{\partial y}\left[\nu\frac{\partial U}{\partial y} - \overline{u'v'}\right] \qquad (3.95)$$

Because the convective terms are negligible in the log layer, the sum of the viscous and Reynolds shear stress must be constant. Hence, we can say

$$\nu\frac{\partial U}{\partial y} - \overline{u'v'} \approx \nu\left(\frac{\partial U}{\partial y}\right)_w = \frac{\tau_w}{\rho} = u_\tau^2 \qquad (3.96)$$

where subscript $w$ denotes value at the wall and $u_\tau = \sqrt{\tau_w/\rho}$ is known as the **friction velocity**. As noted above, the Reynolds stress is much larger than the viscous stress in the log layer. Consequently, according to the mixing-length model,

$$\ell_{mix}^2\left(\frac{\partial U}{\partial y}\right)^2 \approx u_\tau^2 \qquad (3.97)$$

If we say that the mixing length is given by

$$\ell_{mix} = \kappa y \qquad (3.98)$$

where $\kappa$ is a constant, Equation (3.97) can be integrated immediately to yield

$$U \approx \frac{u_\tau}{\kappa}\ell n y + \text{constant} \qquad (3.99)$$

Finally, recall the dimensionless velocity and normal distance defined in Equation (1.21), which we repeat here for convenience, viz.,

$$u^+ \equiv \frac{U}{u_\tau} \quad \text{and} \quad y^+ \equiv \frac{u_\tau y}{\nu} \tag{3.100}$$

Introducing Equation (1.21) into Equation (3.99) yields the classical **law of the wall**, viz.,

$$U^+ \approx \frac{1}{\kappa}\ell n y^+ + C \tag{3.101}$$

The coefficient $\kappa$ is known as the **Kármán constant**, and $C$ is a dimensionless constant. Coles and Hirst (1969) found from correlation of experimental data for a large number of attached, incompressible boundary layers with and without pressure gradient that

$$\kappa \approx 0.41 \tag{3.102}$$

$$C \approx 5.0 \tag{3.103}$$

Note that the mixing-length formula, Equation (3.97) with Equation (3.98), yields the same result as given by dimensional analysis alone [cf. Equations (1.18) and (1.19)].

Using Equation (3.98) all the way from $y = 0$ to $y = \delta$, the mixing-length model fails to provide close agreement with measured skin friction for boundary layers. Of course, not even Prandtl expected that $\ell_{mix} = \kappa y$ throughout the boundary layer. Since the mixing length was first postulated, considerable effort has been made aimed at finding a suitable prescription for boundary-layer computations. Several key modifications to Equation (3.98) have evolved, three of which deserve our immediate attention. See Schlichting (1979) or Hinze (1975) for a more-complete history of the mixing-length model's evolution.

**The first key modification** was devised by Van Driest (1956) who proposed that the mixing length should be multiplied by a damping function. Specifically, Van Driest proposed, with some theoretical support but mainly as a good fit to data, that the mixing length should behave according to

$$\ell_{mix} = \kappa y \left[1 - e^{-y^+/A_o^+}\right] \tag{3.104}$$

where the constant $A_o^+$ is

$$A_o^+ = 26 \tag{3.105}$$

Aside from the primary need to improve predictive accuracy, the Van Driest modification improves our description of the Reynolds stress in the limit $y \to 0$. With $\ell_{mix}$ given by Equation (3.98), the asymptotic behavior of the Reynolds shear stress is $\tau_{xy} \sim y^2$ as $y \to 0$. However, the no-slip boundary condition tells us that $u' = 0$ at $y = 0$. Since there is no a priori reason for $\partial u'/\partial y$ to vanish at the surface, we conclude that $u' \sim y$ as $y \to 0$. Since the fluctuating velocity satisfies the continuity equation, we also conclude that $v' \sim y^2$. Hence, the Reynolds shear stress must go to zero as $y^3$. Results of DNS studies (Chapter 8) indicate that indeed $\tau_{xy} \sim y^3$ as $y \to 0$. However, as noted by Hinze (1975), the coefficient of the $y^3$ term in a Taylor series expansion for $\tau_{xy}$ must be very small as measurements are as close to $\tau_{xy} \sim y^4$ as they are to $\tau_{xy} \sim y^3$ when $y \to 0$. In the limit of small $y$ the Van Driest mixing length implies $\tau_{xy}$ goes to zero as $y^4$ approaching the surface.

**The second key modification** was made by Clauser (1956) who addressed the proper form of the eddy viscosity in the defect layer. Similar to Prandtl's special form of the eddy viscosity for wake flows given in Equation (3.25), Clauser specifies that

$$\nu_{T_o} = \alpha U_e \delta^* \tag{3.106}$$

where $\nu_{T_o}$ is the kinematic eddy viscosity in the outer part of the layer, $\delta^*$ is the displacement thickness, $U_e$ is the velocity at the edge of the layer, and $\alpha$ is a closure coefficient.

In a similar vein, Escudier (1966) found that predictive accuracy is improved by limiting the peak value of the mixing length according to

$$(\ell_{mix})_{max} = 0.09\delta \tag{3.107}$$

where $\delta$ is boundary-layer thickness. Escudier's modification is the same approximation we used in analyzing free shear flows [Equation (3.34)], although the value 0.09 is half the value we found for the far wake.

Using an eddy viscosity appropriate to wake flow in the outer portion of the boundary layer also improves our physical description of the turbulent boundary layer. Measurements indeed indicate that the turbulent boundary layer exhibits wake-like characteristics in the defect layer. As pointed out by Coles and Hirst (1969), **"a typical boundary layer flow can be viewed as a wake-like structure which is constrained by a wall."** Figure 3.8 illustrates Coles' notion that the defect layer resembles a wake flow while the wall constraint is felt primarily in the sublayer and log layer. Strictly speaking, turbulence structure differs a lot between a boundary layer and a wake. Hence, the terminology "wake component" is misleading from a conceptual point of view. Nevertheless, the mathematical approximations that yield accurate predictions for a wake and for the outer portion of a turbulent boundary layer in zero pressure gradient are remarkably similar.

**The third key modification** is due to Corrsin and Kistler (1954) and Klebanoff (1954) as a corollary result of their experimental studies of **intermittency**. They found that approaching the freestream from within the boundary layer, the flow is not always turbulent. Rather, it is sometimes laminar and sometimes turbulent, i.e., it is **intermittent**. Their measurements indicate that for smooth

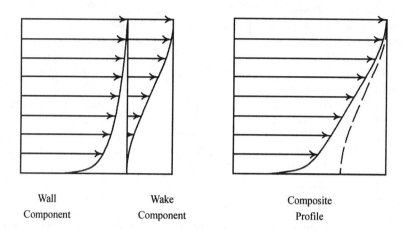

Wall                    Wake                              Composite

Component               Component                         Profile

Figure 3.8: *Coles' description of the turbulent boundary layer. [From Coles and Hirst (1969) — Used with permission.]*

walls, the eddy viscosity should be multiplied by

$$F_{Kleb}(y;\delta) = \left[1 + 5.5\left(\frac{y}{\delta}\right)^6\right]^{-1} \tag{3.108}$$

where $\delta$ is the boundary-layer thickness. This provides a measure of the effect of intermittency on the flow.

All of these modifications have evolved as a result of the great increase in power and accuracy of computing equipment and measurement techniques since the 1940's. The next two subsections introduce the two most noteworthy models in use today that are based on the mixing-length concept. Both include variants of the Van Driest, Clauser, and Klebanoff modifications. Although it is not used in these two models, the Escudier modification has also enjoyed great popularity.

As a final comment, we have introduced two new closure coefficients, $A_o^+$ and $\alpha$, and an empirical function, $F_{Kleb}$. As we continue in our journey through this book, we will find that the number of such coefficients increases as we attempt to describe more and more features of the turbulence.

### 3.4.1  Cebeci-Smith Model

The Cebeci-Smith model [Smith and Cebeci (1967)] is a two-layer model with $\nu_T$ given by separate expressions in each layer. The eddy viscosity is

$$\nu_T = \begin{cases} \nu_{T_i}, & y \leq y_m \\ \nu_{T_o}, & y > y_m \end{cases} \tag{3.109}$$

where $y_m$ is the smallest value of $y$ for which $\nu_{T_i} = \nu_{T_o}$. The values of $\nu_T$ in the inner layer, $\nu_{T_i}$, and the outer layer, $\nu_{T_o}$, are computed as follows.

**Inner Layer:**

$$\nu_{T_i} = \ell_{mix}^2 \left[ \left( \frac{\partial U}{\partial y} \right)^2 + \left( \frac{\partial V}{\partial x} \right)^2 \right]^{1/2} \qquad (3.110)$$

$$\ell_{mix} = \kappa y \left[ 1 - e^{-y^+/A^+} \right] \qquad (3.111)$$

**Outer layer:**

$$\nu_{T_o} = \alpha U_e \delta_v^* F_{Kleb}(y; \delta) \qquad (3.112)$$

**Closure Coefficients:**

$$\kappa = 0.40, \qquad \alpha = 0.0168, \qquad A^+ = 26 \left[ 1 + y \frac{dP/dx}{\rho u_\tau^2} \right]^{-1/2} \qquad (3.113)$$

The function $F_{Kleb}$ is the Klebanoff intermittency function given by Equation (3.108), $U_e$ is boundary-layer edge velocity, and $\delta_v^*$ is the **velocity thickness** defined by

$$\delta_v^* = \int_0^\delta (1 - U/U_e) \, dy \qquad (3.114)$$

Note that velocity thickness is identical to displacement thickness for incompressible flow. The coefficient $A^+$ differs from Van Driest's value to improve predictive accuracy for boundary layers with nonzero pressure gradient.[3] The prescription for $\nu_{T_i}$ above is appropriate only for two-dimensional flows; for three-dimensional flows, it should be proportional to a quantity such as the magnitude of the vorticity vector. There are many other subtle modifications to this model for specialized applications including surface mass transfer, streamline curvature, surface roughness, low Reynolds number, etc. Cebeci and Smith (1974) give complete details of their model with all of its variations.

The Cebeci-Smith model is especially elegant and easy to implement. Most of the computational effort, relative to a laminar case, goes into computing the velocity thickness. This quantity is readily available in boundary-layer computations so that a laminar-flow program can usually be converted to a turbulent flow program with just a few extra lines of instructions. Figure 3.9 illustrates a typical eddy viscosity profile constructed by using $\nu_{T_i}$ between $y = 0$ and $y = y_m$, and $\nu_{T_o}$ for the rest of the layer. At Reynolds numbers typical of fully-developed turbulence, matching between the inner and outer layers will occur well into the log layer.

---

[3]However, the Van Driest value should be used in fully-developed pipe flow, for which the $dP/dx$ correction yields imaginary $A^+$.

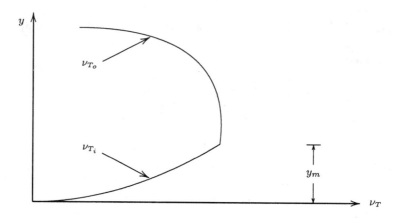

Figure 3.9: *Eddy viscosity for the Cebeci-Smith model.*

We can estimate the value of $y_m^+$ as follows. Since we expect the matching point to lie in the log layer, the exponential term in the Van Driest damping function will be negligible. Also, the law of the wall [Equation (3.99)] tells us $\partial U / \partial y \approx u_\tau / (\kappa y)$. Thus,

$$\nu_{T_i} \approx \kappa^2 y^2 \frac{u_\tau}{\kappa y} \approx \kappa u_\tau y = \kappa \nu y^+ \tag{3.115}$$

Since the matching point also lies close enough to the surface that we can say $y/\delta \ll 1$, the Klebanoff intermittency function will be close to one so that (with $\delta_v^* = \delta^*$):

$$\nu_{T_o} \approx \alpha U_e \delta^* = \alpha \nu Re_{\delta*} \tag{3.116}$$

Hence, equating $\nu_{T_i}$ and $\nu_{T_o}$, we find

$$y_m^+ \approx \frac{\alpha}{\kappa} Re_{\delta*} \approx 0.042 Re_{\delta*} \tag{3.117}$$

Assuming a typical turbulent boundary layer for which $Re_{\delta*} \sim 10^4$, the matching point will lie at $y_m^+ \sim 420$.

### 3.4.2   Baldwin-Lomax Model

The Baldwin-Lomax model [Baldwin and Lomax (1978)] was formulated for use in computations where boundary-layer properties such as $\delta$, $\delta_v^*$ and $U_e$ are difficult to determine. This situation often arises in numerical simulation of separated flows, especially for flows with shock waves. Like the Cebeci-Smith model, this is a two-layer model. The eddy viscosity is given by Equation (3.109), and the inner and outer layer viscosities are as follows.

**Inner Layer:**

$$\nu_{T_i} = \ell_{mix}^2 |\omega| \tag{3.118}$$

$$\ell_{mix} = \kappa y \left[ 1 - e^{-y^+/A_o^+} \right] \tag{3.119}$$

**Outer Layer:**

$$\nu_{T_o} = \alpha C_{cp} F_{wake} F_{Kleb}(y; y_{max}/C_{Kleb}) \tag{3.120}$$

$$F_{wake} = \min \left[ y_{max} F_{max}; C_{wk} y_{max} U_{dif}^2 / F_{max} \right] \tag{3.121}$$

$$F_{max} = \frac{1}{\kappa} \left[ \max_y (\ell_{mix} |\omega|) \right] \tag{3.122}$$

where $y_{max}$ is the value of $y$ at which $\ell_{mix} |\omega|$ achieves its maximum value.

**Closure Coefficients:[4]**

$$\left. \begin{array}{ll} \kappa = 0.40, & \alpha = 0.0168, \quad A_o^+ = 26 \\ C_{cp} = 1.6, & C_{Kleb} = 0.3, \quad C_{wk} = 1 \end{array} \right\} \tag{3.123}$$

The function $F_{Kleb}$ is Klebanoff's intermittency function [Equation (3.108)] with $\delta$ replaced by $y_{max}/C_{Kleb}$, and $\omega$ is the magnitude of the vorticity vector, i.e.,

$$\omega = \left[ \left( \frac{\partial V}{\partial x} - \frac{\partial U}{\partial y} \right)^2 + \left( \frac{\partial W}{\partial y} - \frac{\partial V}{\partial z} \right)^2 + \left( \frac{\partial U}{\partial z} - \frac{\partial W}{\partial x} \right)^2 \right]^{1/2} \tag{3.124}$$

for fully three-dimensional flows. This simplifies to $\omega = |\partial V/\partial x - \partial U/\partial y|$ in a two-dimensional flow. If the boundary layer approximations are used in a two-dimensional flow, then $\omega = |\partial U/\partial y|$.

$U_{dif}$ is the maximum value of $U$ for boundary layers. For free shear layers, $U_{dif}$ is the difference between the maximum velocity in the layer and the value of $U$ at $y = y_{max}$. For more general flows, it is defined by

$$U_{dif} = \left( \sqrt{U^2 + V^2 + W^2} \right)_{max} - \left( \sqrt{U^2 + V^2 + W^2} \right)_{y=y_{max}} \tag{3.125}$$

The primary difference between the Baldwin-Lomax and Cebeci-Smith models is in the outer layer, where the product $C_{cp} F_{wake}$ replaces $U_e \delta_v^*$. To avoid the

---

[4]Personal communication between Dr. Lomax and the author of this text has determined that the original Baldwin-Lomax paper inadvertently: (a) assigns a value of $C_{wk} = 0.25$; (b) defines $U_{dif}$ as the difference between the maximum and minimum velocities.

need to locate the boundary-layer edge, the Baldwin-Lomax model establishes the outer-layer length scale in terms of the vorticity in the layer. On the one hand, in using $F_{wake} = y_{max}F_{max}$, we in effect replace $\delta_v^*$ by $y_{max}^2\omega/U_e$. On the other hand, using $F_{wake} = C_{wk}y_{max}U_{dif}^2/F_{max}$ effectively replaces the shear layer width, $\delta$, in Prandtl's eddy-viscosity model [Equation (3.25)] by $U_{dif}/|\omega|$.

For boundary-layer flows, there is very little difference between the predictions of the Baldwin-Lomax and Cebeci-Smith models. This indicates that the prescription for determining the outer-layer length scale based on the vorticity and distance from the surface [cf. Equations (3.121) and (3.122)] is entirely equivalent to the velocity thickness, $\delta_v*$. For more-complicated flows, such as those involving separation, the Baldwin-Lomax model provides an outer length scale that is well defined for most flows. By contrast, $\delta_v^*$, will generally be negative for a separated flow, and thus is an unsuitable length scale.

However, the Baldwin-Lomax model prescription for computing an outer length scale can fail when the vorticity is nonvanishing above the boundary layer. This will occur, for example, on slender bodies at angle of attack, where regions of crossflow separation dominate [e.g., Degani and Schiff (1986) or Gee, Cummings and Schiff (1992)]. In this type of flow, the function $F(y)$ can exhibit more than one relative maximum as illustrated in Figure 3.10. Using a peak beyond the viscous region can lead to nonphysically large eddy viscosity values that lead to gross distortion of the computed flowfield. To eliminate this problem, Degani and Schiff (1986) have devised a procedure that automatically selects the peak value of $F(y)$ within the viscous region. While the Degani-Schiff modification improves model predictions for separated flows, we will see in Section 3.6 that neither the Cebeci-Smith nor the Baldwin-Lomax model embodies a sufficient physical foundation to warrant application to such flows.

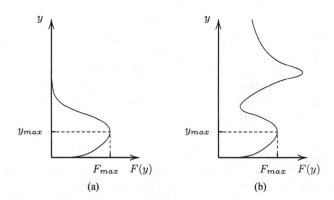

Figure 3.10: *The function $F(y)$ for: (a) a conventional boundary layer; (b) a boundary layer with nonzero freestream vorticity.*

Thus, this type of adjustment to the model reflects no added physical insight, but rather stands as a purely empirical correction.

As a final comment, while Equation (3.123) implies this model has six closure coefficients, there are actually only five. The coefficient $C_{cp}$ appears only in Equation (3.120) where it is multiplied by $\alpha$, so $\alpha C_{cp}$ is actually a single constant.

## 3.5 Application to Wall-Bounded Flows

We turn our attention now to application of the Cebeci-Smith and Baldwin-Lomax models to wall-bounded flows, i.e., to flows with a solid boundary. The no-slip boundary condition must be enforced for wall-bounded flows, and we expect to find a viscous layer similar to that depicted in Figure 3.7. This section first examines two internal flows, viz., channel flow and pipe flow. Then, we consider external flows, i.e., boundary layers growing in a semi-infinite medium.

### 3.5.1 Channel and Pipe Flow

Like the free shear flow applications of Section 3.3, constant-section channel and pipe flow are excellent building-block cases for testing a turbulence model. Although we have the added complication of a solid boundary, the motion can be described with ordinary differential equations and is therefore easy to analyze mathematically. Also, experimental data are abundant for these flows.

The classical problems of flow in a channel, or duct, and a pipe are the idealized case of an infinitely long channel or pipe (Figure 3.11). This approximation is appropriate provided we are not too close to the inlet of the channel/pipe so that the flow has become **fully-developed**. For turbulent flow in a pipe, flow

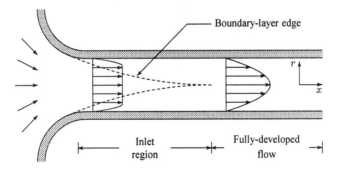

Figure 3.11: *Fully-developed flow in a pipe or channel with the vertical scale magnified.*

asymptotes to full development at a distance $\ell_e$ downstream of the inlet given approximately by [cf. Schlichting (1979)]

$$\frac{\ell_e}{D} = 4.4Re_D^{1/6} \tag{3.126}$$

where $Re_D$ is Reynolds number based on the pipe diameter (or channel half height). Thus, for example, the **entrance length**, $\ell_e$, for flow in a pipe with $Re_D = 10^5$ is about 30 pipe diameters. Because, by definition, properties no longer vary with distance along the channel/pipe, we conclude immediately that

$$\frac{\partial U}{\partial x} = 0 \tag{3.127}$$

Denoting distance from the center of the channel or pipe by $r$, conservation of mass is

$$\frac{\partial U}{\partial x} + \frac{1}{r^j}\frac{\partial}{\partial r}\left[r^j V\right] = 0 \tag{3.128}$$

where $j = 0$ for channel flow and $j = 1$ for pipe flow. In light of Equation (3.127), we see that $V$ does not vary across the channel/pipe. Since $V$ must vanish at the channel/pipe walls, we conclude that $V = 0$ throughout the fully-developed region. Hence, for both channel and pipe flow, the inertial terms are exactly zero, so that the momentum equation simplifies to

$$0 = -\frac{dP}{dx} + \frac{1}{r^j}\frac{d}{dr}\left[r^j\left(\mu\frac{dU}{dr} - \rho\overline{u'v'}\right)\right] \tag{3.129}$$

In fully-developed flow pressure gradient must be independent of $x$, and if $V = 0$, it is also exactly independent of $r$. Hence, we can integrate once to obtain

$$\mu\frac{dU}{dr} - \rho\overline{u'v'} = \frac{r}{j+1}\frac{dP}{dx} \tag{3.130}$$

Now, the Reynolds stress vanishes at the channel/pipe walls, and this establishes a direct relationship between the pressure gradient and the shear stress at the walls. If we let $R$ denote the half-height of the channel or the radius of the pipe, applying Equation (3.130) at $r = R$ tells us that

$$\tau_w = -\frac{R}{j+1}\frac{dP}{dx} \tag{3.131}$$

Hence, introducing the friction velocity, $u_\tau$, the momentum equation for channel/pipe flow simplifies to the following first-order, ordinary differential equation.

$$\mu\frac{dU}{dr} - \rho\overline{u'v'} = -\rho u_\tau^2\frac{r}{R} \tag{3.132}$$

Noting that both channel and pipe flow are symmetric about the centerline, we can obtain the complete solution by solving Equation (3.132) with $r$ varying between 0 and $R$. It is more convenient however to define $y$ as the distance from the wall so that

$$y = R - r \qquad (3.133)$$

Hence, representing the Reynolds stress in terms of the eddy viscosity, i.e., $-\rho \overline{u'v'} = \mu_T dU/dy$, we arrive at the following equation for the velocity.

$$(\mu + \mu_T)\frac{dU}{dy} = \rho u_\tau^2 \left(1 - \frac{y}{R}\right) \qquad (3.134)$$

Finally, we introduce sublayer-scaled coordinates, $U^+$ and $y^+$, from Equation (3.100), as well as $\mu_T^+ = \mu_T/\mu$. This results in the dimensionless form of the momentum equation for channel flow and pipe flow, viz.,

$$(1 + \mu_T^+)\frac{dU^+}{dy^+} = \left(1 - \frac{y^+}{R^+}\right) \qquad (3.135)$$

where

$$R^+ = u_\tau R/\nu \qquad (3.136)$$

Equation (3.135) must be solved subject to the no-slip boundary condition at the channel/pipe wall. Thus, we require

$$U^+(0) = 0 \qquad (3.137)$$

At first glance, this appears to be a standard initial-value problem that can, in principle, be solved using an integration scheme such as the Runge-Kutta method. However, the problem is a bit more difficult, and for both the Cebeci-Smith and Baldwin-Lomax models, the problem must be solved iteratively. That is, for the Cebeci-Smith model, we don't know $U_e$ and $\delta_v^*$ a priori. Similarly, with the Baldwin-Lomax model, we don't know the values of $U_{dif}$ and $y_{max}$ until we have determined the entire velocity profile. This is not a serious complication however, and the solution converges after just a few iterations.

The equations for channel and pipe flow can be conveniently solved using a standard over-relaxation iterative procedure. Appendix C describes a program called **PIPE** that yields a numerical solution for several turbulence models, including the Cebeci-Smith and Baldwin-Lomax models.

Figure 3.12 compares computed two-dimensional channel-flow profiles with Direct Numerical Simulation (DNS) results of Mansour, Kim and Moin (1988) for Reynolds number based on channel height and average velocity of 13,750. As shown, the Cebeci-Smith and Baldwin-Lomax velocity profiles are within 8% and 5%, respectively, of the DNS profiles. Computed Reynolds shear stress profiles for both models differ from the DNS profiles by no more than 2%.

Computed skin friction for both models differs by less than 2% from Halleen
and Johnston's (1967) correlation of experimental data, viz.,

$$c_f = 0.0706 Re_H^{-1/4} \qquad (3.138)$$

where the skin friction and Reynolds number are based on the average velocity
across the channel and the channel height $H$, i.e., $c_f = \tau_w/(\frac{1}{2}\rho U_{avg}^2)$ and on
Reynolds number, $Re_H = U_{avg}H/\nu$.

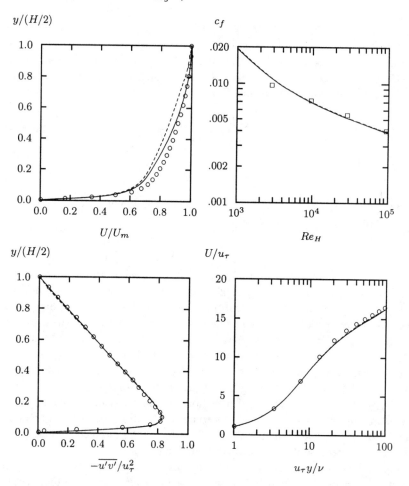

Figure 3.12: *Comparison of computed and measured channel-flow properties,*
$Re_H = 13,750$. ——— *Baldwin-Lomax;* - - - *Cebeci-Smith;* ○ *Mansour et al.*
*(DNS);* □ *Halleen-Johnston correlation.*

Figure 3.13 compares model predicted pipe-flow properties with the experimental data of Laufer (1952) for a Reynolds number based on pipe diameter and average velocity of 40,000. Baldwin-Lomax velocity and Reynolds shear stress differ from measured values by no more than 3%. As with channel flow, the Cebeci-Smith velocity shows greater differences (8%) from the data, while the Reynolds shear stress values are very close to those predicted by the Baldwin-Lomax model.

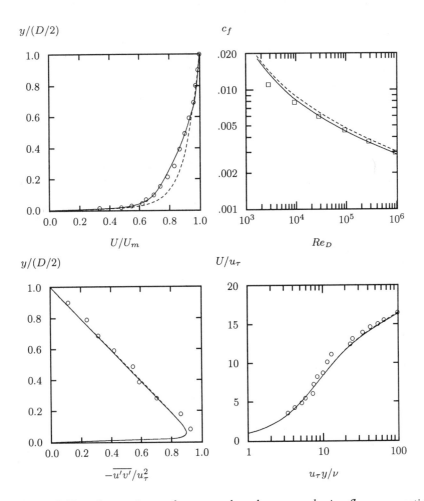

Figure 3.13: *Comparison of computed and measured pipe-flow properties,* $Re_D = 40,000.$ —— *Baldwin-Lomax; - - - Cebeci-Smith; ○ Laufer; □ Prandtl correlation.*

Computed skin friction is within 7% and 1% for the Cebeci-Smith and Baldwin-Lomax models, respectively, of Prandtl's universal law of friction for smooth pipes [see Schlichting (1979)] given by

$$\frac{1}{\sqrt{c_f}} = 4\log_{10}\left(2Re_D\sqrt{c_f}\right) - 1.6 \qquad (3.139)$$

where $c_f$ and $Re_D$ are based on average velocity across the pipe and pipe diameter, D.

These computations illustrate that subtle differences in the Reynolds shear stress can lead to much larger differences in velocity for pipe and channel flow. This means we must determine the Reynolds shear stress very accurately in order to obtain accurate velocity profiles. To some extent this seems odd. The Reynolds stress is a higher-order correlation while velocity is a simple time average. Our natural expectation is for the mean velocity to be determined with great precision while higher-order quantities such as Reynolds stress are determined with a bit less precision. The dilemma appears to stem from the fact that we need the same precision in $\tau_{xy}$ as in $\partial U/\partial y$. As we advance to more complicated turbulence models, we will see this accuracy dilemma repeated, although generally with less severity. As applications go, channel and pipe flow are not very forgiving.

Interestingly, Figure 3.13 shows that for the higher Reynolds number pipe flow, higher velocity is predicted with the Cebeci-Smith model than with the Baldwin-Lomax model. The opposite is true for the lower Reynolds number channel-flow case. Cebeci and Smith (1974) have devised low-Reynolds-number corrections for their model which, presumably, would reduce the differences from the DNS channel-flow results.

## 3.5.2  Boundary Layers

In general, for a typical boundary layer, we must account for pressure gradient. Ignoring effects of normal Reynolds stresses and introducing the eddy viscosity to determine the Reynolds shear stress, the two-dimensional ($j = 0$) and axisymmetric ($j = 1$) boundary-layer equations are as follows.

$$\frac{\partial U}{\partial x} + \frac{1}{y^j}\frac{\partial}{\partial y}\left(y^j V\right) = 0 \qquad (3.140)$$

$$U\frac{\partial U}{\partial x} + V\frac{\partial U}{\partial y} = -\frac{1}{\rho}\frac{dP}{dx} + \frac{1}{y^j}\frac{\partial}{\partial y}\left[y^j(\nu + \nu_T)\frac{\partial U}{\partial y}\right] \qquad (3.141)$$

The appropriate boundary conditions follow from the no slip condition at the surface and from insisting that $U \rightarrow U_e$ as we approach the boundary-layer edge.

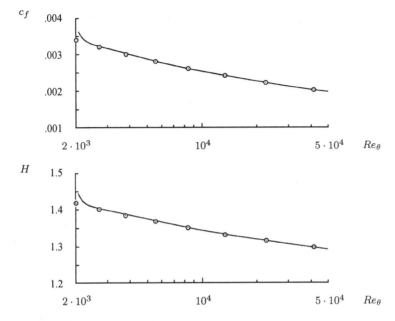

Figure 3.14: *Comparison of computed and correlated shape factor and skin friction for flat-plate boundary layer flow;* ⊙ *Coles;* —— *Cebeci-Smith model. [From Kline et al. (1969) — Used with permission.]*

Consequently, we must solve Equations (3.140) and (3.141) subject to

$$\left.\begin{array}{rcl} U(x,0) & = & 0 \\ V(x,0) & = & 0 \\ U(x,y) & \rightarrow & U_e(x) \quad \text{as} \quad y \rightarrow \delta(x) \end{array}\right\} \tag{3.142}$$

where $\delta(x)$ is the boundary-layer thickness.

The Cebeci-Smith model has been applied to a wide range of boundary-layer flows and has enjoyed a great deal of success. Figure 3.14, for example, compares computed skin friction, $c_f$, and shape factor, $H$, for a constant-pressure (flat-plate) boundary layer with Coles' [Coles and Hirst (1969)] correlation of experimental data. Results are expressed as functions of Reynolds number based on momentum thickness, $Re_\theta$. As shown, model predictions virtually duplicate correlated values.

The model remains reasonably accurate for favorable pressure gradient and for mild adverse pressure gradient. Because the model has been fine tuned for boundary-layer flows, differences between computed and measured velocity profiles generally are small. Typically, integral parameters such as momentum thickness and shape factor show less than 10% differences from measured values.

Figure 3.15 compares computed and measured boundary layer properties for two of the flows considered in the 1968 AFOSR-IFP-Stanford Conference on the Computation of Turbulent Boundary Layers (this conference is often referred to colloquially as Stanford Olympics I). For both cases, computed and measured velocity profiles are nearly identical. Flow 3100 is two dimensional with a mild favorable pressure gradient. Despite the close agreement in velocity profiles overall, differences in shape factor are between 8% and 10%. Flow 3600 is axisymmetric with an adverse pressure gradient. For this flow, shape factors differ by less than 5%. The Baldwin-Lomax model also closely reproduces measured flow properties for these types of boundary layers.

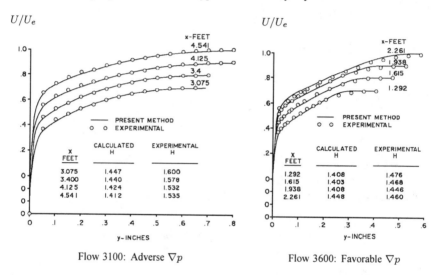

Figure 3.15: *Comparison of computed and measured boundary layer velocity profiles and shape factor for flows with nonzero pressure gradient; Cebeci-Smith model. [From Kline et al. (1969) — Used with permission.]*

Figure 3.16 compares computed and measured skin friction for sixteen incompressible boundary layers subjected to favorable, zero and adverse pressure gradients. For both models, computed and measured $c_f$ generally differ by less than 10%. Fifteen of the sixteen cases considered are from Stanford Olympics I. The lone exception is Flow 0141, which corresponds to a boundary layer in an increasingly adverse pressure gradient. This flow has been studied experimentally by Samuel and Joubert [see Kline et al. (1981)]. It was a key boundary-layer case included in the 1980-81 AFOSR-HTTM-Stanford Conference on Complex Turbulent Flows (known colloquially as Stanford Olympics II). Measurements for all cases satisfy the momentum-integral equation, thus assuring their two-dimensionality and accuracy of the experiments.

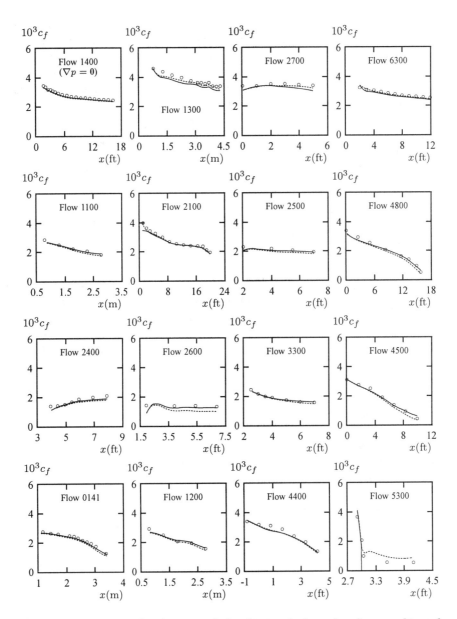

Figure 3.16: *Computed and measured skin friction for boundary layers subjected to a pressure gradient. Top row - favorable $\nabla p$; next to top row - mild adverse $\nabla p$; next to bottom row - moderate adverse $\nabla p$; bottom row - strong adverse $\nabla p$. —— Baldwin-Lomax model; - - - Cebeci-Smith model; ○ measured.*

Table 3.1 summarizes the difference between computed and measured $c_f$ at the final station for the various pressure gradients. This is a sensible measure of the overall accuracy as all transients have settled out, and, with the exception of Flow 2400, the pressure gradient is strongest at the end of the computation. The overall average difference for 15 cases (Flow 5300 has been excluded) is 7% for the Baldwin-Lomax model and 9% for the Cebeci-Smith model.

Table 3.1: *Differences Between Computed and Measured Skin Friction.*

| Pressure Gradient | Flows | Baldwin-Lomax | Cebeci-Smith |
|---|---|---|---|
| Favorable | 1400, 1300, 2700, 6300 | 7.8% | 5.0% |
| Mild Adverse | 1100, 2100, 2500, 4800 | 5.8% | 6.5% |
| Moderate Adverse | 2400, 2600, 3300, 4500 | 9.9% | 15.4% |
| Strong Adverse | 0141, 1200, 4400 | 3.1% | 10.1% |
| All | – | 6.7% | 9.3% |

One noteworthy case is Flow 3300 of the 1968 AFOSR-IFP-Stanford Conference on the Computation of Turbulent Boundary Layers. This flow, also known as Bradshaw Flow C, has a strongly adverse pressure gradient that is gradually relaxed and corresponds to an experiment performed by Bradshaw (1969). It was generally regarded as one of the most difficult to predict of all flows considered in the Conference. As shown, both models predict skin friction very close to the measured value. The Cebeci-Smith value for $c_f$ at the final station ($x = 7$ ft.) is 6% lower than the measured value. The Baldwin-Lomax value exceeds the measured value at $x = 7$ ft. by 3%.

A second case worthy of mention is Flow 0141 of the 1980-81 AFOSR-HTTM-Stanford Conference on Complex Turbulent Flows. The close agreement between theory and experiment for this flow is actually remarkable. The Cebeci-Smith and Baldwin-Lomax values for $c_f$ at $x \approx 3$ m. is 13% and 2% of the measured value, respectively. This boundary layer was presumed to be a "simple" flow for Conference participants. However, as we will discuss further in Chapter 4, it proved to be the Achilles heel of the best turbulence models of the day.

The only case the models fail to predict accurately is Flow 5300, which is known as the Stratford (1959) "incipient-separation" flow. The boundary layer experiences an adverse pressure gradient that is of sufficient strength to drive it to the brink of separation. The Cebeci-Smith model's skin friction at the final station ($x = 4.1$ ft.) is 60% higher than measured. The Baldwin-Lomax model predicts boundary-layer separation at $x \approx 3$ ft.

As a final comment, all sixteen computations have been done using Program **EDDYBL**, a boundary-layer program suitable for two-dimensional and axisymmetric flows. Appendix D describes the program, which is included on the diskette provided with this book.

## 3.6 Separated Flows

All of the applications in the preceding section are for attached boundary layers. We turn now to flows having an adverse pressure gradient of sufficient strength to cause the boundary layer to separate. Separation occurs in many practical applications including stalled airfoils, flow near the stern of a ship, flow through a diffuser, etc. Engineering design would be greatly enhanced if our turbulence model were a reliable analytical tool for predicting separation and its effect on surface pressure, skin friction and heat transfer. Unfortunately, algebraic models are quite unreliable for separated flows.

When a boundary layer separates, the streamlines are no longer nearly parallel to the surface as they are for attached boundary layers. We must solve the full Reynolds-averaged Navier-Stokes equation [Equation (2.24)], which includes all components of the Reynolds-stress tensor. In analogy to Stokes hypothesis for laminar flow, we set

$$\tau_{ij} = 2\nu_T S_{ij} \tag{3.143}$$

where $S_{ij}$ is the mean strain-rate tensor defined by

$$S_{ij} = \frac{1}{2}[U_{i,j} + U_{j,i}] \tag{3.144}$$

Figure 3.17 is typical of separated flow results for an algebraic model. Menter (1992b) applied the Baldwin-Lomax model to an axisymmetric flow with a strong adverse pressure gradient. The experiment was conducted by Driver (1991). Inspection of the skin friction shows that the Baldwin-Lomax model yields a separation bubble nearly twice as long as the experimentally observed bubble.

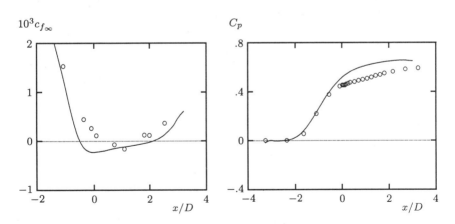

Figure 3.17: *Computed and measured flow properties for Driver's separated flow;* —— *Baldwin-Lomax;* ∘ *Driver.*

The corresponding rise in pressure over the separation region is 15% to 20% higher than measured. As pointed out by Menter, the Cebeci-Smith model yields similar results.

It is not surprising that a turbulence model devoid of any information about flow history will perform poorly for separated flows. On the one hand, the mean strain-rate tensor undergoes rapid changes in a separated flow associated with the curved streamlines over and within the separation bubble. On the other hand, the turbulence adjusts to changes in the flow on a time scale unrelated to the mean rate of strain. Rotta (1962), for example, concludes from analysis of experimental data that when a turbulent boundary layer is perturbed from its equilibrium state, a new equilibrium state is not attained for at least 10 boundary-layer thicknesses downstream of the perturbation. In other words, separated flows are very much out of "equilibrium." The Boussinesq approximation, along with all the "equilibrium" approximations implicit in an algebraic model, can hardly be expected to provide an accurate description for separated flows.

Attempts have been made to remedy the problem of poor separated flow predictions with the Cebeci-Smith model. Shang and Hankey (1975) introduced the notion of a relaxation length, $L$, to account for upstream turbulence history effects. They introduced what they called a **relaxation eddy viscosity model** and determined the eddy viscosity as follows.

$$\mu_T = \mu_{T_{eq}} - (\mu_{T_{eq}} - \mu_{T_1})e^{-(x-x_1)/L} \tag{3.145}$$

The quantity $\mu_{T_{eq}} = \rho\nu_{T_{eq}}$ denotes the equilibrium eddy viscosity corresponding to the value given by Equations (3.109) through (3.112), while $\mu_{T_1}$ is the value of the eddy viscosity at a reference point, $x = x_1$, upstream of the separation region. Typically, the relaxation length is about $5\delta_1$, where $\delta_1$ is the boundary-layer thickness at $x = x_1$. The principal effect of Equation (3.145) is to reduce the Reynolds stress from the "equilibrium" value predicted by the Cebeci-Smith model. This mimics the experimental observation that the Reynolds stress remains nearly frozen at its initial value while it is being convected along streamlines in the separation region, and approaches a new equilibrium state exponentially.

In a similar vein, Hung (1976) proposed a differential form of Shang and Hankey's Equation (3.145), viz.,

$$\frac{d\mu_T}{dx} = \frac{\mu_{T_{eq}} - \mu_T}{L} \tag{3.146}$$

This equation is very similar to the earlier proposal of Reyhner [Kline et al. (1969)]. Hung (1976) exercised these relaxation models in several supersonic shock-separated flows. He was able to force close agreement between computed and measured locations of the separation point and the surface pressure distribution. However, he found that these improvements came at the expense of increased discrepancies between computed and measured skin friction, heat transfer and reattachment-point location.

## 3.7 The 1/2-Equation Model

Johnson and King (1985) [see also Johnson (1987) and Johnson and Coakley (1990)] have devised a "non-equilibrium" version of the algebraic model. Their starting point is a so-called "equilibrium" algebraic model in which the eddy viscosity is

$$\mu_T = \mu_{T_o} \tanh(\mu_{T_i}/\mu_{T_o}) \tag{3.147}$$

where $\mu_{T_i}$ and $\mu_{T_o}$ represent inner-layer and outer-layer eddy viscosity, respectively. The hyperbolic tangent is used to eliminate the discontinuity in $\partial \mu_T/\partial y$ attending the use of Equation (3.109).

**Inner Layer:**
    The inner layer viscosity, $\mu_{T_i}$, is similar to the form used in the Cebeci-Smith and Baldwin-Lomax models. However, the dependence on velocity gradient has been replaced by explicit dependence on distance from the surface, $y$, and two primary velocity scales, $u_\tau$ and $u_m$, as follows:

$$\mu_{T_i} = \rho \left[ 1 - \exp\left( -\frac{u_D y/\nu}{A^+} \right) \right]^2 \kappa u_s y \tag{3.148}$$

$$\sqrt{\rho}\, u_s = (1 - \gamma_2)\sqrt{\tau_w} + \gamma_2 \sqrt{\tau_m} \tag{3.149}$$

$$\gamma_2 = \tanh(y/L_c) \tag{3.150}$$

$$L_c = \frac{\sqrt{\tau_w}}{\sqrt{\tau_w} + \sqrt{\tau_m}} L_m \tag{3.151}$$

$$L_m = \begin{cases} \kappa y_m, & y_m/\delta \leq C_1/\kappa \\ C_1 \delta, & y_m/\delta > C_1/\kappa \end{cases} \tag{3.152}$$

$$u_m = \sqrt{\tau_m/\rho_m} \tag{3.153}$$

$$u_D = \max[u_m, u_\tau] \tag{3.154}$$

where subscript $m$ denotes the value at the point, $y = y_m$, at which the Reynolds shear stress, $\rho\tau_{xy}$, assumes its maximum value denoted by $\tau_m = (\rho\tau_{xy})_{max}$. Additionally $u_\tau$ is the conventional friction velocity and $\rho_w$ is the density at the surface, $y = 0$. In its original form, this model used only the velocity scale $u_m$ in Equation (3.148). This scale proved to provide better predictions of velocity profile shape for separated flows than the velocity-gradient prescription of Prandtl [Equation (3.15)]. Later, the secondary velocity scales $u_s$ and $u_D$ were added to improve predictions for reattaching flows and for flows with nontrivial effects of compressibility.

**Outer Layer:**

The "non-equilibrium" feature of the model comes in through the appearance of a "nonequilibrium parameter," $\sigma(x)$, so that:

$$\mu_{T_o} = \alpha \rho U_e \delta_v^* F_{Kleb}(y; \delta) \sigma(x) \tag{3.155}$$

Comparison of this equation with Equation (3.112) shows that the outer layer viscosity, $\mu_{T_o} = \rho \nu_{T_o}$, is equal to that used in the Cebeci-Smith model multiplied by $\sigma(x)$. The Johnson-King model solves the following **ordinary differential equation** for the maximum Reynolds shear stress, $\tau_m$, in terms of $u_m = \sqrt{\tau_m/\rho_m}$.

$$U_m \frac{d}{dx}\left(u_m^2\right) = a_1 \left[\frac{(u_m)_{eq} - u_m}{L_m}\right] u_m^2 - C_{dif} \left[\frac{u_m^3}{C_2 \delta - y_m}\right] \left|1 - \sigma^{1/2}(x)\right| \tag{3.156}$$

where $U_m$ is mean velocity and $(u_m)_{eq}$ is the value of $u_m$ according to the "equilibrium" algebraic model [$\sigma(x) = 1$]. The first term on the right-hand side of Equation (3.156) is reminiscent of Hung's relaxation model [Equation (3.146)]. The second term is an estimate of the effect of turbulent diffusion on the Reynolds shear stress. Equation (3.156) is solved along with the Reynolds-averaged equations to determine $\tau_m$. As the solution proceeds, the coefficient $\sigma(x)$ is determined so that the maximum Reynolds shear stress is given by

$$\tau_m = (\mu_T)_m \left(\frac{\partial U}{\partial y} + \frac{\partial V}{\partial x}\right)_m \tag{3.157}$$

That is, the $\mu_T$ distribution is adjusted to agree with $\tau_m$. In using this model, computations must be done iteratively since $\sigma(x)$ is unknown a priori, wherefore the value from a previous iteration or an extrapolated value must be used in solving Equation (3.156) for $\tau_m$.

**Closure Coefficients:**

$$\left.\begin{array}{lll} \kappa = 0.40, & \alpha = 0.0168, & A^+ = 17 \\ a_1 = 0.25, & C_1 = 0.09, & C_2 = 0.70 \\ C_{dif} = 0.50 & \text{for} \ \sigma(x) \geq 1; & 0 \ \text{otherwise} \end{array}\right\} \tag{3.158}$$

The general idea of this model is that the Reynolds shear stress adjusts to departures from "equilibrium" at a rate different from that predicted by the algebraic model. The ordinary differential equation for $u_m$ is used to account for the difference in rates. Because this equation is an **ordinary**, as opposed to a **partial**, differential equation, the turbulence community has chosen the curious terminology **1/2-Equation Model** to describe this model. It is unclear whether this means it has half the number of dimensions (but then, it would have to be a 1/3-Equation Model for three-dimensional applications) or if partial differential equations are twice as hard to solve as ordinary differential equations.

Figure 3.18 compares computed and measured skin friction for the sixteen boundary-layer flows of Stanford Olympics I and II discussed earlier. As with the algebraic models, the computations have been done using Program **EDDYBL** (Appendix D). Note that predicted $c_f$ for the constant-pressure case (Flow 1400) is 5% less than measured values. This is a direct consequence of using Equation (3.147). As can be readily verified, using Equation (3.109), computed skin friction matches measured values almost exactly.

As summarized in Table 3.2, overall differences between computed and measured $c_f$ are somewhat larger than corresponding differences for the Baldwin-Lomax model. The overall average difference at the final station for the flows, excluding Flow 5300, is 12% as compared to 7% for the Baldwin-Lomax model (and 9% for the Cebeci-Smith model).

Table 3.2: *Differences Between Computed and Measured Skin Friction.*

| Pressure Gradient | Flows | Johnson-King | Baldwin-Lomax |
|---|---|---|---|
| Favorable | 1400, 1300, 2700, 6300 | 7.8% | 7.8% |
| Mild Adverse | 1100, 2100, 2500, 4800 | 10.0% | 5.8% |
| Moderate Adverse | 2400, 2600, 3300, 4500 | 13.6% | 9.9% |
| Strong Adverse | 0141, 1200, 4400 | 16.4% | 3.1% |
| All | − | 12.0% | 6.7% |

As noted earlier, Bradshaw Flow C (Flow 3300) was one of the most difficult cases in the 1968 AFOSR-IFP-Stanford Conference on the Computation of Turbulent Boundary Layers. The Johnson-King model fares rather poorly on this case with computed skin friction 19% higher than the measured value at the final station. Recall that the Cebeci-Smith and Baldwin-Lomax model predictions were within 6% and 3% of the measured value, respectively.

Also, for the Samuel-Joubert increasingly adverse pressure gradient case (Flow 0141), the computed skin friction is 13% *higher* than the measured value. By contrast, the Cebeci-Smith and Baldwin-Lomax models predict $c_f$ 13% and 2% *lower*, respectively.

As a bit of a surprise, while the predicted boundary layer remains attached for the Stratford incipient-separation case (Flow 5300), the computed skin friction is more than four times the measured value. Results obtained with the Cebeci-Smith model (see Figure 3.16) are quite a bit closer to measurements.

Although the differences are somewhat larger than those of the algebraic models, inspection of Figure 3.18 shows that, with the exception of Flow 5300, the accuracy is satisfactory for most engineering applications. The differences can probably be reduced to the same levels as for the Baldwin-Lomax and Cebeci-Smith models by either recalibrating the closure coefficients or by using Equation (3.109) instead of Equation (3.147).

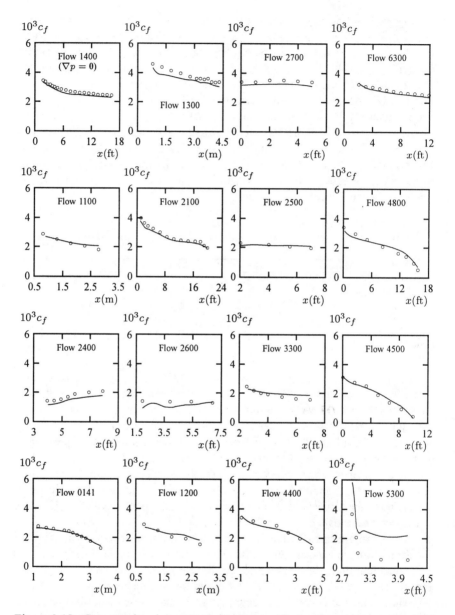

Figure 3.18: *Computed and measured skin friction for boundary layers subjected to a pressure gradient. Top row - favorable $\nabla p$; next to top row - mild adverse $\nabla p$; next to bottom row - moderate adverse $\nabla p$; bottom row - strong adverse $\nabla p$. —— Johnson-King model; ○ measured.*

Menter (1992b) has applied the Johnson-King model to Driver's (1991) separated flow. Figure 3.19 compares computed and measured values; results for the Baldwin-Lomax model are also included. As shown, the Johnson-King model predictions are much closer to measurements, most notably in the size of the separation region.

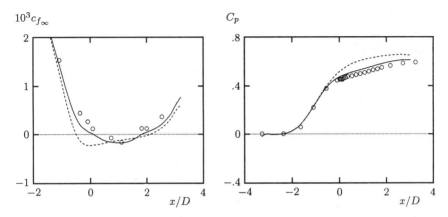

Figure 3.19: *Computed and measured flow properties for Driver's separated flow;* —— *Johnson-King;* - - - *Baldwin-Lomax;* ○ *Driver.*

## 3.8 Range of Applicability

Algebraic models are the simplest and easiest to implement of all turbulence models. They are conceptually very simple and rarely cause unexpected numerical difficulties. Because algebraic models are so easy to use, they should be replaced only where demonstrably superior alternatives are available.

The user must always be aware of the issue of **incompleteness**. These models will work well only for the flows for which they have been fine tuned. There is very little hope of extrapolating beyond the established data base for which an algebraic model is calibrated. We need only recall that for the four free shear flows considered in Section 3.3, four different values for the mixing length are needed—and none of these lengths is appropriate for wall-bounded flows!

On balance, both the Cebeci-Smith and Baldwin-Lomax models faithfully reproduce skin friction and velocity profiles for incompressible turbulent boundary layers provided the pressure gradient is not too strong. Neither model is clearly superior to the other: the accuracy level is about the same for both models. The chief virtue of the Baldwin-Lomax model over the Cebeci-Smith model is its independence of properties such as $\delta_v^*$ that can often be difficult to define

accurately in complex flows. Its other differences from the Cebeci-Smith model
are probably accidental. However, neither model is reliable for separated flows.
Despite this well-known limitation, many incautious researchers have applied the
Baldwin-Lomax model to extraordinarily complex flows where its only virtue is
that it doesn't cause the computation to blow up.

The Johnson-King model offers a promising modification that removes much
of the inadequacy of algebraic models for separated flows. However, like algebraic models, the Johnson-King model provides no information about the turbulence length scale and is thus **incomplete**. Consequently, it shares many of the
shortcomings of the underlying algebraic model. **On the negative side**, the improved agreement between theory and experiment for separated flows has been
gained with a loss of the elegance and simplicity of the Cebeci-Smith model.
The number of ad hoc closure coefficients has increased from three to seven,
and the model inherently requires an iterative solution procedure. The model
is also formulated specifically for wall-bounded flows and is thus restricted to
such flows, i.e., the model is highly geometry dependent. **On the positive side**,
the Johnson-King model has been applied to many transonic flows that tend to
be particularly difficult to predict with modern turbulence models. The model's
track record has been quite good with such flows. Its predictions for attached
boundary layers could be made even closer to measurements by either using
Equation (3.109) instead of Equation (3.147) or simply recalibrating the model's
closure coefficients. On balance, this model appears to be a useful engineering
design tool, within its verified range of applicability.

# Problems

**3.1** For the far wake, verify that the solution to Equations (3.45) is given by Equations (3.46) – (3.48).

**3.2** For the mixing layer, beginning with Equation (3.65), introduce Equations (3.68) through (3.71) and derive Equation (3.73).

**3.3** For the jet, begin with Equation (3.83) and derive Equation (3.86).

**3.4** Using Equation (3.25) to represent the eddy viscosity, generate a similarity solution for the far wake. Obtain the exact closed-form solution, and determine the value of $\chi$ by forcing agreement with the corresponding $u_o(x)$ and $\delta(x)$ derived in this chapter. The following integral will be useful when you apply the integral constraint.

$$\int_0^\infty e^{-\xi^2} d\xi = \frac{\sqrt{\pi}}{2}$$

**3.5** Using Equation (3.25) to represent the eddy viscosity, generate a similarity solution for the plane jet. Obtain the exact closed-form solution, and determine the value of $\chi$ by forcing agreement with the corresponding $u_o(x)$ and $\delta(x)$ derived in this chapter. The following integrals will be useful in deriving the solution.

$$\int \frac{dx}{c^2 - x^2} = \frac{1}{c} \tanh^{-1}\left(\frac{x}{c}\right) + \text{constant}$$

$$\int_0^\infty \left[1 - \tanh^2 \xi\right]^2 d\xi = \frac{2}{3}$$

**3.6** Show that using Equation (3.98) for the mixing length in the viscous sublayer yields a velocity that behaves according to:

$$U^+ \approx y^+ - \frac{\kappa^2}{3}(y^+)^3 + \cdots \quad \text{as} \quad y^+ \to 0$$

**3.7** For a constant-pressure turbulent boundary layer, the skin friction and displacement thickness are approximately

$$c_f \approx 0.045 Re_\delta^{-1/4} \quad \text{and} \quad \delta^* \approx \frac{1}{8}\delta$$

where $Re_\delta = U_e\delta/\nu$ is Reynolds number based on $\delta$. Note also that, by definition, $c_f = 2u_\tau^2/U_e^2$. Assuming the matching point always occurs in the log layer so that $\partial U/\partial y = u_\tau/(\kappa y)$, make a graph of $y_m/\delta$ and $y_m^+$ versus $Re_\delta$ for the Cebeci-Smith model. Let $Re_\delta$ vary between $10^4$ and $10^6$. You should first rewrite the equations for $\nu_{T_i}$ and $\nu_{T_o}$ in terms of $y/\delta$ and $Re_\delta$. Then, solve the resulting equation for $y_m/\delta$ with an iterative procedure such as Newton's method. Compare your numerical results with Equation (3.117).

**3.8** Using a standard numerical integration scheme such as the Runge-Kutta method, determine the constant $C$ in the law of the wall implied by the mixing-length model. That is, solve the following equation for $U^+$.

$$(1 + \mu_T^+)\frac{dU^+}{dy^+} = 1$$

Integrate from $y^+ = 0$ to $y^+ = 500$ and calculate the limiting value of $C$ as $y^+ \to \infty$ from examination of

$$C = U^+ - \frac{1}{\kappa}\ell ny^+ \quad \text{at} \quad y^+ = 250, 300, 350, 400, 450 \text{ and } 500$$

Do the computation with the mixing length given by (i) Equation (3.98) and (ii) Equation (3.104). **NOTE:** To avoid truncation error, verify the following limiting form of the equation for $dU^+/dy^+$.

$$\frac{dU^+}{dy^+} \approx 1 - \left(\ell_{mix}^+\right)^2 + 2\left(\ell_{mix}^+\right)^4 + \cdots \quad \text{as} \quad \ell_{mix}^+ \to 0$$

Use this asymptotic form very close to $y^+ = 0$.

**3.9** Assume the velocity in a boundary layer for $y^+ \gg 1$ is given by

$$U^+ \approx \frac{1}{\kappa}\ell ny^+ + 5.0 + \frac{1}{\kappa}\sin^2\left(\frac{\pi y}{2\delta}\right)$$

Also, assume that $y_{max} \gg 26\nu/u_\tau$ for the Baldwin-Lomax model. Compute the quantities $y_{max}F_{max}$ and $C_{wk}y_{max}U_{dif}^2/F_{max}$ for this boundary layer. Then, noting that skin friction is given by $c_f = 2u_\tau^2/U_e^2$, determine the largest value of $c_f$ for which $F_{wake} = y_{max}F_{max}$. **HINT:** The solution to the transcendental equation $\xi + \tan\xi = 0$ is $\xi \approx 2.03$.

**3.10** For a turbulent boundary layer with surface mass transfer, the momentum equation in the sublayer and log layer simplifies to:

$$v_w\frac{dU}{dy} = \frac{d}{dy}\left[(\nu + \nu_T)\frac{dU}{dy}\right]$$

where $v_w$ is the (constant) vertical velocity at the surface.

(a) Integrate once using the appropriate surface boundary conditions. Introduce the friction velocity, $u_\tau$, in stating your integrated equation.

(b) Focusing now upon the log layer where $\nu_T \gg \nu$, what is the approximate form of the equation derived in Part (a) if we use the Cebeci-Smith model?

(c) Verify that the solution to the simplified equation of Part (b) is

$$2\frac{u_\tau}{v_w}\sqrt{1 + v_wU/u_\tau^2} = \frac{1}{\kappa}\ell ny + \text{constant}$$

**3.11** Generate a solution for channel and pipe flow using a mixing-length model with the mixing length in the inner and outer layers given by

$$
\ell_{mix} = \begin{cases} \kappa y \left[ 1 - e^{-y^+/26} \right] & , \quad \text{Inner Layer} \\ .09R & , \quad \text{Outer Layer} \end{cases}
$$

where $R$ is channel half-height or pipe radius. Use a numerical integration scheme such as the Runge-Kutta method, or modify Program **PIPE** (Appendix C). Compare computed skin friction with Equations (3.138) and (3.139). **See NOTE below.**

**NOTE for 3.11 and 3.12:** To assist in presenting your results, verify that skin friction and Reynolds number are given by $c_f = 2/(U_{avg}^+)^2$ and $Re_D = 2U_{avg}^+ R^+$ where $R^+ = u_\tau R/\nu$ and $U_{avg}$ is the average velocity across the channel/pipe. Also, to avoid truncation error, verify the following limiting form of the equation for $dU^+/dy^+$ in the limit $\ell_{mix}^+ \to 0$.

$$
\frac{dU^+}{dy^+} \approx \left( 1 - \frac{y^+}{R^+} \right) \left[ 1 - \left( 1 - \frac{y^+}{R^+} \right) (\ell_{mix}^+)^2 + 2 \left( 1 - \frac{y^+}{R^+} \right)^2 (\ell_{mix}^+)^4 \right]
$$

Use this asymptotic form very close to $y^+ = 0$.

**3.12** Generate a solution for pipe flow using a mixing-length model with the mixing length given by Nikuradse's formula, i.e.,

$$
\ell_{mix}/R = 0.14 - 0.08(1 - y/R)^2 - 0.06(1 - y/R)^4
$$

where $R$ is pipe radius. Use a numerical integration scheme such as the Runge-Kutta method, or modify Program **PIPE** (Appendix C). Compare computed skin friction with Equation (3.139). **See NOTE above.**

**3.13** Using a standard numerical integration scheme such as the Runge-Kutta method, determine the constant $C$ in the law of the wall implied by the Johnson-King model. That is, solve the following equation for $U^+$.

$$
(1 + \mu_T^+)\frac{dU^+}{dy^+} = 1
$$

Integrate from $y^+ = 0$ to $y^+ = 500$ and calculate the limiting value of $C$ as $y^+ \to \infty$ from examination of

$$
C = U^+ - \frac{1}{\kappa}\ell n y^+ \quad \text{at} \quad y^+ = 250, 300, 350, 400, 450 \text{ and } 500
$$

**NOTE:** To avoid truncation error, verify the following limiting form for $dU^+/dy^+$.

$$
\frac{dU^+}{dy^+} \approx 1 - \left( \ell_{mix}^+ \right)^2 + 2 \left( \ell_{mix}^+ \right)^4 + \cdots \quad \text{as} \quad \ell_{mix}^+ \to 0
$$

Use this asymptotic form very close to $y^+ = 0$.

**3.14** Using Program **PIPE** (Appendix C), compute the skin friction for channel flow according to the Johnson-King model. Compare your results with the Halleen-Johnston correlation [Equation (3.138] for $10^3 \leq Re_H \leq 10^5$. Also, compare the computed velocity profile for $Re_H = 13,750$ with the Mansour et al. DNS data, which are as follows.

| $y/(H/2)$ | $U/U_m$ | $y/(H/2)$ | $U/U_m$ |
|-----------|---------|-----------|---------|
| 0.000     | 0.000   | 0.602     | 0.945   |
| 0.103     | 0.717   | 0.710     | 0.968   |
| 0.207     | 0.800   | 0.805     | 0.984   |
| 0.305     | 0.849   | 0.902     | 0.995   |
| 0.404     | 0.887   | 1.000     | 1.000   |
| 0.500     | 0.917   |           |         |

**3.15** Using Program **PIPE** (Appendix C), compute the skin friction for pipe flow according to the Johnson-King model. Compare your results with the Prandtl correlation [Equation (3.139] for $10^3 \leq Re_D \leq 10^6$. Also, compare the computed velocity profile for $Re_D = 40,000$ with Laufer's data, which are as follows.

| $y/(D/2)$ | $U/U_m$ | $y/(D/2)$ | $U/U_m$ |
|-----------|---------|-----------|---------|
| 0.010     | 0.333   | 0.590     | 0.931   |
| 0.095     | 0.696   | 0.690     | 0.961   |
| 0.210     | 0.789   | 0.800     | 0.975   |
| 0.280     | 0.833   | 0.900     | 0.990   |
| 0.390     | 0.868   | 1.000     | 1.000   |
| 0.490     | 0.902   |           |         |

**3.16** The object of this problem is to compare predictions of algebraic models with measured properties of a turbulent boundary layer with adverse pressure gradient. The experiment to be simulated was conducted by Ludwieg and Tillman [see Coles and Hirst (1969) – Flow 1200]. Use Program **EDDYBL** and its menu-driven setup utility, Program **SETEBL**, to do the computations (see Appendix D).

(a) Using **SETEBL**, change appropriate input parameters to accomplish the following: use SI units (IUTYPE); set freestream conditions to $p_{t_\infty} = 1.01858 \cdot 10^5$ N/m$^2$, $T_{t_\infty}$ = 294 K, $M_\infty$ = 0.08656 (PT1, TT1, XMA); use an initial stepsize of $\Delta s = 0.01$ m (DS); set the initial boundary-layer properties so that $c_f = 0.00292$, $\delta = 0.0224$ m, $H = 1.36$, $Re_\theta = 5454$, $s_i = 0.75$ m (CF, DELTA, H, RETHET, SI); set the maximum arc length to $s_f = 2.782$ m (SSTOP); and, set up for $N = 11$ points to define the pressure (NUMBER).

(b) Use the following data to define the pressure distribution in a file named **presur.dat**. The initial and final pressure gradients are given by $(dp_e/dx)_i = 180.9$ N/m$^3$ and $(dp_e/dx)_f = -15.97$ N/m$^3$, respectively. Also, prepare a file **heater.dat** with constant wall temperature, $T_w$ = 294 K, and zero heat flux.

| $s$ (m) | $p_e$ (N/m$^2$) | $s$ (m) | $p_e$ (N/m$^2$) | $s$ (m) | $p_e$ (N/m$^2$) |
|---------|-----------------|---------|-----------------|---------|-----------------|
| 0.000 | $1.01067 \cdot 10^5$ | 2.282 | $1.01415 \cdot 10^5$ | 3.532 | $1.01554 \cdot 10^5$ |
| 0.782 | $1.01201 \cdot 10^5$ | 2.782 | $1.01491 \cdot 10^5$ | 3.732 | $1.01563 \cdot 10^5$ |
| 1.282 | $1.01271 \cdot 10^5$ | 3.132 | $1.01526 \cdot 10^5$ | 3.932 | $1.01562 \cdot 10^5$ |
| 1.782 | $1.01358 \cdot 10^5$ | 3.332 | $1.01541 \cdot 10^5$ | | |

(c) Do three computations using the Cebeci-Smith, Baldwin-Lomax and Johnson-King models.

(d) Compare computed skin friction with the following measured values.

| $s$ (m) | $c_f$ | $s$ (m) | $c_f$ |
|---------|-------|---------|-------|
| 0.782 | $2.92 \cdot 10^{-3}$ | 2.282 | $1.94 \cdot 10^{-3}$ |
| 1.282 | $2.49 \cdot 10^{-3}$ | 2.782 | $1.55 \cdot 10^{-3}$ |
| 1.782 | $2.05 \cdot 10^{-3}$ | | |

**3.17** The object of this problem is to compare predictions of algebraic models with measured properties of a turbulent boundary layer with adverse pressure gradient. The experiment to be simulated was conducted by Bradshaw [see Coles and Hirst (1969) – Flow 3300]. Use Program **EDDYBL** and its menu-driven setup utility, Program **SETEBL**, to do the computations (see Appendix D).

(a) Using **SETEBL**, change appropriate input parameters to accomplish the following: use USCS units (IUTYPE); set freestream conditions to $p_{t_\infty} = 2148$ lb/ft$^2$, $T_{t_\infty} = 537.6°$ R, $M_\infty = 0.106$ (PT1, TT1, XMA); use an initial stepsize given by $\Delta s = 0.05$ ft and a geometric-progression ratio $k_g = 1.08$ (DS, XK); set initial boundary-layer properties so that $c_f = 0.00225$, $\delta = 0.125$ ft, $H = 1.40$, $Re_\theta = 9216$, $s_i = 2.55$ ft (CF, DELTA, H, RETHET, SI); set the maximum arc length to $s_f = 7.0$ ft (SSTOP); and, set up for $N = 11$ points to define the pressure (NUMBER).

(b) Use the following data to define the pressure distribution in a file named **presur.dat**. The initial and final pressure gradients are given by $(dp_e/dx)_i = 3.410939$ lb/ft$^3$ and $(dp_e/dx)_f = 0.64929$ lb/ft$^3$, respectively. Also, prepare a file **heater.dat** with constant wall temperature, $T_w = 537.6°$ R, and zero heat flux.

| $s$ (ft) | $p_e$ (lb/ft$^2$) | $s$ (ft) | $p_e$ (lb/ft$^2$) | $s$ (ft) | $p_e$ (lb/ft$^2$) |
|----------|-------------------|----------|-------------------|----------|-------------------|
| 2.5 | $2.13123 \cdot 10^3$ | 4.5 | $2.13556 \cdot 10^3$ | 6.5 | $2.13768 \cdot 10^3$ |
| 3.0 | $2.13272 \cdot 10^3$ | 5.0 | $2.13621 \cdot 10^3$ | 7.0 | $2.13806 \cdot 10^3$ |
| 3.5 | $2.13387 \cdot 10^3$ | 5.5 | $2.13677 \cdot 10^3$ | 7.5 | $2.13841 \cdot 10^3$ |
| 4.0 | $2.13480 \cdot 10^3$ | 6.0 | $2.13726 \cdot 10^3$ | | |

(c) Do three computations using the Cebeci-Smith, Baldwin-Lomax and Johnson-King models.

(d) Compare computed skin friction with the following measured values.

| $s$ (ft) | $c_f$ | $s$ (ft) | $c_f$ | $s$ (ft) | $c_f$ |
|----------|-------|----------|-------|----------|-------|
| 2.5 | $2.45 \cdot 10^{-3}$ | 4.00 | $1.91 \cdot 10^{-3}$ | 7.00 | $1.56 \cdot 10^{-3}$ |
| 3.0 | $2.17 \cdot 10^{-3}$ | 5.00 | $1.74 \cdot 10^{-3}$ | | |
| 3.5 | $2.00 \cdot 10^{-3}$ | 6.00 | $1.61 \cdot 10^{-3}$ | | |

**3.18** The object of this problem is to predict the separation point for flow past a circular cylinder with the boundary-layer equations, using the measured pressure distribution. The experiment to be simulated was conducted by Patel (1968). Use Program **EDDYBL** and its menu-driven setup utility, Program **SETEBL**, to do the computations (see Appendix D).

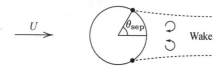

**Problem 3.18**

(a) Using **SETEBL**, change appropriate input parameters to accomplish the following: use USCS units (IUTYPE); set the freestream conditions to $p_{t_\infty} = 2147.7$ lb/ft$^2$, $T_{t_\infty} = 529.6°$ R, $M_\infty = 0.144$ (PT1, TT1, XMA); use an initial stepsize given by $\Delta s = 0.001$ ft (DS); set the initial boundary-layer properties so that $c_f = 0.00600$, $\delta = 0.006$ ft, $H = 1.40$, $Re_\theta = 929$, $s_i = 0.262$ ft (CF, DELTA, H, RETHET, SI); set the maximum arc length to $s_f = 0.785$ ft and maximum number of steps to 300 (SSTOP, IEND1); and, set up for $N = 47$ points to define the pressure (NUMBER).

(b) Use the following data to define the pressure distribution in a file named **presur.dat**. The initial and final pressure gradients are zero. Then, prepare a file **heater.dat** with constant wall temperature, $T_w = 529.4°$ R, and zero heat flux.

| s (ft) | $p_e$ (lb/ft$^2$) | s (ft) | $p_e$ (lb/ft$^2$) | s (ft) | $p_e$ (lb/ft$^2$) |
|--------|-------------------|--------|-------------------|--------|-------------------|
| .0000 | $2.147540 \cdot 10^3$ | .1500 | $2.116199 \cdot 10^3$ | .3500 | $2.055516 \cdot 10^3$ |
| .0025 | $2.147528 \cdot 10^3$ | .1625 | $2.112205 \cdot 10^3$ | .3625 | $2.056591 \cdot 10^3$ |
| .0050 | $2.147491 \cdot 10^3$ | .1750 | $2.107903 \cdot 10^3$ | .3750 | $2.058435 \cdot 10^3$ |
| .0075 | $2.147429 \cdot 10^3$ | .1875 | $2.103448 \cdot 10^3$ | .3875 | $2.061661 \cdot 10^3$ |
| .0100 | $2.147343 \cdot 10^3$ | .2000 | $2.098378 \cdot 10^3$ | .4000 | $2.066423 \cdot 10^3$ |
| .0125 | $2.147233 \cdot 10^3$ | .2125 | $2.093155 \cdot 10^3$ | .4125 | $2.071954 \cdot 10^3$ |
| .0250 | $2.146314 \cdot 10^3$ | .2250 | $2.087317 \cdot 10^3$ | .4250 | $2.079021 \cdot 10^3$ |
| .0375 | $2.144796 \cdot 10^3$ | .2375 | $2.081325 \cdot 10^3$ | .4375 | $2.085473 \cdot 10^3$ |
| .0500 | $2.142688 \cdot 10^3$ | .2500 | $2.075334 \cdot 10^3$ | .4500 | $2.089161 \cdot 10^3$ |
| .0625 | $2.140018 \cdot 10^3$ | .2625 | $2.069189 \cdot 10^3$ | .4625 | $2.091004 \cdot 10^3$ |
| .0750 | $2.136807 \cdot 10^3$ | .2750 | $2.064580 \cdot 10^3$ | .4750 | $2.092080 \cdot 10^3$ |
| .0875 | $2.134021 \cdot 10^3$ | .2875 | $2.060893 \cdot 10^3$ | .4875 | $2.092230 \cdot 10^3$ |
| .1000 | $2.130641 \cdot 10^3$ | .3000 | $2.058588 \cdot 10^3$ | .5000 | $2.092230 \cdot 10^3$ |
| .1125 | $2.127261 \cdot 10^3$ | .3125 | $2.056898 \cdot 10^3$ | .6500 | $2.092230 \cdot 10^3$ |
| .1250 | $2.123881 \cdot 10^3$ | .3250 | $2.055823 \cdot 10^3$ | .7850 | $2.092230 \cdot 10^3$ |
| .1375 | $2.120194 \cdot 10^3$ | .3375 | $2.055362 \cdot 10^3$ | | |

(c) Do three computations using the Cebeci-Smith, Baldwin-Lomax and Johnson-King models. Compare computed separation angle measured from the downstream symmetry axis with the measured value of $\theta_{sep} = 70°$. The radius of the cylinder is $R = 0.25$ ft, so that separation arc length, $s_{sep}$, is related to this angle by $\theta_{sep} = \pi - s_{sep}/R$.

# Chapter 4

# One-Equation and Two-Equation Models

As computers have increased in power since the 1960's, turbulence models based upon the equation for the turbulence kinetic energy have become the cornerstone of modern turbulence modeling research. This chapter discusses two types of eddy-viscosity models, viz., **One-Equation Models** and **Two-Equation Models**, with most of the emphasis on the latter. These models both retain the Boussinesq eddy-viscosity approximation, but differ in one important respect. One-equation models based on the turbulence energy equation are **incomplete** as they relate the turbulence length scale to some typical flow dimension. These models are rarely used. By contrast, one-equation models based on a equation for the eddy viscosity and two-equation models automatically provide the turbulence length scale or its equivalent and are thus **complete**.

The chapter begins with a derivation and discussion of the turbulence kinetic energy equation. We then introduce one-equation models based on this equation and upon a postulated equation for the eddy viscosity. The discussion includes examples of how such models fare for several flows. Next, we introduce two-equation models with specific details of the two most commonly used models. Our first two-equation model applications are to the same free shear flows as were considered in Chapter 3. Then, we focus upon a powerful tool, singular-perturbation theory, that we use to analyze model-predicted features of the turbulent boundary layer. We apply the two-equation model to attached wall-bounded flows and compare to corresponding algebraic-model predictions. We discuss the issue of asymptotic consistency approaching a solid boundary, and the ability of two-equation models to predict transition from laminar to turbulent flow. Our final applications are to separated flows. The concluding section discusses the range of applicability of one- and two-equation models.

## 4.1    The Turbulence Energy Equation

Turbulence energy equation models have been developed to incorporate nonlocal and flow history effects in the eddy viscosity. Prandtl (1945) postulated computing a characteristic velocity scale for the turbulence, $v_{mix}$, thus obviating the need for assuming that $v_{mix} \sim \ell_{mix}|\partial U/\partial y|$ [c.f. Equation (3.15)]. He chose the kinetic energy (per unit mass) of the turbulent fluctuations, $k$, as the basis of his velocity scale, i.e.,

$$k = \frac{1}{2}\overline{u_i' u_i'} = \frac{1}{2}\left(\overline{u'^2} + \overline{v'^2} + \overline{w'^2}\right) \tag{4.1}$$

Thus, in terms a turbulence length scale, $\ell$, and $k$, dimensional arguments dictate that the kinematic eddy viscosity is given by

$$\nu_T = \text{constant} \cdot k^{1/2}\ell \tag{4.2}$$

Note that we drop subscript "mix" in this chapter for convenience, and to avoid confusion with the mixing length used in algebraic models.

The question now arises as to how we determine $k$. The answer is provided by taking the trace of the Reynolds-stress tensor, which yields the following.

$$\tau_{ii} = -\overline{u_i' u_i'} = -2k \tag{4.3}$$

Thus, the trace of the Reynolds-stress tensor is proportional to the kinetic energy of the turbulent fluctuations per unit volume. The quantity $k$ should strictly be referred to as **specific turbulence kinetic energy** ("specific" meaning "per unit mass"), but is often just called **turbulence kinetic energy**.

In Chapter 2 we derived a differential equation describing the behavior of the Reynolds-stress tensor, $\tau_{ij}$, i.e., Equation (2.34). We can derive a corresponding equation for $k$ by taking the trace of the Reynolds-stress equation. Noting that the trace of the tensor $\Pi_{ij}$ vanishes for incompressible flow, contracting Equation (2.34) leads to the following **transport equation** for the turbulence kinetic energy.

$$\frac{\partial k}{\partial t} + U_j \frac{\partial k}{\partial x_j} = \tau_{ij}\frac{\partial U_i}{\partial x_j} - \epsilon + \frac{\partial}{\partial x_j}\left[\nu\frac{\partial k}{\partial x_j} - \frac{1}{2}\overline{u_i' u_i' u_j'} - \frac{1}{\rho}\overline{p' u_j'}\right] \tag{4.4}$$

The quantity $\epsilon$ is the **dissipation per unit mass** and is defined by the following correlation.

$$\epsilon = \nu\,\overline{\frac{\partial u_i'}{\partial x_k}\frac{\partial u_i'}{\partial x_k}} \tag{4.5}$$

The various terms appearing in Equation (4.4) represent physical processes occurring as the turbulence moves about in a given flow. The sum of the two

terms on the left-hand side, i.e., the **unsteady term** and the **convection**, is the
familiar substantial derivative of $k$ that gives the rate of change of $k$ following a
fluid particle. The first term on the right-hand side is known as **Production**, and
represents the rate at which kinetic energy is transferred from the mean flow to
the turbulence. Rewritten as $\tau_{ij} S_{ij}$ (because $\tau_{ij}$ is symmetric), this term is seen
to be the rate at which work is done by the mean strain rate against the turbulent
stresses. **Dissipation** is the rate at which turbulence kinetic energy is converted
into thermal internal energy, equal to the mean rate at which work is done by the
fluctuating part of the strain rate against the fluctuating viscous stresses. The term
involving $\nu \partial k / \partial x_j$ is called **Molecular Diffusion**, and represents the diffusion
of turbulence energy caused by the fluid's natural molecular transport process.
We refer to the triple velocity correlation term as **Turbulent Transport**, and
regard it as the rate at which turbulence energy is transported through the fluid
by turbulent fluctuations. The last term on the right-hand side of the equation
is called **Pressure Diffusion**, another form of turbulent transport resulting from
correlation of pressure and velocity fluctuations.

The quantity $\epsilon$ as defined in Equation (4.5) differs from the classical definition
of dissipation given in the preceding paragraph. From the latter, it follows that
[cf. Townsend (1976) or Hinze (1975)] the true dissipation, $\epsilon_{true}$, is proportional
to the square of the fluctuating strain-rate tensor, $s'_{ik}$, viz.,

$$\epsilon_{true} = 2\nu \overline{s'_{ik} s'_{ik}}, \qquad s'_{ik} = \frac{1}{2} \left( \frac{\partial u'_i}{\partial x_k} + \frac{\partial u'_k}{\partial x_i} \right) \tag{4.6}$$

Hence, the quantity $\epsilon$ is given by (for incompressible flow):

$$\epsilon = \epsilon_{true} - \frac{\partial}{\partial x_k} \left( \overline{\nu u'_i \frac{\partial u'_k}{\partial x_i}} \right) \tag{4.7}$$

In practice, the difference between $\epsilon$ and $\epsilon_{true}$ is small and should be expected
to be significant only in regions of strong gradients, e.g., shock waves or the
viscous wall region. In the latter case, Bradshaw and Perot (1993) have shown
that the maximum difference is just 2%, and can thus be ignored.

The unsteady term, convection and molecular diffusion are exact while production, dissipation, turbulent transport and pressure diffusion involve unknown
correlations. To **close** this equation, we must specify $\tau_{ij}$, dissipation, turbulent
transport and pressure diffusion.

The conventional approach to closure of the $k$ equation was initiated by
Prandtl (1945) who established arguments for each term in the equation. This
term-by-term modeling approach amounts to performing **drastic surgery** on the
exact equation, replacing unknown correlations with closure approximations.
This process is by no means rigorous. The closure approximations are no better
than the turbulence data upon which they are based. Our hope is that we can find

closure approximations that make accurate solutions possible. We will discuss this point in greater detail when we introduce two-equation models.

**Reynolds-Stress Tensor:** For the class of turbulence models considered in this chapter, we assume the Boussinesq approximation is valid. Thus, we say that the specific Reynolds-stress tensor is given by

$$\tau_{ij} = 2\nu_T S_{ij} - \frac{2}{3}k\delta_{ij} \tag{4.8}$$

where $S_{ij}$ is the mean strain-rate tensor. Note that the second term on the right-hand side of Equation (4.8) is needed to obtain the proper trace of $\tau_{ij}$. That is, since $S_{ii} = 0$ for incompressible flow, contracting Equation (4.8) yields $\tau_{ii} = -2k$ in accord with Equation (4.3).

Strictly, we should regard Equation (4.8) as the definition of $\nu_T$. In this spirit, no approximation is implied, provided we don't explicitly say $\nu_T$ is a scalar. However, for the purposes of this chapter, we do in fact assume $\nu_T$ is a scalar so that the term "approximation" is appropriate.

**Turbulent Transport and Pressure Diffusion:** The standard approximation made to represent turbulent transport of scalar quantities in a turbulent flow is that of gradient-diffusion. In analogy to molecular transport processes, we say that $-\overline{u'_j\phi'} \sim \nu_T \partial\Phi/\partial x_j$. Unfortunately, there is no corresponding straightforward analog for the pressure-diffusion term. In the absence of definitive experimental data, the pressure-diffusion term has generally been grouped with the turbulent-transport term, and the sum assumed to behave as a gradient-transport process. Fortunately, recent DNS results [e.g., Mansour, Kim and Moin (1988)] indicate that the term is quite small for simple flows. Thus, we assume that

$$\frac{1}{2}\overline{u'_iu'_iu'_j} + \frac{1}{\rho}\overline{p'u'_j} = -\frac{\nu_T}{\sigma_k}\frac{\partial k}{\partial x_j} \tag{4.9}$$

where $\sigma_k$ is a closure coefficient. Assuming the vectors on the left- and right-hand sides of Equation (4.9) are parallel (a somewhat optimistic assumption!), this equation defines $\sigma_k$. As stressed by Bradshaw (1994), this statement applies to all turbulence closure coefficients. At this point, no approximation has entered although, of course, we hope the model is realistic enough that $\sigma_k$ can be chosen to be constant.

**Dissipation:** The manner in which we determine the dissipation is not unique amongst turbulence energy equation models. It suffices at this point to note that we still have two unknown parameters, which are the turbulence length scale, $\ell$, and the dissipation, $\epsilon$. If both properties are assumed to be strictly functions of the turbulence independent of natural fluid properties such as molecular viscosity, purely dimensional arguments [Taylor (1935)] show that

$$\epsilon \sim k^{3/2}/\ell \qquad (4.10)$$

Hence, we still need a prescription for the length scale of the turbulence in order to close our system of equations. In the following sections, we will explore the various methods that have been devised to determine the length scale.

Combining Equations (4.4) and (4.9), we can write the modeled version of the turbulence kinetic energy equation that is used in virtually all turbulence energy equation models. The equation assumes the following form,

$$\frac{\partial k}{\partial t} + U_j \frac{\partial k}{\partial x_j} = \tau_{ij} \frac{\partial U_i}{\partial x_j} - \epsilon + \frac{\partial}{\partial x_j}\left[(\nu + \nu_T/\sigma_k)\frac{\partial k}{\partial x_j}\right] \qquad (4.11)$$

where $\tau_{ij}$ is given by Equation (4.8).

## 4.2 One-Equation Models

To complete the closure of the turbulence kinetic energy equation, Prandtl (1945) postulated that the dissipation assumes the form quoted in Equation (4.10). Introducing a closure coefficient that we will call $C_D$, the dissipation is

$$\epsilon = C_D k^{3/2}/\ell \qquad (4.12)$$

and the turbulence length scale remains the only unspecified part of the model. Given twenty years of experience with the mixing-length model, Prandtl had sufficient confidence that he could generalize established prescriptions for the turbulence length scale $\ell$. [Of course, $\ell \propto \ell_{mix}$ only if the ratio of production to dissipation is constant. To see this, note that in a thin shear layer, Equation (3.18) gives $\partial U/\partial y = (-\overline{u'v'})^{1/2}/\ell_{mix}$. Hence, balancing production and dissipation means $-\overline{u'v'}\partial U/\partial y = (-\overline{u'v'})^{3/2}/\ell_{mix} = C_D k^{3/2}/\ell$ so that $\ell \propto \ell_{mix}$ if $-\overline{u'v'}/k = $ constant.] As we will discuss further below, measurements show that the constant is about 0.3 for many thin shear layers. Thus, Prandtl's **One-Equation Model** is as follows:

$$\frac{\partial k}{\partial t} + U_j \frac{\partial k}{\partial x_j} = \tau_{ij} \frac{\partial U_i}{\partial x_j} - C_D \frac{k^{3/2}}{\ell} + \frac{\partial}{\partial x_j}\left[(\nu + \nu_T/\sigma_k)\frac{\partial k}{\partial x_j}\right] \qquad (4.13)$$

where $\tau_{ij}$ is given by Equation (4.8) and the kinematic eddy viscosity is

$$\nu_T = k^{1/2}\ell = C_D k^2/\epsilon \qquad (4.14)$$

Note that at this point we make an implicit assumption regarding the "constant" in Equation (4.2), which has been set equal to one in Equation (4.14).

That is, there is no a priori reason why $\nu_T$ should depend only upon $k$ and $\ell$, i.e., no reason why "constant" should really be constant. In reality, $\nu_T$ is the ratio of a turbulence quantity (e.g., $-\overline{u'v'}$) to a mean flow quantity (e.g., $\partial U/\partial y + \partial V/\partial x$). Consequently, $\nu_T$ will not, in general, precisely follow mean-flow scales such as $U_e$ and $\delta^*$ or turbulence scales such as $k$ and $\ell$. Only in **equilibrium flows** for which production and dissipation balance are mean-flow and turbulence scales proportional — and then either can be used for $\nu_T$. Otherwise, an unknown mix of scales is needed.

Emmons (1954) independently proposed essentially the same model. Before the model can be used in applications, the length scale, $\ell$, and the closure coefficients $\sigma_k$ and $C_D$ must be specified. Emmons (1954) and Glushko (1965) applied this model to several flows with some degree of success using Equation (4.14) with $\sigma_k = 1$ and $C_D$ ranging between 0.07 and 0.09. Their length scale distributions were similar to those used for the mixing-length model. Wolfshtein (1967) found that by introducing damping factors in the dissipation and eddy viscosity similar to the Van Driest factor [Equation (3.104)], more satisfactory results can be obtained with this model for low-Reynolds-number flows. More recently, Goldberg (1991) has refined the model even further.

Although it is clearly more complex than an algebraic model, the Prandtl-Emmons-Glushko one-equation model is certainly straightforward and elegant. As originally postulated it involves two closure coefficients and one closure function (the length scale). Even with Wolfshtein's low-Reynolds-number corrections, the number of closure coefficients increases by only two so that the model actually has fewer closure coefficients than the Baldwin-Lomax and Johnson-King models. For attached flows, the Goldberg model has five closure coefficients, two damping functions, and a closure function for the length scale. Goldberg's number of closure coefficients and empirical functions more than doubles for separated flows.

Bradshaw, Ferriss and Atwell (1967) formulated a one-equation model that avoids introducing a gradient-diffusion approximation. Rather than introduce the Boussinesq approximation, they argue that for a wide range of flows, the ratio of the Reynolds shear stress, $\tau_{xy}$, to the turbulence kinetic energy, $k$, is constant. Measurements [Townsend (1976)] indicate that for boundary layers, wakes and mixing layers the ratio is nearly the same and given by

$$\tau_{xy} \approx \beta_r k, \qquad \beta_r = 0.3 \qquad (4.15)$$

The stress/energy ratio, i.e., the constant $\beta_r$, is often referred to as **Bradshaw's constant**, and sometimes as **Townsend's constant**.[1] Building upon this presumably universal result, Bradshaw, Ferriss and Atwell formulated a one-equation model based on the turbulence kinetic energy. A novel feature of their formulation is that the equations are hyperbolic for boundary layers rather than parabolic.

---

[1]The notation $\tau_{xy} = 2a_1 k$ is sometimes used where $a_1 \approx 0.15$.

This is a direct consequence of modeling the $k$ equation's turbulent transport term by a "bulk-convection" process rather than a gradient-diffusion approximation as in Equation (4.11). The resulting equations are thus solved by using the method of characteristics.

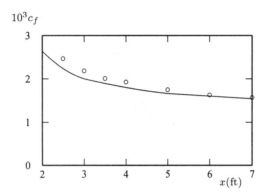

Figure 4.1: *Comparison of computed and measured skin friction for Bradshaw Flow C;* —— *Bradshaw-Ferriss-Atwell;* ○ *Bradshaw.*

Figure 4.1 compares computed and measured skin friction for Flow 3300 of the 1968 AFOSR-IFP-Stanford Conference on the Computation of Turbulent Boundary Layers. As shown, the differences between theory and experiment are even less than those obtained using the Cebeci-Smith and Baldwin-Lomax models [see Figure 3.16]. Overall, the Bradshaw-Ferriss-Atwell model's skin friction for boundary layers in adverse pressure gradient was closest of the various models tested in the 1968 Conference to measured values.

One-equation models have been formulated that are based on something other than the turbulence energy equation. Nee and Kovasznay (1968), for example, postulated a phenomenological transport equation for the kinematic eddy viscosity, $\nu_T$. The equation involves terms similar to those appearing in Equation (4.13). The model has four closure coefficients and requires prescription of the turbulence length scale. Sekundov (1971) developed a similar model that has generated considerable interest in the Russian research community, but that has rarely been referenced in Western journals. The English-language report of Gulyaev et al. (1993) summarizes work on the Sekundov model in its 1992 version. The paper by Vasiliev et al. (1997) shows that the 1971 version, although very simple, is complete and quite capable.

More recently, Baldwin and Barth (1990) and Spalart and Allmaras (1992) have devised even more elaborate model equations for the eddy viscosity. The Baldwin-Barth model, for example, includes seven closure coefficients and three empirical damping functions. The Baldwin-Barth model is as follows.

**Kinematic Eddy Viscosity:**

$$\nu_T = C_\mu \nu \tilde{R}_T D_1 D_2 \qquad (4.16)$$

**Turbulence Reynolds Number:**

$$\frac{\partial}{\partial t}\left(\nu \tilde{R}_T\right) + U_j \frac{\partial}{\partial x_j}\left(\nu \tilde{R}_T\right) = (C_{\epsilon 2} f_2 - C_{\epsilon 1})\sqrt{\nu \tilde{R}_T P}$$

$$+ (\nu + \nu_T/\sigma_\epsilon)\frac{\partial^2(\nu \tilde{R}_T)}{\partial x_k \partial x_k} - \frac{1}{\sigma_\epsilon}\frac{\partial \nu_T}{\partial x_k}\frac{\partial(\nu \tilde{R}_T)}{\partial x_k} \qquad (4.17)$$

**Closure Coefficients and Auxiliary Relations:**

$$C_{\epsilon 1} = 1.2, \quad C_{\epsilon 2} = 2.0, \quad C_\mu = 0.09, \quad A_o^+ = 26, \quad A_2^+ = 10 \qquad (4.18)$$

$$\frac{1}{\sigma_\epsilon} = (C_{\epsilon 2} - C_{\epsilon 1})\frac{\sqrt{C_\mu}}{\kappa^2}, \qquad \kappa = 0.41 \qquad (4.19)$$

$$P = \nu_T\left[\left(\frac{\partial U_i}{\partial x_j} + \frac{\partial U_j}{\partial x_i}\right)\frac{\partial U_i}{\partial x_j} - \frac{2}{3}\frac{\partial U_k}{\partial x_k}\frac{\partial U_k}{\partial x_k}\right] \qquad (4.20)$$

$$D_1 = 1 - e^{-y^+/A_o^+} \qquad \text{and} \qquad D_2 = 1 - e^{-y^+/A_2^+} \qquad (4.21)$$

$$f_2 = \frac{C_{\epsilon 1}}{C_{\epsilon 2}} + \left(1 - \frac{C_{\epsilon 1}}{C_{\epsilon 2}}\right)\left(\frac{1}{\kappa y^+} + D_1 D_2\right) \cdot$$
$$\left[\sqrt{D_1 D_2} + \frac{y^+}{\sqrt{D_1 D_2}}\left(\frac{D_2}{A_o^+}e^{-y^+/A_o^+} + \frac{D_1}{A_2^+}e^{-y^+/A_2^+}\right)\right] \qquad (4.22)$$

The Baldwin-Barth model is **complete** as it involves no adjustable functions or coefficients. While this guarantees nothing regarding its suitability for a given application, it does make its implementation convenient. This type of model constitutes the simplest complete model of turbulence.

Note that the Baldwin-Barth model circumvents the need to specify a dissipation length such as the quantity $\ell$ in Equation (4.13) by expressing the decay, or dissipation, of the eddy viscosity in terms of spatial gradients. That is, the dissipation term in Equation (4.17), $\epsilon_\nu$, is

$$\epsilon_\nu = \frac{1}{\sigma_\epsilon}\frac{\partial \nu_T}{\partial x_k}\frac{\partial(\nu \tilde{R}_T)}{\partial x_k} \qquad (4.23)$$

As a consequence of this closure approximation, $\epsilon_\nu = 0$ when spatial gradients vanish. Thus, rather than decaying with streamwise distance, the eddy viscosity will remain constant in a uniform stream. This incorrect feature can produce nonphysical diffusion in a numerical computation, for example, in a multi-element airfoil.

The Spalart-Allmaras model is also written in terms of the eddy viscosity. The model includes eight closure coefficients and three closure functions. Its defining equations are as follows.

**Kinematic Eddy Viscosity:**

$$\nu_T = \tilde{\nu} f_{v1} \qquad (4.24)$$

**Eddy Viscosity Equation:**

$$\frac{\partial \tilde{\nu}}{\partial t} + U_j \frac{\partial \tilde{\nu}}{\partial x_j} = c_{b1} \tilde{S} \tilde{\nu} - c_{w1} f_w \left(\frac{\tilde{\nu}}{d}\right)^2 + \frac{1}{\sigma} \frac{\partial}{\partial x_k} \left[(\nu + \tilde{\nu}) \frac{\partial \tilde{\nu}}{\partial x_k}\right] + \frac{c_{b2}}{\sigma} \frac{\partial \tilde{\nu}}{\partial x_k} \frac{\partial \tilde{\nu}}{\partial x_k}$$
$$(4.25)$$

**Closure Coefficients and Auxiliary Relations:**

$$c_{b1} = 0.1355, \quad c_{b2} = 0.622, \quad c_{v1} = 7.1, \quad \sigma = 2/3 \qquad (4.26)$$

$$c_{w1} = \frac{c_{b1}}{\kappa^2} + \frac{(1 + c_{b2})}{\sigma}, \quad c_{w2} = 0.3, \quad c_{w3} = 2, \quad \kappa = 0.41 \qquad (4.27)$$

$$f_{v1} = \frac{\chi^3}{\chi^3 + c_{v1}^3}, \quad f_{v2} = 1 - \frac{\chi}{1 + \chi f_{v1}}, \quad f_w = g \left[\frac{1 + c_{w3}^6}{g^6 + c_{w3}^6}\right]^{1/6} \qquad (4.28)$$

$$\chi = \frac{\tilde{\nu}}{\nu}, \quad g = r + c_{w2}(r^6 - r), \quad r = \frac{\tilde{\nu}}{\tilde{S} \kappa^2 d^2} \qquad (4.29)$$

$$\tilde{S} = S + \frac{\tilde{\nu}}{\kappa^2 d^2} f_{v2}, \quad S = \sqrt{2 \Omega_{ij} \Omega_{ij}} \qquad (4.30)$$

The tensor $\Omega_{ij} = \frac{1}{2}(\partial U_i/\partial x_j - \partial U_j/\partial x_i)$ is the rotation tensor and $d$ is distance from the closest surface. Although not listed here, the model even includes a transition correction that introduces four additional closure coefficients and two more empirical functions. Finally, note that the source terms for the eddy viscosity equation depend upon the distance from the closest surface, $d$, as well as upon the gradient of $\tilde{\nu}$. Since $d \to \infty$ far from solid boundaries, this model also predicts no decay of the eddy viscosity in a uniform stream.

To determine how close one-equation model predictions are to measurements, we turn first to the four free shear flow applications considered in Section 3.3. We also include the **radial jet**, which occurs when two jets of equal strength collide and spread radially. Since the Baldwin-Barth and Spalart-Allmaras models are complete, the turbulence scales are automatically defined, i.e., neither model involves an adjustable closure coefficient such as $\alpha$ in Equation (3.34). Comparing computed and measured **spreading rate** provides a straightforward, and concise, gauge of how well the models reproduce measured flow properties.

The conventional definition of **spreading rate for the wake** is the value of the similarity variable, $\eta = y\sqrt{\rho U_\infty^2/(Dx)}$ (see Subsection 3.3.1), where the velocity defect is half its maximum value. Similarly **for the plane jet, round jet and radial jet, the spreading rate** is the value of $y/x$ where the velocity is half its centerline value. **For the mixing layer, the spreading rate** is usually defined as the difference between the values of $y/x$ where $(U - U_2)^2/(U_1 - U_2)^2$ is 9/10 and 1/10.

Table 4.1: *Free Shear Flow Spreading Rates for One-Equation Models*

| Flow | Baldwin-Barth | Spalart-Allmaras | Measured |
|------|---------------|------------------|----------|
| Far Wake | .315 | .341 | .365 |
| Mixing Layer | – | .108 | .115 |
| Plane Jet | – | .156 | .100-.110 |
| Round Jet | – | .246 | .086-.096 |
| Radial Jet | – | .166 | .096-.110 |

Table 4.1 compares computed and measured spreading rates for the Baldwin-Barth and Spalart-Allmaras models. The numerical results for the Spalart-Allmaras model have been obtained using Programs **WAKE**, **JET** and **MIXER**, which are described in Appendix C. The table includes only the far-wake spreading rate inferred from the computations of Baldwin and Barth (1990) for the Baldwin-Barth model.

Attempts at incorporating the Baldwin-Barth model in the Appendix C programs have proven unsuccessful because the tridiagonal matrix corresponding to the discretized form of the equation for $\nu\tilde{R}_T$ proves to be particularly ill conditioned. Contrary to the comments of Baldwin and Barth (1990), who warn of possible numerical difficulties, the problem does not stem from poor grid resolution. Rather, the model predicts a sharp discontinuity in the eddy viscosity just inside the edge of the shear layer that destabilizes the computation, independent of grid size.

As summarized in the table, Spalart-Allmaras and Baldwin-Barth models predict a far-wake spreading rate within 6% and 14%, respectively, of the measured rate. The Spalart-Allmaras model's mixing-layer spreading rate is also within 6% of the measured value. However, its predicted plane-jet and radial-jet spreading rates are more than 40% higher than measured, while the round-jet value is nearly triple the corresponding experimental value. These results are entirely consistent with the fact that these models have been optimized for aerodynamic applications, most notably for flow past a wing. The mixing layer and far wake are salient in this context, while the plane and round jet are not. These applications do show one of the limitations of the Spalart-Allmaras model, i.e., it is inappropriate to applications involving jet-like free shear regions.

Figure 4.2 compares computed and measured skin friction for the sixteen boundary layers with pressure gradient used to assess algebraic models in Chapter 3. All computations have been done with Program **EDDYBL** (Appendix D). Table 4.2 summarizes overall differences between computed and measured $c_f$ for the Baldwin-Barth and Spalart-Allmaras models. The overall average difference at the final station for the flows, excluding Flow 5300, is 26% for the Baldwin-Barth model and 7% for the Spalart-Allmaras model.

Table 4.2: *Differences Between Computed and Measured Skin Friction.*

| Pressure Gradient | Flows | Baldwin-Barth | Spalart-Allmaras |
|---|---|---|---|
| Favorable | 1400, 1300, 2700, 6300 | 8.2% | 1.4% |
| Mild Adverse | 1100, 2100, 2500, 4800 | 22.3% | 9.9% |
| Moderate Adverse | 2400, 2600, 3300, 4500 | 32.2% | 11.0% |
| Strong Adverse | 0141, 1200, 4400 | 40.3% | 7.2% |
| All | – | 25.8% | 7.4% |

The Baldwin-Barth model's predicted skin friction is consistently smaller than measured for boundary layers with adverse pressure gradient. As an example, for the Samuel-Joubert increasingly adverse pressure gradient case (Flow 0141), the computed skin friction is 35% lower than the measured value. Although all twelve adverse-pressure-gradient flows are attached, the Baldwin-Barth model predicts separation for three cases, viz., Flows 4800 (mild adverse $\nabla p$), 4500 (moderate adverse $\nabla p$) and 5300 (strong adverse $\nabla p$). This clearly illustrates the model's tendency to respond too strongly to adverse pressure gradient, relative to measurements, in the sense that it always predicts too large a decrease in skin friction.

As shown by Sai and Lutfy (1995), the Baldwin-Barth model is extremely sensitive to the freestream value of the eddy viscosity. While using nonphysically large values for $\tilde{R}_T$ reduces differences between computed and measured $c_f$, no freestream values have been found that can prevent separation for Flows 4800, 4500 and 5300.

By contrast, aside from transients near the beginning of several of the computations, the Spalart-Allmaras $c_f$ is as close to corresponding measured values as the Baldwin-Lomax algebraic model. For the Samuel-Joubert case, the computed skin friction is 10% higher than measured. The sole case with large differences between computed and measured flow properties is the "incipient-separation" case of Stratford (1959), i.e., Flow 5300. As shown in Figure 4.2, the predicted value of $c_f$ at the end of the computation is 3.4 times the measured value. Recall that the Johnson-King 1/2-equation model (see Figure 3.18) exhibits similar behavior for this flow, while being much closer to measurements for the other fifteen cases.

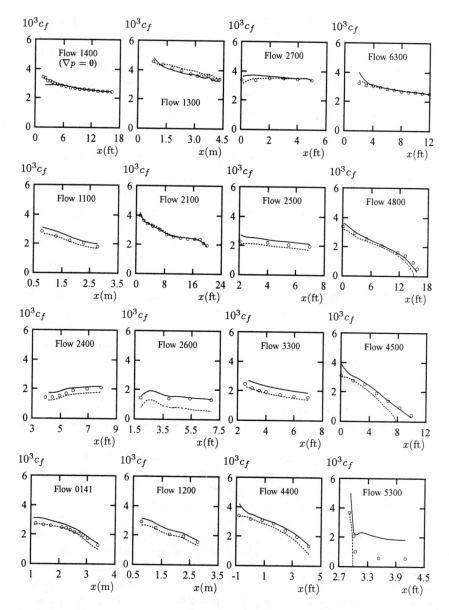

Figure 4.2: *Computed and measured skin friction for boundary layers subjected to a pressure gradient. Top row - favorable* $\nabla p$; *next to top row - mild adverse* $\nabla p$; *next to bottom row - moderate adverse* $\nabla p$; *bottom row - strong adverse* $\nabla p$. —— *Spalart-Allmaras model;* - - - *Baldwin-Barth model;* ∘ *measured.*

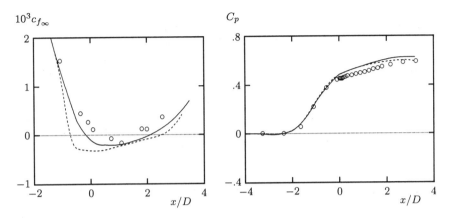

Figure 4.3: *Computed and measured flow properties for Driver's separated flow;* —— *Spalart-Allmaras;* - - - *Baldwin-Barth;* ∘ *Driver.*

Figure 4.3 shows how the Spalart-Allmaras and Baldwin-Barth models fare for Driver's separated flow as demonstrated by Menter (1992b, 1994). The Spalart-Allmaras model predicts a separation bubble that is about 60% larger than measured. The Baldwin-Barth model skin friction deviates from measured values even more than the Baldwin-Lomax model (see Figure 3.17), with a predicted separation bubble region that is more than twice the size measured by Driver. The results for the Baldwin-Barth model are not surprising in light of how poorly the model fares for attached boundary layers in adverse $\nabla p$.

The backward-facing step (Figure 4.4) is a popular test case for turbulence models because the geometry is simple. Additionally, separation occurs at the

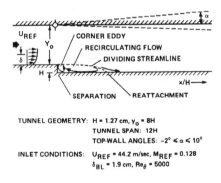

Figure 4.4: *Backward-facing step flow geometry and inlet conditions for the Driver-Seegmiller (1985) experiments. [From Driver and Seegmiller (1985) — Copyright © AIAA 1985 — Used with permission.]*

sharp corner so the flow is easier to predict than a flow for which the separation point is unknown a priori. Figure 4.5 compares computed and measured [Driver and Seegmiller (1985)] skin friction for backstep flow with the upper channel wall inclined to the lower wall at 0°. The Spalart-Allmaras model predicts reattachment at 6.1 step heights, $H$, downstream of the step. This is within 2% of the measured value of $6.2H$. Although not shown here, the model predicts reattachment at $8.6H$ when the upper wall is inclined at 6°, which is within 6% of the measured value of $8.1H$.

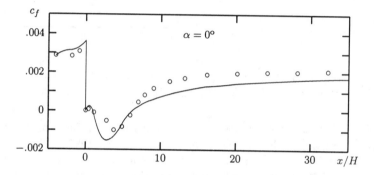

Figure 4.5: *Computed and measured skin friction for flow past a backward-facing step; —— Spalart-Allmaras model; • Driver-Seegmiller data.*

Thus, on balance, Spalart-Allmaras predictions are satisfactory for many engineering applications. It is especially attractive for airfoil and wing applications, for which it has been calibrated. Its failure to accurately reproduce jet spreading rates is a cause for concern, and should serve as a warning that the model has some shortcomings. Nevertheless, the model appears to be a valuable engineering tool.

By contrast, the Baldwin-Barth model predicts much larger discrepancies between computed and measured $c_f$ than the Spalart-Allmaras model and the much simpler algebraic models. The discrepancies are so large (an average of 26% for the 15 boundary-layer cases excluding Flow 5300) that its use for boundary-layer flows is inadvisable. It is also extremely sensitive to the freestream value of the eddy viscosity and can be very difficult to cast in finite-difference form (e.g., by yielding ill-conditioned matrices). Given all of these flaws, the model should be abandoned in favor of the Spalart-Allmaras model.

In light of these facts, we have not yet arrived at a universal turbulence model. In general, one-equation models share a few of the failures as well as most of the successes of the mixing-length model. While there is a smaller need for adjustment from flow to flow than with the mixing-length model, the Spalart-Allmaras model, as good as it is, is unable to predict spreading rates for plane and round jets that are consistent with measurements. Also, while the model's predictions for attached boundary layers are usually as close to measurements as those of algebraic models, its skin friction for the Stratford incipient-separation case (Flow 5300) is several times higher than measured. Finally, while it provides close agreement with measured reattachment length for the backward-facing step and airfoils with small separation bubbles, its predicted separation bubble for the Driver flow is significantly larger than measured. This erratic pattern is a bit discomforting, and suggests that something better is needed for general turbulent-flow applications. To reach a more-nearly universal model, especially for separated flows, we must seek a model in which transport effects for the velocity and length scales are accounted for separately. The rest of this chapter is devoted to investigating such models.

## 4.3   Two-Equation Models

**Two-Equation Models** of turbulence have served as the foundation for much of the turbulence model research during the past two decades. For example, almost all of the computations done for the 1980-81 AFOSR-HTTM-Stanford Conference on Complex Turbulent Flows used two-equation turbulence models. These models provide not only for computation of $k$, but also for the turbulence length scale or equivalent. Consequently, two-equation models are **complete**, i.e., can be used to predict properties of a given turbulent flow with no prior knowledge of the turbulence structure.

In the following discussion of two-equation models, arguments are often presented in terms of the $k$-$\omega$ model rather than the more-commonly used $k$-$\epsilon$ model. This, in no way, constitutes a campaign to popularize the former model. Rather, it usually reflects either the author's greater familiarity with the $k$-$\omega$ model (as one of its developers) or its analytical simplicity relative to the $k$-$\epsilon$ model. Except in cases where conclusions are obvious, every attempt has been made to leave the reader to make judgments regarding the superiority of any model described in this book.

The starting point for virtually all two-equation models is the Boussinesq approximation, Equation (4.8), and the turbulence kinetic energy equation in the form of Equation (4.11). As pointed out at the end of Section 4.1, there is an arbitrariness in the way we define the turbulence length scale, $\ell$, to go with the velocity scale, $k^{1/2}$.

Kolmogorov (1942), for example, pointed out that a second transport equation is needed to compute the so-called **specific dissipation rate**, $\omega$. This quantity has dimensions of (time)$^{-1}$. On dimensional grounds, the eddy viscosity, turbulence length scale and dissipation can be determined from

$$\nu_T \sim k/\omega, \qquad \ell \sim k^{1/2}/\omega, \qquad \epsilon \sim \omega k \qquad (4.31)$$

Chou (1945) proposed modeling the exact equation for $\epsilon$. In terms of this formulation, the kinematic eddy viscosity and turbulence length scale are

$$\nu_T \sim k^2/\epsilon, \quad \ell \sim k^{3/2}/\epsilon \tag{4.32}$$

Rotta (1951) first suggested a transport equation for the turbulence length scale and later (1968) proposed an equation for the product of $k$ and $\ell$. In either case,

$$\nu_T \sim k^{1/2}\ell, \quad \epsilon \sim k^{3/2}/\ell \tag{4.33}$$

More recently, Zeierman and Wolfshtein (1986) introduced a transport equation for the product of $k$ and a **turbulence dissipation time**, $\tau$, which is essentially the reciprocal of Kolmogorov's $\omega$. Also, Speziale, Abid and Anderson (1990) have postulated an equation for $\tau$. For these models,

$$\nu_T \sim k\tau, \quad \ell \sim k^{1/2}\tau, \quad \epsilon \sim k/\tau \tag{4.34}$$

Regardless of the choice of the second variable in our two-equation model, we see a recurring theme. Specifically, the dissipation, eddy viscosity and length scale are all related on the basis of dimensional arguments. Historically, dimensional analysis has been one of the most powerful tools available for deducing and correlating properties of turbulent flows. However, we should always be aware that while dimensional analysis is extremely useful, it unveils nothing about the physics underlying its implied scaling relationships. The physics is in the choice of variables.

One of the key conclusions of the 1980-81 AFOSR-HTTM-Stanford Conference on Complex Turbulent Flows was that the greatest amount of uncertainty about two-equation models lies in the second transport equation complementing the equation for $k$. Further, it was even unclear about what the most appropriate choice of the second dependent variable is. In the two decades following the Conference, interesting developments have occurred, most notably with the $k$-$\omega$ model, that help clear up most of the uncertainty.

Before proceeding to details of two-equation models, it is worthwhile to pause and note the following. As with one-equation models, there is no fundamental reason that $\nu_T$ should depend only upon turbulence parameters such as $k$, $\ell$, $\epsilon$ or $\omega$. In general, the ratio of individual Reynolds stresses to mean strain rate components depends upon both mean-flow and turbulence scales. Thus, two-equation turbulence models are no more likely than one-equation models to apply universally to turbulent flows, and can be expected to be inaccurate for many non-equilibrium turbulent flows.

Additionally, some researchers even argue that the addition of another differential equation invites unexpected numerical difficulties and miscellaneous unintended mathematical anomalies. We will indeed see some of this behavior as we investigate two-equation turbulence models, and what has been done to deal with the additional complexities attending their implementation.

## 4.3.1 The $k$-$\omega$ Model

As noted above, Kolmogorov (1942) proposed the first two-equation model of turbulence. Kolmogorov chose the kinetic energy of the turbulence as one of his turbulence parameters and, like Prandtl (1945), modeled the differential equation governing its behavior. His second parameter was the dissipation per unit turbulence kinetic energy, $\omega$. In his $k$-$\omega$ model, $\omega$ satisfies a differential equation similar to the equation for $k$. With no prior knowledge of Kolmogorov's work, Saffman (1970) formulated a $k$-$\omega$ model that would prove superior to the Kolmogorov model. As part of the Imperial College efforts on two-equation models, Spalding [see Launder and Spalding (1972)] offered an improved version of the Kolmogorov model that removed some of its flaws. Shortly after formulation of Saffman's model and continuing to the present time, Wilcox et al. [Wilcox and Alber (1972), Saffman and Wilcox (1974), Wilcox and Traci (1976), Wilcox and Rubesin (1980), and Wilcox (1988a)] have pursued further development and application of $k$-$\omega$ turbulence models in earnest. Coakley (1983) has developed a $k^{1/2}$-$\omega$ model. Speziale, Abid and Anderson (1990), Menter (1992c) and Peng, Davidson and Holmberg (1997) have also devised $k$-$\omega$ models. Robinson, Harris and Hassan (1995) have developed a $k$-$\omega^2$ model.

In formulating his model, Kolmogorov referred to $\omega$ as "the rate of dissipation of energy in unit volume and time." To underscore its physical relation to the " 'external scale' of turbulence, $\ell$," he also called it "some mean 'frequency' determined by $\omega = ck^{1/2}/\ell$, where $c$ is a constant." On the one hand, the reciprocal of $\omega$ is the time scale on which dissipation of turbulence energy occurs. While the actual process of dissipation takes place in the smallest eddies, the rate of dissipation is the transfer rate of turbulence kinetic energy to the smallest eddies. Hence, it is set by the properties of the large eddies, and thus scales with $k$ and $\ell$, wherefore $\omega$ is indirectly associated with dissipative processes. On the other hand, in analogy to molecular viscosity, we expect the eddy viscosity to be proportional to the product of length and velocity scales characteristic of turbulent fluctuations, which is consistent with Kolmogorov's argument that $\omega \sim k^{1/2}/\ell$. Of course, we should keep in mind that analogies between molecular and turbulent processes are not trustworthy, and Kolmogorov's argument is essentially an exercise in dimensional analysis, not fundamental physics.

The development of the Kolmogorov model (1942) is quite brief and doesn't even establish values for all of the closure coefficients. Since little formal development of the equations is given, we can only speculate about how this great turbulence researcher may have arrived at his model equations. Since he makes no specific reference to any exact equations, it seems unlikely that he attempted to close the $k$ equation or other moments of the Navier-Stokes equation term by term. Rather, as the great believer in the power of dimensional analysis that he was, it is easy to imagine that Kolmogorov's original reasoning may have gone something like this.

- Since $k$ already appears in the postulated constitutive relation [Equation (4.8)], it is plausible that $\nu_T \propto k$.

- The dimensions of $\nu_T$ are (length)$^2$/(time) while the dimensions of $k$ are (length)$^2$/(time)$^2$.

- Consequently $\nu_T/k$ has dimensions (time).

- Turbulence dissipation $\epsilon$ has dimensions (length)$^2$/(time)$^3$.

- Consequently $\epsilon/k$ has dimensions 1/(time).

- We can close Equations (4.8) and (4.11) by introducing a variable with dimensions (time) or 1/(time).

The next step is to postulate an equation for $\omega$. In doing so, the avenue that Kolmogorov took was to recognize that there is a fairly small number of physical processes commonly observed in the motion of a fluid. The most common processes are unsteadiness, convection (often referred to as advection), diffusion, dissipation, dispersion and production. Combining physical reasoning with dimensional analysis, Kolmogorov postulated the following equation for $\omega$.

$$\frac{\partial \omega}{\partial t} + U_j \frac{\partial \omega}{\partial x_j} = -\beta \omega^2 + \frac{\partial}{\partial x_j}\left[\sigma \nu_T \frac{\partial \omega}{\partial x_j}\right] \tag{4.35}$$

We have taken some notational liberties in writing Equation (4.35), and $\beta$ and $\sigma$ are two new closure coefficients. This equation has four particularly noteworthy features. **First**, there is no analog to the $k$-equation's turbulence production term. The absence of a production term is consistent with Kolmogorov's notion that $\omega$ is associated with the smallest scales of the turbulence, and thus has no direct interaction with the mean motion. His logic is flawed on this issue as the large-scale, energy-bearing eddies are primarily responsible for determining the appropriate time scale of the turbulence, and the rate of dissipation itself. **Second**, the equation is written in terms of $\omega$ rather than $\omega^2$. As will be shown when we analyze the defect layer in Subsection 4.6.2, Kolmogorov's decision to write his equation in terms of $\omega$ was a somewhat prophetic choice. **Third**, there is no molecular diffusion term so that this equation applies strictly to high-Reynolds-number flows and cannot be integrated through the viscous sublayer as it stands. **Fourth**, it is entirely empirical, guided by physical reasoning.

The interpretation of $\omega$ has behaved a bit like the turbulent fluctuations it is intended to describe. Saffman (1970) described $\omega$ as "a frequency characteristic of the turbulence decay process under its self-interaction." He stated further, "The rough idea is that $\omega^2$ is the mean square vorticity of the 'energy containing eddies' and [$k$] is the kinetic energy of the motion induced by this vorticity." Spalding [Launder and Spalding (1972)], Wilcox and Alber (1972) and Robinson, Harris and Hassan (1995) identify $\omega$ as the RMS fluctuating vorticity, so that $\omega^2$ is twice the **enstrophy**. Wilcox and Rubesin (1980), Wilcox (1988a) and Speziale et al. (1990) regard $\omega$ simply as the ratio of $\epsilon$ to $k$.

The $\omega$ equation has changed as the $k$-$\omega$ model has evolved over the past five decades. A production term has been added by all model developers subsequent to Kolmogorov. Like Kolmogorov, Wilcox (1988a), Speziale et al. (1990) and Peng et al. (1997) write the equation for $\omega$ in terms of $\omega$, while most other models use an equation for $\omega^2$. The following version of the $k$-$\omega$ model dramatically improves predictive accuracy of the Wilcox (1988a) model for free shear flows.[2]

**Kinematic Eddy Viscosity:**

$$\nu_T = k/\omega \tag{4.36}$$

**Turbulence Kinetic Energy:**

$$\frac{\partial k}{\partial t} + U_j \frac{\partial k}{\partial x_j} = \tau_{ij} \frac{\partial U_i}{\partial x_j} - \beta^* k \omega + \frac{\partial}{\partial x_j}\left[(\nu + \sigma^* \nu_T) \frac{\partial k}{\partial x_j}\right] \tag{4.37}$$

**Specific Dissipation Rate:**

$$\frac{\partial \omega}{\partial t} + U_j \frac{\partial \omega}{\partial x_j} = \alpha \frac{\omega}{k} \tau_{ij} \frac{\partial U_i}{\partial x_j} - \beta \omega^2 + \frac{\partial}{\partial x_j}\left[(\nu + \sigma \nu_T) \frac{\partial \omega}{\partial x_j}\right] \tag{4.38}$$

**Closure Coefficients and Auxiliary Relations:**

$$\alpha = \frac{13}{25}, \quad \beta = \beta_o f_\beta, \quad \beta^* = \beta_o^* f_{\beta^*}, \quad \sigma = \frac{1}{2}, \quad \sigma^* = \frac{1}{2} \tag{4.39}$$

$$\beta_o = \frac{9}{125}, \quad f_\beta = \frac{1 + 70\chi_\omega}{1 + 80\chi_\omega}, \quad \chi_\omega \equiv \left|\frac{\Omega_{ij}\Omega_{jk}S_{ki}}{(\beta_o^* \omega)^3}\right| \tag{4.40}$$

$$\beta_o^* = \frac{9}{100}, \quad f_{\beta^*} = \begin{cases} 1, & \chi_k \leq 0 \\ \dfrac{1 + 680\chi_k^2}{1 + 400\chi_k^2}, & \chi_k > 0 \end{cases}, \quad \chi_k \equiv \frac{1}{\omega^3} \frac{\partial k}{\partial x_j} \frac{\partial \omega}{\partial x_j} \tag{4.41}$$

$$\epsilon = \beta^* \omega k \quad \text{and} \quad \ell = k^{1/2}/\omega \tag{4.42}$$

---

[2]These equations can be used for simple compressible flows by replacing $\nu$ and $\nu_T$ by $\mu = \rho\nu$ and $\mu_T = \rho\nu_T$, respectively, and multiplying all but the diffusion terms by $\rho$. Also, a mean-energy and equation of state must be added — see Chapter 5 for more details.

To avoid confusion, from this point on, we will refer to Equations (4.36) through (4.42) as the **Wilcox (1998) $k$-$\omega$ model**.

The tensors $\Omega_{ij}$ and $S_{ij}$ appearing in Equation (4.40) are the mean-rotation and mean-strain-rate tensors, respectively defined by

$$\Omega_{ij} = \frac{1}{2}\left(\frac{\partial U_i}{\partial x_j} - \frac{\partial U_j}{\partial x_i}\right), \qquad S_{ij} = \frac{1}{2}\left(\frac{\partial U_i}{\partial x_j} + \frac{\partial U_j}{\partial x_i}\right) \qquad (4.43)$$

As can be easily verified, the quantity $\chi_\omega$ is zero for two-dimensional flows. The dependence of $\beta$ on $\chi_\omega$, patterned after the work of Pope (1978), has a significant effect for round and radial jets.

The most important differences between this version of the $k$-$\omega$ model and the Wilcox (1988a) version is in the coefficients of the dissipation terms, $\beta^*$ and $\beta$. The functions $f_{\beta^*}$ and $f_\beta$, which depend upon $\chi_k$ and $\chi_\omega$ and are defined in Equations (4.40) and (4.41), do not appear in the Wilcox (1988a) model. Also, the values of $\alpha = 0.52$ and $\beta_o = 0.072$ for the new model are a bit different from the values used for the Wilcox (1988a) model ($\alpha = 0.56$ and $\beta_o = 0.075$).

On the one hand, the new dissipation coefficients have a very small effect on boundary-layer flows, mainly because $\omega$ is generally very large close to a solid boundary so that $\chi_k$ and $\chi_\omega$ are small. *This is important* because the Wilcox (1988a) model predicts boundary-layer properties that are in very close agreement with measurements. The new model is nearly identical for all boundary-layer computations attempted to date. On the other hand, $\chi_k$ and $\chi_\omega$ are significantly larger for free shear flows. Consequently, the new $k$-$\omega$ model predicts more dissipation in free shear flows than the Wilcox (1988a) version, which results in a reduction of predicted spreading rates. *This is important* because the previous version of the model predicts free shear flow spreading rates that are significantly larger than measured. As we will discuss at greater length in Section 4.5, the functions $f_{\beta^*}$ and $f_\beta$ and the values of $\alpha$ and $\beta_o$ have been calibrated to provide spreading rates that are consistent with measurements for far wakes, mixing layers and plane, round and radial jets. We will also show how these modifications eliminate confusion regarding the $k$-$\omega$ model's sensitivity to freestream boundary conditions.

Thus, the new $k$-$\omega$ model is as accurate as the Wilcox (1988a) model for boundary layers. However, its predicted free shear flow spreading rates are much closer to measurements, so that the model defined in Equations (4.36) through (4.42) is applicable to both wall-bounded and free shear flows. *Since most complex turbulent flows include both types of regions, this is a minimum requirement for any turbulence model that is proposed for use in complex flows.* As we will see in this section, the new $k$-$\omega$ model satisfies this minimum requirement. With the exception of the enstrophy-equation model developed by Robinson, Harris and Hassan (1995) — using 11 closure coefficients and 2 closure functions — no other two-equation model known to this author satisfies this requirement.

## 4.3.2 The $k$-$\epsilon$ Model

By far, the most popular two-equation model is the $k$-$\epsilon$ model. The earliest development efforts based on this model were those of Chou (1945), Davidov (1961) and Harlow and Nakayama (1968). The central paper however, is that by Jones and Launder (1972) that, in the turbulence modeling community, has nearly reached the status of the Boussinesq and Reynolds papers. That is, the model is so well known that it is referred to as the **Standard $k$-$\epsilon$ model** and reference to the Jones-Launder paper is often omitted. Actually, Launder and Sharma (1974) "retuned" the model's closure coefficients and most researchers use the form of the model presented in the 1974 paper.

Again, we begin with Equations (4.8) and (4.11). In formulating the $k$-$\epsilon$ model, the idea is to derive the exact equation for $\epsilon$ and to find suitable closure approximations for the exact equation governing its behavior. Recall that $\epsilon$ is defined by Equation (4.5). The exact equation for $\epsilon$ is derived by taking the following moment of the Navier-Stokes equation:

$$2\nu \overline{\frac{\partial u_i'}{\partial x_j} \frac{\partial}{\partial x_j} [\mathcal{N}(u_i)]} = 0 \tag{4.44}$$

where $\mathcal{N}(u_i)$ is the Navier-Stokes operator defined in Equation (2.26). After a considerable amount of algebra, the following equation for $\epsilon$ results.

$$\frac{\partial \epsilon}{\partial t} + U_j \frac{\partial \epsilon}{\partial x_j} = -2\nu \left[ \overline{u_{i,k}' u_{j,k}'} + \overline{u_{k,i}' u_{k,j}'} \right] \frac{\partial U_i}{\partial x_j} - 2\nu \overline{u_k' u_{i,j}'} \frac{\partial^2 U_i}{\partial x_k \partial x_j}$$
$$- 2\nu \overline{u_{i,k}' u_{i,m}' u_{k,m}'} - 2\nu^2 \overline{u_{i,km}' u_{i,km}'}$$
$$+ \frac{\partial}{\partial x_j} \left[ \nu \frac{\partial \epsilon}{\partial x_j} - \nu \overline{u_j' u_{i,m}' u_{i,m}'} - 2\frac{\nu}{\rho} \overline{p_{,m}' u_{j,m}'} \right] \tag{4.45}$$

This equation is far more complicated than the turbulence kinetic energy equation and involves several new unknown double and triple correlations of fluctuating velocity, pressure and velocity gradients. The terms on the three lines of the right-hand side of Equation (4.45) are generally regarded as **Production of Dissipation, Dissipation of Dissipation**, and the sum of **Molecular Diffusion of Dissipation** and **Turbulent Transport of Dissipation**, respectively. These correlations are essentially impossible to measure with any degree of accuracy so that there is presently little hope of finding reliable guidance from experimentalists regarding suitable closure approximations. Recent DNS studies such as the work of Mansour, Kim and Moin (1988) have helped gain some insight to the exact $\epsilon$ transport equation for low-Reynolds-number flows. However, the database for establishing closure approximations similar to those used for the $k$ equation remains very sparse.

Many researchers have proceeded undaunted by the lack of a rational basis for establishing closure approximations with a feeling that using Equation (4.45) as their foundation adds rigor to their approach. The strongest claim that can actually be made is that conventional closure approximations used for Equation (4.45) are dimensionally correct. This is not very different from the Kolmogorov (1942) and Saffman (1970) approaches that are guided almost exclusively by physical reasoning and dimensional analysis. An important point we should keep in mind is to **avoid modeling the differential equations rather than the physics of turbulence**. That is not to say we should avoid any reference to the differential equations, for then we might formulate a model that violates a fundamental physical feature of the Navier-Stokes equation. Rather, we should avoid deluding ourselves by thinking that the **drastic surgery** approach to something as complex as Equation (4.45) is any more rigorous than dimensional analysis.

Even if we had demonstrably accurate closure approximations for the exact $\epsilon$ transport equation, there is a serious question of their relevance to our basic closure problem. That is, the length or time scale required is that of the energy-containing, Reynolds-stress-bearing eddies rather than the dissipating eddies represented by the exact $\epsilon$ equation. So, we must ask whether the modeled $\epsilon$ equation represents the dissipation as such [as Equation (4.45) does], or whether it is actually an empirical equation for the rate of energy transfer from the large eddies (equal, of course, to the rate of dissipation in the small eddies). The answer seems clear since the closure approximations normally used parameterize the various terms in the modeled $\epsilon$ equation as functions of large-eddy scales (our use of dimensional analysis does this implicitly). As a consequence, the relation between the modeled equation for $\epsilon$ and the exact equation is so tenuous as not to need serious consideration. The **Standard $k$-$\epsilon$ model** is as follows.

**Kinematic Eddy Viscosity:**

$$\nu_T = C_\mu k^2 / \epsilon \tag{4.46}$$

**Turbulence Kinetic Energy:**

$$\frac{\partial k}{\partial t} + U_j \frac{\partial k}{\partial x_j} = \tau_{ij} \frac{\partial U_i}{\partial x_j} - \epsilon + \frac{\partial}{\partial x_j} \left[ (\nu + \nu_T/\sigma_k) \frac{\partial k}{\partial x_j} \right] \tag{4.47}$$

**Dissipation Rate:**

$$\frac{\partial \epsilon}{\partial t} + U_j \frac{\partial \epsilon}{\partial x_j} = C_{\epsilon 1} \frac{\epsilon}{k} \tau_{ij} \frac{\partial U_i}{\partial x_j} - C_{\epsilon 2} \frac{\epsilon^2}{k} + \frac{\partial}{\partial x_j} \left[ (\nu + \nu_T/\sigma_\epsilon) \frac{\partial \epsilon}{\partial x_j} \right] \tag{4.48}$$

**Closure Coefficients and Auxiliary Relations:**

$$C_{\epsilon 1} = 1.44, \quad C_{\epsilon 2} = 1.92, \quad C_\mu = 0.09, \quad \sigma_k = 1.0, \quad \sigma_\epsilon = 1.3 \tag{4.49}$$

$$\omega = \epsilon/(C_\mu k) \quad \text{and} \quad \ell = C_\mu k^{3/2}/\epsilon \tag{4.50}$$

Strictly speaking, the Launder-Sharma (1974) model is the Standard $k$-$\epsilon$ model. In addition to the equations quoted here, it involves viscous damping functions, which are discussed in Section 4.9.

A more recent version of the $k$-$\epsilon$ model has been developed by Yakhot and Orszag (1986) [see also Yakhot et al. (1992)]. Using techniques from renormalization group theory, they have developed what is known as the RNG $k$-$\epsilon$ model. The eddy viscosity, $k$ and $\epsilon$ are still given by Equations (4.46), (4.47) and (4.48). However, the model uses a modified coefficient, $C_{\epsilon 2}$, defined by

$$C_{\epsilon 2} \equiv \tilde{C}_{\epsilon 2} + \frac{C_\mu \lambda^3 (1 - \lambda/\lambda_o)}{1 + \beta \lambda^3}, \qquad \lambda \equiv \frac{k}{\epsilon} \sqrt{2 S_{ij} S_{ji}} \qquad (4.51)$$

The closure coefficients for the RNG $k$-$\epsilon$ model are[3]

$$C_{\epsilon 1} = 1.42, \quad \tilde{C}_{\epsilon 2} = 1.68, \quad C_\mu = 0.085, \quad \sigma_k = 0.72, \quad \sigma_\epsilon = 0.72 \quad (4.52)$$

$$\beta = 0.012, \quad \lambda_o = 4.38 \qquad (4.53)$$

### 4.3.3   Other Two-Equation Models

Two-equation models based on the turbulence length scale, $\ell$, and the turbulence time scale, $\tau$, have received less attention than the $k$-$\omega$ and $k$-$\epsilon$ models. Generally speaking, the level of agreement between measurements and predictions made with other models is comparable to $k$-$\omega$ and $k$-$\epsilon$ predictions for simple constant-pressure flows, but these models have not been pursued to any great extent. This subsection presents a brief overview of some of the length-scale and time-scale models. More details can be found in the various papers referenced in the discussion.

The proposed foundation for Rotta's (1968) $k$-$k\ell$ model is the **two-point velocity correlation tensor** defined in Equation (2.49), viz.,

$$R_{ij}(\mathbf{x}, t; \mathbf{r}) = \overline{u_i'(\mathbf{x}, t) \, u_j'(\mathbf{x} + \mathbf{r}, t)} \qquad (4.54)$$

As discussed in Subsection 2.5.2, the turbulence kinetic energy is simply one half the trace of $R_{ij}$ with a displacement $\mathbf{r} = \mathbf{0}$. Rotta's second variable is the product of $k$ and the **integral length scale**, $\ell$, which is the integral of $R_{ii}$ over all displacements, $r = |\mathbf{r}|$. Thus Rotta's variables are given by

---

[3]This version of the RNG $k$-$\epsilon$ model has been gleaned from the open literature. A proprietary improved version exists, but is available only in commercial computer programs for general turbulent-flow applications.

$$k = \frac{1}{2}R_{ii}(\mathbf{x}, t; \mathbf{0}) \quad \text{and} \quad k\ell = \frac{3}{16} \int_{-\infty}^{\infty} R_{ii}(\mathbf{x}, t; r) \, dr \qquad (4.55)$$

As with attempts to model the exact dissipation equation, no particular advantage has been gained by introducing the two-point velocity correlation tensor. While an exact equation for $k\ell$ can indeed be derived, Rotta (1968) still had to perform drastic surgery on the exact equation. For example, using standard closure approximations based largely on the strength of dimensional analysis, the following modeled version of the exact $k\ell$ equation results.

$$\frac{\partial}{\partial t}(k\ell) + U_j \frac{\partial}{\partial x_j}(k\ell) = C_{L1} \ell \tau_{ij} \frac{\partial U_i}{\partial x_j} - C_{L2} k^{3/2}$$

$$+ \frac{\partial}{\partial x_j} \left[ \nu \frac{\partial}{\partial x_j}(k\ell) + (\nu_T/\sigma_{L1}) \ell \frac{\partial k}{\partial x_j} + (\nu_T/\sigma_{L2}) k \frac{\partial \ell}{\partial x_j} \right] \qquad (4.56)$$

For this model, $k$ and $\nu_T$ are given by Equations (4.13) and (4.14). Rodi and Spalding (1970) and Ng and Spalding (1972) developed this model further. More recently, Smith (1990) has pursued development of a $k$-$k\ell$ model. Smith (1994) has also developed a $k$-$\ell$ model for which the dependent variable is $\ell$ rather than $k\ell$. Ng and Spalding found that for wall-bounded flows, the closure coefficient $C_{L2}$ must vary with distance from the surface. They propose the following set of closure coefficients.

$$C_{L1} = 0.98, \quad C_{L2} = 0.059 + 702(\ell/y)^6, \quad C_D = 0.09, \quad \sigma_k = \sigma_{L1} = \sigma_{L2} = 1 \qquad (4.57)$$

On a similar tack, Zeierman and Wolfshtein (1986) base their model upon the **autocorrelation tensor** defined in Equation (2.43), i.e.,

$$\mathcal{R}_{ij}(\mathbf{x}, t; t') = \overline{u_i'(\mathbf{x}, t) u_j'(\mathbf{x}, t + t')} \qquad (4.58)$$

The turbulence kinetic energy is half the trace of $\mathcal{R}_{ij}$ with $t' = 0$, while the **integral time scale** is proportional to the integral of $\mathcal{R}_{ii}$ over all possible values of $t'$. Thus,

$$k = \frac{1}{2}\mathcal{R}_{ii}(\mathbf{x}, t; 0) \quad \text{and} \quad k\tau = \frac{1}{2} \int_0^{\infty} \mathcal{R}_{ii}(\mathbf{x}, t; t') \, dt' \qquad (4.59)$$

The Zeierman-Wolfshtein $k$-$k\tau$ model is as follows.

**Kinematic Eddy Viscosity:**

$$\nu_T = C_\mu k\tau \qquad (4.60)$$

**Turbulence Kinetic Energy:**

$$\frac{\partial k}{\partial t} + U_j \frac{\partial k}{\partial x_j} = \tau_{ij} \frac{\partial U_i}{\partial x_j} - \frac{k}{\tau} + \frac{\partial}{\partial x_j}\left[(\nu + \nu_T/\sigma_k)\frac{\partial k}{\partial x_j}\right] \quad (4.61)$$

**Integral Time Scale:**

$$\frac{\partial}{\partial t}(k\tau) + U_j \frac{\partial}{\partial x_j}(k\tau) = C_{\tau 1}\tau\tau_{ij}\frac{\partial U_i}{\partial x_j} - C_{\tau 2}k$$

$$+ \frac{\partial}{\partial x_j}\left[(\nu + \nu_T/\sigma_\tau)\frac{\partial}{\partial x_j}(k\tau)\right] \quad (4.62)$$

**Closure Coefficients and Auxiliary Relations:**

$$C_{\tau 1} = 0.173, \quad C_{\tau 2} = 0.225, \quad C_\mu = 0.09, \quad \sigma_k = 1.46, \quad \sigma_\tau = 10.8 \quad (4.63)$$

$$\omega = 1/(C_\mu\tau), \quad \epsilon = k/\tau \quad \text{and} \quad \ell = C_\mu k^{1/2}\tau \quad (4.64)$$

Note that because the eddy viscosity is proportional to $k\tau$, Equation (4.62) can also be regarded as an equation for $\nu_T$.

Speziale, Abid and Anderson (1990) have taken a different approach in devising a $k$-$\tau$ model. Specifically, they introduce the formal change of dependent variables $\epsilon = k/\tau$ and transform the Standard $k$-$\epsilon$ model. The resulting equation for $\tau$ is as follows.

$$\frac{\partial \tau}{\partial t} + U_j \frac{\partial \tau}{\partial x_j} = (1 - C_{\epsilon 1})\frac{\tau}{k}\tau_{ij}\frac{\partial U_i}{\partial x_j} + (C_{\epsilon 2} - 1)$$

$$+ \frac{\partial}{\partial x_j}\left[(\nu + \nu_T/\sigma_{\tau 2})\frac{\partial \tau}{\partial x_j}\right]$$

$$+ \frac{2}{k}(\nu + \nu_T/\sigma_{\tau 1})\frac{\partial k}{\partial x_k}\frac{\partial \tau}{\partial x_k} - \frac{2}{\tau}(\nu + \nu_T/\sigma_{\tau 2})\frac{\partial \tau}{\partial x_k}\frac{\partial \tau}{\partial x_k} \quad (4.65)$$

Speziale, Abid and Anderson use the following revised set of closure coefficient values for their $k$-$\tau$ model that make it a bit different from the Standard $k$-$\epsilon$ model.

$$C_{\epsilon 1} = 1.44, \quad C_{\epsilon 2} = 1.83, \quad C_\mu = 0.09, \quad \sigma_k = \sigma_{\tau 1} = \sigma_{\tau 2} = 1.36 \quad (4.66)$$

In summary, the models listed above are representative of the various two-equation models that have been devised since Kolmogorov's (1942) $k$-$\omega$ model. While other models have been created, the intent of this text is to study models in a generic sense, as opposed to creating an encyclopedia of turbulence models. In the following sections we investigate several aspects of two-equation models including: (a) details on how the closure-coefficient values are chosen; (b) surface boundary conditions for wall-bounded flows; and, (c) applications to a variety of flows.

## 4.4   Closure Coefficients

All of the two-equation models have closure coefficients that have been introduced in replacing unknown double and triple correlations with algebraic expressions involving known turbulence and mean-flow properties. The $k$-$\omega$ model, for example, has five, viz., $\alpha$, $\beta$, $\beta^*$, $\sigma$ and $\sigma^*$. If our theory were exact, we could set the values of these coefficients from first principles much as we use the kinetic theory of gases to determine the viscosity coefficient in Stokes' approximation for laminar flows. Unfortunately, the theory is not exact, but rather a model developed mainly on the strength of dimensional analysis. Consequently, the best we can do is to set the values of the closure coefficients to assure agreement with observed properties of turbulence.

This section describes the manner in which the closure coefficients have been determined for the $k$-$\omega$ model. There is no loss of generality in doing this however, since *these same general arguments have been used in establishing the values of the closure coefficients in most two-equation models.* The problems section at the end of the chapter examines some of the (relatively minor) differences among the various models.

We can establish the ratio of $\beta_o^*$ to $\beta_o$ by applying the model to decaying homogeneous, isotropic turbulence. In this kind of turbulence, there are no spatial gradients of any mean-flow properties wherefore Equations (4.37) and (4.38) simplify to

$$\frac{dk}{dt} = -\beta_o^* \omega k \quad \text{and} \quad \frac{d\omega}{dt} = -\beta_o \omega^2 \qquad (4.67)$$

where we note that, because $\chi_k = \chi_\omega = 0$, we have $f_{\beta^*} = f_\beta = 1$ [see Equations (4.40) and (4.41)]. The asymptotic solution for $k$ is readily found to be

$$k \sim t^{-\beta_o^*/\beta_o} \qquad (4.68)$$

Experimental observations [see Townsend (1976)] indicate that $k \sim t^{-n}$ where $n = 1.25 \pm 0.06$ for decaying homogeneous, isotropic turbulence. Choosing $\beta_o^*/\beta_o = 5/4$ sets the ratio at the center of the range of accepted values.

Values for the coefficients $\alpha$ and $\beta_o^*$ can be established by examining the **log layer**. Recall from Section 3.4 that the log layer is defined as the portion of the boundary layer sufficiently distant from the surface that molecular viscosity is negligible relative to eddy viscosity, yet close enough for convective effects to be negligible. In the limiting case of an incompressible constant-pressure boundary layer, the mean-momentum equation and the equations for $k$ and $\omega$ simplify to the following.

$$\left.\begin{array}{l} 0 = \dfrac{\partial}{\partial y}\left[\nu_T \dfrac{\partial U}{\partial y}\right] \\[3mm] 0 = \nu_T \left(\dfrac{\partial U}{\partial y}\right)^2 - \beta_o^* \omega k + \sigma^* \dfrac{\partial}{\partial y}\left[\nu_T \dfrac{\partial k}{\partial y}\right] \\[3mm] 0 = \alpha \left(\dfrac{\partial U}{\partial y}\right)^2 - \beta_o \omega^2 + \sigma \dfrac{\partial}{\partial y}\left[\nu_T \dfrac{\partial \omega}{\partial y}\right] \end{array}\right\} \qquad (4.69)$$

We will justify the limiting form of these equations when we use perturbation methods to analyze the log layer in Subsection 4.6.1. We seek the conditions for which these simplified equations yield a solution consistent with the law of the wall. As can be easily verified, Equations (4.69) possess such a solution, viz.,

$$U = \frac{u_\tau}{\kappa}\ell n y + \text{constant}, \qquad k = \frac{u_\tau^2}{\sqrt{\beta_o^*}}, \qquad \omega = \frac{u_\tau}{\sqrt{\beta_o^*}\kappa y} \qquad (4.70)$$

where $u_\tau$ is the conventional friction velocity and $\kappa$ is Kármán's constant. There is one constraint imposed in the solution to Equations (4.69), namely, a unique relation exists between the implied value of Kármán's constant and the various closure coefficients. Specifically, the following equation must hold.

$$\alpha = \beta_o/\beta_o^* - \sigma\kappa^2/\sqrt{\beta_o^*} \qquad (4.71)$$

Additionally, according to our solution the Reynolds shear stress, $\tau_{xy}$, is constant and equal to $u_\tau^2$. Inspection of Equations (4.70) shows that this implies $\tau_{xy} = \sqrt{\beta_o^*}\, k$ in the log layer. A variety of measurements [Townsend (1976)] indicate the ratio of $\tau_{xy}$ to $k$ is about 3/10 (i.e., Bradshaw's constant) in the log layer. This is the same ratio Bradshaw, Ferriss and Atwell (1967) used in formulating their one-equation model [c.f. Equation (4.15)]. Thus, the predicted log-layer solution is consistent with experimental observations provided we select $\beta_o^* = 9/100$. Since we have concluded above that $\beta_o^*/\beta_o = 5/4$, necessarily $\beta_o = 9/125$.

We must work a bit harder to determine the values of $\sigma$ and $\sigma^*$. As we will see in Subsections 4.6.2 and 4.6.3, detailed analysis of the defect layer and the sublayer indicates that the optimum choice is $\sigma = \sigma^* = 1/2$. Finally, Equation (4.71) shows that selecting $\alpha = 13/25$ is consistent with Coles' value for the Kármán constant of 0.41. Thus, in summary, the values of the five primary closure coefficients in the $k$-$\omega$ model are

$$\alpha = \frac{13}{25}, \qquad \beta_o = \frac{9}{125}, \qquad \beta_o^* = \frac{9}{100}, \qquad \sigma = \frac{1}{2}, \qquad \sigma^* = \frac{1}{2} \qquad (4.72)$$

These are the values quoted in Equations (4.39), (4.40) and (4.41).

Other arguments have been used to determine closure coefficients prior to any applications or computer optimization. Saffman (1970), for example, uses estimates based on vortex-stretching processes in simple shear and pure extension to effectively establish bounds on a coefficient similar to $\alpha$. He also requires that the length scale, $\ell$, be discontinuous at a turbulent/nonturbulent interface and finds that his model requires $\sigma = \sigma^* = 1/2$ to guarantee such behavior.

Zeierman and Wolfshtein (1986) use the fact that very close to separation, measurements [Townsend (1976)] indicate the law of the wall is replaced by

$$U \rightarrow \frac{1}{0.24}\sqrt{\frac{y}{\rho}\frac{dP}{dx}} \qquad \text{as} \qquad y \rightarrow 0 \qquad (4.73)$$

They also observe from measurements of Laufer (1950) and Clark (1968) that, for flow near the center of a channel, the turbulence kinetic energy and velocity are closely approximated by

$$\left.\begin{array}{rcl} k/k_o & \approx & 1 + 6.67(y/R)^2 \\ U/U_o & \approx & 1 - 0.242(y/R)^2 \\ u_\tau^2 & \approx & 0.048 U_o k_o^{1/2} \end{array}\right\} \qquad \text{as} \qquad y \rightarrow R \qquad (4.74)$$

Briggs et al. (1996) provide another simple argument that can be used to establish closure-coefficient values. They have done a Large Eddy Simulation (LES – Chapter 8) of a shear-free mixing layer, an idealized flow that is relevant for geophysical studies. In this flow, the turbulent-transport (diffusion) terms balance dissipation terms in the $k$, $\omega$, $\epsilon$, etc. equations. For example, the $k$-$\omega$ model simplifies to

$$\left.\begin{array}{rcl} \dfrac{\partial}{\partial y}\left(\sigma^* \nu_T \dfrac{\partial k}{\partial y}\right) & = & \beta_o^* \omega k \\[4mm] \dfrac{\partial}{\partial y}\left(\sigma \nu_T \dfrac{\partial \omega}{\partial y}\right) & = & \beta_o \omega^2 \end{array}\right\} \qquad (4.75)$$

Briggs et al. conclude that the asymptotic behavior of $k$ and $\nu_T$ is

$$k \sim K y^{-2.45} \quad \text{and} \quad \nu_T \sim V y^{-0.42} \qquad \text{as} \quad y \rightarrow \infty \qquad (4.76)$$

where $K$ and $V$ are constants. Based on the Briggs et al. results, a short calculation yields the following constraint on the $k$-$\omega$ model's closure coefficients.

$$\sigma \beta_o^* = \sigma^* \beta_o \qquad (4.77)$$

On the one hand, this condition would be satisfied, for example, by changing $\sigma^*$ to 5/8 and leaving all other $k$-$\omega$ closure-coefficient values unchanged. On the other hand, for the values given in Equation (4.72), the $k$-$\omega$ model predicts $k \sim K y^{-2.68}$ and $\nu_T \sim V y^{-0.34}$, which is fairly close to the LES behavior.

Briggs et al. also show that the $k$-$\epsilon$ model predicts $k \sim K y^{-4.9}$ and $\nu_T \sim V y^{-1.5}$, which bears no resemblance to the LES results.

In conclusion, the specific flows selected for determination of the closure coefficients are a free choice of the developer. For example, using data for homogeneous turbulence *and* boundary layers assumes we have a degree of universality that may be grossly optimistic. That is, we are implicitly assuming our model is valid for grid turbulence, boundary layers, and many flows in between. Dropping homogeneous turbulence in favor of more boundary-layer data may yield a model optimized for boundary layers but restricted to such flows. Ideally, we would find flows that isolate each closure coefficient. Often, more than one is involved [e.g., Equation (4.71)]. In any event, for the sake of clarity, the arguments should be as simple as possible.

# 4.5 Application to Free Shear Flows

Our first applications will be for free shear flows. In this section, we seek similarity solutions to determine far-field behavior for the plane wake, mixing layer, plane jet, round jet and radial jet. In addition to developing the similarity solutions for the $k$-$\omega$ and $k$-$\epsilon$ models, we also discuss several aspects of the solutions and differences between the $k$-$\omega$ and $k$-$\epsilon$ models. These include: solution sensitivity to freestream boundary conditions; (b) cross diffusion; and, (c) the round-jet/plane-jet anomaly.

Solution sensitivity to freestream boundary conditions is an issue that previously has not been completely understood. We will find that solutions for two-equation turbulence models are sensitive to the freestream value of $\omega$, $\epsilon$, etc. even when boundary conditions are chosen so that $k$ and $\nu_T$ are both very small in the freestream. Cross diffusion is a term appearing in the $\omega$, $\epsilon$ or other second transport equation that results from making a formal change of variables in transforming from one set of turbulence parameters (e.g., $k$ and $\omega$) to another (e.g., $k$ and $\epsilon$ or $k$ and $\ell$). We will see how cross diffusion affects free-shear-layer predictions. The round-jet/plane-jet anomaly is a classical problem that has plagued turbulence models. Many models predict that the round jet grows faster than the plane jet, while measurements show the opposite to be true. We will see which models suffer from the anomaly and which do not.

## 4.5.1 Developing the Similarity Solution

There are two noteworthy changes in our approach to obtaining a solution for free shear flows. **First**, for the mixing layer and the jets we can choose our similarity variable to be $\eta = y/x$. That is, with no loss of generality, we can set all scaling constants such as $A$ in Equations (3.70) and (3.71) equal to one.

We had to carry such scaling coefficients for the mixing-length model because, by hypothesis, the mixing length is proportional to the width of the layer, which is proportional to the coefficient $A$. With two-equation models, the turbulence length scale is determined as part of the solution so that the way in which we scale the similarity variable $\eta$ is of no consequence. **Second**, while the rest of the methodology is the same, the addition of two extra differential equations complicates the problem somewhat. Because they are the most widely used two-equation models, we confine our attention to the $k$-$\omega$ and $k$-$\epsilon$ models. With the standard boundary-layer/shear-layer approximations, the equations of motion become:

$$\frac{\partial U}{\partial x} + \frac{1}{y^j}\frac{\partial}{\partial y}\left[y^j V\right] = 0 \tag{4.78}$$

$$U\frac{\partial U}{\partial x} + V\frac{\partial U}{\partial y} = \frac{1}{y^j}\frac{\partial}{\partial y}\left[y^j \tau_{xy}\right] \tag{4.79}$$

$$\tau_{xy} = \nu_T \frac{\partial U}{\partial y} \tag{4.80}$$

**$k$-$\omega$ Model:**

$$\left.\begin{aligned}
U\frac{\partial k}{\partial x} + V\frac{\partial k}{\partial y} &= \tau_{xy}\frac{\partial U}{\partial y} - \beta^*\omega k + \frac{1}{y^j}\frac{\partial}{\partial y}\left[y^j \sigma^* \nu_T \frac{\partial k}{\partial y}\right] \\
U\frac{\partial \omega}{\partial x} + V\frac{\partial \omega}{\partial y} &= \alpha\frac{\omega}{k}\tau_{xy}\frac{\partial U}{\partial y} - \beta\omega^2 + \frac{1}{y^j}\frac{\partial}{\partial y}\left[y^j \sigma \nu_T \frac{\partial \omega}{\partial y}\right] \\
\nu_T &= k/\omega
\end{aligned}\right\} \tag{4.81}$$

**$k$-$\epsilon$ Model:**

$$\left.\begin{aligned}
U\frac{\partial k}{\partial x} + V\frac{\partial k}{\partial y} &= \tau_{xy}\frac{\partial U}{\partial y} - \epsilon + \frac{1}{y^j}\frac{\partial}{\partial y}\left[y^j \frac{\nu_T}{\sigma_k}\frac{\partial k}{\partial y}\right] \\
U\frac{\partial \epsilon}{\partial x} + V\frac{\partial \epsilon}{\partial y} &= C_{\epsilon 1}\frac{\epsilon}{k}\tau_{xy}\frac{\partial U}{\partial y} - C_{\epsilon 2}\frac{\epsilon^2}{k} + \frac{1}{y^j}\frac{\partial}{\partial y}\left[y^j \frac{\nu_T}{\sigma_\epsilon}\frac{\partial \epsilon}{\partial y}\right] \\
\nu_T &= C_\mu k^2/\epsilon
\end{aligned}\right\} \tag{4.82}$$

In Equations (4.78) through (4.82), $j = 1$ for the round jet and $j = 0$ for the other four cases.

The similarity solution for the various free shear flows can be written in the following compact form.

**Far Wake:**

$$U(x,y) = U_\infty - \sqrt{\frac{D}{\rho x}}\,\mathcal{U}(\eta), \quad k(x,y) = \frac{D}{\rho x}K(\eta)$$

$$\omega(x,y) = \frac{U_\infty}{x}W(\eta) \quad \epsilon(x,y) = \frac{DU_\infty}{\rho x^2}E(\eta) \tag{4.83}$$

$$\eta = y\sqrt{\frac{\rho U_\infty^2}{Dx}}$$

**Mixing Layer:**

$$U(x,y) = U_1\mathcal{U}(\eta), \quad k(x,y) = U_1^2 K(\eta)$$

$$\omega(x,y) = \frac{U_1}{x}W(\eta), \quad \epsilon(x,y) = \frac{U_1^3}{x}E(\eta) \tag{4.84}$$

$$\eta = \frac{y}{x}$$

**Jet:**

$$U(x,y) = \frac{J^{1/2}}{x^{(m+1)/2}}\mathcal{U}(\eta), \quad k(x,y) = \frac{J}{x^{(m+1)}}K(\eta)$$

$$\omega(x,y) = \frac{J^{1/2}}{x^{(m+3)/2}}W(\eta), \quad \epsilon(x,y) = \frac{J^{3/2}}{x^{(3m+5)/2}}E(\eta) \tag{4.85}$$

$$\eta = \frac{y}{x}$$

where $m = 0$ for the plane jet and $m = 1$ for the round and radial jets.

Substituting these self-similar representations into the mean-momentum equation yields the general form

$$\mathcal{V}\frac{d\mathcal{U}}{d\eta} - \frac{1}{\eta^j}\frac{d}{d\eta}\left[\eta^j N \frac{d\mathcal{U}}{d\eta}\right] = S_u\mathcal{U} \tag{4.86}$$

where the functions $N(\eta)$ and $\mathcal{V}(\eta)$ are the transformed eddy viscosity and normal velocity-like function, respectively. The two terms on the left-hand side of Equation (4.86) are essentially vertical convection and diffusion. The term on the right-hand side is a source term that originates from the streamwise convection of momentum.

Table 4.3 lists the coefficient $S_u$ and the normal-velocity function, $\mathcal{V}(\eta)$, for each of the free shear flows considered. The transformed $k$, $\omega$ and $\epsilon$ equations are:

**$k$-$\omega$ Model:**

$$
\left.
\begin{aligned}
&\mathcal{V}\frac{dK}{d\eta} - \frac{1}{\eta^j}\frac{d}{d\eta}\left[\eta^j\sigma^* N\frac{dK}{d\eta}\right] = S_k K + N\left(\frac{d\mathcal{U}}{d\eta}\right)^2 - \beta^* W K \\[2mm]
&\mathcal{V}\frac{dW}{d\eta} - \frac{1}{\eta^j}\frac{d}{d\eta}\left[\eta^j\sigma N\frac{dW}{d\eta}\right] = S_w W + \alpha\frac{W}{K}N\left(\frac{d\mathcal{U}}{d\eta}\right)^2 - \beta W^2 \\[2mm]
&N = \frac{K}{W}
\end{aligned}
\right\} \quad (4.87)
$$

**$k$-$\epsilon$ Model:**

$$
\left.
\begin{aligned}
&\mathcal{V}\frac{dK}{d\eta} - \frac{1}{\eta^j}\frac{d}{d\eta}\left[\eta^j\frac{N}{\sigma_k}\frac{dK}{d\eta}\right] = S_k K + N\left(\frac{d\mathcal{U}}{d\eta}\right)^2 - E \\[2mm]
&\mathcal{V}\frac{dE}{d\eta} - \frac{1}{\eta^j}\frac{d}{d\eta}\left[\eta^j\frac{N}{\sigma_\epsilon}\frac{dE}{d\eta}\right] = S_e E + C_{\epsilon 1}\frac{E}{K}N\left(\frac{d\mathcal{U}}{d\eta}\right)^2 - C_{\epsilon 2}\frac{E^2}{K} \\[2mm]
&N = C_\mu\frac{K^2}{E}
\end{aligned}
\right\} \quad (4.88)
$$

Table 4.3: *Free Shear Flow Parameters*

| Flow | $S_u$ | $S_k$ | $S_w$ | $S_e$ | $j$ | $m$ | $\mathcal{V}(\eta)$ |
|------|-------|-------|-------|-------|-----|-----|---------------------|
| Far Wake | $\frac{1}{2}$ | 1 | 1 | 2 | 0 | 0 | $-\frac{1}{2}\eta$ |
| Mixing Layer | 0 | 0 | $\mathcal{U}$ | $\mathcal{U}$ | 0 | 0 | $-\int_0^\eta \mathcal{U}(\eta')d\eta'$ |
| Plane Jet | $\frac{1}{2}\mathcal{U}$ | $\mathcal{U}$ | $\frac{3}{2}\mathcal{U}$ | $\frac{5}{2}\mathcal{U}$ | 0 | 0 | $-\frac{1}{2}\int_0^\eta \mathcal{U}(\eta')d\eta'$ |
| Round Jet | $\mathcal{U}$ | $2\mathcal{U}$ | $2\mathcal{U}$ | $4\mathcal{U}$ | 1 | 1 | $-\frac{1}{\eta}\int_0^\eta \mathcal{U}(\eta')\eta'd\eta'$ |
| Radial Jet | $\mathcal{U}$ | $2\mathcal{U}$ | $2\mathcal{U}$ | $4\mathcal{U}$ | 1 | 0 | $-\int_0^\eta \mathcal{U}(\eta')d\eta'$ |

The $k$, $\omega$ and $\epsilon$ equations contain convective terms, diffusion terms, and additional source terms corresponding to streamwise convection, production and dissipation. Table 4.3 lists the convective source-term coefficients, $S_k$, $S_w$ and $S_e$. The table also lists the exponents $j$ and $m$ for each free shear flow.

To complete the solution, we must specify the parameter $\lambda$ appearing in the RNG $k$-$\epsilon$ model [Equation (4.51)]. In terms of similarity variables, $\lambda$ is

$$\lambda = \frac{K}{E} \left| \frac{d\mathcal{U}}{d\eta} \right| \tag{4.89}$$

For the $k$-$\omega$ model, the parameter $\chi_k$ needed to specify the function $f_{\beta^*}$ in the $k$-$\omega$ model [see Equation (4.41)] transforms to:

$$\chi_k = \frac{1}{W^3} \frac{dK}{d\eta} \frac{dW}{d\eta} \tag{4.90}$$

The $k$-$\omega$ model also uses the vortex-stretching parameter $\chi_\omega$ to specify the function $f_\beta$ [see Equation (4.40)]. Because evaluation of this parameter involves matrix multiplication, it is worthwhile to illustrate details of the mathematics. For two-dimensional shear flows, the strain-rate and rotation tensors are

$$[S_{ij}] \approx \begin{bmatrix} \frac{\partial U}{\partial x} & \frac{1}{2}\frac{\partial U}{\partial y} & 0 \\ \frac{1}{2}\frac{\partial U}{\partial y} & \frac{\partial V}{\partial y} & 0 \\ 0 & 0 & 0 \end{bmatrix} \quad \text{and} \quad [\Omega_{ij}] \approx \begin{bmatrix} 0 & \frac{1}{2}\frac{\partial U}{\partial y} & 0 \\ -\frac{1}{2}\frac{\partial U}{\partial y} & 0 & 0 \\ 0 & 0 & 0 \end{bmatrix} \tag{4.91}$$

where $x$ and $y$ denote streamwise and normal directions, respectively. Hence, for two-dimensional incompressible flow, we have

$$\begin{aligned} \Omega_{ij}\Omega_{jk}S_{ki} &= \Omega_{12}\Omega_{21}S_{11} + \Omega_{21}\Omega_{12}S_{22} \\ &\approx -\frac{1}{4}\left(\frac{\partial U}{\partial y}\right)^2\left(\frac{\partial U}{\partial x} + \frac{\partial V}{\partial y}\right) = 0 \end{aligned} \tag{4.92}$$

where we use the fact that the divergence of the velocity vanishes for incompressible flow. This corresponds to the fact that vortex stretching is exactly zero in two-dimensional flows.

By contrast, for axisymmetric shear flows, these two tensors are

$$[S_{ij}] \approx \begin{bmatrix} \frac{\partial U}{\partial x} & \frac{1}{2}\frac{\partial U}{\partial r} & 0 \\ \frac{1}{2}\frac{\partial U}{\partial r} & \frac{\partial V}{\partial r} & 0 \\ 0 & 0 & \frac{V}{r} \end{bmatrix} \quad \text{and} \quad [\Omega_{ij}] \approx \begin{bmatrix} 0 & \frac{1}{2}\frac{\partial U}{\partial r} & 0 \\ -\frac{1}{2}\frac{\partial U}{\partial r} & 0 & 0 \\ 0 & 0 & 0 \end{bmatrix} \tag{4.93}$$

where $x$ and $r$ denote streamwise and radial directions, respectively. So, for incompressible shear flows, there follows

$$\begin{aligned} \Omega_{ij}\Omega_{jk}S_{ki} &= \Omega_{12}\Omega_{21}S_{11} + \Omega_{21}\Omega_{12}S_{22} \\ &\approx -\frac{1}{4}\left(\frac{\partial U}{\partial r}\right)^2\left(\frac{\partial U}{\partial x} + \frac{\partial V}{\partial r}\right) = \frac{1}{4}\left(\frac{\partial U}{\partial r}\right)^2\frac{V}{r} \end{aligned} \tag{4.94}$$

where we use the fact that the continuity equation in axisymmetric flows is

$$\frac{\partial U}{\partial x} + \frac{\partial V}{\partial r} + \frac{V}{r} = 0 \quad \Longrightarrow \quad \frac{V}{r} = -\left(\frac{\partial U}{\partial x} + \frac{\partial V}{\partial r}\right) \qquad (4.95)$$

Hence, $\chi_\omega$ is nonzero for axisymmetric flows. This reflects the fact that rings of vorticity with an axis parallel to the direction of flow can be stretched as the flow spreads radially. Therefore, the parameter $\chi_\omega$ is

$$\chi_\omega = \begin{cases} 0, & \text{Plane jet} \\ \dfrac{1}{4}\dfrac{(\partial U/\partial r)^2}{(\beta_o^*\omega)^3}\dfrac{V}{r}, & \text{Round and radial jets} \end{cases} \qquad (4.96)$$

In terms of the similarity solution, we have

$$\chi_\omega = \begin{cases} 0, & \text{Plane jet} \\ \dfrac{1}{4}\left(\dfrac{d\mathcal{U}}{d\eta}\right)^2 \dfrac{\mathcal{V}}{\eta\,(\beta_o^*W)^3}, & \text{Round jet} \\ \dfrac{1}{4}\left(\dfrac{d\mathcal{U}}{d\eta}\right)^2 \dfrac{\mathcal{U}}{(\beta_o^*W)^3}, & \text{Radial jet} \end{cases} \qquad (4.97)$$

Boundary conditions on the velocity are the same as in Chapter 3. We must also specify boundary conditions for $K$, $W$ and $E$. Solutions for two-equation models often feature (nonphysical) sharp turbulent/nonturbulent interfaces for free shear flows, i.e., interfaces at which derivatives of flow properties are discontinuous (see Subsection 7.2.2). Consequently, the most sensible boundary conditions in the freestream are those corresponding to nonturbulent flow, i.e., $K(\eta)$, $W(\eta)$ and $E(\eta)$ all vanish approaching the edge of the shear layer. As it turns out, two-equation-model solutions are affected by finite values of $K$, $W$ and $E$ in the freestream, and are sensitive to the freestream value of $E$ or $W$. Subsection 4.5.3 focuses in more detail on this sensitivity. The most appropriate boundary conditions for $K$, $W$ and $E$ are as follows.

**Wake and Jet:**
$$K'(0) = W'(0) = E'(0) = 0 \qquad (4.98)$$

**Wake, Jet and Mixing Layer:**

$$K(\eta) \to 0, \quad W(\eta) \to 0, \quad \text{and} \quad E(\eta) \to 0 \quad \text{as} \quad |\eta| \to \infty \quad (4.99)$$

This completes formulation of the similarity solution for the $k$-$\omega$ and $k$-$\epsilon$ models. We have demonstrated that all pertinent equations and boundary conditions transform to a set of equations and boundary conditions that can be written in terms of the similarity variable, $\eta$. In so doing, we have formulated a nonlinear, two-point boundary-value problem that obviously cannot be solved in closed form. In the next section, we discuss the numerical solution.

## 4.5.2 Numerical Solution

As in Section 4.2, we use the conventional definition of **spreading rate for the wake**, which is the value of $\eta$ given in Equation (4.83), where the velocity defect is half its maximum value. Similarly **for the plane, round and radial jet, the spreading rate** is the value of $y/x$ where the velocity is half its centerline value. **For the mixing layer, the spreading rate** is the difference between the values of $y/x$ where $(U - U_2)^2/(U_1 - U_2)^2$ is 9/10 and 1/10. Table 4.4 compares computed (using Programs **WAKE, MIXER** and **JET** — see Appendix C) and measured spreading rates for the $k$-$\omega$, $k$-$\epsilon$ and RNG $k$-$\epsilon$ models. Figures 4.6 through 4.9 compare computed and measured velocity profiles for these three models.

Table 4.4: *Free Shear Flow Spreading Rates for Two-Equation Models*

| Flow | $k$-$\omega$ Model | $k$-$\epsilon$ Model | RNG $k$-$\epsilon$ Model | Measured |
|------|------|------|------|------|
| Far Wake | .339 | .256 | .290 | .365 |
| Mixing Layer | .105 | .098 | .099 | .115 |
| Plane Jet | .101 | .108 | .146 | .100-.110 |
| Round Jet | .088 | .120 | .185 | .086-.096 |
| Radial Jet | .099 | .094 | .110 | .096-.110 |

Of the three models, the $k$-$\omega$ model is closest to measured spreading rates, with differences between computed and measured values ranging from 3% to 9% — the average difference is 5%. The $k$-$\epsilon$ model predicts a spreading rate that is 30% lower than measured for the far wake, 15% lower than measured for the mixing layer, and between 25% to 40% higher than measured for the round jet. Only for the plane jet and the radial jet does its predicted spreading rate fall within the range of measured values. The average difference between computed and measured spreading rates for the $k$-$\epsilon$ model is 18%. The RNG $k$-$\epsilon$ model yields even larger differences (an average of 29%), including a predicted round-jet spreading rate that is double the measured value.

Figures 4.6 and 4.7 reveal an especially noteworthy feature of the $k$-$\omega$ solutions. The figures show the smooth variation of the velocity profiles approaching the freestream for the far wake and the mixing layer, respectively. By contrast, both the $k$-$\epsilon$ and RNG $k$-$\epsilon$ models predict a discontinuous slope in the velocity profile at the edge of the shear layer for these two cases. Interestingly, Figures 4.8 and 4.9 show that all three models predict a smooth approach to the freestream velocity for round and radial jets. While the $k$-$\epsilon$ model's free shear flow velocity profiles exhibit larger differences from corresponding measurements for the far wake, mixing layer and round jet, its predicted profile for the plane jet is the closest of the three models to the measured profile.

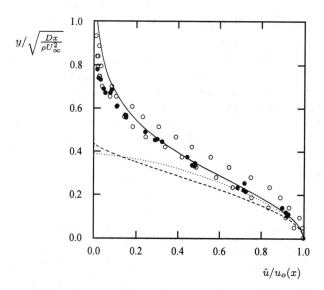

Figure 4.6: *Comparison of computed and measured velocity profiles for the far wake;* —— *$k$-$\omega$; - - - $k$-$\epsilon$; · · · · RNG $k$-$\epsilon$; • Fage and Falkner (1932); ∘ Weygandt and Mehta (1995).*

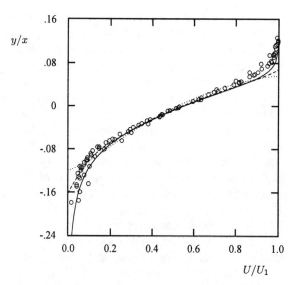

Figure 4.7: *Comparison of computed and measured velocity profiles for the mixing layer;* —— *$k$-$\omega$; - - - $k$-$\epsilon$; · · · · RNG $k$-$\epsilon$; ∘ Liepmann and Laufer (1947).*

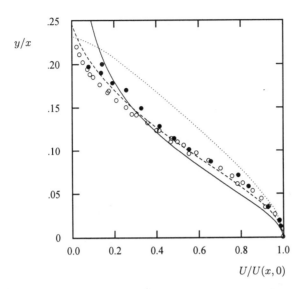

Figure 4.8: *Comparison of computed and measured velocity profiles for the plane jet; —— k-ω; - - - k-ε; ···· RNG k-ε; ○ Bradbury (1965); ● Heskestad (1965).*

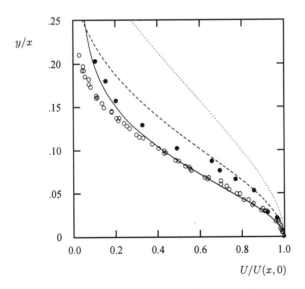

Figure 4.9: *Comparison of computed and measured velocity profiles for the round jet; —— k-ω; - - - k-ε; ···· RNG k-ε; ○ Wygnanski and Fiedler (1969); ● Rodi (1975).*

Table 4.5: *More Two-Equation Model Free Shear Flow Spreading Rates*

| Flow | Robinson et al., $k$-$\zeta$ | Speziale et al., $k$-$\tau$ | Peng et al. $k$-$\omega$ | Wilcox (1988a) $k$-$\omega$ | Measured |
|------|------|------|------|------|------|
| Far Wake | .313 | .221 | .206 | .496 | .365 |
| Mixing Layer | .112 | .082 | .071 | .141 | .115 |
| Plane Jet | .115 | .089 | – | .135 | .100-.110 |
| Round Jet | .091 | .102 | .096 | .369 | .086-.096 |
| Radial Jet | .097 | .073 | .040 | .317 | .096-.110 |

Table 4.5 lists computed spreading rates[4] for four other models that provide a measure of how difficult it has proven to be to develop a model that adequately describes free shear flows. As indicated in the table, the Robinson et al. (1995) enstrophy-equation ($k$-$\zeta$) model predicts spreading rates that are quite close to measured values for all five free shear flows. By contrast, both the Speziale et al. (1990) $k$-$\tau$ model and the Peng et al. (1997) $k$-$\omega$ model predict spreading rates that are significantly smaller than measured for all but the round jet. Finally, the Wilcox (1988a) $k$-$\omega$ model predicts spreading rates that are larger than measured for all five cases.

The latter results provide a definitive measure of how much improvement of the $k$-$\omega$ model attends the modifications to $\beta^*$ and $\beta$ defined in Equations (4.39) through (4.42). The variation of $\beta^*$ with $\chi_k$ produces more dissipation relative to production, which reduces computed spreading rates for free shear flows in general. The variation of $\beta$ with $\chi_\omega$ reduces dissipation relative to production in the $\omega$ equation for round and radial jets, which, in turn, increases $\omega$ and thus the dissipation in the $k$ equation. Hence, both modifications counter the Wilcox (1988a) model's excess production, relative to dissipation, for free shear flows.

### 4.5.3   Sensitivity to Finite Freestream Boundary Conditions

Two equation models have a unique, and unexpected feature when nonzero freestream boundary conditions are specified for $k$, $\omega$, $\epsilon$, etc. Specifically, even if we select $k$ and the second turbulence property ($\omega$, $\epsilon$, etc.) to be sufficiently small that both $k$ and $\nu_T$ are negligible, the solution is sensitive to our choice of the second turbulence property's freestream value. This is an important consideration since most computations are done with these assumptions.

Figure 4.10 shows how the spreading rate, $\delta'$, varies with the freestream value of $\omega$ for the $k$-$\omega$ model defined in Equations (4.36) through (4.42) and the Wilcox (1988a) $k$-$\omega$ model for the far wake, the mixing layer and the plane jet.

---

[4]Values listed for the $k$-$\zeta$ model are from Robinson et al. (1995). All other values have been obtained using modified versions of Programs **WAKE**, **MIXER** and **JET** – see problems section.

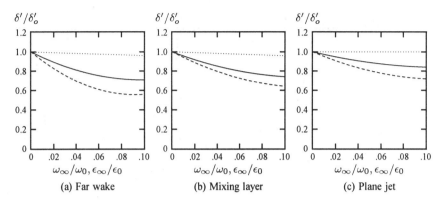

Figure 4.10: *Sensitivity of free shear flow spreading rates to freestream conditions:* $\cdots$ *$k$-$\epsilon$;* —— *$k$-$\omega$;* - - - *Wilcox (1988a) $k$-$\omega$. $\omega_0$ and $\epsilon_0$ are for $\eta = 0$.*

It also shows the variation of $\delta'$ with the freestream value of $\epsilon$ for the Standard $k$-$\epsilon$ model defined in Equations (4.46) through (4.50). In all three graphs, $\delta'_o$ is the predicted spreading rate for the limiting case $\omega_\infty \to 0$ for the $k$-$\omega$ models and $\epsilon_\infty \to 0$ for the $k$-$\epsilon$ model. All computations have been done with the dimensionless eddy viscosity, $N(\infty)$, equal to $10^{-6}$.

All three models predict a decrease in spreading rate as the freestream value of $\omega$ or $\epsilon$ increases. In all three graphs, the freestream value is scaled with respect to the value at $\eta = 0$, which is very close to the maximum value for each flow. As shown, the $k$-$\omega$ models display a much stronger sensitivity than the $k$-$\epsilon$ model. The graphs also show that if the freestream value is less than 0.1% of the maximum value $[\omega_\infty/\omega_0 < 0.001]$ there is virtually no effect on the predicted spreading rate. Certainly this is not an unreasonable constraint as $k$, $\omega$ and $\epsilon$ all vary by many orders of magnitude in typical turbulent shear flows. Using a freestream value of $\omega$ or $\epsilon$ in excess of even 0.01% of the peak value would very likely correspond to using a physically unrealistic value.

There is no mystery about why the solution should be sensitive to freestream boundary conditions. We are, after all, solving a two-point boundary-value problem, which requires freestream boundary conditions on all variables, including $\omega$ and $\epsilon$. In light of this, it is clear that there must be some range of boundary values that affect the solution. Figure 4.10 shows that there is a well defined limiting form of the solution for vanishing freestream boundary values, further validating the claim that Equations (4.98) and (4.99) are the proper freestream boundary conditions.

Another idealized flow that demonstrates how farfield boundary conditions affect two-equation model predictions is the one-dimensional propagation of a turbulent front into a quiescent fluid. This problem has been analyzed by several

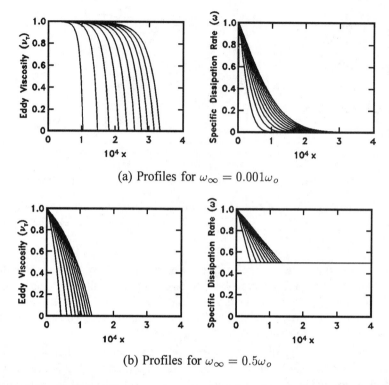

(a) Profiles for $\omega_\infty = 0.001\omega_o$

(b) Profiles for $\omega_\infty = 0.5\omega_o$

Figure 4.11: *Turbulent front propagation — Wilcox (1988a) k-ω model. The 10 curves displayed in each graph are computed profiles at 10 different times as the front advances to the right.*

authors, including Lele (1985) and Wilcox (1995b). Briefly, we imagine a planar source of turbulence at $x = 0$ where we maintain constant values of $k = k_o$ and $\omega = \omega_o$ or $\epsilon = \epsilon_o$ for all time. The turbulence source is instantaneously "turned on" at time $t = 0$, and a front propagates into the fluid at a finite rate.

Figure 4.11 shows computed dimensionless $\nu_T$ and $\omega$ profiles for farfield values of $\omega$ equal to $0.001\omega_o$ and $0.5\omega_o$ based on the Wilcox (1988a) k-ω model. Both computations have been done with the farfield value of $k$ chosen so that the farfield eddy viscosity is $10^{-6}$ times the value at $x = 0$. The graphs all include a family of curves corresponding to 10 different times, with the front advancing to the right. The motion of the front is clearly indicated by the $\nu_T$ curves, which exhibit the sharp interface between the spreading turbulence and the nonturbulent fluid.

Inspection of the curves shows that when the freestream value of $\omega$ is $0.001\omega_o$, the $\omega$ curves all tend smoothly to the farfield value as the front advances. By

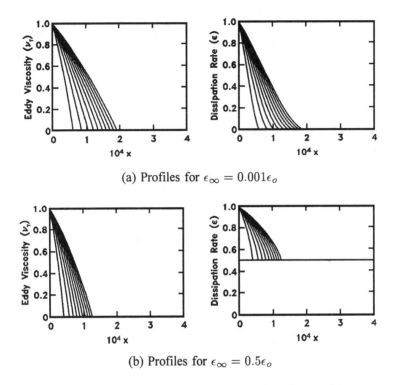

(a) Profiles for $\epsilon_\infty = 0.001\epsilon_o$

(b) Profiles for $\epsilon_\infty = 0.5\epsilon_o$

Figure 4.12: *Turbulent front propagation — $k$-$\epsilon$ model. The 10 curves displayed in each graph are computed profiles at 10 different times as the front advances to the right.*

contrast, when $\omega = 0.5\omega_o$, the farfield value has a strong effect on the solution. It places a large lower bound on $\omega$, and causes the solution to have discontinuous slope at the front. It also retards the rate at which the front advances. Specifically, when $\omega = 0.5\omega_o$, the rate of advance of the front is only about 40% of the rate for $\omega = 0.001\omega_o$.

Figure 4.12 includes similar graphs for the Standard $k$-$\epsilon$ model, corresponding to farfield values of $\epsilon$ equal to $0.001\epsilon_o$ and $0.5\epsilon_o$. Again, both computations have been done with $k_\infty$ chosen so that the farfield eddy viscosity is $10^{-6}$ times the value at the origin. The rate of advance of the turbulent front for $\epsilon = 0.5\epsilon_o$ is 65% of the rate for the smaller farfield value. Thus, while the effect of the farfield condition is smaller for the $k$-$\epsilon$ model than for the $k$-$\omega$ model, it is nevertheless very substantial.

Clearly, some degree of care must be exercised when selecting freestream or farfield boundary conditions for two-equation models. It is not sufficient to simply select small values for $k$ and $\nu_T$, as the choice can imply a nonphysically large value of the second turbulence parameter, viz., $\omega$, $\epsilon$ or $\ell$. In complex flows, estimates should be made regarding the peak value of the second variable in regions of intense shear, to be sure the freestream value is small enough. To be certain appropriately small values are used in the freestream, the values can always be adjusted as the computation proceeds.

### 4.5.4   Cross Diffusion

There is an interesting relationship between the $k$-$\epsilon$ and $k$-$\omega$ models (or any pair of models whose second variable is $k^m \epsilon^n$ for some $m$ and $n$) that helps delineate some of the key differences. Specifically, if we let $\epsilon = C_\mu k \omega$ define a change of dependent variables from $\epsilon$ to $\omega$, it is a straightforward matter to demonstrate that the resulting equation for $\omega$ is

$$\frac{\partial \omega}{\partial t} + U_j \frac{\partial \omega}{\partial x_j} = \alpha \frac{\omega}{k} \tau_{ij} \frac{\partial U_i}{\partial x_j} - \beta \omega^2 + \frac{\partial}{\partial x_j} \left[ (\nu + \sigma \nu_T) \frac{\partial \omega}{\partial x_j} \right]$$
$$+ 2 \frac{(\nu + \sigma \nu_T)}{k} \frac{\partial k}{\partial x_j} \frac{\partial \omega}{\partial x_j} + \frac{\omega}{k} \frac{\partial}{\partial x_j} \left[ (\sigma - \sigma^*) \nu_T \frac{\partial k}{\partial x_j} \right] \quad (4.100)$$

where $\beta^* = C_\mu$. Also, $\alpha$, $\beta$, $\sigma$ and $\sigma^*$ are simple functions of the $k$-$\epsilon$ model's closure coefficients (see problems section). Focusing on free shear flows, we can ignore molecular viscosity, $\nu$. Also, if we assume $\sigma = \sigma^*$ for simplicity, Equation (4.100) simplifies to

$$\frac{\partial \omega}{\partial t} + U_j \frac{\partial \omega}{\partial x_j} = \alpha \frac{\omega}{k} \tau_{ij} \frac{\partial U_i}{\partial x_j} - \beta \omega^2 + \sigma_d \frac{1}{\omega} \frac{\partial k}{\partial x_j} \frac{\partial \omega}{\partial x_j} + \frac{\partial}{\partial x_j} \left[ \sigma \nu_T \frac{\partial \omega}{\partial x_j} \right] \quad (4.101)$$

where $\sigma_d = 2\sigma$. The term proportional to $\sigma_d$ in Equation (4.101) is referred to as **cross diffusion**, depending upon gradients of both $k$ and $\omega$.

In free shear flows the cross-diffusion term enhances production of $\omega$, which in turn increases dissipation of $k$ (assuming $\sigma_d > 0$). This occurs for small freestream values of $k$ and $\omega$, for which both quantities decrease approaching the shear-layer edge. The overall effect is to reduce the net production of $k$, which reduces the predicted spreading rates from the values listed in Table 4.5.

Several authors, including Speziale et al. (1990), Menter (1992c), Wilcox (1993a) and Peng et al. (1997) have attempted to improve the $k$-$\omega$ model by adding cross diffusion. Speziale et al. (1990) and Peng et al. (1997) include cross diffusion in their $\omega$ equations, and achieve some success in a limited number of applications. However, the differential equations are far more difficult to handle numerically with cross diffusion included. As an example, while Programs **WAKE** and **MIXER** will converge with cross diffusion included, the timestep must be reduced to about a tenth of the value that can be used when there is no cross diffusion. The same is true for Program **JET** for round and radial jets. However, the program fails to converge for any timestep in the plane-jet case. Inspection of Table 4.5 shows that the Peng et al. model's spreading rates are significantly lower than measured, The spreading rates predicted by the Speziale et al. $k$-$\omega$ model are identical to those of the Speziale et al. $k$-$\tau$ model, which are also much smaller than measured.

The numerical difficulties arise because the cross-diffusion term becomes quite large approaching the freestream. Standard numerical algorithms are generally optimized for equations dominated by convective and diffusive terms, and numerical stability is particularly sensitive to the treatment of the convective terms. Cross diffusion augments convection (even if it is treated as a source term), replacing the normal convection term, $V \partial\omega/\partial y$, by $(V - \sigma_d\omega^{-1}\partial k/\partial y)\partial\omega/\partial y$. In practice, the cross-diffusion term dominates approaching the freestream, and reduces the timestep for which stable computation is possible.

Menter (1992c) has enjoyed more success with cross diffusion than Speziale et al. and Peng et al. Menter also introduces a "blending function" that makes $\sigma_d = 0$ close to solid boundaries, while $\sigma_d \to 2\sigma$ away from such boundaries. Additionally, his blending function causes all of the model's closure coefficients to assume the $k$-$\omega$ model values near solid boundaries, and to asymptotically approach values similar to those used with the $k$-$\epsilon$ model otherwise. The net result is a model that behaves very much like the Wilcox (1988a) $k$-$\omega$ model for wall-bounded flows, and that is nearly identical to the $k$-$\epsilon$ model for free shear flows. As with the Speziale et al and Peng et al. models, it is very difficult to obtain numerical solutions with Menter's model using any but the most advanced numerical methods such as those described in Section 7.5.

Wilcox (1993a) has tried a similar concept with

$$
\sigma_d = \begin{cases} 0, & \dfrac{\partial k}{\partial x_j}\dfrac{\partial \omega}{\partial x_j} \leq 0 \\[2ex] \sigma, & \dfrac{\partial k}{\partial x_j}\dfrac{\partial \omega}{\partial x_j} > 0 \end{cases}
\tag{4.102}
$$

Additionally, the value of $\sigma^*$ assumes a value larger than 0.5. As we will see in Subsection 4.6.2, it is important to suppress this cross-diffusion term close to solid boundaries for wall-bounded flows. Just as Menter's blending function causes $\sigma_d$ to approach 0 near a solid boundary, so does Equation (4.102) since $k$ increases and $\omega$ decreases in the viscous sublayer. While simpler than Menter's blending function approach, this straightforward modification yields free shear layer spreading rates that are farther from measurements than those predicted by the $k$-$\epsilon$ model. Other values of the $k$-$\omega$ model's closure coefficients exist that yield closer agreement with measured spreading rates, but that also compromise the model's accuracy for wall-bounded flows.

We can gain further insight into the role of cross diffusion by rewriting Equation (4.101) in terms of the dimensionless grouping $\chi_k$ defined by

$$\chi_k = \frac{1}{\omega^3} \frac{\partial k}{\partial x_j} \frac{\partial \omega}{\partial x_j} \qquad (4.103)$$

The equation then becomes

$$\frac{\partial \omega}{\partial t} + U_j \frac{\partial \omega}{\partial x_j} = \alpha \frac{\omega}{k} \tau_{ij} \frac{\partial U_i}{\partial x_j} - [\beta - \sigma_d \chi_k]\, \omega^2 + \frac{\partial}{\partial x_j} \left[ \sigma \nu_T \frac{\partial \omega}{\partial x_j} \right] \qquad (4.104)$$

On the one hand, since the cross-diffusion term is positive in both free shear flows and boundary layers for most of the flow, its primary role is to increase $\omega$. This, of course, makes the dissipation term in the $k$ equation, $\beta^* \omega k$, larger. Because the net effect is to reduce the $k$-$\omega$ model's predicted spreading rates to desired levels, there has been great interest in examining its importance. On the other hand, as we will see when we examine boundary layers in Section 4.6, it has an undesirable effect on boundary layers. Since we will also see that the $k$-$\omega$ model predicts boundary-layer properties that are quite close to measurements, addition of cross diffusion inevitably leads to an impasse.

However, the dimensionless parameter, $\chi_k$, has a very interesting property. Specifically, it is significantly larger in free shear flows than in boundary layers, primarily because $\omega$ assumes much larger values due to the presence of the solid boundary. In the far wake, for example, it reaches values as large as $5 \cdot 10^6$. By contrast, for the constant-pressure boundary layer, its maximum value is less than 0.4 and a typical value is 0.006. This observation motivates the modification to the coefficient $\beta^*$ in the $k$-$\omega$ model specified in Equations (4.39) through (4.42).

Given that our analysis has led to a parameter that is much larger for free shear flows than for boundary layers, by making the $k$-$\omega$ model's closure coefficients functions of $\chi_k$, we can adjust them for free shear flows without impacting their values appreciably for boundary layers. There is no a priori reason to limit any modifications to the $\omega$ equation or, for that matter, to change the $\omega$ equation at all. The cross-diffusion term as listed in Equation (4.101) appears only because we started with the $k$-$\epsilon$ model. To argue that the cross-diffusion term is "missing" from the $k$-$\omega$ model, as several authors have done, assumes the modeled $\epsilon$ equation is in some sense more fundamental than the modeled $\omega$ equation. Given how poorly the $k$-$\epsilon$ model fares in predicting turbulent flows, especially wall-bounded flows (see Sections 4.6 through 4.10), the argument is obviously non sequitur.

Thus, we choose to introduce the parameter $\chi_k$ in the $k$ equation rather than in the $\omega$ equation. Since the objective is to increase the amount of dissipation in the $k$ equation, making $\beta^*$ rather than $\beta$ a function of $\chi_k$ is a more direct approach. Inspection of Equations (4.39) – (4.42) shows that the modification amounts to saying the dissipation, $\epsilon$, is related to $k$ and $\omega$ by

$$\epsilon = \beta_o^* f_{\beta^*} \omega k \tag{4.105}$$

where

$$\beta_o^* = \frac{9}{100}, \qquad f_{\beta^*}(\chi_k) \equiv \begin{cases} 1, & \chi_k \leq 0 \\ \dfrac{1 + 680\chi_k^2}{1 + 400\chi_k^2}, & \chi_k > 0 \end{cases} \tag{4.106}$$

Although it makes little difference for free shear flows, it is important to use the cross-diffusion parameter only when it is positive. This is so because it has undesirable effects in the viscous sublayer. The function $f_{\beta^*}(\chi_k)$ has been chosen to yield agreement, as closely as possible, with measurements for the far wake, mixing layer and plane jet.

As will be discussed in Section 4.6, boundary-layer predictions are insensitive to the function $f_{\beta^*}(\chi_k)$. To illustrate why this is so, Figure 4.13 shows $f_{\beta^*}$ for the far wake, the constant-pressure boundary layer and the Samuel-Joubert [see Kline et al. (1981)] adverse pressure-gradient boundary layer. The quantity $\eta_e$ denotes the value of $\eta = y/x$ at the outer edge of the wake, while $\delta$ denotes the boundary-layer thickness. As indicated, $f_{\beta^*}$ assumes its asymptotic value of 1.7 for the outer 60% of the wake, corresponding to $\chi_k > 0.5$. For both boundary-layer cases, $\chi_k$ is too small to cause $f_{\beta^*}$ to differ appreciably from 1 in the inner 70% of the layer. It differs from 1 only in the outer 30%, where it has a negligible effect on the overall solution.

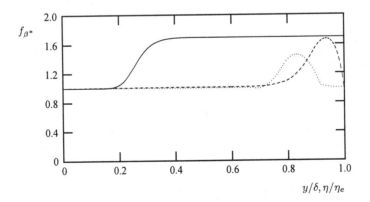

Figure 4.13: *Variation of the function* $f_{\beta^*}$ *for a far wake (——), a constant-pressure boundary layer (- - -) and an adverse-$\nabla p$ boundary layer (·····).*

### 4.5.5    The Round-Jet/Plane-Jet Anomaly

Inspection of Tables 4.1, 4.4 and 4.5 shows that all but two of the turbulence models listed predict that the round jet spreads more rapidly than the plane jet. The two exceptions are the $k$-$\omega$ model and the Robinson et al. (1995) enstrophy-equation model. However, measurements indicate the opposite trend, with the round-jet spreading rate being about 10% lower than that of the plane jet. This shortcoming, common to most turbulence models, is known as the **round-jet/plane-jet anomaly**.

Pope (1978) has proposed a modification to the $\epsilon$ equation that resolves the round-jet/plane-jet anomaly for the $k$-$\epsilon$ model. In Pope's modification, the *dissipation of dissipation* term in the $\epsilon$ equation is replaced by

$$C_{\epsilon2}\frac{\epsilon^2}{k} \rightarrow [C_{\epsilon2} - C_{\epsilon3}\chi_p]\,\frac{\epsilon^2}{k} \tag{4.107}$$

where $\chi_p$ is a "nondimensional measure of vortex stretching" defined as

$$\chi_p \equiv \frac{\Omega_{ij}\Omega_{jk}S_{ki}}{(\epsilon/k)^3} \tag{4.108}$$

The tensors $\Omega_{ij}$ and $S_{ij}$ are the mean-rotation and mean-strain-rate tensors defined in Equation (4.43).

Pope's reasoning is that the primary mechanism for transfer of energy from large to small eddies is vortex stretching. Any mechanism that enhances vortex stretching will increase this rate of transfer. Since the energy is being transferred to the smallest eddies where dissipation occurs, necessarily the dissipation, $\epsilon$, must increase. Because mean-flow vortex lines cannot be stretched in a two-dimensional flow, $\chi_p$ is zero for the plane jet. By contrast, as shown earlier [see Equations (4.91) through (4.96)], the vortex-stretching parameter is nonzero for an axisymmetric mean flow. As discussed earlier, this corresponds to the fact that vortex rings are being stretched radially. Thus, we expect to have $\chi_p \neq 0$ for a round jet.

Using $C_{\epsilon3} = 0.79$ reduces the $k$-$\epsilon$ model's predicted spreading rate to 0.86, consistent with measurements. However, as pointed out by Rubel (1985), the Pope correction has an adverse effect on model predictions for the radial jet, which also has nonzero $\chi_p$. Without the Pope correction, the $k$-$\epsilon$ model predicts a radial-jet spreading rate of 0.094 which is close to the measured range of 0.096 to 0.110 [see Tanaka and Tanaka (1976) and Witze and Dwyer (1976)]. Using the Pope correction for the radial jet reduces the $k$-$\epsilon$ model-predicted spreading rate to 0.040. Hence, as noted by Rubel, "the round jet/plane jet anomaly has been exchanged for a round jet/radial jet anomaly."

In contrast to the $k$-$\epsilon$ model, as indicated in Table 4.5, the Wilcox (1988a) $k$-$\omega$ model predicts comparable spreading rates for both the round and radial jets, both larger than the predicted plane-jet spreading rate. The same is true of the $k$-$\omega$ model defined in Equations (4.36) to (4.42). When a constant value of $\beta = 0.072$ is used for the latter, the predicted round- and radial-jet spreading rates are 0.139 and 0.149, respectively. Numerical experimentation shows that if $\beta$ is reduced to about 7/8 of this value, the model's spreading rates for both the round and radial jets are close to the measured values. Since Pope's argument implies nothing regarding the functional dependence of the modification upon the vortex-stretching parameter, $\chi_p$, it is completely consistent to propose that $\beta$ depend upon this parameter in a manner that reduces the value of $\beta$ as needed for both flows. Thus, as a generalization of the Pope modification, the $k$-$\omega$ model uses the following prescription for $\beta$.

$$\beta = \beta_o f_\beta \tag{4.109}$$

where

$$\beta_o = \frac{9}{125}, \qquad f_\beta = \frac{1 + 70\chi_\omega}{1 + 80\chi_\omega} \tag{4.110}$$

and

$$\chi_\omega \equiv \left| \frac{\Omega_{ij}\Omega_{jk}S_{ki}}{(\beta_o^*\omega)^3} \right| \tag{4.111}$$

Comparison of Equations (4.108) and (4.111) shows that $\chi_\omega = |\chi_p|$. Also, the functional form of $f_\beta$ is such that its asymptotic value is 7/8 for large values of $\chi_\omega$. Finally, as with the cross-diffusion parameter, $\chi_k$ [see Equation (4.41)], the vortex-stretching parameter is very small in axisymmetric boundary layers because $\omega$ is very large.

Interestingly, the Robinson et al. (1995) enstrophy-equation model contains a term similar to the Pope modification. The vortex-stretching mechanism that it represents plays an important role in the model's ability to predict the measured spreading rates for all three jets within a few percent of measurements. Although the usefulness of Pope's correction as represented by Equations (4.107) and (4.108) is limited by a flaw in the $k$-$\epsilon$ model, the concepts underlying the formulation are not. We can reasonably conclude that Pope's analysis provides a sensible reflection of the physics of turbulent jets, at least in the context of $\omega$-based two-equation models.

Our analysis of free shear flows is now complete. In the following sections we turn our attention to wall-bounded flows. To demonstrate how two-equation models fare for such flows, we are going to use a powerful mathematical tool to analyze fine details of model-predicted structure of the turbulent boundary layer. In particular, we will use **perturbation methods** to analyze the various regions in the turbulent boundary layer.

# 4.6    Perturbation Analysis of the Boundary Layer

The differential equations for all but the simplest turbulence models are sufficiently complicated for most flows that closed-form solutions do not exist. This is especially true for boundary layers because of nonlinearity of the convection terms and the turbulent diffusion terms attending introduction of the eddy viscosity. Our inability to obtain closed-form solutions is unfortunate because such solutions are invaluable in design studies and for determining trends with a parameter such as Reynolds number, or more generally, for establishing laws of similitude. Furthermore, without analytical solutions, our ability to check the accuracy of numerical solutions is limited.

There is a powerful mathematical tool available to us to generate approximate solutions that are valid in special limiting cases, viz., **perturbation analysis**. The idea of perturbation analysis is to develop a solution in the form of an **asymptotic expansion** in terms of a parameter, the error being small for sufficiently small values of the parameter. Our desire in developing such an expansion is for the first few terms of the expansion to illustrate all of the essential physics of the problem and to provide a close approximation to the exact solution. Fortunately, this is usually the case in fluid mechanics. This section shows how perturbation analysis can be used to dissect model-predicted structure of the turbulent boundary layer. Appendix B introduces basic concepts of perturbation theory for the reader with no prior background in the field.

## 4.6.1    The Log Layer

We direct our focus to the turbulent boundary layer. Experimental observations provide a strong argument for using perturbation analysis. Specifically, Coles' description of the turbulent boundary layer as a "wake-like structure constrained by a wall" (see Figure 3.8) suggests that different scales and physical processes are dominant in the inner (near-wall) and outer (main body) of the layer. These are concepts upon which perturbation analysis is based. Coles [see Coles and Hirst (1969)] makes an explicit connection with perturbation theory in saying:

> *"The idea that there are two distinct scales in a turbulent boundary layer is an old one, although quantitative expressions of this idea have evolved very slowly... To the extent that the outer velocity boundary condition for the inner (wall) profile is the same as the inner velocity boundary condition for the outer (wake) profile, the turbulent boundary layer is a singular perturbation problem of classical type. In fact, we can claim to have discovered the first two terms in a composite expansion, complete with logarithmic behavior."*

Often perturbation solutions are guided by dimensional considerations and a knowledge of physical aspects of the problem. For the turbulent boundary layer, we can draw from empirically established laws to aid us in developing our perturbation solution. We observe that close to a solid boundary, the **law of the wall** holds. As discussed in Subsection 1.3.5, we can write this symbolically as

$$U(x, y) = u_\tau(x) f(u_\tau y/\nu), \qquad u_\tau = \sqrt{\tau_w/\rho} \qquad (4.112)$$

Similarly, the main body of the turbulent boundary layer behaves according to Clauser's (1956) well-known **defect law**, viz.,

$$U(x, y) = U_e(x) - u_\tau(x) F[y/\Delta(x)], \qquad \Delta(x) = U_e \delta^*/u_\tau \qquad (4.113)$$

The reader should keep in mind that Equation (4.113) only applies to a special class of boundary layers, i.e., boundary layers that are self preserving. Thus, we seek solutions where $F(y/\Delta)$ is independent of $x$. As we will see, the model equations predict existence of such solutions under precisely the same conditions Clauser discovered experimentally.

We develop the leading terms in a perturbation solution for the turbulent boundary layer in the following subsections. There are two small parameters in our problem, the first being the reciprocal of the Reynolds number. This is consistent with the standard boundary-layer approximations. The second small parameter is $u_\tau/U_e$. Clauser's defect law suggests this parameter since the velocity is expressed as a (presumably) small deviation from the freestream velocity that is proportional to $u_\tau$. The analysis will lead to a relation between these two parameters.

The analysis in this section, which is patterned after the work of Bush and Fendell (1972) and Fendell (1972), shows in Subsection 4.6.3 that the inner expansion is of the form quoted in Equation (4.112) and is valid in the **viscous sublayer** (see Figure 3.7). We also show in Subsection 4.6.2 that the outer expansion is identical in form to Equation (4.113) and holds in the **defect layer**. Formal matching of the sublayer and defect-layer solutions occurs in an overlap region that is often described as the **log layer**. In fact, the common part of the inner and outer expansions is precisely the law of the wall. Thus, although it is not formally a separate layer, establishing flow properties in the log layer permits independent analysis of the sublayer and defect layer. It also forms the basis of surface boundary conditions for many two-equation turbulence models. We discuss the log layer in this subsection.

Before performing any analysis, we anticipate that we will be solving a **singular-perturbation problem**. We expect this, but not because of a reduction in order of the differential equations. Rather, we have no hope of satisfying the no-slip condition with our outer solution because of the assumed form in the defect layer, i.e., velocity being a small perturbation from the freestream value.

Likewise, the sublayer solution, if it is consistent with measurements, predicts velocity increasing logarithmically with distance from the surface as $y \to \infty$ so that we cannot satisfy the freestream boundary condition with our inner solution. This is the irregular behavior near boundaries alluded to in Appendix B where we define a singular-perturbation problem.

We begin our analysis with the incompressible boundary-layer equations. Conservation of mass and momentum are sufficient for establishing the form of the expansions, so that we have no need to introduce the model equations now. For two-dimensional flow, we have

$$\frac{\partial U}{\partial x} + \frac{\partial V}{\partial y} = 0 \tag{4.114}$$

$$U\frac{\partial U}{\partial x} + V\frac{\partial U}{\partial y} = -\frac{1}{\rho}\frac{dP}{dx} + \frac{\partial}{\partial y}\left[(\nu + \nu_T)\frac{\partial U}{\partial y}\right] \tag{4.115}$$

The easiest way to arrive at the **log-layer** equations is to derive the **sublayer** equations and then to determine the limiting form of the sublayer equations for $y^+ \to \infty$. Consistent with the normal boundary-layer concept that variations in the streamwise ($x$) direction are much less rapid than those in the normal ($y$) direction, we scale $x$ and $y$ differently. Letting $L$ denote a dimension characteristic of distances over which flow properties change in the $x$ direction, we scale $x$ and $y$ according to

$$\xi = x/L \quad \text{and} \quad y^+ = u_\tau y/\nu \tag{4.116}$$

The appropriate expansions for the streamfunction and kinematic eddy viscosity are

$$\psi_{inner}(x,y) \sim \nu[f_0(\xi,y^+) + \phi_1 f_1(\xi,y^+) + O(\phi_2)] \tag{4.117}$$

$$\nu_{T_{inner}}(x,y) \sim \nu[N_0(\xi,y^+) + \phi_1 N_1(\xi,y^+) + O(\phi_2)] \tag{4.118}$$

where the asymptotic sequence $\{1, \phi_1, \phi_2, \ldots\}$ is to be determined. Consequently, the streamwise velocity becomes

$$U(x,y) \sim u_\tau\left[\hat{u}_0(\xi,y^+) + \phi_1\hat{u}_1(\xi,y^+) + O(\phi_2)\right], \quad \hat{u}_n \equiv \frac{\partial f_n}{\partial y^+} \tag{4.119}$$

Substituting into the momentum equation, we obtain

$$\frac{\partial}{\partial y^+}\left[(1 + N_0)\frac{\partial\hat{u}_0}{\partial y^+}\right] + O(\phi_1) = \frac{\nu}{u_\tau\delta^*}\left[\beta_T + O\left(\frac{\delta^*}{L}\right)\right] \tag{4.120}$$

where the quantity $\beta_T$ is the so-called **equilibrium parameter** [see Coles and Hirst (1969)] defined by

$$\beta_T \equiv \frac{\delta^*}{\tau_w}\frac{dP}{dx} \tag{4.121}$$

In general, we regard $\beta_T$ as being of order one. In fact, when we analyze the defect layer, this will be the key parameter quantifying the effect of pressure gradient on our solution. Additionally, $u_\tau \delta^*/\nu \gg 1$ and $\delta^* \ll L$. Hence, we conclude that

$$\phi_1 = \frac{\nu}{u_\tau \delta^*} \tag{4.122}$$

and

$$\frac{\partial}{\partial y^+}\left[(1 + N_0)\frac{\partial \hat{u}_0}{\partial y^+}\right] = 0 \tag{4.123}$$

To enhance physical understanding of what we have just proven, it is worthwhile to return to dimensional variables. We have shown that, to leading order, the convective terms and the pressure gradient are small compared to the other terms in the sublayer so that the momentum equation simplifies to

$$\frac{\partial}{\partial y}\left[(\nu + \nu_T)\frac{\partial U}{\partial y}\right] = 0 \tag{4.124}$$

Integrating once tells us that the sum of the specific molecular and Reynolds shear stress is constant in the sublayer, i.e.,

$$(\nu + \nu_T)\frac{\partial U}{\partial y} = \frac{\tau_w}{\rho} \tag{4.125}$$

Equation (4.124) or (4.125) is the equation for the leading-order term in the inner expansion for a turbulent boundary layer. As we will demonstrate in greater detail in Subsection 4.6.3, we can satisfy the no-slip condition ($U = 0$) at $y = 0$ while the solution as $y^+ \to \infty$ asymptotes to the law of the wall, i.e., velocity increasing logarithmically with distance from the surface. Another feature of the solution is that the eddy viscosity increases linearly with $y^+$ as $y^+ \to \infty$ so that the eddy viscosity becomes very large compared to the molecular viscosity. Consistent with this behavior, the molecular viscosity can be neglected in Equation (4.124) or (4.125) for the limiting case $y^+ \to \infty$. As noted above, we refer to the form of the differential equations in this limit as the **log-layer equations**. Thus, we conclude that in the log layer we can neglect convection, pressure gradient and molecular diffusion. The momentum equation thus simplifies to the following:

$$0 = \frac{\partial}{\partial y}\left[\nu_T \frac{\partial U}{\partial y}\right] \tag{4.126}$$

To the same degree of approximation, in the log layer, the $k$-$\omega$ model equations for two-dimensional flow (so that $\chi_\omega = 0 \Rightarrow \beta = \beta_o$) simplify to:

**$k$-$\omega$ Model:**

$$
\left.
\begin{aligned}
&0 = \nu_T \left(\frac{\partial U}{\partial y}\right)^2 - \beta^* \omega k + \sigma^* \frac{\partial}{\partial y}\left[\nu_T \frac{\partial k}{\partial y}\right] \\[2mm]
&0 = \alpha \left(\frac{\partial U}{\partial y}\right)^2 - \beta_o \omega^2 + \sigma \frac{\partial}{\partial y}\left[\nu_T \frac{\partial \omega}{\partial y}\right] \\[2mm]
&\nu_T = k/\omega
\end{aligned}
\right\} \tag{4.127}
$$

As can be shown by direct substitution, the solution to Equations (4.126) and (4.127) is

$$
U = \frac{u_\tau}{\kappa}\ell n y + C, \qquad k = \frac{u_\tau^2}{\sqrt{\beta_o^*}}, \qquad \omega = \frac{u_\tau}{\sqrt{\beta_o^*}\kappa y} \tag{4.128}
$$

where $C$ is a constant and the implied value of the Kármán constant, $\kappa$, is given by

$$
\kappa^2 = \sqrt{\beta_o^*}(\beta_o/\beta_o^* - \alpha)/\sigma \tag{4.129}
$$

Note that the term proportional to $\sigma^*$ disappears because $\partial k/\partial y = 0$, while $\chi_k = 0$ and $\beta^* = \beta_o^*$ for the same reason. The closure coefficient values specified in Equation (4.39) have been chosen to give $\kappa = 0.41$. We discussed the log-layer solution in Section 4.4 to illustrate how values for some of the closure coefficients have been selected. There are additional features of the solution worthy of mention. For example, the eddy viscosity varies linearly with distance from the surface and is given by

$$
\nu_T = \kappa u_\tau y \tag{4.130}
$$

This variation is equivalent to the mixing-length variation, $\ell_{mix} = \kappa y$. Also, the ratio of the Reynolds shear stress to the turbulence energy is constant, i.e.,

$$
\tau_{xy} = \sqrt{\beta_o^*}\, k \tag{4.131}
$$

In a similar way, the $k$-$\epsilon$ model equations simplify to the following:

**$k$-$\epsilon$ Model:**

$$
\left.
\begin{aligned}
&0 = \nu_T \left(\frac{\partial U}{\partial y}\right)^2 - \epsilon + \frac{\partial}{\partial y}\left[\frac{\nu_T}{\sigma_k}\frac{\partial k}{\partial y}\right] \\[2mm]
&0 = C_{\epsilon 1} C_\mu k \left(\frac{\partial U}{\partial y}\right)^2 - C_{\epsilon 2}\frac{\epsilon^2}{k} + \frac{\partial}{\partial y}\left[\frac{\nu_T}{\sigma_\epsilon}\frac{\partial \epsilon}{\partial y}\right] \\[2mm]
&\nu_T = C_\mu k^2/\epsilon
\end{aligned}
\right\} \tag{4.132}
$$

The solution to Equations (4.126) and (4.132) is

$$U = \frac{u_\tau}{\kappa}\ell n y + \text{constant}, \qquad k = \frac{u_\tau^2}{\sqrt{C_\mu}}, \qquad \epsilon = \frac{u_\tau^3}{\kappa y} \qquad (4.133)$$

where we again find an implied value for the Kármán constant, $\kappa$, viz.,

$$\kappa^2 = \sqrt{C_\mu}(C_{\epsilon 2} - C_{\epsilon 1})\sigma_\epsilon \qquad (4.134)$$

Using the closure coefficient values for the Standard $k$-$\epsilon$ model [Equation (4.49)], $\kappa$ assumes a somewhat large value of 0.433. For the RNG $k$-$\epsilon$ model [Equations (4.51) – (4.53)], we find $\kappa = 0.399$.

Keep in mind that the turbulent boundary layer consists of the sublayer and the defect layer. The sublayer is a thin near-wall region, while the defect layer constitutes most of the boundary layer. In the spirit of matched asymptotic expansions, the log layer is the overlap region which, in practice, is usually much thicker than the sublayer (see Figure 3.7). Part of our reason for focusing on this region of the boundary layer is of historical origin. Aside from the $k$-$\omega$ model, most two-equation models fail to agree satisfactorily with experiment in the viscous sublayer unless the coefficients are made empirical functions of an appropriate turbulence Reynolds number (which we discuss in Subsection 4.9.1). Consequently, the log-layer solution has often been used as a replacement for the no-slip boundary condition. Early $k$-$\epsilon$ model solutions, for example, were generated by enforcing the asymptotic behavior given in Equation (4.133). We must postpone further discussion of surface boundary conditions pending detailed analysis of the sublayer. Analysis of the log layer can also prove useful in determining leading-order effects of complicating factors such as surface curvature, coordinate-system rotation, and compressibility. As our most immediate goal, we have, in effect, done our matching in advance. Thus, we are now in a position to analyze the defect layer and the sublayer independent of one another. We turn first to the defect layer.

## 4.6.2 The Defect Layer

In this subsection, we make use of singular-perturbation methods to analyze model-predicted structure of the classical defect layer, including effects of pressure gradient. Our analysis includes three turbulence models, viz.: the Wilcox $k$-$\omega$ model; the Standard $k$-$\epsilon$ model; and the RNG $k$-$\epsilon$ model. First, we generate the perturbation solution. Next, we compare solutions for the three models in the absence of pressure gradient. Then, we examine effects of pressure gradient for the models. Finally, as promised in Section 4.4, we justify the values chosen for $\sigma$ and $\sigma^*$ in the $k$-$\omega$ model.

To study the defect layer, we continue to confine our analysis to incompressible flow so that we begin with Equations (4.114) and (4.115). The perturbation expansion for the defect layer proceeds in terms of the ratio of friction velocity to the boundary-layer-edge velocity, $u_\tau/U_e$, and the dimensionless coordinates, $\xi$ and $\eta$, defined by

$$\xi = x/L \quad \text{and} \quad \eta = y/\Delta(x), \quad \Delta = U_e \delta^*/u_\tau \tag{4.135}$$

where $\delta^*$ is displacement thickness and $L$ is a characteristic streamwise length scale that is presumed to be very large compared to $\delta^*$. As in our approach to the log layer, we first establish the general form of the solution for the mean momentum equation. We expand the streamfunction and kinematic eddy viscosity as follows.

$$\psi_{outer}(x,y) \quad \sim \quad U_e \Delta \left[\eta - \frac{u_\tau}{U_e} F_1(\xi,\eta) + o\left(\frac{u_\tau}{U_e}\right)\right] \tag{4.136}$$

$$\nu_{T_{outer}}(x,y) \quad \sim \quad U_e \delta^* \left[N_0(\xi,\eta) + o(1)\right] \tag{4.137}$$

Observe that, as is so often the case in perturbation analysis, we needn't continue the expansions beyond the first one or two terms to capture most of the important features of the solution. For the specified streamfunction, the velocity becomes:

$$U(x,y) \sim U_e \left[1 - \frac{u_\tau}{U_e} U_1(\xi,\eta) + o\left(\frac{u_\tau}{U_e}\right)\right], \quad U_1 = \frac{\partial F_1}{\partial \eta} \tag{4.138}$$

Substituting Equations (4.135) – (4.138) into the mean conservation equations [Equations (4.114) and (4.115)] yields the transformed momentum equation, viz.,

$$2\sigma_T \xi \frac{\partial U_1}{\partial \xi} = (\alpha_T - 2\beta_T - 2\omega_T)\eta \frac{\partial U_1}{\partial \eta} + (\beta_T - 2\omega_T)U_1 + \frac{\partial}{\partial \eta}\left[N_0 \frac{\partial U_1}{\partial \eta}\right] \tag{4.139}$$

where the parameters $\alpha_T$, $\beta_T$, $\sigma_T$ and $\omega_T$ are defined in terms of $\delta^*$, $u_\tau$ and skin friction, $c_f = 2(u_\tau/U_e)^2$, i.e.,

$$\alpha_T \equiv \frac{2}{c_f}\frac{d\delta^*}{dx}, \quad \beta_T \equiv \frac{\delta^*}{\tau_w}\frac{dP}{dx}, \quad \sigma_T \equiv \frac{\delta^*}{c_f x}, \quad \omega_T \equiv \frac{\delta^*}{c_f u_\tau}\frac{du_\tau}{dx} \tag{4.140}$$

Equation (4.139) must be solved subject to two boundary conditions. First, to satisfy the requirement that $U \to U_e$ as $y \to \infty$, necessarily

$$U_1 \to 0 \quad \text{as} \quad \eta \to \infty \tag{4.141}$$

Also, we must asymptote to the log-layer solution as $\eta \to 0$. One way to insure this is to insist that

$$\frac{\partial U_1}{\partial \eta} \to -\frac{1}{\kappa \eta} \quad \text{as} \quad \eta \to 0 \tag{4.142}$$

At this point, we have not greatly simplified our problem. Equation (4.139), like the original momentum equation, is a partial differential equation. The only simplification thus far is that molecular viscosity is negligible relative to the eddy viscosity. However, even this is not necessarily advantageous since the no-slip velocity boundary condition has been replaced by singular behavior approaching the surface. And, of course, we are now working in a transformed coordinate system $(\xi, \eta)$ rather than the familiar Cartesian coordinate system $(x, y)$. So why go to all this trouble? The answer is, we have only just begun.

Reexamination of the steps we have taken thus far should reveal a familiar tack; specifically, we appear to be developing a similarity solution. Indeed this is intentional, and inspection of Clauser's defect law [Equation (4.113)] shows that there has been method in our madness. Comparison of Equation (4.113) with the assumed form of our perturbation expansion for $U$ given in Equation (4.138) shows that $U_1$ must be a function only of $\eta$. Thus, we now pose the question as to what conditions must be satisfied in order for a similarity solution to exist.

Clearly, the coefficients $\alpha_T$, $\beta_T$ and $\omega_T$ must be independent of $x$, for then the coefficients of all terms on the right-hand side of Equation (4.139) will be independent of $x$. The coefficient $\sigma_T$ is of no consequence since, if $U_1$ is independent of $x$, the left-hand side of Equation (4.139) vanishes regardless of the value of $\sigma_T$.

The coefficients $\alpha_T$ and $\omega_T$ are simple algebraic functions of $\beta_T$. To show this, we begin by performing the formal matching of the defect-layer and sublayer solutions. As shown in the preceding section,

$$U_{inner}(\xi, y^+) \sim u_\tau \left[ \frac{1}{\kappa} \ell n y^+ + C \right] \quad \text{as} \quad y^+ \to \infty \qquad (4.143)$$

Assuming that a similarity solution exists so that $U_1$ depends only upon $\eta$, straightforward substitution into Equation (4.139) with a vanishing left-hand side shows that

$$U_1 \sim \frac{1}{\kappa} [-\ell n \eta + u_0 - u_1 \eta \ell n \eta + \cdots] \quad \text{as} \quad \eta \to 0 \qquad (4.144)$$

where the constants $u_0, u_1, \ldots$ depend upon the complete solution which, in turn, depends upon what turbulence model is used. We now do a formal matching of the inner and outer expansions noting that $y^+ = \eta Re_{\delta^*}$ and the outer solution is $U_{outer}(\xi, \eta) \sim [U_e - u_\tau U_1(\eta) + \cdots]$. To match through first order, we require the following:

$$\left[ \frac{1}{\kappa} \ell n y^+ + C \right] - \left[ \frac{U_e}{u_\tau} + \frac{1}{\kappa} \ell n \eta - \frac{u_0}{\kappa} \right] \to 0 \quad \text{as} \quad y^+ \to \infty, \quad \eta \to 0 \quad (4.145)$$

Hence, we conclude from matching that:

$$\frac{U_e}{u_\tau} = \left( C + \frac{u_0}{\kappa} \right) + \frac{1}{\kappa} \ell n Re_{\delta^*} \qquad (4.146)$$

This is a useful result that enables us to compute the skin friction from our defect-layer solution, a point we will return to later. For our present purpose, Equation (4.146) enables us to determine $\omega_T$. That is, since

$$u_\tau = \frac{U_e}{\left(C + \frac{u_0}{\kappa}\right) + \frac{1}{\kappa}\ell n Re_{\delta*}} \tag{4.147}$$

Differentiating with respect to $x$ yields

$$
\begin{aligned}
\frac{du_\tau}{dx} &= \frac{dU_e/dx}{\left(C + \frac{u_0}{\kappa}\right) + \frac{1}{\kappa}\ell n Re_{\delta*}} - \frac{U_e dRe_{\delta*}/dx}{\kappa Re_{\delta*}\left[\left(C + \frac{u_0}{\kappa}\right) + \frac{1}{\kappa}\ell n Re_{\delta*}\right]^2} \\
&= \frac{u_\tau}{U_e}\frac{dU_e}{dx} - \frac{u_\tau^2}{\kappa U_e Re_{\delta*}}\frac{dRe_{\delta*}}{dx} \tag{4.148}
\end{aligned}
$$

Substituting Equation (4.148) into the definition of $\omega_T$ [see Equation (4.140)] and using the fact that $u_\tau^2 = \frac{1}{2}U_e^2 c_f$, we find

$$
\begin{aligned}
\omega_T &= \frac{\delta^*}{c_f U_e}\frac{dU_e}{dx} - \frac{\delta^*}{c_f u_\tau}\frac{\frac{1}{2}U_e^2 c_f}{\kappa U_e^2 \delta^*/\nu}\frac{d}{dx}\left(\frac{U_e \delta^*}{\nu}\right) \\
&= \frac{\delta^*}{c_f U_e}\frac{dU_e}{dx} - \frac{1}{2\kappa u_\tau}\frac{d}{dx}(U_e \delta^*) \\
&= \frac{\delta^*}{c_f U_e}\left[1 - \frac{1}{\kappa}\frac{c_f}{2}\frac{U_e}{u_\tau}\right]\frac{dU_e}{dx} - \frac{1}{2\kappa}\frac{U_e}{u_\tau}\frac{d\delta^*}{dx} \\
&= \frac{\delta^*}{c_f U_e}\left[1 - \frac{1}{\kappa}\frac{u_\tau}{U_e}\right]\frac{dU_e}{dx} - \frac{1}{2\kappa}\frac{U_e}{u_\tau}\frac{d\delta^*}{dx} \tag{4.149}
\end{aligned}
$$

We can compute $d\delta^*/dx$ and $dU_e/dx$ from the definitions of $\alpha_T$ and $\beta_T$ given in Equation (4.140), i.e.,

$$\frac{d\delta^*}{dx} = \frac{c_f \alpha_T}{2} \quad \text{and} \quad \frac{dU_e}{dx} = -\frac{1}{\rho U_e}\frac{dP}{dx} = -\frac{\tau_w}{\rho U_e \delta^*}\beta_T \tag{4.150}$$

Combining Equations (4.149) and (4.150), we have

$$
\begin{aligned}
\omega_T &= \frac{\delta^*}{c_f U_e}\left[1 - \frac{1}{\kappa}\frac{u_\tau}{U_e}\right]\left(-\frac{\tau_w}{\rho U_e \delta^*}\right)\beta_T - \frac{1}{2\kappa}\frac{U_e}{u_\tau}\left(\frac{c_f \alpha_T}{2}\right) \\
&= -\frac{\tau_w}{\underbrace{\rho U_e^2 c_f}_{=2\tau_w}}\left[1 - \frac{1}{\kappa}\frac{u_\tau}{U_e}\right]\beta_T - \frac{1}{4\kappa}\,c_f\underbrace{\frac{U_e}{u_\tau}}_{=2u_\tau/U_e}\alpha_T \\
&= -\frac{1}{2}\beta_T\left[1 - \frac{1}{\kappa}\frac{u_\tau}{U_e}\right] - \frac{1}{2\kappa}\frac{u_\tau}{U_e}\alpha_T \tag{4.151}
\end{aligned}
$$

Therefore, regrouping terms, we conclude that

$$\omega_T = -\frac{1}{2}\beta_T + \frac{1}{2\kappa}(\beta_T - \alpha_T)\frac{u_\tau}{U_e} \tag{4.152}$$

Finally, since we seek a solution valid in the limit $u_\tau/U_e \to 0$, we have

$$\omega_T = -\frac{1}{2}\beta_T + O\left(\frac{u_\tau}{U_e}\right) \tag{4.153}$$

Note that Bush and Fendell (1972) incorrectly argue that $\omega_T = o(1)$ in the limit $u_\tau/U_e \to 0$, an error repeated in the first edition of the present book. Using arguments similar to those above, Tennekes and Lumley (1983) and Henkes (1998a) also show that $\omega_T$ is given by Equation (4.153).

This leaves us with the reduced requirement for existence of a similarity solution that only $\alpha_T$ and $\beta_T$ are independent of $x$. However, we can also show that $\alpha_T$ and $\beta_T$ are uniquely related to leading order. To see this, we examine the classical momentum-integral equation that follows from integrating the mean-momentum equation across the boundary layer [c.f., Schlichting (1979)], viz.,

$$\frac{c_f}{2} = \frac{d\theta}{dx} - (2+H)\frac{\theta}{\rho U_e^2}\frac{dP}{dx} \tag{4.154}$$

where $\theta$ is momentum thickness and $H = \delta^*/\theta$ is the shape factor. In terms of $\alpha_T$ and $\beta_T$, the momentum-integral equation can be rewritten as

$$\alpha_T\frac{d\theta}{dx} = \left[1 + \frac{(2+H)}{H}\beta_T\right]\frac{d\delta^*}{dx} \tag{4.155}$$

If we evaluate the displacement and momentum thickness using our perturbation solution we find two important facts. First, evaluating the displacement thickness integral yields an integral constraint on our solution for $U_1$, $U_2$, etc. Second, we find to leading order that $\delta^*$ and $\theta$ are equal, i.e., the shape factor approaches 1 as $Re_{\delta^*} \to \infty$ and/or $u_\tau/U_e \to 0$. The proof of these facts is straightforward and thus left for the problems section; the results are:

$$\int_0^\infty U_1(\eta)\,d\eta = 1, \qquad \int_0^\infty U_n(\eta)\,d\eta = 0, \quad n \geq 2 \tag{4.156}$$

$$H \sim 1 + O\left(\frac{u_\tau}{U_e}\right) \qquad \text{as} \qquad Re_{\delta^*} \to \infty, \quad \frac{u_\tau}{U_e} \to 0 \tag{4.157}$$

The perturbation solution for $U_1(\eta)$ provides sufficient information to determine the $O(u_\tau/U_e)$ term[5] for $H$ (see problems section). Hence, Equation (4.155) yields the following relationship between $\alpha_T$ and $\beta_T$.

$$\alpha_T = 1 + 3\beta_T \tag{4.158}$$

---

[5]The coefficient of this term is generally large, and realistic shape factors (e.g., $H \approx 1.3$ for a flat-plate) follow from the perturbation solution.

Thus, we see that the requirement for existence of a similarity solution to Equation (4.139) for large Reynolds number is simply that the **equilibrium parameter**, $\beta_T$, be constant. This is a very satisfactory state of affairs because it is consistent with experimental observations at finite (laboratory-scale) Reynolds numbers. That is, Clauser found that, outside the viscous sublayer, turbulent boundary layers assume a self-similar form when the equilibrium parameter is constant.

Appealing to Equations (4.153) and (4.158), the coefficients appearing in Equation (4.139) are

$$\alpha_T - 2\beta_T - 2\omega_T = 1 + 2\beta_T \quad \text{and} \quad \beta_T - 2\omega_T = 2\beta_T \qquad (4.159)$$

Then, the problem we must solve to determine $U_1(\eta)$ is:

$$\frac{d}{d\eta}\left[N_0 \frac{dU_1}{d\eta}\right] + (1 + 2\beta_T)\eta \frac{dU_1}{d\eta} + 2\beta_T U_1 = 0 \qquad (4.160)$$

$$\frac{dU_1}{d\eta} \to -\frac{1}{\kappa\eta} \quad \text{as} \quad \eta \to 0 \quad \text{and} \quad U_1(\eta) \to 0 \quad \text{as} \quad \eta \to \infty \qquad (4.161)$$

The integral constraint, Equation (4.156), must also be enforced. The dimensionless eddy viscosity, $N_0(\eta)$, depends upon the turbulence model selected. For our purposes, we will consider three different turbulence models, viz.: the $k$-$\omega$ model [Equations (4.36) – (4.42)]; the Standard $k$-$\epsilon$ model [Equations (4.46) – (4.49)]; and the RNG $k$-$\epsilon$ model [Equations (4.46) – (4.48) and (4.51) – (4.53)].

Making standard boundary-layer approximations for the model equations, we seek a perturbation solution for $k$, $\omega$ and $\epsilon$ of the following form.

$$\left.\begin{aligned}
k &\sim \frac{u_\tau^2}{\sqrt{\beta_o^*}} \left[K_0(\eta) + o(1)\right] \\[2ex]
\omega &\sim \frac{u_\tau}{\sqrt{\beta_o^*}\Delta} \left[W_0(\eta) + o(1)\right] \\[2ex]
\epsilon &\sim \frac{u_\tau^3}{\Delta} \left[E_0(\eta) + o(1)\right]
\end{aligned}\right\} \qquad (4.162)$$

Note that for the $k$-$\epsilon$ models, we make the identification $\beta_o^* = C_\mu$. For all three turbulence models, the transformed equation for $k$ can be written as

**All Models:**

$$\sigma^* \frac{d}{d\eta}\left[N_0 \frac{dK_0}{d\eta}\right] + (1 + 2\beta_T)\eta \frac{dK_0}{d\eta} + 2\beta_T K_0$$
$$+ \sqrt{\beta_o^*}\left[N_0 \left(\frac{dU_1}{d\eta}\right)^2 - E_0\right] = 0 \qquad (4.163)$$

where, for the $k$-$\epsilon$ models, we note that $\sigma^* = 1/\sigma_k$. The second equation and auxiliary equations are specific to each model. The transformed equations are:

**$k$-$\omega$ Model:**

$$\left.\begin{array}{l} \sigma\dfrac{d}{d\eta}\left[N_0\dfrac{dW_0}{d\eta}\right] + (1+2\beta_T)\eta\dfrac{dW_0}{d\eta} + (1+4\beta_T)W_0 \\[3mm] +\sqrt{\beta_o^*}\left[\alpha\left(\dfrac{dU_1}{d\eta}\right)^2 - \dfrac{\beta_o}{\beta_o^*}W_0^2\right] = 0 \\[3mm] N_0 = K_0/W_0 \quad \text{and} \quad E_0 = f_{\beta^*}K_0W_0 \end{array}\right\} \quad (4.164)$$

**$k$-$\epsilon$ Model:**

$$\left.\begin{array}{l} \sigma_\epsilon^{-1}\dfrac{d}{d\eta}\left[N_0\dfrac{dE_0}{d\eta}\right] + (1+2\beta_T)\eta\dfrac{dE_0}{d\eta} + (1+6\beta_T)E_0 \\[3mm] +\sqrt{C_\mu}\left[C_{\epsilon 1}K_0\left(\dfrac{dU_1}{d\eta}\right)^2 - C_{\epsilon 2}\dfrac{E_0^2}{K_0}\right] = 0 \\[3mm] N_0 = K_0^2/E_0 \quad \text{and} \quad E_0 = K_0W_0 \end{array}\right\} \quad (4.165)$$

We must specify boundary conditions on the dimensionless functions $K_0$, $W_0$ and $E_0$ both in the freestream and approaching the surface. For nonturbulent flow in the freestream, we require that the turbulence parameters all vanish as $\eta \to \infty$. However, we also stipulate that these quantities approach zero in such a way that $N_0$ vanishes. Thus, the freestream boundary conditions are:

$$K_0(\eta) \to 0, \quad W_0(\eta) \to 0, \quad E_0(\eta) \to 0, \quad U_1(\eta) \to 0 \quad \text{as} \quad \eta \to \infty \quad (4.166)$$

Approaching the surface, we must formally match to the law of the wall. Matching is a bit different for each model but is nevertheless straightforward; we omit details of the algebra in the interest of brevity. The limiting forms used for $\eta \to 0$ follow.

$$\left.\begin{array}{l} K_0(\eta) \sim [1 + k_1\eta \ell n\eta + \cdots] \\[3mm] E_0(\eta) \sim \dfrac{1}{\kappa\eta}[1 + e_1\eta \ell n\eta + \cdots] \\[3mm] W_0(\eta) \sim \dfrac{1}{\kappa\eta}[1 + w_1\eta \ell n\eta + \cdots] \\[3mm] U_1(\eta) \sim \dfrac{1}{\kappa}[-\ell n\eta + u_0 - u_1\eta \ell n\eta + \cdots] \end{array}\right\} \quad (4.167)$$

The coefficients $k_1$, $u_1$, $w_1$ and $e_1$ are as follows, where for notational consistency, we define

$$\alpha^* \equiv \sqrt{\beta_o^*} = \sqrt{C_\mu} \qquad (4.168)$$

Also, we again write some of the results in terms of $\sigma^*$ with the understanding that $\sigma^* = 1/\sigma_k$ for the $k$-$\epsilon$ models.

**All Models:**

$$k_1 = \frac{2\beta_T/\kappa}{\sigma^*\kappa^2/(2\alpha^*) - 1} \qquad (4.169)$$

**$k$-$\omega$ Model:**

$$\left.\begin{array}{l} u_1 = \dfrac{\left[\beta_o/\left(\alpha\beta_o^*\right)\right]\left[\sigma^*\kappa^2/\left(2\alpha^*\right)\right]}{1 - \beta_o/\left(\alpha\beta_o^*\right)} k_1 \\[4mm] w_1 = \dfrac{\sigma^*\kappa^2/\left(2\alpha^*\right)}{1 - \beta_o/\left(\alpha\beta_o^*\right)} k_1 \end{array}\right\} \qquad (4.170)$$

**Standard $k$-$\epsilon$ Model:**

$$\left.\begin{array}{l} u_1 = \dfrac{\left(1 + \sigma^*\kappa^2/\alpha^*\right)C_{\epsilon 2} - C_{\epsilon 1}}{2\left(C_{\epsilon 1} - C_{\epsilon 2}\right)} k_1 \\[4mm] e_1 = \dfrac{\left(1 + \sigma^*\kappa^2/\alpha^*\right)C_{\epsilon 1} - C_{\epsilon 2}}{2\left(C_{\epsilon 1} - C_{\epsilon 2}\right)} k_1 \end{array}\right\} \qquad (4.171)$$

**RNG $k$-$\epsilon$ Model:**

$$\left.\begin{array}{l} u_1 = \left\{(\sigma_\epsilon/2)\left[1/\sigma_\epsilon - \sigma^*\left(C_{\epsilon 1} - C'_{\epsilon 2}\right)\right] - \left[1 + \sigma^*\kappa^2/\left(2\alpha^*\right)\right]\right\} k_1 \\[3mm] e_1 = (\sigma_\epsilon/2)\left[1/\sigma_\epsilon - \sigma^*\left(C_{\epsilon 1} - C'_{\epsilon 2}\right)\right] k_1 \\[3mm] C'_{\epsilon 2} = \dfrac{1 - 1/\left(\lambda_o\alpha^*\right)}{2\alpha^*\left[1 + \beta/(\alpha^*)^3\right]}\left[3 - \dfrac{1}{\lambda_o\alpha^* - 1} - \dfrac{3\beta/(\alpha^*)^3}{1 + 3\beta/(\alpha^*)^3}\right] \end{array}\right\} \qquad (4.172)$$

Additionally, the coefficient $u_0$ is determined from the integral constraint for mass conservation, which is guaranteed by the integral constraint in the first of Equations (4.156). Table 4.6 summarizes the equations for the leading-order terms in the defect-layer solution.

Before proceeding to discussion of the defect-layer similarity solution, there are two quantities of interest that follow from the leading order solution, viz., the skin friction, $c_f$, and Coles' **wake-strength parameter**, $\Pi$. Recall that from matching defect-layer and sublayer velocity profiles, we deduced Equation (4.146). Noting that $c_f = 2(u_\tau/U_e)^2$, we conclude that

$$\sqrt{\frac{2}{c_f}} = \left(C + \frac{u_0}{\kappa}\right) + \frac{1}{\kappa}\ell n Re_{\delta^*} \qquad (4.173)$$

Table 4.6: *Summary of the Defect-Layer Equations*

| | |
|---|---|
| Mass (Integral Constraint) | Equation (4.156) |
| Momentum | Equation (4.160) |
| Velocity Boundary Conditions | Equation (4.161) |
| Turbulence Kinetic Energy | Equation (4.163) |
| Specific Dissipation ($k$-$\omega$ Model) | Equation (4.164) |
| Dissipation ($k$-$\epsilon$ Model) | Equation (4.165) |
| $k$, $\omega$, $\epsilon$ Boundary Conditions for $\eta \to \infty$ | Equation (4.166) |
| $k$, $\omega$, $\epsilon$ Boundary Conditions for $\eta \to 0$ | Equation (4.167) |

The **composite law of the wall, law of the wake** profile according to Coles' meticulous correlation of experimental data [see Coles and Hirst (1969)] is given by

$$\frac{U}{u_\tau} = \frac{1}{\kappa}\ell n\left(\frac{u_\tau y}{\nu}\right) + C + \frac{2\Pi}{\kappa}\sin^2\left(\frac{\pi}{2}\frac{y}{\delta}\right) \qquad (4.174)$$

The $\sin^2$ function is purely a curve fit: several other functions have been suggested, including forms that yield $\partial U/\partial y = 0$ at $y = \delta$ [which is not the case for Equation (4.174)]. At the boundary-layer edge, $y = \delta$, we have

$$\frac{U_e}{u_\tau} = \frac{1}{\kappa}\ell n\left(\frac{u_\tau \delta}{\nu}\right) + C + \frac{2\Pi}{\kappa} \qquad (4.175)$$

Combining Equations (4.146) and (4.175) and canceling the constant $C$ yields

$$\frac{1}{\kappa}\ell n\left(\frac{u_\tau \delta}{\nu}\right) + \frac{2\Pi}{\kappa} = \frac{u_0}{\kappa} + \frac{1}{\kappa}\ell n Re_{\delta^*} \qquad (4.176)$$

Hence, solving for $\Pi$, we find

$$\Pi = \frac{1}{2}u_0 + \frac{1}{2}\ell n\left(\frac{U_e \delta^*}{\nu}\right)\left(\frac{\nu}{u_\tau \delta}\right) = \frac{1}{2}u_0 - \frac{1}{2}\ell n\left(\frac{u_\tau \delta}{U_e \delta^*}\right) \qquad (4.177)$$

Finally, defect-layer solutions include sharp (nonphysical) turbulent/nonturbulent interfaces so that the edge of the defect-layer lies at a finite value $\eta = \eta_e$, i.e.,

$$\eta_e = \frac{\delta}{\Delta} = \frac{u_\tau \delta}{U_e \delta^*} \qquad (4.178)$$

Therefore, combining Equations (4.177) and (4.178) leads to the following expression for the wake-strength parameter.

$$\Pi = \frac{1}{2}(u_0 - \ell n \eta_e) \qquad (4.179)$$

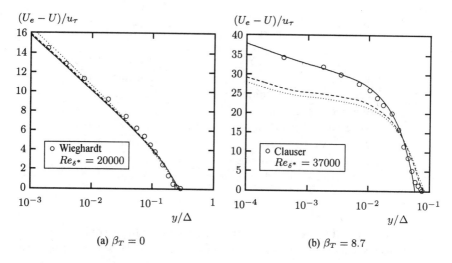

(a) $\beta_T = 0$                              (b) $\beta_T = 8.7$

Figure 4.14: *Comparison of computed and measured defect-layer velocity profiles; ——— k-ω model; - - - Standard k-ε model; ···· RNG k-ε model.*

Figure 4.14(a) compares the defect-layer solution for the three models with corresponding experimental data of Wieghardt as tabulated by Coles and Hirst (1969). The experimental data presented are those at the highest Reynolds number for which data are reported. This is consistent with the defect-layer solution that is formally valid for very large Reynolds number. As shown, all three models predict velocity profiles that differ from measured values by no more than about three percent of scale. Thus, based on analysis of the constant-pressure defect layer, there is little difference amongst the three models.

Turning now to the effect of pressure gradient, we consider defect-layer solutions for the equilibrium parameter, $\beta_T$, ranging from -0.5 to +9.0, where positive $\beta_T$ corresponds to an adverse pressure gradient. The choice of this range of $\beta_T$ has been dictated by the requirement of the perturbation solution that $\beta_T$ be constant. This appears to be the maximum range over which experimental data have been taken with $\beta_T$ more-or-less constant.

Figure 4.14(b) compares computed velocity profiles with experimental data of Clauser [see Coles and Hirst (1969)] for $\beta_T = 8.7$. As shown, the $k$-$\omega$ model yields a velocity profile that is within 3% of measurements while the $k$-$\epsilon$ models show much larger differences.

Figure 4.15 compares computed wake strength, $\Pi$, with values inferred by Coles and Hirst (1969) from experimental data. In addition to results for the two-equation models, the figure includes predicted $\Pi$ according to the Baldwin-Barth (1990) one-equation model. Inspection of Figure 4.15 reveals provocative differences amongst the four models. Most notably, the $k$-$\omega$ model yields wake

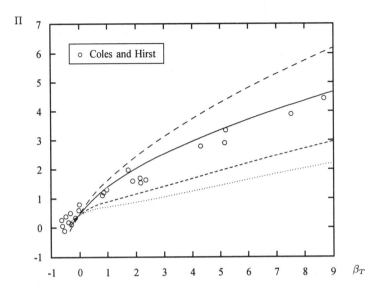

Figure 4.15: *Computed and measured wake-strength parameter; —— k-ω model; - - - Standard k-ε model; ···· RNG k-ε model; – – Baldwin-Barth model.*

strengths closest to values inferred from data over the complete range considered. Consistent with the velocity-profile discrepancies shown in Figure 4.14(b), the $k$-$\epsilon$ models exhibit much larger differences, with predicted wake strength 30%-40% lower than inferred values when $\beta_T$ is as small as two! Also, the Baldwin-Barth model predicts values of $\Pi$ that are typically 30% higher than measured.

To appreciate the significance of these results, observe that since we know $U_e/u_\tau = \sqrt{2/c_f}$, we can rewrite Equation (4.175) as follows.

$$\sqrt{\frac{2}{c_f}} = \frac{1}{\kappa}\ell n\left(\frac{u_\tau\delta}{\nu}\right) + C + \frac{2\Pi}{\kappa} \qquad (4.180)$$

Although this is not an explicit equation for $c_f$ as a function of $\Pi$, if we assume the logarithmic term varies more slowly than the term proportional to $\Pi$, reducing the value of $\Pi$ increases the value of $c_f$, and vice versa. This indeed turns out to be the case.

Consequently, if a model's predicted values of $\Pi$ are smaller than measured, its predicted skin friction is larger than observed. As we will see in Sections 4.8 through 4.10, the $k$-$\epsilon$ model consistently predicts values of skin friction that are significantly larger than measured. Similarly, if the values of $\Pi$ are larger than measured, predicted skin friction is smaller than observed. Inspection of Figure 4.2 confirms that the Baldwin-Barth model predicts skin friction values that are substantially below corresponding measured values. Finally, if a turbulence model predicts values of $\Pi$ similar to measured values over the entire range of $\beta_T$, its skin-friction (and other boundary-layer property) predictions will be consistent with measurements. We will see in subsequent sections and chapters that the $k$-$\omega$ model accurately predicts boundary-layer properties, including effects of pressure gradient. Although we have not shown the results here, the Baldwin-Lomax, Cebeci-Smith, Johnson-King and Spalart-Allmaras models all predict $\Pi$ versus $\beta_T$ curves that are much closer to the $k$-$\omega$ curve than the Baldwin-Barth and $k$-$\epsilon$ models. Correspondingly, they all predict boundary-layer features within a few percent of measurements (cf. Figures 3.16, 3.18 and 4.2).

Thus, we see that using perturbation methods to analyze the defect layer provides an excellent test of how well any turbulence model will ultimately perform for boundary layers. Although the analysis is confined to equilibrium boundary layers, in the sense that $\beta_T$ is constant (and is strictly valid only in the limit of very large Reynolds number), it is nevertheless an objective and important test. This is true because, if the boundary layer is not changing too rapidly, its properties will be consistent with those of the equilibrium case corresponding to the local value of $\beta_T$.

The explanation of the $k$-$\epsilon$ models' poor performance for adverse pressure gradient can be developed from inspection of the asymptotic behavior of solutions as $\eta \to 0$. For the models analyzed, the velocity behaves as

$$\frac{U_e - U}{u_\tau} \sim -\frac{1}{\kappa}\ell n\eta + A - \beta_T B\eta\ell n\eta + \cdots \qquad \text{as} \qquad \eta \to 0 \qquad (4.181)$$

where Table 4.7 summarizes the constants $A$ and $B$, defined by

$$A \equiv \frac{u_0}{\kappa} \qquad \text{and} \qquad B \equiv \frac{u_1}{\beta_T \kappa} \qquad (4.182)$$

Note that, while the coefficient $A$ is determined as part of the solution (from the integral constraint that mass be conserved), the coefficient $B$ follows directly from the limiting form of the solution as $\eta \to 0$. As seen from Table 4.7, $B$ is largest for the RNG $k$-$\epsilon$ model and is smallest for the $k$-$\omega$ model. The presence of the $\eta\ell n\eta$ term gives rise to an inflection in the velocity profile as $\eta \to 0$ that is most pronounced for the $k$-$\epsilon$ models. In terms of turbulence properties, the turbulence length scale, $\ell$, behaves according to

$$\ell \sim (\beta_o^*)^{1/4}\kappa\eta[1 + \beta_T L\eta\ell n\eta + \cdots] \qquad \text{as} \qquad \eta \to 0 \qquad (4.183)$$

Table 4.7 also includes the coefficient $L$ for each model. Again, we see that the contribution of the $\eta\ell n\eta$ term is much larger for the $k$-$\epsilon$ models than it is for the $k$-$\omega$ model. Thus, for adverse pressure gradient, the $k$-$\epsilon$ models' turbulence length scales tend to be too large in the near-wall region. Note, of course, that this shortcoming is not evident in the constant-pressure case, which has $\beta_T = 0$.

Table 4.7: *Coefficients A, B and L for $\beta_T = 8.7$*

| Model | $A$ | $B$ | $L$ |
|---|---|---|---|
| $k$-$\omega$ | 15.42 | 5.53 | -4.31 |
| Standard $k$-$\epsilon$ | 7.53 | 30.51 | -13.02 |
| RNG $k$-$\epsilon$ | 4.64 | 36.07 | -15.39 |

The manner in which the $k$-$\omega$ model achieves smaller values of $\ell$ than the $k$-$\epsilon$ models can be seen by changing dependent variables. That is, starting with the $k$-$\omega$ formulation and defining $\epsilon = \beta_o^* \omega k$, we can deduce the following incompressible equation for $\epsilon$ implied by the $k$-$\omega$ model (with constant $\beta^*$).

$$U\frac{\partial \epsilon}{\partial x} + V\frac{\partial \epsilon}{\partial y} = (1+\alpha)k\left(\frac{\partial U}{\partial y}\right)^2 - (1+\beta_o/\beta_o^*)\frac{\epsilon^2}{k} + \frac{\partial}{\partial y}\left[\sigma \nu_T \frac{\partial \epsilon}{\partial y}\right]$$
$$- 2\sigma \nu_T \frac{\partial k}{\partial y}\frac{\partial(\epsilon/k)}{\partial y} \qquad (4.184)$$

All terms except the last on the right-hand side of Equation (4.184) are identical in form to those of the Standard $k$-$\epsilon$ model [see Equation (4.48)]. The **cross-diffusion term**, which is discussed in detail in Subsection 4.5.4, is negligibly small as $\eta \to 0$ for constant-pressure boundary layers because $k \to$ constant as $\eta \to 0$. However, $\partial k/\partial y$ is nonvanishing when $\beta_T \neq 0$ and $\partial(\epsilon/k)/\partial y$ generally is quite large as $\eta \to 0$. The net effect of this additional term is to suppress the rate of increase of $\ell$ close to the surface.

We can draw an important conclusion from these observations about cross diffusion. Specifically, with a change of dependent variables to $\omega$ from the second parameter being used (e.g., $\epsilon$, $\ell$, $\tau$), any two-equation model can be rewritten as a $k$-$\omega$ model. In general, the implied equation for $\omega$ includes a cross-diffusion term. Since cross diffusion has such an undesirable effect on boundary-layer predictions, additional corrections to the model will be needed to counter the undesirable effects of the term. Rodi and Scheuerer (1986), Yap (1987) and Henkes (1998b), for example, have proposed corrections to the $k$-$\epsilon$ model which implicitly counter the effects of cross diffusion (relative to the $k$-$\omega$ model) with varying degrees of success.

As with free shear flows, the freestream value of $\omega$ has an effect on $k$-$\omega$ model defect-layer solutions even when the freestream eddy viscosity is negligibly small. However, the sensitivity is far less significant than it is for free shear cases (see Subsection 4.5.3). By contrast, the freestream value of $\epsilon$ has virtually no effect for the $k$-$\epsilon$ model. We can use Program **DEFECT** (see Appendix C) to demonstrate the sensitivity. All computations have been done with $\beta_T = 0$ and

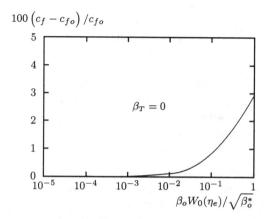

Figure 4.16: *Effect of the freestream value of $\omega$ on $k$-$\omega$ model skin friction.*

have a freestream eddy viscosity of $N(\eta_e) = 10^{-4}$. As it turns out, a self-similar solution exists for $W_0(\eta_e) = \sqrt{\beta_o^*/\beta_o} = 4.166$, which is a relatively large value. Regarding this as the upper bound on $W_0(\eta_e)$, computations have been performed to determine the sensitivity of skin friction, $c_f$, to the freestream value of $W_0$. Figure 4.16 shows the variation of $c_f$ with $W_0(\eta_e)$. The quantity $c_{fo}$ denotes the value of $c_f$ for zero freestream conditions. As shown, the effect is small. For the largest value of $W_0(\eta_e)$, the change in $c_f$ is less than 3%. Note that, in typical numerical computations, large values of $\omega$ diffuse from the wall toward the boundary-layer edge, so that $W_0(\eta_e)$ tends toward $\sqrt{\beta_o^*/\beta_o}$, which corresponds to $U\,d\omega/dx = -\beta_o\omega^2$ in the freestream.

Finally, unlike the three closure coefficients discussed in Section 4.4, simple arguments have not been found to establish the values of $\sigma$ and $\sigma^*$ for the $k$-$\omega$ model. In Subsection 4.6.3, we will find that using $\sigma = 1/2$ yields an excellent solution in the viscous sublayer, almost independent of the value of $\sigma^*$. Equation (4.170) shows that the coefficient $B = u_1/(\beta_T\kappa)$ is proportional to $\sigma^*$, so that smaller values of $\sigma^*$ should improve the model's predictions for boundary layers with variable pressure. The computed variation of $\Pi$ with $\beta_T$ (Figure 4.15) closely matches experimental results when $\sigma^* = 1/2$, and this is the value that has been chosen for the $k$-$\omega$ model.

### 4.6.3   The Viscous Sublayer

In order to facilitate integration of the model equations through the viscous sublayer, we must, at a minimum, have molecular diffusion terms in the equations of motion. Potentially, we might also have to allow the various closure coefficients to be functions of viscosity (i.e., turbulence Reynolds number) as well. This should come as no surprise since even the mixing-length model requires the Van Driest damping factor and one-equation models need similar viscous damping [Wolfshtein (1967), Baldwin and Barth (1990), Spalart and Allmaras (1992)]. In this section, we use perturbation methods to analyze viscous-sublayer structure predicted by several two-equation models. As we will see, with the exception of some $k$-$\omega$ models, virtually all two-equation models require Reynolds-number dependent corrections in order to yield a realistic sublayer solution.

We have already derived the sublayer solution in Subsection 4.6.1 when we discussed the **log layer**. Recapping the highlights of the expansion procedure, the velocity is given by an expansion of the form

$$U(x,y) \sim u_\tau[\hat{u}_0(y^+) + Re_{\delta^*}^{-1}\hat{u}_1(\xi, y^+) + o(Re_{\delta^*}^{-1})] \tag{4.185}$$

To leading order, the convective terms and pressure gradient are negligible. Thus, for example, the leading-order equations for the $k$-$\omega$ model expressed in terms of dimensional quantities are given by

$$\left.\begin{aligned} &(\nu + \nu_T)\frac{dU}{dy} = u_\tau^2 \\ &\frac{d}{dy}\left[(\nu + \sigma^*\nu_T)\frac{dk}{dy}\right] + \nu_T\left(\frac{dU}{dy}\right)^2 - \beta^*\omega k = 0 \\ &\frac{d}{dy}\left[(\nu + \sigma\nu_T)\frac{d\omega}{dy}\right] + \alpha\left(\frac{dU}{dy}\right)^2 - \beta_o\omega^2 = 0 \\ &\nu_T = \frac{k}{\omega} \end{aligned}\right\} \tag{4.186}$$

Because the Reynolds shear stress is constant, the viscous sublayer is often referred to as the **constant-stress layer**. Five boundary conditions are needed for this fifth-order system, two of which follow from matching to the law of the wall as $y^+ \to \infty$, viz.,

$$k \to \frac{u_\tau^2}{\sqrt{\beta_o^*}} \quad \text{and} \quad \omega \to \frac{u_\tau}{\sqrt{\beta_o^*}\,\kappa y} \quad \text{as} \quad y^+ \to \infty \tag{4.187}$$

where $y^+ \equiv u_\tau y/\nu$. Two more boundary conditions follow from **no slip** at the surface, which implies that $U$ and $k$ vanish at $y = 0$. Thus,

$$U = k = 0 \quad \text{at} \quad y^+ = 0 \tag{4.188}$$

The final condition follows from examination of the differential equations for $k$ and $\omega$ approaching the surface. The $k$-$\omega$ model possesses two kinds of solutions. The first type of solution has a finite value of $\omega$ at the surface. This fact was first observed by Saffman (1970) who speculated that the constant in the law of the wall, $C$, would vary with the surface value of $\omega$. This feature is unique to $k$-$\omega$ and $k$-$\omega^2$ models and will be explored in detail in Section 4.7. The second type of solution is common to all two-equation models and this is the one we will focus on now. Examination of the differential equations approaching $y = 0$ shows that for all two-equation models,

$$k \sim y^n \quad \text{and} \quad \beta_o^* y^2 \omega / \nu \sim \text{constant} \quad \text{as} \quad y \to 0 \quad (4.189)$$

Table 4.8 lists the values of $n$ and the constant for several models. As shown, none of the models predicts the exact theoretical value of 2 for both $n$ and $\beta_o^* y^2 \omega / \nu$. This can only be accomplished with additional modification of the model equations.

Table 4.8: *Sublayer Behavior Without Viscous Damping*

| Model | Type | $C$ | $n$ | $\beta_o^* y^2 \omega / \nu$ |
|---|---|---|---|---|
| Wilcox-Rubesin (1980) | $k$-$\omega^2$ | 7.1 | 4.00 | 12.00 |
| Saffman (1970) | $k$-$\omega^2$ | 6.0 | 3.7-4.0 | 12.00 |
| Launder-Spalding (1972) | $k$-$\omega^2$ | 5.7 | 3.79 | 12.00 |
| Wilcox (1998) | $k$-$\omega$ | 5.1 | 3.28 | 7.50 |
| Wilcox (1988a) | $k$-$\omega$ | 5.1 | 3.23 | 7.20 |
| Kolmogorov (1942) | $k$-$\omega$ | 3.1 | 3.62 | 7.20 |
| Launder-Sharma (1974) | $k$-$\epsilon$ | -2.2 | 1.39 | 0.53 |
| Speziale (1990) | $k$-$\tau$ | -2.2 | 1.39 | 0.53 |
| Exact/Measured | | 5.0 | 2.00 | 2.00 |

The exact values follow from expanding the fluctuating velocity in Taylor series near a solid boundary. That is, we know that the fluctuating velocity satisfies the no-slip boundary condition and also satisfies conservation of mass (see Section 2.3). Consequently, the three velocity components must behave as follows.

$$\left.\begin{array}{rcl} u' & \sim & A(x, z, t)y & + & O(y^2) \\ v' & \sim & B(x, z, t)y^2 & + & O(y^3) \\ w' & \sim & C(x, z, t)y & + & O(y^2) \end{array}\right\} \quad \text{as} \quad y \to 0 \quad (4.190)$$

Hence, the turbulence energy and dissipation are given by

$$k \sim \frac{1}{2}\left(\overline{A^2 + C^2}\right)y^2 + O(y^3) \quad \text{and} \quad \epsilon \sim \nu\left(\overline{A^2 + C^2}\right) + O(y) \quad (4.191)$$

Assuming that $\epsilon = \beta_o^* \omega k$, Equation (4.191) tells us that

$$k \sim y^2 \quad \text{and} \quad \beta_o^* y^2 \omega / \nu \sim 2 \quad \text{as} \quad y \to 0 \quad (4.192)$$

$U^+$

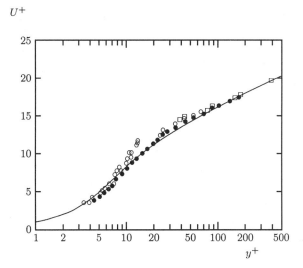

Figure 4.17: *Computed and measured sublayer velocity:* ○ *Laufer;* ● *Andersen et al.;* □ *Wieghardt;* —— *k-ω model.*

Hence, using the asymptotic behavior of $\omega$ for $y \rightarrow 0$ appropriate to each model as the fifth boundary condition, we can solve the sublayer equations (see Subsection 7.2.1 for an explanation of how to handle the singular behavior of $\omega$ numerically). One of the most interesting features of the solution is the constant in the law of the wall, $C$, that is evaluated from the following limit.

$$C = \lim_{y^+ \rightarrow \infty} \left[ U^+ - \frac{1}{\kappa} \ell n y^+ \right] \qquad (4.193)$$

In practice, integrating from $y^+ = 0$ to $y^+ = 500$ is sufficient for numerical solution of the sublayer equations. Program **SUBLAY** (see Appendix C) can be used to solve the sublayer equations for the $k$-$\omega$ model.

Table 4.8 also lists the computed value of $C$ for the various two-equation models. As shown, the Spalding (1972) $k$-$\omega^2$ model, the Wilcox (1988a) $k$-$\omega$ model and the $k$-$\omega$ model defined in Equations (4.36) – (4.42) are sufficiently close to the standard value of 5.0 to be used with no additional viscous modifications. The Standard $k$-$\epsilon$ model and the Speziale et al. $k$-$\tau$ model are farthest from the generally accepted value for $C$.

Figure 4.17 compares $k$-$\omega$ model velocity profiles with corresponding measurements of Laufer (1952), Andersen, Kays and Moffat (1972), and Wieghardt [as tabulated by Coles and Hirst (1969)]. As shown, computed velocities generally fall within experimental data scatter for all values of $y^+$ considered.

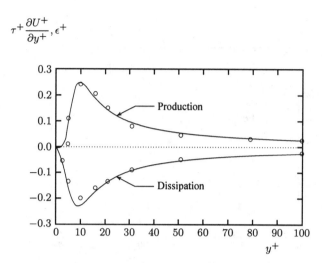

Figure 4.18: *Computed and measured production and dissipation:* ∘ *Laufer;* —— *k-ω model.*

Figure 4.18 compares computed production and dissipation with Laufer's (1952) near-wall pipe-flow measurements. Again, predictions are close to measurements. However, note that Laufer's dissipation data are incorrect for values of $y^+$ less than 10, a point we will discuss further in Subsection 4.8.1.

This concludes our perturbation analysis of the turbulent boundary layer. As we have seen, using perturbation analysis, we have been able to dissect model-predicted structure of the defect layer, log layer and sublayer, never having to solve more than an ordinary differential equation. This is a great advantage in testing a turbulence model in light of the ease and accuracy with which ordinary differential equations can be solved. The equations are not trivial to solve however since we are dealing with two-point boundary-value problems, and the resulting systems of equations are of sixth order for the defect layer and fifth order for the sublayer. However, this is far easier to handle than the partial differential equations we started with, and parametric studies (e.g., varying the equilibrium parameter, $\beta_T$) are much simpler. As a final comment, results obtained in this section should make the following statement obvious.

*Given the demonstrated power and utility of perturbation analysis in analyzing the turbulent boundary layer, this type of analysis can, and should, be used in developing all turbulence models. Failure to use these methods is the primary reason so many turbulence models have been devised that fail to accurately predict properties of incompressible, equilibrium boundary layers.*

## 4.7 Surface Boundary Conditions

In order to apply a two-equation turbulence model to wall-bounded flows, we must specify boundary conditions appropriate to a solid boundary for the velocity and the two turbulence parameters. As shown in the preceding section, many two-equation models fail to predict a satisfactory value of the constant $C$ in the law of the wall (see Table 4.8). Consequently, for these models, applying the no-slip boundary condition and integrating through the viscous sublayer yields unsatisfactory results. One approach we can take to remove this deficiency is to introduce viscous damping factors analogous to the Van Driest correction for the mixing-length model. Since introduction of damping factors accomplishes much more than improving predictions of the velocity profile in the sublayer, we defer detailed discussion of such modifications to Section 4.9. An alternative approach is to circumvent the inability to predict a satisfactory log-layer solution by simply matching to the law of the wall using a suitable value for $C$. This is what we did in analyzing the defect layer, and the procedure is equally valid for general wall-bounded flows.

### 4.7.1 Wall Functions

Historically, researchers implementing this matching procedure have referred to the functional forms used in the limit $y \to 0$ as **wall functions**. This procedure uses the law of the wall as the constitutive relation between velocity and surface shear stress. That is, in terms of the velocity at the mesh point closest to the surface (the "matching point"), we can regard the law of the wall, viz.,

$$U = u_\tau \left[ \frac{1}{\kappa} \ell n \left( \frac{u_\tau y}{\nu} \right) + C \right] \tag{4.194}$$

as a transcendental equation for the friction velocity and, hence, the shear stress. Once the friction velocity is known, we use Equations (4.128) for the $k$-$\omega$ model or Equations (4.133) for the $k$-$\epsilon$ model to define the values of $k$ and $\omega$ or $\epsilon$ at the grid points closest to the surface. Because $\omega$ and $\epsilon$ are odd functions of $u_\tau$ and both quantities are positive definite, care must be taken for separated flows. We can either use the absolute value of $u_\tau$ or combine the equations for $k$ and $\omega$ or $k$ and $\epsilon$ so that the **wall functions** for $k$, $\omega$ and $\epsilon$ become:

$$k = \frac{u_\tau^2}{\sqrt{\beta_o^*}}, \qquad \omega = \frac{k^{1/2}}{(\beta_o^*)^{1/4} \kappa y}, \qquad \epsilon = (\beta_o^*)^{3/4} \frac{k^{3/2}}{\kappa y} \tag{4.195}$$

The wall-function approach is not entirely satisfactory for several reasons. Most importantly, numerical solutions generally are sensitive to the point above the surface where the wall functions are used, i.e., the point where the matching occurs (see Subsection 7.2.1 for an in-depth discussion of this problem). Furthermore, the law of the wall doesn't always hold for flow near solid boundaries, most notably for separated flows.

There is a more subtle danger attending the use of wall functions. Specifically, when poor results are obtained with a two-equation model, researchers sometimes mistakenly blame their difficulties on the use of non-optimum wall functions. This assessment is too often made when the wall functions are not the real cause of the problem. For example, the $k$-$\epsilon$ model just doesn't perform well for boundary layers with adverse pressure gradient, even when accurately matched to the log law. Many articles have appeared claiming that discrepancies between the $k$-$\epsilon$ model's predicted skin friction and corresponding measurements for such flows are caused by the wall functions. This incorrectly assumes that the surface shear is a localized force that depends only upon sublayer structure. As shown in the defect-layer solution of the preceding section, no viscous modification is likely to remove the curious inflection [Figure 4.14(b)] in the $k$-$\epsilon$ model's velocity profile unless viscous effects (unrealistically) penetrate far above the viscous sublayer. We must not lose sight of the fact that the momentum flux through a boundary layer affects the surface shear stress and vice versa [see Equation (4.154)]. Hence, inaccurate skin friction predictions can be caused by inaccuracies in the velocity profile anywhere in the layer.

As a final comment on wall functions, Wilcox (1989) demonstrates that pressure gradient must be included in order to achieve solutions independent of the matching point. Retaining pressure gradient in the log-layer equations [i.e., retaining the term $\beta_T/Re_{\delta*}$ in Equation (4.120)], then the asymptotic behavior for the $k$-$\omega$ model [as defined in Equations (4.36) – (4.42)] approaching the surface is given by the following equations:

$$
\left.
\begin{aligned}
U &= u_\tau \left[ \frac{1}{\kappa} \ell n \left( \frac{u_\tau y}{\nu} \right) + C - 1.13 \frac{u_\tau y}{\nu} P^+ + O(P^+)^2 \right] \\[2mm]
k &= \frac{u_\tau^2}{\sqrt{\beta_o^*}} \left[ 1 + 1.16 \frac{u_\tau y}{\nu} P^+ + O(P^+)^2 \right] \\[2mm]
\omega &= \frac{u_\tau}{\sqrt{\beta_o^*}\,\kappa y} \left[ 1 - 0.30 \frac{u_\tau y}{\nu} P^+ + O(P^+)^2 \right]
\end{aligned}
\right\} \quad (4.196)
$$

where $P^+$ is the dimensionless pressure-gradient parameter defined by

$$
P^+ = \frac{\nu}{\rho u_\tau^3} \frac{dP}{dx} \tag{4.197}
$$

The expansions in Equation (4.196) have been derived assuming that $P^+$ is a small parameter.

## 4.7.2  Surface Roughness

As noted in the preceding section, a key advantage of the $k$-$\omega^2$ and $k$-$\omega$ formulations over the $k$-$\epsilon$ formulation is the fact that $\omega$-oriented equations possess
solutions in which the value of $\omega$ may be arbitrarily specified at the surface.
This is an advantage because it provides a natural way to incorporate effects
of surface roughness through surface boundary conditions. This feature of the
equations was originally recognized by Saffman (1970). If we write the surface
boundary condition on $\omega$ as

$$\omega = \frac{u_\tau^2}{\nu} S_R \quad \text{at} \quad y = 0 \qquad (4.198)$$

we can generate sublayer solutions for arbitrary $S_R$, including the limiting cases
$S_R \to 0$ and $S_R \to \infty$. Figure 4.19 shows the computed value of $C$ for a wide
range of values of $S_R$. As shown, in the limit $S_R \to \infty$, $C$ tends to 5.13. In the
limit $S_R \to 0$, an excellent correlation of the numerical predictions is given by

$$C \to 8.4 + \frac{1}{\kappa} \ell n (S_R/100) \quad \text{as} \quad S_R \to 0 \qquad (4.199)$$

Nikuradse [see Schlichting (1979)] found experimentally that for flow over
very rough surfaces (see Figure 1.6),

$$C \to 8.5 + \frac{1}{\kappa} \ell n \left( 1/k_s^+ \right), \qquad k_s^+ = u_\tau k_s/\nu \qquad (4.200)$$

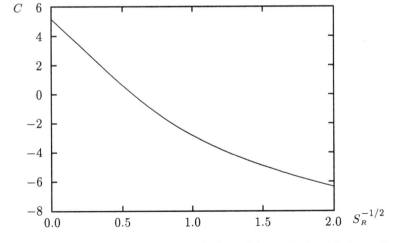

Figure 4.19: *Variation of the constant in the law of the wall, C, with the surface
value of the specific dissipation rate.*

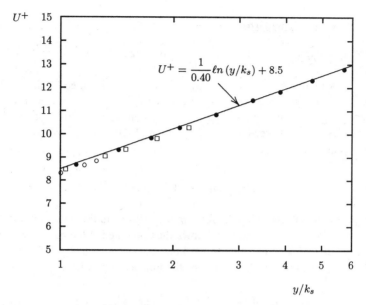

Figure 4.20: *Comparison of computed sublayer velocity profiles for "completely rough" surfaces with Nikuradse's correlation:* ○ *computed,* $k_s^+ = 400;$ □ *computed,* $k_s^+ = 225;$ ● *computed,* $k_s^+ = 50.$

where $k_s$ is the average height of sand-grain roughness elements. (Note that the computations use $\kappa = 0.41$ while Nikuradse found $\kappa = 0.40$.) Thus, if we make the correlation

$$S_R = 100/k_s^+, \qquad k_s^+ \gg 1 \qquad (4.201)$$

then Equations (4.199) and (4.200) are nearly identical. Figure 4.20 compares computed velocity profiles with Nikuradse's correlation, which is obtained by using Equation (4.200) in the law of the wall, viz.,

$$U^+ = \frac{1}{\kappa} \ell n \left( y/k_s \right) + 8.5, \qquad \kappa = 0.40 \qquad (4.202)$$

for three values of $k_s^+$. Computed velocities are very close to the correlation. The most remarkable fact about this correlation is that Equation (4.202) is the form the law of the wall assumes for flow over "completely-rough" surfaces [recall Equation (1.31)], including the value of the additive constant (8.4 and 8.5 differ by one percent).

By making a qualitative argument based on flow over a wavy wall, Wilcox and Chambers (1975) [see problems section] show that for small roughness heights, we should expect to have

$$S_R \sim (1/k_s^+)^2 \qquad \text{as} \qquad k_s^+ \to 0 \qquad (4.203)$$

Comparison with Nikuradse's data shows that the following correlation between $S_R$ and $k_s^+$ reproduces measured effects of sand-grain roughness for values of $k_s^+$ up to about 400.

$$S_R = \begin{cases} (50/k_s^+)^2, & k_s^+ < 25 \\ \\ 100/k_s^+, & k_s^+ \geq 25 \end{cases} \tag{4.204}$$

As a final comment, the solution for $k_s^+ \to 0$ is identical to the sublayer solution discussed in Subsection 4.6.3 [see Equation (4.189)]. The analysis of this section shows that the singular case corresponds to the perfectly-smooth surface. In practice, Equation (4.204) should be used rather than Equation (4.189) even if a perfectly-smooth surface is desired. Specifically, we can combine Equations (4.198) and the first of Equations (4.204) to arrive at the **slightly-rough-surface boundary condition** on $\omega$, viz.,

$$\omega = \frac{2500\nu_w}{k_s^2} \quad \text{at} \quad y = 0 \tag{4.205}$$

It is important to select a small enough value of $k_s$ to insure that $k_s^+ < 5$, corresponding to a "hydraulically-smooth surface" as defined by Schlichting (1979). If too large a value is selected, the skin friction values will be larger than smooth-wall values.

The advantage in using either Equation (4.204) or Equation (4.205) is obvious for several reasons.

- Local geometry (e.g., distance normal to the surface) does not appear so it can be applied even in three-dimensional geometries.

- $k_s$ need only be small enough to have a hydraulically smooth surface, i.e., $u_\tau k_s/\nu < 5$. Resulting surface values of $\omega$ are rarely ever large enough to cause numerical error provided a sensible finite-difference grid is used (see Subsection 7.2.1).

- Experience shows that Equation (4.204) works well for separated flows.

### 4.7.3 Surface Mass Injection

For boundary layers with surface mass injection, the introduction of an additional velocity scale ($v_w$ = area-averaged normal flow velocity through the porous surface) suggests that the scaling for $\omega$ at the surface may differ from Equation (4.198). Andersen, Kays and Moffat (1972) provide further evidence that the specific-dissipation-rate boundary condition must be revised when mass injection is present by showing, from correlation of their experimental data, that both $\kappa$ and $C$ are functions of $v_w^+ = v_w/u_\tau$. Because rough-surface computations show that the value of $C$ is strongly affected by the surface value of the specific dissipation rate, this suggests that the surface value of $\omega$ will depend in some manner upon $v_w$. Examination of the limiting form of the model equations for $y^+ \to \infty$ (i.e., in the log layer) shows immediately that the effective Kármán "constant", $\kappa_v$, varies with $v_w^+$ according to

$$\kappa_v = \frac{\kappa}{1 + \Xi v_w^+} \tag{4.206}$$

where $\Xi$ is given by

$$\Xi = \frac{1}{2}\left(C_o - \frac{3\sigma - 2}{2\sigma\kappa}\right) + \frac{1}{4\kappa}\ell n y^+ = 3.17 + 0.61\ell n y^+ \tag{4.207}$$

Note that $C_o = 5.13$ is the $k$-$\omega$ model-predicted constant in the law of the wall for a perfectly-smooth wall with no surface mass transfer.[6]

The variation of $\kappa_v$ predicted in Equations (4.206) and (4.207) is consistent with the Andersen et al. data. Including appropriate convective terms in Equations (4.186), we can use Program **SUBLAY** to perform sublayer computations for the cases experimentally documented by Andersen et al. In each case, the surface value of $\omega$ is given by

$$\omega = \frac{u_\tau^2}{\nu}S_B \quad \text{at} \quad y = 0 \tag{4.208}$$

Following Wilcox (1988a), we vary the value of $S_B$ to achieve optimum agreement between measured and computed velocities. The correlation between $S_B$ and $v_w^+$ is given in analytical form as

$$S_B = \frac{25}{v_w^+(1 + 5v_w^+)} \tag{4.209}$$

Figure 4.21 compares measured velocities with values computed using Equations (4.208) and (4.209).

## 4.8    Application to Wall-Bounded Flows

Using the surface boundary conditions devised in Section 4.7, we can now apply two-equation turbulence models to wall-bounded flows. Because of their relative simplicity, we consider pipe and channel flow first using the $k$-$\omega$ model. Then, we will consider several incompressible boundary-layer applications. In these applications we exercise the $k$-$\omega$ model and the Standard $k$-$\epsilon$ model.

---

[6]For boundary layers with suction, i.e., for $v_w < 0$, the $k$-$\omega$ model provides close agreement with measured velocity profiles treating the wall as being smooth. That is, $\omega$ should be given either by Equation (4.189) or by Equation (4.205) with $k_s^+ < 5$.

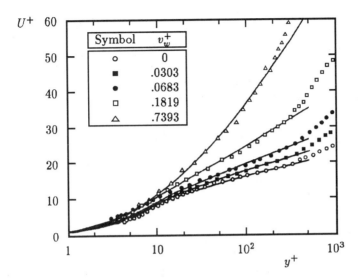

Figure 4.21: *Sublayer velocity profiles for boundary layers with surface mass injection;* —— *$k$-$\omega$ model;* ∘ □ • □ △ *Andersen et al.*

## 4.8.1 Channel and Pipe Flow

Figures 4.22 and 4.23 compare computed and measured channel and pipe flow properties, respectively. Six different comparisons are shown in each figure, including mean velocity, skin friction, Reynolds shear stress, turbulence kinetic energy, turbulence-energy production and dissipation rate.

Figure 4.22 compares $k$-$\omega$ model channel-flow predictions with the Direct Numerical Simulation (DNS) computations performed by Mansour, Kim and Moin (1988). Reynolds number based on channel height and average velocity is 13,750. Velocity profiles and Reynolds shear stress profiles differ by less than 3%. Computed skin friction differs from Halleen and Johnston's (1967) correlation [Equation (3.137)] by less than about 2% except at the lowest Reynolds number shown. Although the model fails to predict the peak value of $k$ near the channel wall, the computed $k$ profile differs from the DNS profile by less than 5% over 80% of the channel. Despite the fact that the model is not asymptotically consistent (Subsection 4.9.1) approaching the surface, even the dimensionless turbulence-energy production, $\mathcal{P}^+ = \nu\tau_{xy}(\partial U/\partial y)/u_\tau^4$, and dissipation, $\epsilon^+ = \nu\epsilon/u_\tau^4$, nearly duplicate the DNS results except very close to the surface (see discussion of pipe flow below). On balance, the $k$-$\omega$ results are a bit closer to the DNS results than either the Cebeci-Smith or Baldwin-Lomax models (Subsection 3.5.1).

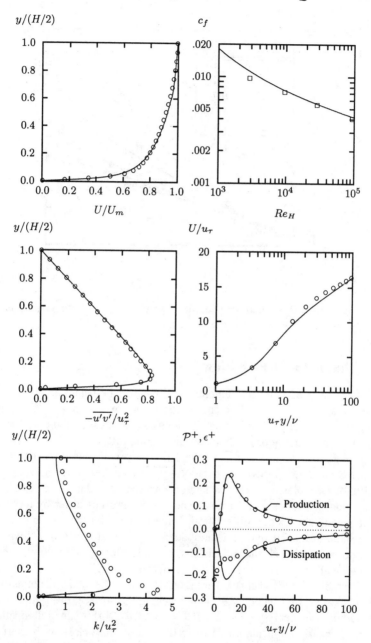

Figure 4.22: *Comparison of computed and measured channel-flow properties,* $Re_H = 13,750$. —— *k-ω model;* ○ *Mansour et al. (DNS);* □ *Halleen-Johnston correlation.*

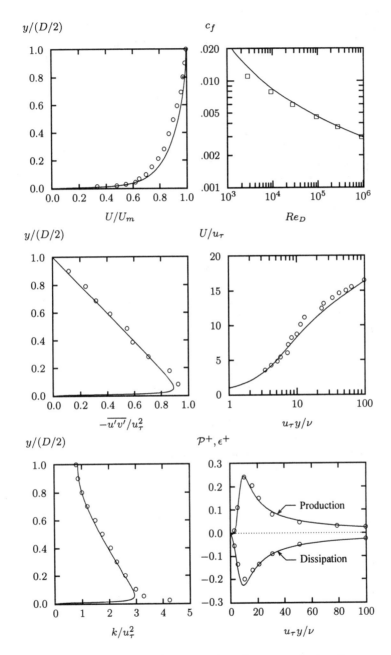

Figure 4.23: *Comparison of computed and measured pipe-flow properties,* $Re_D = 40,000$. —— *k-ω model;* ○ *Laufer;* □ *Prandtl correlation.*

Figure 4.23 compares $k$-$\omega$ model pipe-flow results with Laufer's (1952) measurements at a Reynolds number based on pipe diameter and average velocity of 40,000. As shown, computed and measured velocity and Reynolds shear stress profiles differ by less than 6%. As with channel flow, computed and measured turbulence kinetic energy differ by about 4% except close to the surface where the sharp peak occurs. Computed production and dissipation differ from measured values by less than 5%. However, dissipation is really nonzero at the surface (see the DNS results in Figure 4.22). Laufer's dissipation measurements are certainly incorrect as $y^+ \to 0$, and the model is reproducing erroneous results! Finally, computed skin friction is within 4% of Prandtl's universal law of friction [Equation (3.139)]. Overall, velocity and Reynolds-stress predictions are as close to measurements as those of the Cebeci-Smith and Baldwin-Lomax models.

It is interesting, and perhaps illuminating, that the most important flow properties are accurately predicted even though the sharp peak in turbulence energy is underestimated by 40% and 25%, respectively, for channel and pipe flow. That is, for engineering applications, the most important quantity is the skin friction. The next most important quantity typically is the velocity profile. Only for specialized applications is a subtle feature such as the peak value of $k$ important. Thus, we see that even though the $k$-$\omega$ model fails to predict this subtle feature, this is apparently of little consequence for most engineering applications.

## 4.8.2  Boundary Layers

We turn now to application of the $k$-$\omega$ and $k$-$\epsilon$ model equations to the same 16 incompressible boundary layers considered for algebraic (Figure 3.16), 1/2-equation (Figure 3.18) and one-equation models (Figure 4.2). All of the $k$-$\omega$ model results use the surface boundary conditions described in Subsection 4.7.2. The $k$-$\epsilon$ model computations were done using the Launder-Sharma (1974) low-Reynolds-number version subject to appropriate surface boundary conditions [see Subsection 4.9.1, Equations (4.213) – (4.217), (4.219) and (4.223)].

**Favorable Pressure Gradient.** The top row of graphs in Figure 4.24 compares computed and measured $c_f$ for the constant-pressure boundary layer (Flow 1400) and three boundary layers with favorable $\nabla p$ (Flows 1300, 2700 and 6300). As shown, for the $k$-$\omega$ model, computed $c_f$ virtually duplicates measurements for all four cases — differences between computed and measured $c_f$ are no more than 4%. The $k$-$\epsilon$ predictions are also quite close to measurements for Flows 1400 and 6300. However, $k$-$\epsilon$ skin friction is 10% below measured values for Flows 1300 and 2700. Thus, as no great surprise, the $k$-$\omega$ and $k$-$\epsilon$ models are quite accurate for the flat-plate boundary layer and boundary layers with favorable pressure gradient. The average difference between computed and measured $c_f$ at the final station is 2.6% and 7.2% for the $k$-$\omega$ and $k$-$\epsilon$ models, respectively.

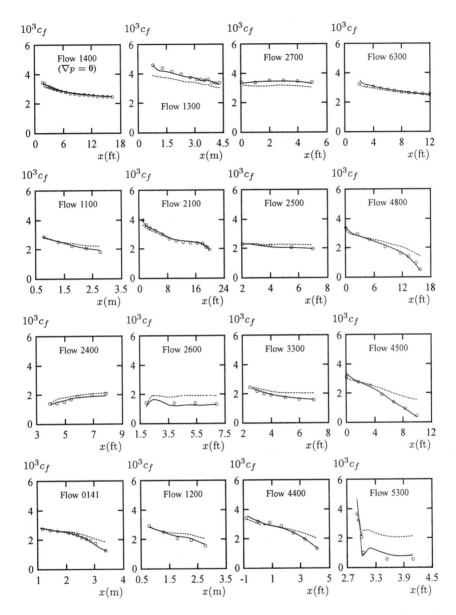

Figure 4.24: *Computed and measured skin friction for boundary layers subjected to a pressure gradient. Top row - favorable $\nabla p$; next to top row - mild adverse $\nabla p$; next to bottom row - moderate adverse $\nabla p$; bottom row - strong adverse $\nabla p$. —— $k$-$\omega$ model; - - - Launder-Sharma (1974) $k$-$\epsilon$ model; o measured.*

**Mild Adverse Pressure Gradient.** The second row of graphs in Figure 4.24 compares computed and measured $c_f$ for boundary layers with "mild" adverse pressure gradient. These flows (1100, 2100, 2500 and 4800) correspond to values of the equilibrium parameter, $\beta_T$, less than about 2. The $k$-$\omega$ predictions are again very close to measurements, even for Flow 4800, which is approaching separation. By contrast, the $k$-$\epsilon$ model's skin friction is close to corresponding measured values only for Flow 2100. The model's predicted skin friction is almost three times the measured value for Flow 4800, and the average difference between computed and measured $c_f$ for the four cases is 27%.

**Moderate Adverse Pressure Gradient.** Turning to "moderate" adverse $\nabla p$ ($\beta_T$ between about 2 and 4), we focus on the next to bottom row of graphs in Figure 4.24, i.e., Flows 2400, 2600, 3300 and 4500. As shown, there is no significant increase in differences between computed and measured $c_f$ for the $k$-$\omega$ model even for the nearly-separated Flow 4500, with the average difference being 6%. However, the $k$-$\epsilon$ model's predictions show even greater deviations from measured $c_f$, with the computed value being nearly 4 times the measured value for Flow 4500. The average difference at the end of each computation is 40%. Flow 3300, Bradshaw (1969) Flow C, was one of the most difficult cases considered in Stanford Olympics I. Throughout the flow, the $k$-$\omega$ model's $c_f$ is within 5% of measurements, while the $k$-$\epsilon$ model predicts a final value of $c_f$ that exceeds the measured value by 30%. The difference can be reduced to about 20% using wall functions [Chambers and Wilcox (1977)]. Because the equilibrium parameter $\beta_T \approx 2$ for this flow, the poor results for the $k$-$\epsilon$ model are unsurprising.

**Strong Adverse Pressure Gradient.** The bottom row of graphs in Figure 4.24 correspond to "strong" adverse pressure gradient, which corresponds to $\beta_T > 4$. Inspection of Figure 4.15 suggests that the $k$-$\omega$ model should be expected to continue predicting boundary-layer properties close to measurements, while differences between $k$-$\epsilon$ predictions and measurements should continue to increase. This is indeed the case. For example, Flow 0141 has increasingly adverse pressure gradient, the experimental data being those of Samuel and Joubert [see Kline et al. (1981)]. For the $k$-$\omega$ model, computed and measured skin friction differ by less than 5% of scale. Since $\beta_T$ exceeds 9 toward the end of the computation, the poor performance of the $k$-$\epsilon$ model (computed $c_f$ exceeds measured values by as much as 46%) is consistent with the defect-layer analysis of Subsection 4.6.3. The difference can be reduced to about 30% using wall functions [Rodi and Scheuerer (1986)]. While the $k$-$\omega$ model's skin friction is 55% higher than measured for the Stratford (1959) "incipient-separation" flow, this prediction is closer to the measured $c_f$ than any of the algebraic, 1/2-equation and one-equation models considered in Chapters 3 and 4. The $k$-$\epsilon$ model's $c_f$ is 4 times the measured value at the end of the computation.

Table 4.9 summarizes differences between computed and measured $c_f$ at the final station for the various pressure gradients. The overall average difference for 15 cases (Flow 5300 has been excluded) is a little less than 4% for the $k$-$\omega$ model and 29% for the $k$-$\epsilon$ model.

Table 4.9: *Differences Between Computed and Measured Skin Friction.*

| Pressure Gradient | Flows | $k$-$\omega$ | $k$-$\epsilon$ |
|---|---|---|---|
| Favorable | 1400, 1300, 2700, 6300 | 2.6% | 7.2% |
| Mild Adverse | 1100, 2100, 2500, 4800 | 3.2% | 27.2% |
| Moderate Adverse | 2400, 2600, 3300, 4500 | 5.9% | 39.8% |
| Strong Adverse | 0141, 1200, 4400 | 2.4% | 41.8% |
| All | – | 3.5% | 29.0% |

# 4.9 Low-Reynolds-Number Effects

Thus far, the turbulence models we have considered are restricted to high-Reynolds number applications. Even in the case of the $k$-$\omega$ model, while we have been able to obtain acceptably accurate results by integrating through the viscous sublayer, we have paid no attention to low-Reynolds-number effects. For example, the model fails to predict the sharp peak in turbulence kinetic energy close to the surface for pipe and channel flow (see Figures 4.22 and 4.23). Most two-equation models fail to predict a realistic value of the additive constant, $C$, in the law of the wall, and require viscous damping in order to do so. Finally, there are applications for which viscous effects must be accurately represented, and this section will discuss commonly used low-Reynolds-number corrections.

## 4.9.1 Asymptotic Consistency

In formulating viscous corrections for two-equation models, we can obtain some guidance from looking at the limiting behavior of the fluctuating velocities approaching a solid boundary. That is, we assume standard Taylor-series expansions for each of the fluctuating velocities and substitute into the exact equations of motion, viz., the instantaneous continuity and Navier-Stokes equations. We did this in Subsection 4.6.3 when we were formulating surface boundary conditions for the viscous-sublayer perturbation solution. Thus, we again begin by assuming

$$\left. \begin{array}{ccccc} u' & \sim & f_x(x,z,t)y & + & O(y^2) \\ v' & \sim & f_y(x,z,t)y^2 & + & O(y^3) \\ w' & \sim & f_z(x,z,t)y & + & O(y^2) \end{array} \right\} \quad \text{as} \quad y \to 0 \qquad (4.210)$$

where $f_x(x, z, t)$, $f_y(x, z, t)$ and $f_z(x, z, t)$ must have zero time average and satisfy the equations of motion. Note that the no-slip surface boundary condition dictates the fact that $\mathbf{u}'$ must go to zero as $y \to 0$. Since we expect Navier-Stokes solutions to be analytic everywhere, we conclude that the fluctuating velocity components $u'$ and $w'$ vary linearly with $y$. Also, substituting Equations (4.210) into the continuity equation shows that $v'$ varies quadratically with $y$. While we don't know the precise values of $f_x$, $f_y$ and $f_z$ without solving the complete Navier-Stokes equation, we can still use the exact asymptotic variations of $u'$, $v'$ and $w'$ with $y$ to deduce the limiting behavior of time-averaged properties approaching the surface. For example, the turbulence kinetic energy and dissipation are

$$k \sim \frac{1}{2}(\overline{f_x^2 + f_z^2})y^2 + O(y^3) \quad \text{and} \quad \epsilon \sim \nu\,(\overline{f_x^2 + f_z^2}) + O(y) \quad (4.211)$$

Also, the Reynolds shear stress is given by

$$\tau_{xy} \sim -\overline{f_x f_y}\, y^3 + O(y^4) \tag{4.212}$$

A model that duplicates the power-law forms of $k$, $\epsilon$ and $\tau_{xy}$ given in Equations (4.211) and (4.212) is said to be **asymptotically consistent** with the near-wall behavior of the exact equations of motion.

Many researchers have attempted to devise viscous corrections for $k$-$\epsilon$ and other two-equation models to permit their integration through the viscous sublayer. All have achieved some degree of **asymptotic consistency**. Jones and Launder (1972) were the first to propose viscous modifications for the $k$-$\epsilon$ model. Other proposals have been made by Launder and Sharma (1974), Hoffmann (1975), Reynolds (1976), Hassid and Poreh (1978), Lam and Bremhorst (1981), Dutoya and Michard (1981), Chien (1982), Myong and Kasagi (1990), Speziale, Abid and Anderson (1990), Shih and Hsu (1991), Durbin (1991), Zhang, So, Speziale and Lai (1993), Yang and Shih (1993), and Fan, Lakshminarayana and Barnett (1993). For steady, incompressible boundary layers, all of these models can be written compactly as follows:

$$U\frac{\partial k}{\partial x} + V\frac{\partial k}{\partial y} = \nu_T\left(\frac{\partial U}{\partial y}\right)^2 - \epsilon + \frac{\partial}{\partial y}\left[(\nu + \nu_T/\sigma_k)\frac{\partial k}{\partial y}\right] \tag{4.213}$$

$$U\frac{\partial\tilde{\epsilon}}{\partial x} + V\frac{\partial\tilde{\epsilon}}{\partial y} = C_{\epsilon 1}f_1\frac{\tilde{\epsilon}}{k}\nu_T\left(\frac{\partial U}{\partial y}\right)^2 - C_{\epsilon 2}f_2\frac{\tilde{\epsilon}^2}{k} + E + \frac{\partial}{\partial y}\left[(\nu + \nu_T/\sigma_\epsilon)\frac{\partial\tilde{\epsilon}}{\partial y}\right] \tag{4.214}$$

where the dissipation, $\epsilon$, is related to the quantity $\tilde{\epsilon}$ by

$$\epsilon = \epsilon_o + \tilde{\epsilon} \tag{4.215}$$

The quantity $\epsilon_o$ is the value of $\epsilon$ at $y = 0$, and is defined differently for each model. The eddy viscosity is defined as

$$\nu_T = C_\mu f_\mu k^2 / \tilde{\epsilon} \qquad (4.216)$$

Equations (4.213) – (4.216) contain five empirical **damping functions**, $f_1$, $f_2$, $f_\mu$, $\epsilon_o$ and $E$. These functions depend upon one or more of the following three dimensionless parameters.

$$Re_T = \frac{k^2}{\tilde{\epsilon}\nu}, \qquad R_y = \frac{k^{1/2}y}{\nu}, \qquad y^+ = \frac{u_\tau y}{\nu} \qquad (4.217)$$

The models devised by Jones and Launder (1972), Launder and Sharma (1974), Lam and Bremhorst (1981), and Chien (1982) exemplify most of the features incorporated in $k$-$\epsilon$ model viscous damping functions. The damping functions and closure coefficients for these four low-Reynolds-number $k$-$\epsilon$ models are as follows.

**Jones-Launder Model**

$$\left.\begin{array}{l} f_\mu = e^{-2.5/(1+Re_T/50)} \\[4pt] f_1 = 1 \\[4pt] f_2 = 1 - 0.3e^{-Re_T^2} \\[4pt] \epsilon_o = 2\nu \left(\dfrac{\partial\sqrt{k}}{\partial y}\right)^2 \\[12pt] E = 2\nu\nu_T \left(\dfrac{\partial^2 U}{\partial y^2}\right)^2 \\[12pt] C_{\epsilon 1} = 1.55, \quad C_{\epsilon 2} = 2.00, \quad C_\mu = 0.09, \quad \sigma_k = 1.0, \quad \sigma_\epsilon = 1.3 \end{array}\right\} \quad (4.218)$$

**Launder-Sharma Model**

$$\left.\begin{array}{l} f_\mu = e^{-3.4/(1+Re_T/50)^2} \\[4pt] f_1 = 1 \\[4pt] f_2 = 1 - 0.3e^{-Re_T^2} \\[4pt] \epsilon_o = 2\nu \left(\dfrac{\partial\sqrt{k}}{\partial y}\right)^2 \\[12pt] E = 2\nu\nu_T \left(\dfrac{\partial^2 U}{\partial y^2}\right)^2 \\[12pt] C_{\epsilon 1} = 1.44, \quad C_{\epsilon 2} = 1.92, \quad C_\mu = 0.09, \quad \sigma_k = 1.0, \quad \sigma_\epsilon = 1.3 \end{array}\right\} \quad (4.219)$$

**Lam-Bremhorst Model**

$$
\left.
\begin{aligned}
&f_\mu = \left(1 - e^{-0.0165R_y}\right)^2 \left(1 + 20.5/Re_T\right) \\
&f_1 = 1 + (0.05/f_\mu)^3 \\
&f_2 = 1 - e^{-Re_T^2} \\
&\epsilon_o = 0 \\
&E = 0 \\
&C_{\epsilon 1} = 1.44, \quad C_{\epsilon 2} = 1.92, \quad C_\mu = 0.09, \quad \sigma_k = 1.0, \quad \sigma_\epsilon = 1.3
\end{aligned}
\right\} \quad (4.220)
$$

**Chien Model**

$$
\left.
\begin{aligned}
&f_\mu = 1 - e^{-0.0115y^+} \\
&f_1 = 1 \\
&f_2 = 1 - 0.22e^{-(Re_T/6)^2} \\
&\epsilon_o = 2\nu \frac{k}{y^2} \\
&E = -2\nu \frac{\tilde{\epsilon}}{y^2} e^{-y^+/2} \\
&C_{\epsilon 1} = 1.35, \quad C_{\epsilon 2} = 1.80, \quad C_\mu = 0.09, \quad \sigma_k = 1.0, \quad \sigma_\epsilon = 1.3
\end{aligned}
\right\} \quad (4.221)
$$

By examining the limiting behavior of each of these models close to a solid boundary where $y = 0$, it is easy to demonstrate that, consistent with Equation (4.211), all four models guarantee

$$
k \sim y^2 \quad \text{and} \quad \epsilon/k \to 2\nu/y^2 \quad \text{as} \quad y \to 0 \qquad (4.222)
$$

Additionally, the Lam-Bremhorst model predicts $\tau_{xy} \sim y^4$ while the other three models predict $\tau_{xy} \sim y^3$. Thus, all except the Lam-Bremhorst model are consistent with Equation (4.212) as well.

Surface boundary conditions for low-Reynolds-number $k$-$\epsilon$ models are not entirely straightforward. On the one hand, the no-slip boundary condition tells us that $k$ must vanish at a solid boundary. On the other hand, the strongest thing we can say about the surface value of $\epsilon$ is the second of Equations (4.222). That is, we invariably must tie the surface value of $\epsilon$ to the second derivative of $k$ at the surface. The Jones-Launder, Launder-Sharma and Chien models build in the proper asymptotic behavior through introduction of the function $\epsilon_o$. Consequently, the boundary conditions appropriate at the surface are

$$
k = \tilde{\epsilon} = 0 \quad \text{at} \quad y = 0 \qquad (4.223)
$$

By contrast, Lam and Bremhorst deal directly with $\epsilon$ and specify the surface boundary condition on $\epsilon$ by requiring

$$
\epsilon = \nu \frac{\partial^2 k}{\partial y^2} \quad \text{at} \quad y = 0 \qquad (4.224)
$$

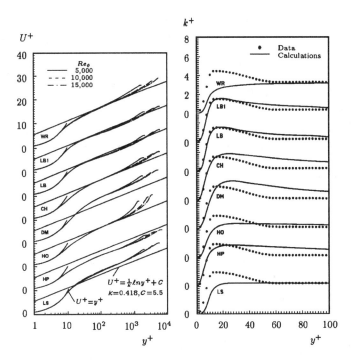

Figure 4.25: *Flat-plate boundary layer properties.  CH = Chien; DM = Dutoya-Michard; HO = Hoffman; HP = Hassid-Poreh; LB = Lam-Bremhorst with $\epsilon = \nu \partial^2 k/\partial y^2$ at $y = 0$; LB1 = Lam-Bremhorst with $\partial \epsilon/\partial y = 0$ at $y = 0$; LS = Launder-Sharma; WR = Wilcox-Rubesin. [From Patel, Rodi and Scheuerer (1985) — Copyright © AIAA 1985 — Used with permission.]*

As an alternative, Lam and Bremhorst propose using

$$\frac{\partial \epsilon}{\partial y} = 0 \quad \text{at} \quad y = 0 \tag{4.225}$$

While Equation (4.225) is easier to implement than Equation (4.224), there is no a priori reason to expect that the next term in the Taylor-series expansion for $\epsilon$ should vanish.

In a review article, Patel, Rodi and Scheuerer (1985) compare seven low-Reynolds-number variants of the $k$-$\epsilon$ model and the Wilcox-Rubesin (1980) $k$-$\omega^2$ model. Figure 4.25 compares computed and measured velocity and dimensionless turbulence energy ($k^+ = k/u_\tau^2$) profiles for the flat-plate boundary layer. As shown, several models fail to provide accurate velocity profiles for the incompressible flat-plate boundary layer.

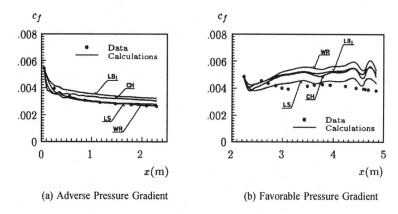

(a) Adverse Pressure Gradient          (b) Favorable Pressure Gradient

Figure 4.26: *Comparison of computed and measured skin friction for low-Reynolds-number flows with pressure gradient.  CH = Chien; LB1 = Lam-Bremhorst with $\partial\epsilon/\partial y = 0$ at $y = 0$; LS = Launder-Sharma; WR = Wilcox-Rubesin. [From Patel, Rodi and Scheuerer (1985) — Copyright © AIAA 1985 — Used with permission.]*

Figure 4.26(a) shows that for adverse pressure gradient, the Wilcox-Rubesin model (which was not designed with low-Reynolds-number applications in mind) most faithfully matches measured [Andersen et al. (1972)] skin friction. Figure 4.26(b) shows that none of the models reproduces the measured skin friction for the low-Reynolds-number, favorable pressure gradient flow of Simpson and Wallace (1975). This further demonstrates that the only thing low-Reynolds-number modifications do is fix the $k$-$\epsilon$ model's problems in predicting the constant $C$ in the law of the wall.

There is a popular misconception that low-Reynolds-number modifications to the $k$-$\epsilon$ model can remove its deficiencies for adverse pressure gradient flows. This mistaken notion overlooks the volumes of data on and physical understanding of turbulent boundary layers established during the twentieth century, most notably by Clauser and Coles. Recall from Subsection 4.6.1 that Coles describes the turbulent boundary layer as a "wake-like structure constrained by a wall" and notes that different scales and physical processes are dominant in the sublayer and defect layer. As noted above, since perturbation analysis shows that the $k$-$\epsilon$ model predicts defect-layer data rather poorly, we cannot reasonably expect viscous corrections (which are negligible in the physical defect layer) to correct the problem.

Figure 4.27 clearly illustrates this point. The figure compares computed and measured skin friction for the 12 incompressible boundary layers with adverse pressure gradient considered earlier (see Figure 4.24). Results are presented

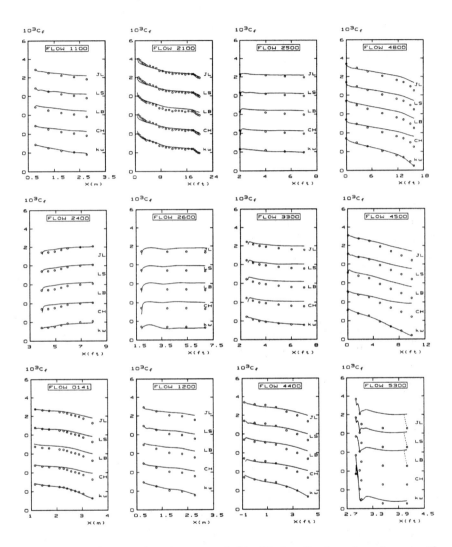

Figure 4.27: *Computed and measured skin friction for boundary layers with adverse pressure gradient; CH = Chien; JL = Jones-Launder; LB = Lam-Bremhorst; LS = Launder-Sharma; kw = k-ω.*

for the Jones-Launder, Launder-Sharma, Lam-Bremhorst and Chien $k$-$\epsilon$ models and for the Wilcox (1998) $k$-$\omega$ model. Discrepancies between computed and measured $c_f$ increase for all four $k$-$\epsilon$ models as the strength of the pressure gradient increases. As discussed in the last section, $k$-$\omega$ results are close to measured values for all twelve cases, including the nearly separated Flow 5300 (the Chien model predicts separation for this case). In terms of the final values of $c_f$, the average difference between computation and measurement for the 11 cases (as before, we exclude Flow 5300) is 4% for the $k$-$\omega$ model, 34% for the Jones-Launder model, 36% for the Launder-Sharma model, 44% for the Chien model and 48% for the Lam-Bremhorst model.

These results confirm the defect-layer perturbation solution presented in Subsection 4.6.2, which shows that [see Equation (4.181)]:

$$\frac{U_e - U}{u_\tau} \sim -\frac{1}{\kappa}\ell n\eta + A - \beta_T B\eta\ell n\eta + O\left(\eta^2\ell n\eta\right) \qquad \text{as} \qquad \eta \to 0 \quad (4.226)$$

where the coefficient $B$ is given in Table 4.7. Combining Equation (4.226) with Equation (4.173), the effective law of the wall predicted by the $k$-$\epsilon$ model is

$$U^+ \sim \frac{1}{\kappa}\ell ny^+ + C + \beta_T B\eta\ell n\eta \qquad \text{as} \qquad y^+ \to \infty \qquad (4.227)$$

Since $\eta < 1$, the term $\beta_T B\eta\ell n\eta$ is negative, so that we should expect the computed velocity profile to lie below the classical law-of-the-wall line on a semi-log plot. Figure 4.28 compares the computed Launder-Sharma model velocity profile with experimental data, the standard law of the wall and a defect-layer solution for $\beta_T = 5.4$, which is the value computed at $x = 3.40$ m in the Samuel-Joubert flow according to the $k$-$\epsilon$ model. Examination of the numerical solution shows that the implied constant in the law of the wall, $C$, is 5.5. As shown, the numerical solution indeed lies below the law-of-the-wall line, while the defect-layer profile shape is similar to the computed profile. We should not expect exact agreement between the computed profile and the defect-layer profile since $\beta_T$ varies quite rapidly with $x$ for the Samuel-Joubert flow. However, the similarity of their shapes is striking. The important point to note is the impact of the term in Equation (4.227) proportional to the equilibrium parameter, $\beta_T$. Its effect is to distort the velocity profile throughout the defect layer, including its asymptotic form approaching the viscous sublayer from above.

As a final comment on low-Reynolds-number corrections for the $k$-$\epsilon$ model, using the dimensionless parameters $R_y$ and $y^+$ [Equation (4.217)] is ill advised. Both depend on distance normal to the surface, which can cause difficulty in complex geometries such as a wing-fuselage junction. Also, it is ironic that several additional closure coefficients and functions are needed for the $k$-$\epsilon$ model to behave properly in the near-wall region of a turbulent boundary layer. Dissipation is, after all, a phenomenon that occurs in the smallest eddies, which is all we find

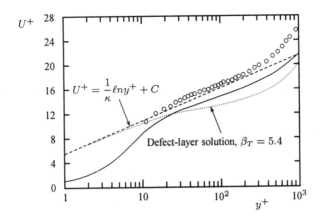

Figure 4.28: *Computed and measured near-wall velocity profiles for Samuel and Joubert's adverse pressure gradient flow, $x = 3.40$ m:* —— *Launder-Sharma model with $\kappa = 0.43$ and $C = 5.5$; $\circ$ Samuel-Joubert.*

in the near-wall region. This further underscores the fact that there is virtually no connection between the exact equation for $\epsilon$ and its modeled counterpart.

## 4.9.2 Transition

Turbulence model equations can be integrated through transition from laminar to turbulent flow, although most models predict transition at Reynolds numbers that are at least an order of magnitude too low. The following discussion focuses mostly on the $k$-$\omega$ model, whose behavior through transition is easiest to understand. The discussion also demonstrates why the $k$-$\epsilon$ model is so much harder to implement for transitional flows. To understand why and how the $k$-$\omega$ model predicts transition, consider the flat-plate boundary layer. For the $k$-$\omega$ model, the incompressible, two-dimensional boundary-layer form of the equations for $k$ and $\omega$ is as follows.

$$U\frac{\partial U}{\partial x} + V\frac{\partial U}{\partial y} = \frac{\partial}{\partial y}\left[(\nu + \nu_T)\frac{\partial U}{\partial y}\right] \tag{4.228}$$

$$U\frac{\partial k}{\partial x} + V\frac{\partial k}{\partial y} = \nu_T\left(\frac{\partial U}{\partial y}\right)^2 - \beta^*\omega k + \frac{\partial}{\partial y}\left[(\nu + \sigma^*\nu_T)\frac{\partial k}{\partial y}\right] \tag{4.229}$$

$$U\frac{\partial \omega}{\partial x} + V\frac{\partial \omega}{\partial y} = \alpha\frac{\omega}{k}\nu_T\left(\frac{\partial U}{\partial y}\right)^2 - \beta\omega^2 + \frac{\partial}{\partial y}\left[(\nu + \sigma\nu_T)\frac{\partial \omega}{\partial y}\right] \tag{4.230}$$

$$\nu_T = \alpha^* k/\omega \tag{4.231}$$

With one exception, all notation and closure coefficients are as defined in Equations (4.36) to (4.42). The only difference is the appearance of an additional closure coefficient $\alpha^*$ in Equation (4.231). This coefficient is equal to unity for the standard high-Reynolds-number version of the $k$-$\omega$ model. We can most clearly illustrate how the model equations predict transition by rearranging terms in Equations (4.229) and (4.230) as follows.

$$U\frac{\partial k}{\partial x} + V\frac{\partial k}{\partial y} = P_k\beta^*\omega k + \frac{\partial}{\partial y}\left[(\nu + \sigma^*\nu_T)\frac{\partial k}{\partial y}\right] \tag{4.232}$$

$$U\frac{\partial \omega}{\partial x} + V\frac{\partial \omega}{\partial y} = P_\omega\beta\omega^2 + \frac{\partial}{\partial y}\left[(\nu + \sigma\nu_T)\frac{\partial \omega}{\partial y}\right] \tag{4.233}$$

The **net production per unit dissipation** terms in the two equations, $P_k$ and $P_\omega$, are defined by:

$$P_k = \frac{\alpha^*}{\beta^*}\left(\frac{\partial U/\partial y}{\omega}\right)^2 - 1 \tag{4.234}$$

$$P_\omega = \frac{\alpha\alpha^*}{\beta}\left(\frac{\partial U/\partial y}{\omega}\right)^2 - 1 \tag{4.235}$$

There are two important observations worthy of mention at this point. **First**, if the turbulence energy is zero, Equation (4.233) has a well-behaved solution. That is, when $k = 0$, the eddy viscosity vanishes and the $\omega$ equation uncouples from the $k$ equation. Consequently, the $k$-$\omega$ model has a nontrivial laminar-flow solution, with $\nu_T = 0$, for $\omega$. **Second**, the signs of $P_k$ and $P_\omega$ determine whether $k$ and $\omega$ are amplified or reduced in magnitude. However, it is not obvious from Equations (4.234) and (4.235) how the signs of these terms vary with Reynolds number as we move from the plate leading edge to points downstream. We can make the variation obvious by rewriting Equations (4.234) and (4.235) in terms of the Blasius transformation for a laminar boundary layer.

Before we introduce the Blasius transformation, we must determine the appropriate scaling for $\omega$. To do this, we note that close to the surface of a flat-plate boundary layer (laminar or turbulent), the specific dissipation rate behaves according to[7] [see Equation (4.189) and Table 4.8]:

$$\omega \to \frac{6\nu}{\beta_o y^2} \quad \text{as} \quad y \to 0 \tag{4.236}$$

In terms of the Blasius similarity variable, $\eta$, defined by

$$\eta = \frac{y}{\sqrt{\nu x/U_\infty}} \tag{4.237}$$

---

[7]Keep in mind that dissipation is $\epsilon = \beta^*\omega k$ so that $\omega$ can be finite even when $k$ and $\epsilon$ vanish.

where $U_\infty$ is freestream velocity, the asymptotic behavior of $\omega$ approaching the surface is

$$\omega \to \frac{U_\infty}{x} \frac{6}{\beta_o \eta^2} \quad \text{as} \quad \eta \to 0 \quad (4.238)$$

Since $U_\infty/x$ has dimensions of 1/time, we conclude that the appropriate scaling for $\omega$ in the Blasius boundary layer is given by

$$\omega = \frac{U_\infty}{x} W(x, \eta) \quad (4.239)$$

where $W(x, \eta)$ is a dimensionless function to be determined as part of the solution. Also, we write the velocity in terms of dimensionless velocity, $\mathcal{U}(x, \eta)$, according to

$$U = U_\infty \mathcal{U}(x, \eta) \quad (4.240)$$

Noting that $f_{\beta^*} \to 1$ for very small values of $k$ and $f_\beta = 1$ for two-dimensional flows, the **net production-per-unit-dissipation** terms become

$$P_k = \frac{\alpha^*}{\beta_o^*} Re_x \left( \frac{\partial \mathcal{U}/\partial \eta}{W} \right)^2 - 1 \quad (4.241)$$

$$P_\omega = \frac{\alpha \alpha^*}{\beta_o} Re_x \left( \frac{\partial \mathcal{U}/\partial \eta}{W} \right)^2 - 1 \quad (4.242)$$

Thus, both $P_k$ and $P_\omega$ increase linearly with Reynolds number, $Re_x$. From the exact laminar solution for $\mathcal{U}(\eta)$ and $W(\eta)$ [the $x$ dependence vanishes for the Blasius boundary layer], the maximum value (with respect to $\eta$) of the ratio of $\partial \mathcal{U}/\partial \eta$ to $W$ is

$$\left( \frac{\partial \mathcal{U}/\partial \eta}{W} \right)_{max} \approx \frac{1}{300} \quad (4.243)$$

The precise value of this ratio is a weak function of the freestream value of $\omega$, ranging between 0.0025 and 0.0040. The maximum occurs about midway through the boundary layer ($y/\delta = 0.56$), a point above which the exact near-wall behavior of $\omega$ [Equation (4.238)] does not hold. Hence, a complete boundary-layer solution is needed to determine the maximum ratio of $\partial \mathcal{U}/\partial \eta$ to $W$.

As long as the eddy viscosity remains small compared to the molecular viscosity, we can specify the precise points where $P_k$ and $P_\omega$ change sign, which impact the beginning and end of transition, respectively. Using Equation (4.243), we conclude that the sign changes occur at the following Reynolds numbers.

$$(Re_x)_k = 9 \cdot 10^4 \frac{\beta_o^*}{\alpha^*} \quad (4.244)$$

$$(Re_x)_\omega = 9 \cdot 10^4 \frac{\beta_o}{\alpha \alpha^*} \quad (4.245)$$

With no viscous modifications, the closure coefficients $\alpha$, $\alpha^*$, $\beta_o$ and $\beta_o^*$ are 13/25, 1, 9/125 and 9/100, respectively. Using these **fully-turbulent** values, we find $(Re_x)_k = 8,100$ and $(Re_x)_\omega = 12,462$. Thus, starting from laminar flow at the leading edge of a flat plate (see Figure 4.29), the following sequence of events occurs.

1. The computation starts in a laminar region with $k = 0$ in the boundary layer and a small freestream value of $k$.

2. Initially, because $P_k < 0$ and $P_\omega < 0$, dissipation of both $k$ and $\omega$ exceeds production. Turbulence energy is entrained from the freestream and spreads through the boundary layer by molecular diffusion. Neither $k$ nor $\omega$ is amplified and the boundary layer remains laminar.

3. At the **critical Reynolds number**, $Re_{x_c} = 8,100$, production overtakes dissipation in the $k$ equation. Downstream of $x_c$, production exceeds dissipation in the $k$ equation and turbulence energy is amplified. At some point in this process, the eddy viscosity grows rapidly and this corresponds to the onset of transition.

4. $k$ continues to be amplified and, beyond $Re_x = 12,462$ production overtakes dissipation in the $\omega$ equation. $\omega$ is now amplified and continues growing until a near balance between production and dissipation is achieved in the $k$ equation. When this near balance is achieved, transition from laminar to turbulent flow is complete.

Consistent with experimental measurements, the entire process is very sensitive to the freestream value of $k$. There is also a sensitivity to the freestream value of $\omega$, although the sensitivity is more difficult to quantify.

Three key points are immediately obvious. **First**, $k$ begins growing at a Reynolds number of 8,100. By contrast, linear-stability theory tells us that Tollmien-Schlichting waves begin forming in the Blasius boundary layer at a Reynolds number of 90,000. This is known as the **minimum critical Reynolds number** for infinitesimal disturbances. Correspondingly, we find that the model predicts transition at much too low a Reynolds number. **Second**, inspection of Equations (4.244) and (4.245) shows that $(Re_x)_w$, and hence the width of the transition region, is controlled by the ratio of $\beta_o$ to $\alpha\alpha^*$. **Third**, transition will never occur if $P_\omega$ reaches zero earlier than $P_k$. Thus, occurrence of transition requires

$$\alpha\alpha^* < \alpha^*\beta_o/\beta_o^* \qquad \text{as} \qquad Re_T \to 0 \qquad (4.246)$$

where the quantity $Re_T$ is turbulence Reynolds number defined by

$$Re_T = \frac{k}{\omega\nu} \qquad (4.247)$$

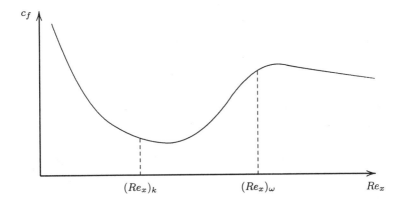

Figure 4.29: *Skin friction variation for a boundary layer undergoing transition from laminar to turbulent flow.*

This fact must be preserved in any viscous modification to the model. Our goal is to devise viscous modifications that depend only upon $Re_T$. As noted in the preceding subsection, this quantity is independent of flow geometry and thus preserves the universal nature of the model. We also proceed with two key objectives in mind. The most important objective is to match the **minimum critical Reynolds number**. Reference to Equation (4.244) shows that, in order to have $(Re_x)_k = 90,000$, we must require

$$\beta_o^*/\alpha^* \to 1 \quad \text{as} \quad Re_T \to 0 \qquad (4.248)$$

Our secondary objective is to achieve **asymptotic consistency** with the exact behavior of $k$ and dissipation, $\epsilon = \beta_o^* k \omega$, approaching a solid boundary where $\chi_k < 0$ so that $f_{\beta^*} = 1$. That is, we would like to have $k/y^2 \to$ constant and $\epsilon/k \to 2\nu/y^2$ as $y \to 0$. Close to a solid boundary, the dissipation and molecular-diffusion terms balance in both the $k$ and $\omega$ equations. The very-near-wall solution for $\omega$ is given by Equation (4.236). The solution for $k$ is of the form

$$k/y^n \to \text{constant} \quad \text{as} \quad y \to 0 \qquad (4.249)$$

where n is given by

$$n = \frac{1}{2}\left[1 + \sqrt{1 + 24\frac{\beta_o^*}{\beta_o}}\right] \qquad (4.250)$$

Noting that dissipation is related to $k$ and $\omega$ by

$$\epsilon = \beta_o^* k \omega \qquad (4.251)$$

we can achieve the desired asymptotic behavior of $k$ provided

$$\beta_o^*/\beta_o \to 1/3 \quad \text{as} \quad Re_T \to 0 \quad (4.252)$$

Requiring this limiting behavior as $Re_T \to 0$ is sufficient to achieve the desired asymptotic behavior as $y \to 0$ since the eddy viscosity, and hence, $Re_T$ vanishes at a solid boundary.

If we choose to have $\beta_o$ constant for all values of $Re_T$, Equations (4.246), (4.248) and (4.252) are sufficient to determine the limiting values of $\alpha^*$ and $\beta_o^*$ and an upper bound for $\alpha\alpha^*$ as turbulence Reynolds number becomes vanishingly small. Specifically, we find

$$\left. \begin{array}{rcl} \alpha\alpha^* & < & \beta_o \\ \alpha^* & \to & \beta_o/3 \\ \beta_o^* & \to & \beta_o/3 \end{array} \right\} \quad \text{as} \quad Re_T \to 0 \quad (4.253)$$

Wilcox and Rubesin (1980) make the equivalent of $\alpha\alpha^*$ and $\alpha^*$ in their $k$-$\omega^2$ model approach the same limiting value and obtain excellent agreement with measured transition width for incompressible boundary layers. Numerical experimentation with the $k$-$\omega$ model indicates the optimum choice for incompressible boundary layers is $\alpha\alpha^* \to 0.80\beta_o$, or

$$\alpha\alpha^* \to 0.0576 \quad \text{as} \quad Re_T \to 0 \quad (4.254)$$

Following Wilcox (1992a), we postulate the following functional dependencies upon $Re_T$ that guarantee the limiting values in Equations (4.253) and (4.254), as well as the original fully-turbulent values for $Re_T \to \infty$.

$$\alpha^* = \frac{\alpha_o^* + Re_T/R_k}{1 + Re_T/R_k} \quad (4.255)$$

$$\alpha = \frac{13}{25} \cdot \frac{\alpha_o + Re_T/R_\omega}{1 + Re_T/R_\omega} \cdot (\alpha^*)^{-1} \quad (4.256)$$

$$\beta^* = \frac{9}{100} \cdot \frac{4/15 + (Re_T/R_\beta)^4}{1 + (Re_T/R_\beta)^4} \cdot f_{\beta^*} \quad (4.257)$$

$$\beta = \frac{9}{125} f_\beta, \quad \sigma^* = \sigma = \frac{1}{2}, \quad \alpha_o^* = \frac{1}{3}\beta_o, \quad \alpha_o = \frac{1}{9} \quad (4.258)$$

$$R_\beta = 8, \quad R_k = 6, \quad R_\omega = 2.95 \quad (4.259)$$

The three coefficients $R_\beta$, $R_k$ and $R_\omega$ control the rate at which the closure coefficients approach their fully-turbulent values. We can determine their values

by using perturbation methods to analyze the viscous sublayer.[8] Using the procedure discussed in Subsection 4.6.3, we can solve for the constant in the law of the
wall, $C$. For given values of $R_\beta$ and $R_k$, there is a unique value of $R_\omega$ that yields
a constant in the law of the wall of 5.0. For small values of $R_\beta$ the peak value of
$k$ near the surface is close to the value achieved without viscous corrections, viz.,
$u_\tau^2/\sqrt{\beta_o^*}$. As $R_\beta$ increases, the maximum value of $k$ near the surface increases.
Comparison of computed sublayer structure with Direct Numerical Simulation
(DNS) results of Mansour, Kim and Moin (1988) indicates the optimum choice
for these three coefficients is as indicated in Equation (4.259).

The only flaw in the model's asymptotic consistency occurs in the Reynolds
shear stress, $\tau_{xy}$. While the exact asymptotic behavior is $\tau_{xy} \sim y^3$, the model
predicts $\tau_{xy} \sim y^4$. This discrepancy could easily be removed with another
viscous modification. However, as will be shown later in this subsection, this is of
no significant consequence. It has no obvious bearing on either the model's ability
to predict transition or properties of interest in turbulent boundary layers. The
additional complexity and uncertainty involved in achieving this subtle feature
of the very-near-wall behavior of $\tau_{xy}$ does not appear to be justified.

It is a simple matter to explain why little progress has been made in predicting
transition with the $k$-$\epsilon$ model. The primary difficulties can be easily demonstrated
by focusing upon incompressible boundary layers. If we use the standard form
of the $k$-$\epsilon$ model, Equations (4.229) through (4.231) are replaced by

$$U\frac{\partial k}{\partial x} + V\frac{\partial k}{\partial y} = \nu_T\left(\frac{\partial U}{\partial y}\right)^2 - \epsilon + \frac{\partial}{\partial y}\left[(\nu + \nu_T/\sigma_k)\frac{\partial k}{\partial y}\right] \qquad (4.260)$$

$$U\frac{\partial \epsilon}{\partial x} + V\frac{\partial \epsilon}{\partial y} = C_{\epsilon 1}\frac{\epsilon}{k}\nu_T\left(\frac{\partial U}{\partial y}\right)^2 - C_{\epsilon 2}\frac{\epsilon^2}{k} + \frac{\partial}{\partial y}\left[(\nu + \nu_T/\sigma_\epsilon)\frac{\partial \epsilon}{\partial y}\right] \qquad (4.261)$$

$$\nu_T = C_\mu k^2/\epsilon \qquad (4.262)$$

A critical difference from the $k$-$\omega$ model is obvious from Equations (4.260) –
(4.262), viz., if $k$ is zero, $\epsilon$ must also be zero. We cannot simply drop the eddy
viscosity in the $\epsilon$ equation, because of the presence of $k$ in the denominator of the
$\epsilon$ equation's dissipation term. The model does possess a laminar-flow solution
for the ratio of $\epsilon$ to $k$. If we make the formal change of variables

$$\epsilon = C_\mu k\omega = \beta_o^* k\omega \qquad (4.263)$$

and assume $\nu_T \ll \nu$, the following laminar-flow equation for $\omega$ results.

---

[8]Note that this approach reflects a degree of optimism that the same viscous corrections can be
expected to capture the physics of the viscous sublayer *and* transitional flows.

$$U\frac{\partial\omega}{\partial x} + V\frac{\partial\omega}{\partial y} = (C_{\epsilon 1} - 1)f_\mu\left(\frac{\partial U}{\partial y}\right)^2 - (C_{\epsilon 2} - 1)C_\mu\omega^2 + \nu\frac{\partial^2\omega}{\partial y^2} + \frac{2\nu}{k}\frac{\partial k}{\partial y}\frac{\partial\omega}{\partial y}$$
(4.264)

Equation (4.264) is nearly identical to the limiting form of Equation (4.230) for $\nu_T/\nu \to 0$. The only significant difference is the last term on the right-hand side of Equation (4.264). Except close to the surface where $k$ must be exactly zero, this term is unlikely to have a significant effect on the solution for small nonzero values of $k$. However, in a numerical solution, products of dependent-variable gradients are generally destabilizing, and the problem can only be aggravated by having a coefficient inversely proportional to $k$. This is not an insurmountable problem. However, establishing starting conditions is clearly more difficult with the $k$-$\epsilon$ model than with the $k$-$\omega$ model.

Given the diverse nature of viscous modifications that have been proposed for the $k$-$\epsilon$ model, it is impossible to make any universal statements about why a specific model fails to predict realistic transition Reynolds numbers. Perhaps the strongest statement that can be made is, **few researchers have approached the problem from the transition point of view**. Most have sought only to achieve asymptotic consistency as $y \to 0$ (Subsection 4.9.1) and attempted transition predictions only as an afterthought. We can gain some insight by examining the net production per unit dissipation terms for the $k$ and $\epsilon$ equations that are analogous to Equations (4.241) and (4.242), viz.,

$$P_k = \frac{f_\mu}{C_\mu}Re_x\left(\frac{\partial U/\partial\eta}{W}\right)^2 - 1$$
(4.265)

$$P_\epsilon = \frac{C_{\epsilon 1}f_\mu}{C_{\epsilon 2}C_\mu}Re_x\left(\frac{\partial U/\partial\eta}{W}\right)^2 - 1$$
(4.266)

**On the one hand**, without viscous damping, if we assume Equation (4.243) is valid, we find $(Re_x)_k = 8,100$ and $(Re_x)_\epsilon = 10,800$. Consequently, as with the high-Reynolds-number version of the $k$-$\omega$ model, transition will occur at too low a Reynolds number. **On the other hand**, because $C_\mu$, $C_{\epsilon 2}$ and sometimes $C_{\epsilon 1}$ are multiplied by functions of distance from the surface and/or functions of $Re_T$ (c.f. $f_\mu$, $f_1$ and $f_2$ in Subsection 4.9.1) in low-Reynolds-number $k$-$\epsilon$ models, we cannot simply use Equation (4.243). Furthermore, as discussed in the preceding subsection, some modelers add terms to the $k$ and $\epsilon$ equations in addition to damping the closure coefficients. Each set of values for the closure coefficients and additional terms must be used in solving Equation (4.264) to determine the laminar-flow solution for $\epsilon/k$. While it is clearly impossible to make a quantitative evaluation of all variants of the $k$-$\epsilon$ model, we can nevertheless make two general observations.

First, from the analysis of the $k$-$\omega$ model, it is obvious that having $f_\mu < 1$ will tend to delay transition. Virtually all modelers implement an $f_\mu$ that accomplishes this end. However, the modifications of Jones and Launder (1972), Chien (1982), and Lam and Bremhorst (1981), for example, damp $C_{\epsilon 2}$ to the extent that $(Re_x)_\epsilon$ is smaller than $(Re_x)_k$. This is the opposite of what is needed and can actually prevent transition.

Second, although this discussion is not intended as an exhaustive survey of the numerous low-Reynolds-number versions of the $k$-$\epsilon$ model, it does illustrate how difficult it can be to apply the model to the transition problem. Given enough additional closure coefficients and damping functions, the $k$-$\epsilon$ model can probably be modified to permit satisfactory transition predictions. However, even if this is done, establishing starting conditions will ultimately require a solution to Equation (4.264). That is, to initialize the computation, we must effectively transform to the $k$-$\omega$ model. Since this is the natural starting point, it seems illogical to perform subsequent computations in terms of $k$ and $\epsilon$.

### 4.9.3  Channel and Pipe Flow

Figure 4.30 compares computed channel-flow skin friction, $c_f$, with the Halleen and Johnston (1967) correlation [see Equation (3.138)] for Reynolds number based on channel height, $H$, and average velocity ranging from $10^3$ to $10^5$ using the $k$-$\omega$ model. As shown, computed $c_f$ differs from the correlation by less than 3% except at the lowest Reynolds number shown where the correlation probably is inaccurate. Velocity, Reynolds shear stress, and turbulence kinetic energy profiles differ by less than 7%. Most notably, the model predicts the peak value of $k$ near the channel wall to within 4% of the DNS value. The low-Reynolds-number modifications have been designed to capture this feature. Additionally, approaching the surface, the dimensionless turbulence-energy production, $\mathcal{P}^+ = \nu \tau_{xy} (\partial U / \partial y) / u_\tau^4$, and dissipation, $\epsilon^+ = \nu \epsilon / u_\tau^4$, are within 10% of the DNS results except very close to the surface.

Figure 4.31 compares computed pipe flow $c_f$ with Prandtl's universal law of friction [see Equation (3.139)]. Reynolds number based on pipe diameter, $D$, and average velocity varies from $10^3$ to $10^6$. As with channel flow, computed $c_f$ falls within 5% of the correlation except at the lowest Reynolds number shown where the correlation is likely to be in error. Computed and measured velocity and Reynolds shear stress profiles differ by less than 8%. As with channel flow, computed and measured turbulence kinetic energy differ by about 5% including the region close to the surface where the sharp peak occurs. Note that, at this high a Reynolds number, the $k$ profile has a spike near $y = 0$ and this feature is captured in the computations. Computed turbulence-energy production, $\mathcal{P}^+$, and dissipation, $\epsilon^+$, differ from measured values by less than 10% except where Laufer's measurements are inaccurate close to the surface.

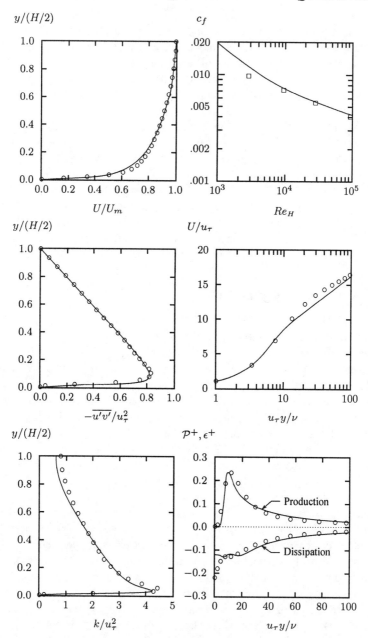

Figure 4.30: *Comparison of computed and measured channel-flow properties,*
$Re_H = 13,750$. ——— *Low-Reynolds-number $k$-$\omega$ model;* ○ *Mansour et al.
(DNS);* □ *Halleen-Johnston correlation.*

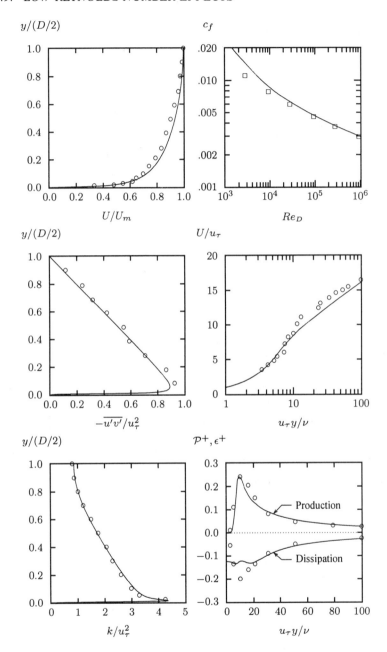

Figure 4.31: *Comparison of computed and measured pipe-flow properties,* $Re_D = 40,000.$ —— *Low-Reynolds-number k-ω model;* ○ *Laufer;* □ *Prandtl correlation.*

## 4.9.4   Boundary-Layer Applications

Figure 4.32 compares computed and measured skin friction for the 16 baseline test cases considered in Chapters 3 and 4. As in all previous applications, computations have been done using Program **EDDYBL** (Appendix D). Additionally, Table 4.10 summarizes average differences between computed and measured $c_f$ at the end of each computation. As indicated in the table, both the low-Reynolds-number (Low-$Re$) and high-Reynolds-number (High-$Re$) versions of the $k$-$\omega$ model reproduce measured skin friction to well within measurement error.

Table 4.10: *Differences Between Computed and Measured Skin Friction.*

| Pressure Gradient | Flows | Low-$Re$ $k$-$\omega$ | High-$Re$ $k$-$\omega$ |
|---|---|---|---|
| Favorable | 1400, 1300, 2700, 6300 | 3.3% | 2.6% |
| Mild Adverse | 1100, 2100, 2500, 4800 | 4.0% | 3.2% |
| Moderate Adverse | 2400, 2600, 3300, 4500 | 9.1% | 5.9% |
| Strong Adverse | 0141, 1200, 4400 | 3.6% | 2.4% |
| All | – | 5.1% | 3.5% |

Differences between the Low-$Re$ and High-$Re$ versions of the $k$-$\omega$ model are almost imperceptible. This is expected since the low-Reynolds-number modifications are confined almost exclusively to the viscous sublayer. The skin friction, by contrast, is controlled by the overall balance of forces (pressure gradient and surface shear stress) and the momentum flux through the entire boundary layer.

The two most noteworthy differences between Low-$Re$ and High-$Re$ model predictions are for the nearly separated Flows 4500 and 5300. For Flow 4500, the Low-$Re$ $k$-$\omega$ $c_f$ approaches zero more rapidly than indicated by measurements. Examination of the figure shows that the High-$Re$ version predicts $c_f$ closer to measurements. This is the primary reason Table 4.10 lists the Low-$Re$ version with 9.1% average differences between computed and measured $c_f$ as compared to 5.9% differences for the High-$Re$ version. For Flow 5300, the Low-$Re$ model provides a solution that is within 9% of the measured $c_f$ at the final station. This is much closer than any of the turbulence models considered earlier. This is not reflected in Table 4.10 because, consistent with the established precedent, the flow is excluded in computing the average difference in $c_f$.

Results obtained for Flows 4500 and 5300 are likely due to the fact that, approaching separation, the specific dissipation rate is reduced to smaller levels than those prevailing in attached boundary layers. Recall that in the sublayer, $\omega$ scales with $u_\tau^2/\nu$. Consequently, since the viscous modification to the closure coefficient $\alpha$ directly impacts the production of $\omega$, the percentage change will be much greater when $\omega$ is small. This can, in turn, have a nontrivial effect throughout the boundary layer, and thus have a larger impact on the skin friction.

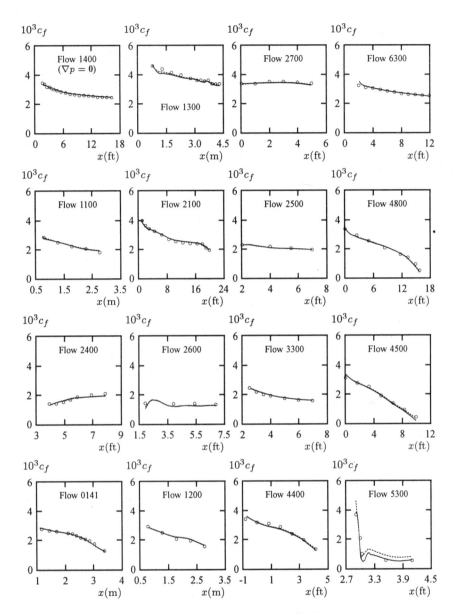

Figure 4.32: *Computed and measured skin friction for boundary layers subjected to a pressure gradient. Top row - favorable* $\nabla p$; *next to top row - mild adverse* $\nabla p$; *next to bottom row - moderate adverse* $\nabla p$; *bottom row - strong adverse* $\nabla p$. —— *Low-Reynolds-number $k$-$\omega$ model;* - - - *High-Reynolds-number $k$-$\omega$ model;* ○ *measured.*

Turning now to transition, Figure 4.33 compares computed and measured transition Reynolds number, $Re_{\theta_t}$, for an incompressible flat-plate boundary layer. We define the transition Reynolds number as the point where the skin friction achieves its minimum value. Results are displayed as a function of freestream turbulence intensity, $T'$, defined by

$$T' = 100\sqrt{\frac{2}{3}\frac{k_e}{U_e^2}} \qquad (4.267)$$

where subscript $e$ denotes the value at the boundary-layer edge. As shown, consistent with the data compiled by Dryden (1959), $Re_{\theta_t}$ increases as the freestream intensity decreases. Because $\omega$ can be thought of as an averaged frequency of the freestream turbulence, it is reasonable to expect the predictions to be sensitive to the freestream value of $\omega$. To assess the effect, the freestream value of the turbulence length scale $\ell = k^{1/2}/\omega$ has been varied from $.001\delta$ to $.100\delta$ where $\delta$ is boundary-layer thickness. As shown, computed $Re_{\theta_t}$ values bracket most of the data. Unlike the situation for free shear flows, the $k$-$\omega$ model's sensitivity to the freestream value of $\omega$ is a *desirable feature* for transition applications. Physical transition location is not simply a function of $T'$, but rather is frequency dependent. While it is unclear how the freestream value of $\omega$ should be specified, consistent with measurements, the model is not confined to a single transition location for a given $T'$ regardless of the frequency of the disturbance.

Figure 4.34 compares computed width of the transition region with measurements of Dhawan and Narasimha (1958), Schubauer and Skramstad (1948), and Fisher and Dougherty (1982). We define transition width, $\Delta x_t$, as the distance between minimum and maximum skin-friction points. The computed width, $Re_{\Delta x_t}$, falls within experimental data scatter for $10^4 < Re_{x_t} < 10^7$. $\Delta x_t$ is unaffected by the freestream value of $\omega$.

While these transition results are interesting, keep in mind that transition is a complicated phenomenon. Transition is triggered by a disturbance in a boundary layer only if the frequency of the disturbance falls in a specific band. Reynolds averaging has masked all spectral effects, and all the model can represent with $k$ and $\omega$ is the intensity of the disturbance and an average frequency. Hence, it is possible for the turbulence model to predict transition when it shouldn't occur. The model equations thus are sensible in the transition context only if the triggering disturbance is broad band, i.e., contains all frequencies.

Additionally, we have only guaranteed that the point where $k$ is first amplified matches the minimum critical Reynolds number for the incompressible, flat-plate boundary layer. To simulate transition with complicating effects such as pressure gradient, surface heat transfer, surface roughness, compressibility, etc., the values of $\alpha_o^*$ and $\alpha_o$ must change [see Wilcox (1977)]. Their values can be deduced from linear-stability theory results, or perhaps from a correlation based

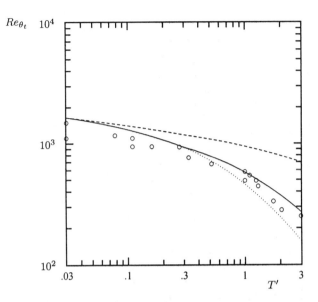

Figure 4.33: *Transition location for an incompressible flat-plate boundary layer:* --- $\ell/\delta = .001$; —— $\ell/\delta = .010$; ··· $\ell/\delta = .100$; ○ *Dryden.*

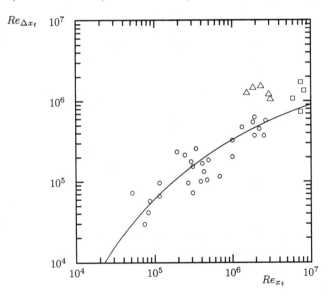

Figure 4.34: *Transition width for an incompressible flat-plate boundary layer:* —— *k-ω model;* ○ *Dhawan and Narasimha;* △ *Schubauer and Skramstad;* □ *Fisher and Dougherty.*

on stability theory. Nevertheless, some information must be provided regarding the minimum critical Reynolds number for each new application.

For any flow, we can always match the measured transition point by adjusting the freestream value of $k$. This is satisfactory when the transition point occurs at a large Reynolds number, which requires $k_\infty$ to be small relative to $U_\infty^2$. However, for a high-speed flow in which transition has been triggered at a relatively small Reynolds number, often unreasonably large values of $k_\infty$ are needed to cause transition, so large as to affect the total energy in the freestream in a physically unrealistic manner. Thus, a new method for triggering transition is needed.

Wilcox (1994) offers an alternative to depending upon the model to predict the onset of transition, known as the **numerical roughness strip**. The foundation of the concept rests upon the fact that by using a finite value for $\omega$ at the surface, the model simulates surface roughness (Subsection 4.7.2). Since increasing the surface roughness height corresponds to decreasing the surface value of $\omega$ (and thus the dissipation in the $k$ equation), the model predicts that roughness will have a destabilizing effect. This is consistent with measurements, and patches of surface roughness are often used to trigger transition in experiments.

Using Equations (4.198) and (4.204) to simulate a roughness strip, Wilcox (1994) has run more than 20 transitional boundary layer cases to test this idea. In all cases, computation begins at the plate leading edge, and the turbulence kinetic energy is initially set to an extremely small value, viz., $10^{-15}U_\infty^2$, throughout the boundary layer. This value is too small to trigger transition naturally. The initial $\omega$ profile is given by the exact laminar-flow solution to the model equations. Using this approach, the numerical roughness strip triggers transition at the desired location for all of the cases considered using a roughness strip with $k_s$ and the streamwise extent of the strip, $\Delta s$, given by the following correlations.

$$\frac{k_s}{\delta_t} = \max\left\{ \frac{5000}{\sqrt{Re_{x_t}}}, 3 \right\} \tag{4.268}$$

$$\frac{\Delta s}{\delta_t} = 0.015\sqrt{Re_{x_t}} \tag{4.269}$$

The quantities $\delta_t$ and $Re_{x_t}$ are the boundary-layer thickness and transition Reynolds number based on arc length.

Figure 4.35 compares computed and measured [Blair and Werle (1981), Blair (1983)] Stanton number, $St$, for transitional boundary layers with surface heat transfer. According to Equations (4.268) and (4.269), the dimensions of the roughness strip required to match the measured transition point for the case with favorable pressure gradient are $(k_s/\delta_t, \Delta s/\delta_t) = (8.5, 8.7)$. As shown, differences between computed and measured Stanton numbers are no more than 15% for the two cases shown.

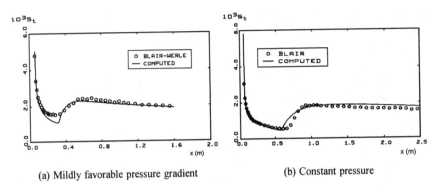

(a) Mildly favorable pressure gradient          (b) Constant pressure

Figure 4.35: *Computed and measured Stanton number for transitional boundary layers with surface heat transfer. [From Wilcox (1994) — Copyright © AIAA 1994 — Used with permission.]*

Perhaps the most practical way to use the model for transitional flows is in describing the transitional region, as opposed to predicting transition onset. Of course, the question of sensitivity to spectral effects in the transition region must be raised. Using linear-stability computations, Wilcox (1981a) shows that after the initial disturbance has grown to a factor of $e^4$ times its initial value, the turbulence model closure coefficients lose all memory of spectral effects. Thus, we can conclude that not far downstream of the minimum critical Reynolds number, Reynolds averaging is sensible. This tells us that, if the point at which the transition begins is known, using a numerical roughness strip is a practical and accurate way of simulating transitional boundary layers.

Low-Reynolds-number corrections increase the complexity of two-equation models significantly. The high-$Re$ $k$-$\omega$ model has just 5 closure coefficients and 2 closure functions. The low-$Re$ version described in this subsection has 10 closure coefficients and 5 closure functions. The various low-Reynolds-number models discussed in Subsection 4.9.1 involve a similar increase in the number of closure coefficients and damping functions. The Launder-Sharma (1974) model, for example, has 9 closure coefficients and 4 closure functions.

If viscous effects are insignificant for a given application, it is advisable to use the simpler high-Reynolds-number version of the model. In the case of the $k$-$\epsilon$ model, if you need to integrate through the viscous sublayer, you have no choice but to use one of the low-Reynolds-number models, preferably one that yields a satisfactory solution for simple flows such as the incompressible flat-plate boundary layer. In the case of the $k$-$\omega$ model, integration through the sublayer can be done without introducing viscous corrections, and there is virtually no difference in model-predicted skin friction and velocity profiles with and without viscous corrections for turbulent boundary layers.

## 4.10  Separated Flows

Turning to separated flows, we first consider the axisymmetric flow with strong adverse pressure gradient that has been experimentally investigated by Driver (1991). Figure 4.36 compares the data with Menter's (1992b) computed skin friction and surface pressure for the Wilcox (1988a) $k$-$\omega$ model. Although the computation has not yet been done with the Wilcox (1998) version of the model defined in Equations (4.36) through (4.42), there are two observations that suggest the new model's predictions will be very similar. First, as indicated in Figure 4.13, the cross-diffusion modification as represented by the function $f_{\beta^*}$ [Equation (4.41)] is relatively minor for wall-bounded flows. Second, the two models' predictions for boundary layers are very similar quantitatively, even for attached cases approaching separation such as Flows 4500, 4800 and 5300 (see Figure 4.24). These facts strongly suggest that the new $k$-$\omega$ model is as accurate as the Wilcox (1988a) version for separated flows.

As shown, the $k$-$\omega$ model yields a separation bubble of about the measured length, with the separation point a bit upstream of the measured location. Also, pressure downstream of reattachment is somewhat higher than measured. Results are clearly much closer to measurements than those obtained with the Baldwin-Lomax and Baldwin-Barth models (see Figures 3.17 and 4.3).

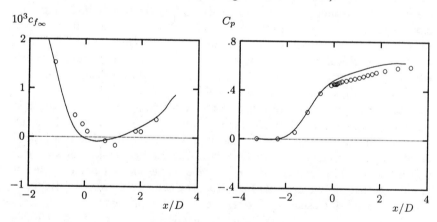

Figure 4.36: *Computed and measured flow properties for Driver's separated flow; —— Wilcox (1988a) $k$-$\omega$ model; o Driver.*

Next, we consider the backward-facing step (Figure 4.4). Figure 4.37 compares computed and measured [Driver and Seegmiller (1985)] skin friction for backstep flow with the upper channel wall inclined to the lower wall at angles of $0°$ and $6°$. Computed results are shown for the Wilcox (1988a) $k$-$\omega$ model and for the Standard $k$-$\epsilon$ model with wall functions; neither model includes viscous

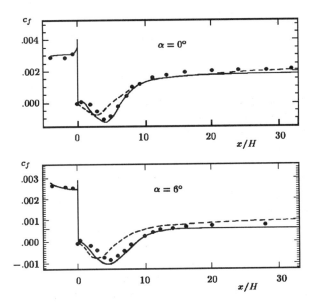

Figure 4.37: *Computed and measured skin friction for flow past a backward-facing step; —— Wilcox (1988a) $k$-$\omega$ model; - - - $k$-$\epsilon$ model; • Driver-Seegmiller data. [From Menter (1992c).]*

corrections. As summarized in Table 4.11, the $k$-$\epsilon$ model predicts reattachment well upstream of the measured point for both cases, while the $k$-$\omega$ model is within 4% of the measured location for both cases. The Wilcox (1998) version has been applied to this flow with $\alpha = 0°$. All computed flow properties are nearly identical to those indicated in Figure 4.37 and Table 4.11.

Many researchers have proposed modifications to the $k$-$\epsilon$ model aimed at improving its predictions for this flow. Driver and Seegmiller (1985), for example, compare four different versions of the model with their experimental data. By contrast, the $k$-$\omega$ model's solution for flow past the backward-facing step is satisfactory with no special modifications.

Table 4.11: *Backstep Reattachment Length*

| Model | Reference | $\alpha = 0°$ | $\alpha = 6°$ |
|---|---|---|---|
| $k$-$\epsilon$ | Launder-Sharma (1974) | 5.20 | 5.50 |
| $k$-$\omega$ | Wilcox (1988a) | 6.40 | 8.45 |
| Measured | Driver-Seegmiller (1985) | 6.20 | 8.10 |

Han (1989) has applied the $k$-$\epsilon$ model with wall functions to flow past a simplified three-dimensional bluff body with a ground plane. The object considered is known as Ahmed's body [Ahmed et al. (1984)] and serves as a simplified automobile-like geometry. In his computations, Han considers a series of afterbody slant angles. Figure 4.38(a) illustrates the shape of Ahmed's body with a 30° slant angle afterbody. Figure 4.38(b) compares computed and measured surface pressure contours on the rear-end surface for a 12.5° slant angle. As shown, computed pressure contours are similar on the slanted surface, but quite different on the vertical base. For slant angles up to 20°, the computed base pressures are significantly lower than measured. Consequently, the computed drag coefficient is about 30% higher than measured. Considering how poorly the $k$-$\epsilon$ model performs for boundary layers in adverse pressure gradient and for the two-dimensional backward-facing step, it is not surprising that the model would predict such a large difference from the measured drag in this extremely complicated three-dimensional, massively-separated flow.

*This is a quintessential example of how important turbulence modeling is to Computational Fluid Dynamics.* Recall that there are three key elements to CFD, viz., the numerical algorithm, the grid and the turbulence model. Han uses an efficient numerical procedure and demonstrates grid convergence of his solutions. Han's computational tools also include state-of-the-art grid-generation procedures. Han's research efforts on this problem are exemplary on both counts. However, using the $k$-$\epsilon$ model undermines the entire computation for the following reasons. Because the model fails to respond in a physically realistic manner to the adverse pressure gradient on the rear-end surface, the predicted skin friction is too high. This means the vorticity at the surface is too large, so that too much vorticity diffuses from the surface. This vorticity is swept into the main flow and too strong a vortex forms when the flow separates. This, of course, reduces the base pressure. Thus, the $k$-$\epsilon$ model's inability to accurately respond to adverse pressure gradient distorts the entire flowfield.

Our final separated-flow application is particularly difficult to simulate, i.e., blood flow in an **arterial stenosis**. The word stenosis, common in bioengineering literature, means "narrowing of a passage." Thus, we consider the flow of blood through an artery that has a narrowing due to the deposit of plaque caused by excess cholesterol in the blood stream. One feature characteristic of blood flow is the low Reynolds number, $Re$, associated with the human body. For example, $Re$ ranges from about 400 in the common carotid artery to 1,500 in the ascending aorta. In the absence of stenosis, the flow is laminar since fully-developed pipe flow does not experience transition to turbulence until the Reynolds number based on diameter and average flow speed exceeds about 2,300. However, the obstruction presented by stenosis leads to flow separation, which in turn causes transition to turbulence. Thus, the problem we address is a low-Reynolds-number flow that includes transition, separation and, ultimately, reattachment.

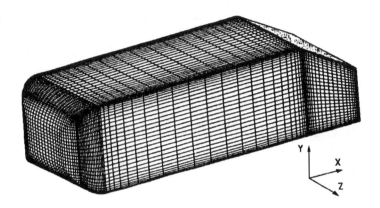

(a) Body geometry and surface grid

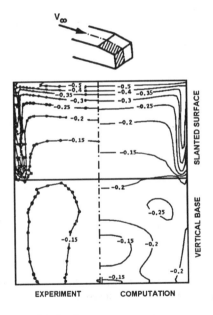

(b) Static-pressure contours

Figure 4.38: *Flow past Ahmed's body – high-Re k-ε computations. [From Han (1989) — Copyright © AIAA 1989 — Used with permission.]*

Figure 4.39 shows the geometry and streamlines of arterial stenosis computations performed by Ghalichi et al. (1998). The Ghalichi et al. computations
have been done using the Wilcox (1994) low-Reynolds-number version of the
$k$-$\omega$ model, which is very similar to the low-$Re$ $k$-$\omega$ model described in Subsection 4.9.2. The flows indicated in Figures 4.39(a) and (b) correspond to a
reduction in cross-sectional area of 50% and 75%, respectively. In both cases, a
separation bubble is present downstream of the stenosis.

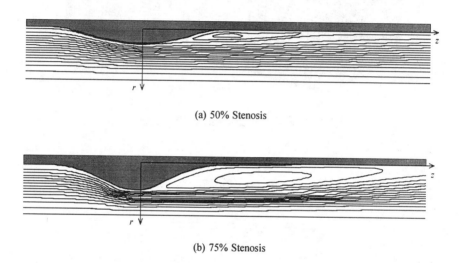

(a) 50% Stenosis

(b) 75% Stenosis

Figure 4.39: *Computed streamlines for blood flow through arteries with 50%
and 75% stenosis;* $Re = 1,000$.

Figure 4.40(a) compares computed and measured [Saad and Giddens (1983)]
reattachment length, $L_a$, for the two different stenoses — the quantity $D$ denotes
the diameter of the unobstructed artery. The largest difference between theory
and experiment is 10% of scale (for 75% stenosis and $Re = 500$). One of
the most remarkable features of the computed flowfield is the critical Reynolds
number at which transition to turbulence occurs. Consistent with measurements,
the $k$-$\omega$ model predicts transition at about $Re = 1,100$ for a 50% stenosis and
at $Re = 400$ for a 75% stenosis.

Figure 4.40(b) compares computed and measured static pressure at the surface
in a 50% stenosis for one of the Ghalichi et al. $k$-$\omega$ based computations, and for
results obtained in an earlier study by Zijlema et al. (1995) using the Standard
$k$-$\epsilon$ model. While differences between computed and measured pressures for the
$k$-$\omega$ model are as large as 20% downstream of the stenosis, the Standard $k$-$\epsilon$
model predicts pressures that bear no resemblance to measured values.

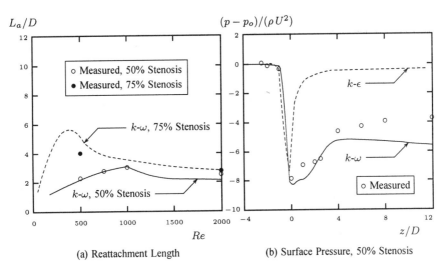

Figure 4.40: *Computed and measured flow properties for blood flow in arterial stenoses.*

## 4.11 Range of Applicability

Early one-equation models were based on the turbulence kinetic energy equation, and were **incomplete**. As discussed in Section 4.2, only a modest advantage is gained in using such models rather than an algebraic model. The primary difficulty is the need to specify the length scale for each new application. There is no natural way to accommodate an abrupt change from a wall-bounded flow to a free shear flow such as near an airfoil trailing edge or beyond the trunk lid of an automobile. The only real advantage of using this type of one-equation model rather than a two-equation model is that numerical solution is simpler. One-equation models tend to be nearly as well behaved as algebraic models, while two-equation models, especially low-$Re$ $k$-$\epsilon$ models, can be very difficult to solve.

By contrast, more recent one-equation models based on a postulated equation for eddy viscosity are **complete**. Two of the most commonly used models are those of Baldwin and Barth (1990) and of Spalart and Allmaras (1992).

The Baldwin-Barth model is very inaccurate for attached boundary layers, consistently predicting values of skin friction that are typically 25% below corresponding measurements. The model's predictions are even farther from measurements for separated flows, and its equation often presents serious numerical difficulties. Thus, it is clear that the Baldwin-Barth model is of little value for general turbulent-flow applications.

The Spalart-Allmaras model predicts skin friction for attached boundary layers that is as close to measurements as algebraic models. The model's predictions are far superior to those of algebraic models for separated flows, and the differential equation presents no serious numerical difficulties. Its only shortcoming appears to be in predicting the asymptotic spreading rates for plane, round and radial jets. On balance, results of experience to date indicate that the Spalart-Allmaras model is an excellent engineering tool for predicting properties of turbulent flows.

Two-equation models are **complete**. Certainly the $k$-$\epsilon$ model is the most widely used two-equation model. It has been applied to many flows with varying degrees of success. Unfortunately, it is even more inaccurate than the Baldwin-Barth one-equation model for flows with adverse pressure gradient, and that poses a serious limitation to its general utility. Because of its inability to respond to adverse pressure gradient (see Tables 4.2 and 4.9), the model is inaccurate for separated flows. Its predictions for free shear flows are also a bit erratic. The $k$-$\epsilon$ model is extremely difficult to integrate through the viscous sublayer and requires viscous corrections simply to reproduce the law of the wall for an incompressible flat-plate boundary layer. No consensus has been achieved on the optimum form of the viscous corrections as evidenced by the number of researchers who have created low-Reynolds-number versions of the model (see Subsection 4.9.1). While the model can be fine tuned for a given application, it is not clear that this represents an improvement over algebraic models. The primary shortcoming of algebraic models is their need of fine tuning for each new application. Although saying the $k$-$\epsilon$ model always needs such fine tuning would be a bit exaggerated, it still remains that such tuning is too often needed. Given all of these well-documented flaws, it remains a mystery to this author why the model has such widespread use.

The $k$-$\omega$ model, although not as popular as the $k$-$\epsilon$ model, enjoys several advantages. Most importantly, the model is significantly more accurate for two-dimensional boundary layers with both adverse and favorable pressure gradient. Also, without any special viscous corrections, the model can be easily integrated through the viscous sublayer. The model accurately reproduces measured spreading rates for all five free shear flows (Table 4.4). Finally, for the limited cases tried to date, the model appears to match measured properties of recirculating flows with no changes to the basic model and its closure coefficients. With viscous corrections included, the $k$-$\omega$ model accurately reproduces subtle features of turbulence kinetic energy behavior close to a solid boundary and even describes boundary-layer transition reasonably well.

Other two-equation models have been created, but they have had even less use than the $k$-$\omega$ model. Before such models can be taken seriously, they should be tested for simple incompressible boundary layers with adverse pressure gradient. How many interesting flows are there, after all, with constant pressure?

The use of perturbation methods to dissect model-predicted boundary-layer structure is perhaps the most important diagnostic tool presented in this chapter. Experience has shown that a turbulence model's ability to accurately predict effects of pressure gradient on boundary layers can be assessed by analyzing its defect-layer behavior. Specifically, models that faithfully replicate measured variation of Coles' wake-strength parameter, $\Pi$, with the equilibrium (pressure-gradient) parameter, $\beta_T$, (see Figure 4.15) also closely reproduce boundary-layer properties for non-equilibrium cases. Conversely, models that deviate significantly from the $\Pi$ vs. $\beta_T$ data predict large deviations from measurements for non-equilibrium boundary layers.

While two-equation models, especially the $k$-$\omega$ model, are more general than less complex models, they nevertheless fail in some applications. In Chapter 5, we will see that these models are unreliable for boundary-layer separation induced by interaction with a shock wave. In Chapter 6, we will see that two-equation models are inaccurate for flows over curved surfaces. Also, two-equation models as presented in this chapter cannot predict secondary motions in noncircular duct flow. In all three of these examples, the difficulty can be traced to the Boussinesq eddy-viscosity approximation.

# Problems

**4.1** Verify that the exact equation for the dissipation, $\epsilon$, is given by Equation (4.45). That is, derive the equation that follows from taking the following moment of the Navier-Stokes equation.

$$2\nu \overline{\frac{\partial u_i'}{\partial x_j} \frac{\partial}{\partial x_j} [\mathcal{N}(u_i)]} = 0$$

where $\mathcal{N}(u_i)$ is the Navier-Stokes operator defined in Equation (2.26).

**4.2** Starting with Equations (4.4) and (4.45), define $\epsilon = \beta_o^* \omega k$ and derive an "exact" $\omega$ equation.

**4.3** Derive the exact equation for the **enstrophy**, $\omega^2$, defined by

$$\omega^2 \equiv \frac{1}{2}\overline{\omega_i' \omega_i'} \qquad \text{where} \qquad \omega_i' = \epsilon_{ijk}\partial u_k'/\partial x_j$$

That is, $\omega_i'$ is the fluctuating vorticity. **HINT:** Beginning with the Navier-Stokes equation, derive the equation for the vorticity, multiply by $\omega_i'$, and time average. The vector identity $\mathbf{u}\cdot\nabla\mathbf{u} = \nabla\left(\frac{1}{2}\mathbf{u}\cdot\mathbf{u}\right) - \mathbf{u}\times(\nabla\times\mathbf{u})$ should prove useful in deriving the vorticity equation.

**4.4** Using Program **PIPE** (Appendix C), compute the skin friction for channel flow according to the Baldwin-Barth and Spalart-Allmaras models. Compare your results with the Halleen-Johnston correlation [Equation (3.138) for $10^3 \le Re_H \le 10^5$. Also, compare the computed velocity profiles for $Re_H = 13,750$ with the Mansour et al. DNS data, which are as follows.

| $y/(H/2)$ | $U/U_m$ | $y/(H/2)$ | $U/U_m$ |
|-----------|---------|-----------|---------|
| 0.000 | 0.000 | 0.602 | 0.945 |
| 0.103 | 0.717 | 0.710 | 0.968 |
| 0.207 | 0.800 | 0.805 | 0.984 |
| 0.305 | 0.849 | 0.902 | 0.995 |
| 0.404 | 0.887 | 1.000 | 1.000 |
| 0.500 | 0.917 | | |

**4.5** Using Program **PIPE** (Appendix C), compute the skin friction for pipe flow according to the Baldwin-Barth and Spalart-Allmaras models. Compare your results with the Prandtl correlation [Equation (3.139) for $10^3 \le Re_D \le 10^6$. Also, compare the computed velocity profiles for $Re_D = 40,000$ with Laufer's data, which are as follows.

| $y/(D/2)$ | $U/U_m$ | $y/(D/2)$ | $U/U_m$ |
|-----------|---------|-----------|---------|
| 0.010 | 0.333 | 0.590 | 0.931 |
| 0.095 | 0.696 | 0.690 | 0.961 |
| 0.210 | 0.789 | 0.800 | 0.975 |
| 0.280 | 0.833 | 0.900 | 0.990 |
| 0.390 | 0.868 | 1.000 | 1.000 |
| 0.490 | 0.902 | | |

**4.6** We wish to create a new two-equation turbulence model. Our first variable is turbulence kinetic energy, $k$, while our second variable is the "eddy acceleration," $a$. Assuming $a$ has dimensions (length)/(time)$^2$, use dimensional arguments to deduce plausible algebraic dependencies of eddy viscosity, $\nu_T$, turbulence energy dissipation rate, $\epsilon$, and turbulence length scale, $\ell$, upon $k$ and $a$.

**4.7** Beginning with the $k$-$\omega$ model and with $\sigma = \sigma^*$, make the formal change of variables $\epsilon = \beta^* \omega k$ and derive the implied $k$-$\epsilon$ model. Express your final results in standard $k$-$\epsilon$ model notation and determine the implied values for $C_\mu$, $C_{\epsilon 1}$, $C_{\epsilon 2}$, $\sigma_k$ and $\sigma_\epsilon$ in terms of $\alpha$, $\beta$, $\beta^*$, $\sigma$ and $\sigma^*$. Assume $f_\beta = f_{\beta^*} = 1$.

**4.8** Beginning with the $k$-$\epsilon$ model, make the formal change of variables $\epsilon = C_\mu \omega k$ and derive the implied $k$-$\omega$ model. Express your final results in standard $k$-$\omega$ model notation and determine the implied values for $\alpha$, $\beta$, $\beta^*$, $\sigma$ and $\sigma^*$ in terms of $C_\mu$, $C_{\epsilon 1}$, $C_{\epsilon 2}$, $\sigma_k$ and $\sigma_\epsilon$.

**4.9** Simplify the $k$-$\epsilon$, $k$-$k\ell$, $k$-$k\tau$ and $k$-$\tau$ models for homogeneous, isotropic turbulence. Determine the asymptotic decay rate for $k$ as a function of the closure coefficient values quoted in Equations (4.49), (4.57), (4.63) and (4.66). Make a table of your results and include the decay rate of $t^{-1.25}$ for the $k$-$\omega$ model. (**NOTE:** You can ignore the $(\ell/y)^6$ contribution to $C_{L2}$ for the $k$-$k\ell$ model.)

**4.10** Simplify the $k$-$\epsilon$, $k$-$k\ell$, $k$-$k\tau$ and $k$-$\tau$ models for the log layer. Determine the value of Kármán's constant, $\kappa$, implied by the closure coefficient values quoted in Equations (4.49), (4.57), (4.63) and (4.66). Make a table of your results and include the value 0.41 for the $k$-$\omega$ model. **NOTE:** For all models, assume a solution of the form $dU/dy = u_\tau/(\kappa y)$, $k = u_\tau^2/\sqrt{C_\mu}$ and $\nu_T = \kappa u_\tau y$. Also, $C_\mu = C_D$ for the $k$-$k\ell$ model.

**4.11** Beginning with Equations (4.84), derive the self-similar form of the $k$-$\omega$ model equations for the mixing layer between a fast stream moving with velocity $U_1$ and a slow stream with velocity $U_2$.

    (a) Assuming a streamfunction of the form $\psi(x, y) = U_1 x F(\eta)$, transform the momentum equation, and verify that $V$ is as given in Table 4.3.

    (b) Transform the equations for $k$ and $\omega$.

    (c) State the boundary conditions on $U$ and $K$ for $|\eta| \to \infty$ and for $V(0)$. Assume $k \to 0$ as $|y| \to \infty$.

    (d) Verify that if $\omega \neq 0$ in the freestream, the only boundary conditions consistent with the similarity solution are:

$$W(\eta) \to \begin{cases} \dfrac{1}{\beta_o}, & \eta \to +\infty \\[2mm] \dfrac{U_1/U_2}{\beta_o}, & \eta \to -\infty \end{cases}$$

**4.12** Derive Equation (4.146).

**4.13** Demonstrate the integral constraint on the defect-layer solution, Equation (4.156).

**4.14** Determine the shape factor to $O(u_\tau/U_e)$ according to the defect-layer solution. Express your answer in terms of an integral involving $U_1(\eta)$.

**4.15** For the $k$-$\omega$ model, very close to the surface and deep within the viscous sublayer, dissipation balances molecular diffusion in the $\omega$ equation. Assuming a solution of the form $\omega = \omega_w/(1 + Ay)^2$, solve this equation for $\omega = \omega_w$ at $y = 0$. Determine the limiting form of the solution as $\omega_w \to \infty$.

**4.16** Consider a flow with freestream velocity $U_\infty$ past a wavy wall whose shape is

$$y = \frac{1}{2}k_s \, \sin\left(\frac{2\pi x}{Nk_s}\right)$$

where $k_s$ is the peak to valley amplitude and $Nk_s$ is wavelength. The linearized incompressible solution valid for $N \gg 1$ is $U = U_\infty + u'$, $V = v'$ where

$$u' = \frac{\pi U}{N} \exp\left(-\frac{2\pi y}{Nk_s}\right) \sin\left(\frac{2\pi y}{Nk_s}\right)$$

$$v' = \frac{\pi U}{N} \exp\left(-\frac{2\pi y}{Nk_s}\right) \cos\left(\frac{2\pi y}{Nk_s}\right)$$

Making an analogy between this linearized solution and the fluctuating velocity field in a turbulent flow, compute the specific dissipation rate, $\omega = \epsilon/(\beta^* k)$. Ignore contributions from the other fluctuating velocity component, $w'$.

**4.17** Using Program **SUBLAY** (Appendix C), determine the variation of the constant $C$ in the law of the wall for the $k$-$\omega$ model with the surface value of $\omega$. Do your computations with ($nvisc = 0$) and without ($nvisc = 1$) viscous modifications. Let $\omega_w^+$ assume the values 1, 3, 10, 30, 100, 300, 1000 and $\infty$. Be sure to use the appropriate value for input parameter *iruff*. Present your results in tabular form.

**4.18** Consider incompressible Couette flow with constant pressure, i.e., flow between two parallel plates separated by a distance $H$, the lower at rest and the upper moving with constant velocity $U_w$.

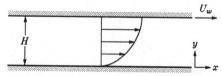

Problems 4.18 and 4.19

(a) Assuming the plates are infinite in extent, simplify the conservation of mass and momentum equations and verify that

$$(\nu + \nu_T)\frac{dU}{dy} = u_\tau^2$$

(b) Now ignore molecular viscosity. What boundary condition on $U$ is appropriate at the lower plate?

(c) Introducing the mixing length given by

$$\ell_{mix} = \kappa y(1 - y/H)$$

solve for the velocity across the channel. **HINT:** Using partial fractions:

$$\frac{1}{y(1 - y/H)} = \frac{1}{y} + \frac{1}{(H - y)}$$

Don't forget to use the boundary condition stated in Part (b).

(d) Develop a relation between friction velocity, $u_\tau$, and the average velocity,

$$U_{avg} = \frac{1}{H} \int_0^H U(y) \, dy$$

(e) Using the $k$-$\omega$ model, simplify the equations for $k$ and $\omega$ with the same assumptions made in Parts (a) and (b).

(f) Deduce the equations for $k$ and $\omega$ that follow from changing independent variables from $y$ to $U$ so that

$$\nu_T \frac{d}{dy} = u_\tau^2 \frac{d}{dU}$$

(g) Assuming $k = u_\tau^2/\sqrt{\beta_o^*}$, simplify the equation for $\omega$. **NOTE:** You might want to use the fact that $(\beta_o - \alpha\beta_o^*) = \sigma\sqrt{\beta_o^*}\kappa^2$.

**4.19** For incompressible, laminar Couette flow, we know that the velocity is given by

$$U = U_w \frac{y}{H}$$

where $U_w$ is the velocity of the moving wall, $y$ is distance form the stationary wall, and $H$ is the distance between the walls.

(a) What is the maximum Reynolds number,

$$Re_{H_c} = U_w H/\nu$$

at which the flow remains laminar according to the high-Reynolds-number version of the $k$-$\omega$ model? To arrive at your answer, you may assume that

$$\omega = \begin{cases} \dfrac{6\nu}{\beta_o y^2}, & 0 \le y \le H/2 \\ \dfrac{6\nu}{\beta_o (H - y)^2}, & H/2 \le y \le H \end{cases}$$

(b) Above what Reynolds number is $\omega$ amplified?

**4.20** This problem studies the effect of viscous-modification closure coefficients for the $k$-$\omega$ model using Program **SUBLAY** (Appendix C).

(a) Modify Subroutine START to permit inputting the values of $R_k$ and $R_\omega$ (program variables $rk$ and $rw$). Determine the value of $R_\omega$ that yields a smooth-wall constant in the law of the wall, $C$, of 5.0 for $R_k$ = 4, 6, 8, 10 and 20.

(b) Now make provision for inputting the value of $R_\beta$ (program variable $rb$). For $R_k$ = 6, determine the value of $R_\omega$ that yields $C$ = 5.0 when $R_\beta$ = 0, 4, 8, and 12. Also, determine the maximum value of $k^+$ for each case.

**4.21** The object of this problem is to compare predictions of one- and two-equation models with measured properties of a turbulent boundary layer with adverse pressure gradient. The experiment to be simulated was conducted by Ludwieg and Tillman [see Coles and Hirst (1969) – Flow 1200]. Use Program **EDDYBL** and its menu-driven setup utility, Program **SETEBL**, to do the computations (see Appendix D).

(a) Using **SETEBL**, change appropriate input parameters to accomplish the following: use SI units (IUTYPE); set freestream conditions to $p_{t_\infty}$ = $1.01858 \cdot 10^5$ N/m$^2$, $T_{t_\infty}$ = 294 K, $M_\infty$ = 0.08656 (PT1, TT1, XMA); use an initial stepsize of $\Delta s$ = 0.01 m (DS); set the initial boundary-layer properties so that $c_f$ = 0.00292, $\delta$ = 0.0224 m, $H$ = 1.36, $Re_\theta$ = 5454, $s_i$ = 0.75 m (CF, DELTA, H, RETHET, SI); set the maximum arc length to $s_f$ = 2.782 m (SSTOP); and, set up for $N$ = 11 points to define the pressure (NUMBER).

(b) Use the following data to define the pressure distribution in a file named **presur.dat**. The initial and final pressure gradients are given by $(dp_e/dx)_i$ = 180.9 N/m$^3$ and $(dp_e/dx)_f$ = $-15.97$ N/m$^3$, respectively. Also, prepare a file **heater.dat** with constant wall temperature, $T_w$ = 294 K, and zero heat flux.

| $s$ (m) | $p_e$ (N/m$^2$) | $s$ (m) | $p_e$ (N/m$^2$) | $s$ (m) | $p_e$ (N/m$^2$) |
|---|---|---|---|---|---|
| 0.000 | $1.01067 \cdot 10^5$ | 2.282 | $1.01415 \cdot 10^5$ | 3.532 | $1.01554 \cdot 10^5$ |
| 0.782 | $1.01201 \cdot 10^5$ | 2.782 | $1.01491 \cdot 10^5$ | 3.732 | $1.01563 \cdot 10^5$ |
| 1.282 | $1.01271 \cdot 10^5$ | 3.132 | $1.01526 \cdot 10^5$ | 3.932 | $1.01562 \cdot 10^5$ |
| 1.782 | $1.01358 \cdot 10^5$ | 3.332 | $1.01541 \cdot 10^5$ | | |

(c) Do three computations using the $k$-$\omega$, Baldwin-Barth and Jones-Launder models.

(d) Compare computed skin friction with the following measured values.

| $s$ (m) | $c_f$ | $s$ (m) | $c_f$ |
|---|---|---|---|
| 0.782 | $2.92 \cdot 10^{-3}$ | 2.282 | $1.94 \cdot 10^{-3}$ |
| 1.282 | $2.49 \cdot 10^{-3}$ | 2.782 | $1.55 \cdot 10^{-3}$ |
| 1.782 | $2.05 \cdot 10^{-3}$ | | |

**4.22** The object of this problem is to compare predictions of one- and two-equation models with measured properties of a turbulent boundary layer with favorable pressure gradient. The experiment to be simulated was conducted by Schubauer and Spangenberg [see Coles and Hirst (1969) – Flow 4800]. Use Program **EDDYBL** and its menu-driven setup utility, Program **SETEBL**, to do the computations (see Appendix D).

(a) Using **SETEBL**, change appropriate input parameters to accomplish the following: use USCS units (IUTYPE); set freestream conditions to $p_{t_\infty} = 2.052 \cdot 10^3$ lb/ft$^2$, $T_{t_\infty} = 528.54°$ R, $M_\infty = 0.0728$ (PT1, TT1, XMA); use an initial stepsize of $\Delta s = 0.008$ ft (DS); set the initial boundary-layer properties so that $c_f = 0.00339$, $\delta = 0.063$ ft, $H = 1.356$, $Re_\theta = 3308$, $s_i = 2.0$ ft (CF, DELTA, H, RETHET, SI); set the maximum arc length to $s_f = 17.833$ ft (SSTOP); and, set up for $N = 16$ points to define the pressure (NUMBER).

(b) Use the following data to define the pressure distribution in a file named **presur.dat**. The initial and final pressure gradients are given by $(dp_e/dx)_i = 0.0$ lb/ft$^3$ and $(dp_e/dx)_f = 1.2 \cdot 10^{-2}$ lb/ft$^3$, respectively. Also, prepare a file **heater.dat** with constant wall temperature, $T_w = 528.5°$ R, and zero heat flux.

| $s$ (ft) | $p_e$ (lb/ft$^2$) | $s$ (ft) | $p_e$ (lb/ft$^2$) | $s$ (ft) | $p_e$ (lb/ft$^2$) |
|---|---|---|---|---|---|
| 1.167 | $2.044405 \cdot 10^3$ | 8.667 | $2.046256 \cdot 10^3$ | 17.833 | $2.049293 \cdot 10^3$ |
| 2.000 | $2.044405 \cdot 10^3$ | 10.333 | $2.046820 \cdot 10^3$ | 18.667 | $2.049478 \cdot 10^3$ |
| 2.833 | $2.044516 \cdot 10^3$ | 12.000 | $2.047399 \cdot 10^3$ | 19.500 | $2.049531 \cdot 10^3$ |
| 3.667 | $2.044662 \cdot 10^3$ | 13.667 | $2.047943 \cdot 10^3$ | 20.333 | $2.049541 \cdot 10^3$ |
| 5.333 | $2.045127 \cdot 10^3$ | 15.333 | $2.048478 \cdot 10^3$ | | |
| 7.000 | $2.045662 \cdot 10^3$ | 17.000 | $2.049045 \cdot 10^3$ | | |

(c) Do three computations using the Baldwin-Barth model, the $k$-$\omega$ model and one of the $k$-$\epsilon$ models.

(d) Compare computed skin friction with the following measured values.

| $s$ (ft) | $c_f$ | $s$ (ft) | $c_f$ | $s$ (ft) | $c_f$ |
|---|---|---|---|---|---|
| 2.000 | $3.39 \cdot 10^{-3}$ | 10.333 | $2.06 \cdot 10^{-3}$ | 17.000 | $0.94 \cdot 10^{-3}$ |
| 4.500 | $2.94 \cdot 10^{-3}$ | 13.667 | $1.61 \cdot 10^{-3}$ | 17.833 | $0.49 \cdot 10^{-3}$ |
| 7.000 | $2.55 \cdot 10^{-3}$ | 15.333 | $1.39 \cdot 10^{-3}$ | | |

**4.23** The object of this problem is to compare predictions of one- and two-equation models with measured properties of a turbulent boundary layer with adverse pressure gradient. The experiment to be simulated was conducted by Schubauer and Spanganberg [see Coles and Hirst (1969) – Flow 4400]. Use Program **EDDYBL** and its menu-driven setup utility, Program **SETEBL**, to do the computations (see Appendix D).

(a) Using **SETEBL**, change the appropriate input parameters to accomplish the following: use USCS units (IUTYPE); set the freestream conditions to $p_{t_\infty} = 2052$ lb/ft$^2$, $T_{t_\infty} = 528.54°$ R, $M_\infty = 0.0728$ (PT1, TT1, XMA); use an initial stepsize $\Delta s = 0.006$ ft (DS); set the initial boundary-layer properties so that $c_f = 0.00340$, $\delta = 0.063$ ft, $H = 1.351$, $Re_\theta = 3066$, $s_i = 1.167$ ft (CF, DELTA, H, RETHET, SI); set the maximum arc length to $s_f = 6.167$ ft and maximum number of steps to 500 (SSTOP, IEND1); and, set up for $N = 15$ points to define the pressure (NUMBER).

(b) Use the following data to define the pressure distribution in a file named **presur.dat**. The initial and final pressure gradients are $(dp_e/dx)_i = 0$ lb/ft$^3$ and $(dp_e/dx)_f = 0.2365$ lb/ft$^3$, respectively. Also, prepare a file **heater.dat** with constant wall temperature, $T_w = 528.54°$ R, and zero heat flux.

| $s$ (ft) | $p_e$ (lb/ft$^2$) | $s$ (ft) | $p_e$ (lb/ft$^2$) | $s$ (ft) | $p_e$ (lb/ft$^2$) |
|---|---|---|---|---|---|
| 1.167 | $2.04441 \cdot 10^3$ | 5.333 | $2.04665 \cdot 10^3$ | 9.500 | $2.04864 \cdot 10^3$ |
| 2.000 | $2.04441 \cdot 10^3$ | 6.167 | $2.04758 \cdot 10^3$ | 10.333 | $2.04870 \cdot 10^3$ |
| 2.833 | $2.04475 \cdot 10^3$ | 7.000 | $2.04826 \cdot 10^3$ | 11.167 | $2.04881 \cdot 10^3$ |
| 3.667 | $2.04516 \cdot 10^3$ | 7.833 | $2.04854 \cdot 10^3$ | 12.000 | $2.04892 \cdot 10^3$ |
| 4.500 | $2.04581 \cdot 10^3$ | 8.667 | $2.04862 \cdot 10^3$ | 12.833 | $2.04911 \cdot 10^3$ |

(c) Do three computations using the $k$-$\omega$ model, one of the $k$-$\epsilon$ models and the Spalart-Allmaras model.

(d) Compare computed skin friction with the following measured values.

| $s$ (ft) | $c_f$ | $s$ (ft) | $c_f$ | $s$ (ft) | $c_f$ |
|---|---|---|---|---|---|
| 1.167 | $3.40 \cdot 10^{-3}$ | 3.667 | $2.86 \cdot 10^{-3}$ | 6.167 | $1.33 \cdot 10^{-3}$ |
| 2.000 | $3.17 \cdot 10^{-3}$ | 4.500 | $2.38 \cdot 10^{-3}$ | | |
| 2.833 | $3.10 \cdot 10^{-3}$ | 5.333 | $1.97 \cdot 10^{-3}$ | | |

**4.24** The object of this problem is to compare predictions of one- and two-equation models with measured properties of a turbulent boundary layer with adverse pressure gradient. The experiment to be simulated was conducted by Stratford [see Coles and Hirst (1969) – Flow 5300]. Use Program **EDDYBL** and its menu-driven setup utility, Program **SETEBL**, to do the computations (see Appendix D).

(a) Using **SETEBL**, change the appropriate input parameters to accomplish the following: use USCS units (IUTYPE); set the freestream conditions to $p_{t_\infty} = 2015$ lb/ft$^2$, $T_{t_\infty}$ = 504° R, $M_\infty$ = 0.05 (PT1, TT1, XMA); use an initial stepsize $\Delta s$ = 0.0001 ft, a geometric-progression ratio $k_g$ = 1.06, and near-wall boundary condition distance $y^+$ = 1 (DS, XK, USTOP); set the initial boundary-layer properties so that $\Lambda$ = 4, $c_f$ = 0.00368, $\delta$ = 0.0609 ft, $H$ = 1.3616, $Re_\theta$ = 2295, $s_i$ = 2.9075 ft (ALAMM, CF, DELTA, H, RETHET, SI); set the maximum arc length to $s_f$ = 4.103 ft and maximum number of steps to 500 (SSTOP, IEND1); and, set up for $N$ = 8 points to define the pressure (NUMBER).

(b) Use the following data to define the pressure distribution in a file named **presur.dat**. The initial and final pressure gradients are $(dp_e/dx)_i$ = 0.406006 lb/ft$^3$ and $(dp_e/dx)_f$ = 1.38916 lb/ft$^3$, respectively. Also, prepare a file **heater.dat** with constant wall temperature, $T_w$ = 504° R, and zero heat flux.

| $s$ (ft) | $p_e$ (lb/ft$^2$) | $s$ (ft) | $p_e$ (lb/ft$^2$) | $s$ (ft) | $p_e$ (lb/ft$^2$) |
|---|---|---|---|---|---|
| 1.4308 | $2.01132 \cdot 10^3$ | 2.9976 | $2.01188 \cdot 10^3$ | 4.4496 | $2.01337 \cdot 10^3$ |
| 2.9075 | $2.01153 \cdot 10^3$ | 3.0580 | $2.01225 \cdot 10^3$ | 6.4407 | $2.01391 \cdot 10^3$ |
| 2.9903 | $2.01181 \cdot 10^3$ | 3.1619 | $2.01249 \cdot 10^3$ | | |

(c) Do three computations using the $k$-$\omega$ model with viscous corrections (MODEL = 0, NVISC = 1), one of the $k$-$\epsilon$ models and the Spalart-Allmaras model.

(d) Compare computed skin friction with the following measured values.

| $s$ (ft) | $c_f$ | $s$ (ft) | $c_f$ |
|---|---|---|---|
| 2.907 | $3.68 \cdot 10^{-3}$ | 3.531 | $0.55 \cdot 10^{-3}$ |
| 2.999 | $2.07 \cdot 10^{-3}$ | 4.103 | $0.53 \cdot 10^{-3}$ |
| 3.038 | $0.99 \cdot 10^{-3}$ | | |

**4.25** The object of this problem is to predict the separation point for flow past a circular cylinder with the boundary-layer equations, using the measured pressure distribution. The experiment to be simulated was conducted by Patel (1968). Use Program **EDDYBL** and its menu-driven setup utility, Program **SETEBL**, to do the computations (see Appendix D).

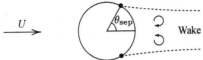

Problem 4.25

(a) Using **SETEBL**, change the appropriate input parameters to accomplish the following: use USCS units (IUTYPE); set the freestream conditions to $p_{t_\infty} = 2147.7$ lb/ft$^2$, $T_{t_\infty} = 529.6°$ R, $M_\infty = 0.144$ (PT1, TT1, XMA); use an initial stepsize $\Delta s = 0.001$ ft (DS); set the initial boundary-layer properties so that $c_f = 0.00600$, $\delta = 0.006$ ft, $H = 1.40$, $Re_\theta = 929$, $s_i = 0.262$ ft (CF, DELTA, H, RETHET, SI); set the maximum arc length to $s_f = 0.785$ ft and maximum number of steps to 300 (SSTOP, IEND1); set up for $N = 47$ points to define the pressure (NUMBER); and, make provision for specifying body curvature (NFLAG).

(b) Use the following data to define the pressure distribution in a file named **presur.dat**. The initial and final pressure gradients are both zero. Then, prepare a file named **heater.dat** with constant wall temperature, $T_w = 529.4°$ R, and zero heat flux. Finally, prepare a file named **blocrv.dat** with constant curvature, $\mathcal{R}^{-1} = 4$ ft$^{-1}$.

| $s$ (ft) | $p_e$ (lb/ft$^2$) | $s$ (ft) | $p_e$ (lb/ft$^2$) | $s$ (ft) | $p_e$ (lb/ft$^2$) |
|---|---|---|---|---|---|
| .0000 | $2.147540 \cdot 10^3$ | .1500 | $2.116199 \cdot 10^3$ | .3500 | $2.055516 \cdot 10^3$ |
| .0025 | $2.147528 \cdot 10^3$ | .1625 | $2.112205 \cdot 10^3$ | .3625 | $2.056591 \cdot 10^3$ |
| .0050 | $2.147491 \cdot 10^3$ | .1750 | $2.107903 \cdot 10^3$ | .3750 | $2.058435 \cdot 10^3$ |
| .0075 | $2.147429 \cdot 10^3$ | .1875 | $2.103448 \cdot 10^3$ | .3875 | $2.061661 \cdot 10^3$ |
| .0100 | $2.147343 \cdot 10^3$ | .2000 | $2.098378 \cdot 10^3$ | .4000 | $2.066423 \cdot 10^3$ |
| .0125 | $2.147233 \cdot 10^3$ | .2125 | $2.093155 \cdot 10^3$ | .4125 | $2.071954 \cdot 10^3$ |
| .0250 | $2.146314 \cdot 10^3$ | .2250 | $2.087317 \cdot 10^3$ | .4250 | $2.079021 \cdot 10^3$ |
| .0375 | $2.144796 \cdot 10^3$ | .2375 | $2.081325 \cdot 10^3$ | .4375 | $2.085473 \cdot 10^3$ |
| .0500 | $2.142688 \cdot 10^3$ | .2500 | $2.075334 \cdot 10^3$ | .4500 | $2.089161 \cdot 10^3$ |
| .0625 | $2.140018 \cdot 10^3$ | .2625 | $2.069189 \cdot 10^3$ | .4625 | $2.091004 \cdot 10^3$ |
| .0750 | $2.136807 \cdot 10^3$ | .2750 | $2.064580 \cdot 10^3$ | .4750 | $2.092080 \cdot 10^3$ |
| .0875 | $2.134021 \cdot 10^3$ | .2875 | $2.060893 \cdot 10^3$ | .4875 | $2.092230 \cdot 10^3$ |
| .1000 | $2.130641 \cdot 10^3$ | .3000 | $2.058588 \cdot 10^3$ | .5000 | $2.092230 \cdot 10^3$ |
| .1125 | $2.127261 \cdot 10^3$ | .3125 | $2.056898 \cdot 10^3$ | .6500 | $2.092230 \cdot 10^3$ |
| .1250 | $2.123881 \cdot 10^3$ | .3250 | $2.055823 \cdot 10^3$ | .7850 | $2.092230 \cdot 10^3$ |
| .1375 | $2.120194 \cdot 10^3$ | .3375 | $2.055362 \cdot 10^3$ | | |

(c) Do three computations using the $k$-$\omega$ model, the Launder-Sharma $k$-$\epsilon$ model and the Spalart-Allmaras model. (Avoid the Lam-Bremhorst, Chien and Yang-Shih $k$-$\epsilon$ models — they are very difficult to implement numerically for this type of flow.)

(d) Compare computed separation angle measured from the downstream symmetry axis with the measured value of $\theta_{sep} = 70°$. The radius of the cylinder is $R = 0.25$ ft, so that separation arc length, $s_{sep}$, is related to this angle by $\theta_{sep} = \pi - s_{sep}/R$.

# Chapter 5

# Effects of Compressibility

For flows in which compressibility effects are important, we must introduce an equation for conservation of energy and an equation of state. Just as Reynolds averaging gives rise to the Reynolds-stress tensor, so we expect that similar averaging will lead to a turbulent heat-flux vector. We should also expect that new compressibility-related correlations will appear throughout the equations of motion. These are important issues that must be addressed in constructing a turbulence model suitable for application to compressible flows, which can be expected to apply to constant-property (low-speed) flows with heat transfer as a special case.

We begin with a brief discussion of common observations pertaining to compressible turbulence. Then, we introduce the Favre mass-averaging procedure and derive the mass-averaged equations of motion. Next, we demonstrate an elegant development in turbulence modeling for the compressible mixing layer. We follow this analysis with an application of perturbation methods to the compressible log layer. We then apply several models to attached compressible boundary layers, including effects of pressure gradient, surface cooling and surface roughness. The chapter concludes with application of various turbulence models to shock-separated flows.

## 5.1 Physical Considerations

By definition, a compressible flow is one in which significant density changes occur, even when pressure changes are small. It includes low-speed flows with large heat-transfer rates. Models for high-speed flows seem to fit the limited data quite well (perhaps with the exception of combusting flows). Generally speaking, compressibility has a relatively small effect on turbulent eddies in

227

wall-bounded flows. This appears to be true for Mach numbers up to about 5 (and perhaps as high as 8), provided the flow doesn't experience large pressure changes over a short distance such as we might have across a shock wave. At subsonic speeds, compressibility effects on eddies are usually unimportant for boundary layers provided $T_w/T_e < 6$. Based on these observations, Morkovin (1962) hypothesized that the effect of density fluctuations on the turbulence is small provided they remain small relative to the mean density. This is a major simplification for the turbulence modeler because it means that, in practice, we need only account for the nonuniform mean density in computing compressible, shock-free, non-hypersonic turbulent flows.

There are limitations to the usefulness of **Morkovin's hypothesis** even at non-hypersonic Mach numbers. For example, it is not useful in flows with significant heat transfer or in flows with combustion because $\rho'/\bar{\rho}$ is typically not small. Also, density fluctuations generally are much larger in free shear flows, and models based on Morkovin's hypothesis fail to predict the measured reduction in spreading rate with increasing freestream Mach number for the compressible mixing layer [e.g., Papamoschou and Roshko (1988)]. As we will see in Section 5.5, the level of $\rho'/\bar{\rho}$ for a boundary layer at Mach 5 is comparable to the level found in a mixing layer at Mach 1. However, in addition, there seem to be qualitative changes in mixing-layer structure as Mach number increases.

On dimensional grounds, we expect the velocity in a turbulent boundary layer to depend, at a minimum, upon basic fluid properties such as Prandtl number, $Pr_L$, and specific-heat ratio, $\gamma$. We also expect it to depend upon the following three dimensionless groupings.

$$\underbrace{y^+ = \frac{u_\tau y}{\nu_w}}_{\substack{Sublayer\ scaled \\ distance}} \quad , \quad \underbrace{q_w^+ = \frac{q_w}{\rho_w c_p u_\tau T_w}}_{\substack{Dimensionless \\ heat\ transfer}}, \quad \underbrace{M_\tau = \frac{u_\tau}{a_w}}_{\substack{Turbulence \\ Mach\ number}} \qquad (5.1)$$

where subscript $w$ denotes surface (wall) value, $q$ is heat flux, $c_p$ is specific-heat coefficient at constant pressure, $T$ is temperature and $a$ is sound speed. Based on the mixing-length model and the assumption that, in analogy to the incompressible case,

$$\frac{\partial U}{\partial y} \approx \frac{\sqrt{\tau_w/\rho}}{\kappa y} \qquad (5.2)$$

where $\rho$ now depends upon $y$. Van Driest (1951) argued[1] that by rescaling the velocity according to

$$\frac{u^*}{u_\tau} = \frac{1}{A}\left[\sin^{-1}\left(\frac{2A^2(U/u_\tau) - B}{\sqrt{B^2 + 4A^2}}\right) + \sin^{-1}\left(\frac{B}{\sqrt{B^2 + 4A^2}}\right)\right] \qquad (5.3)$$

---

[1]The Van Driest argument also requires assuming the turbulent Prandtl number [defined in Equation (5.54)] is constant.

where $A$ and $B$ are functions of $q_w^+$ and $M_\tau$ [see Equation (5.93) below], the velocity is

$$\frac{u^*}{u_\tau} = \frac{1}{\kappa}\ell n y^+ + C \tag{5.4}$$

Equation (5.4) is the **compressible law of the wall**. Correlation of measurements shows that $\kappa$ and $C$ are nearly the same as for incompressible boundary layers [Bradshaw and Huang (1995)]. In principle, however, $C$ is a function of $M_\tau$ and $q_w^+$ since it includes density and viscosity effects in the viscous wall region.

Section 5.6 provides additional detail that explains why we should expect the velocity to scale according to Equation (5.3) in a compressible boundary layer. In general, the compressible law of the wall correlates experimental data for adiabatic walls reasonably well (Section 5.7). It is less accurate for non-adiabatic walls, especially for very cold walls (probably because $C$ varies with $q_w^+$, although data are scarce). An analogous variation of temperature with these parameters can be deduced that is satisfactory for low-speed flows. However, its use is limited because of sensitivity to pressure gradient, even in low-speed flows. Bradshaw and Huang (1995) provide additional detail.

As a final observation, note that the difficulty in predicting properties of the compressible mixing layer is reminiscent of our experience with free shear flows in Chapters 3 and 4. That is, we find again that the seemingly simple free shear flow case is more difficult to model than the wall-bounded case.

## 5.2 Favre Averaging

In addition to velocity and pressure fluctuations, we must also account for density and temperature fluctuations when the medium is a compressible fluid. If we use the standard time-averaging procedure introduced in Chapter 2, the mean conservation equations contain additional terms that have no analogs in the laminar equations. To illustrate this, consider conservation of mass. We write the instantaneous density $\rho$ as the sum of mean, $\bar{\rho}$, and fluctuating, $\rho'$, parts, i.e.,

$$\rho = \bar{\rho} + \rho' \tag{5.5}$$

Expressing the instantaneous velocity in the usual way [Equation (2.4)], substituting into the continuity equation yields

$$\frac{\partial}{\partial t}(\bar{\rho} + \rho') + \frac{\partial}{\partial x_i}(\bar{\rho}U_i + \rho'U_i + \bar{\rho}u_i' + \rho'u_i') = 0 \tag{5.6}$$

After time averaging Equation (5.6), we arrive at the Reynolds-averaged continuity equation for compressible flow, viz.,

$$\frac{\partial\bar{\rho}}{\partial t} + \frac{\partial}{\partial x_i}\left(\bar{\rho}U_i + \overline{\rho'u_i'}\right) = 0 \tag{5.7}$$

Some authors refer to this as the **primitive-variable form** of the continuity equation. Note that in order to achieve closure, an approximation for the correlation between $\rho'$ and $u_i'$ is needed. The problem is even more complicated for the momentum equation where the Reynolds-stress tensor originates from time averaging the product $\rho u_i u_j$ that appears in the convective acceleration. Clearly, a triple correlation involving $\rho'$, $u_i'$, and $u_j'$ appears, thus increasing the complexity of establishing suitable closure approximations.

The problem of establishing the appropriate form of the time-averaged equations can be simplified dramatically by using the density-weighted averaging procedure suggested by Favre (1965). That is, we introduce the **mass-averaged** velocity, $\tilde{u}_i$, defined by

$$\tilde{u}_i = \frac{1}{\bar{\rho}} \lim_{T \to \infty} \frac{1}{T} \int_t^{t+T} \rho(\mathbf{x}, \tau) u_i(\mathbf{x}, \tau) \, d\tau \qquad (5.8)$$

where $\bar{\rho}$ is the conventional Reynolds-averaged density. Thus, in terms of conventional Reynolds averaging, we can say that

$$\bar{\rho}\tilde{u}_i = \overline{\rho u_i} \qquad (5.9)$$

where an overbar denotes conventional Reynolds average. The value of this averaging process, known as **Favre averaging**, becomes obvious when we expand the right-hand side of Equation (5.9). Performing the indicated Reynolds-averaging process, there follows

$$\bar{\rho}\tilde{u}_i = \bar{\rho}U_i + \overline{\rho'u_i'} \qquad (5.10)$$

Inspection of Equation (5.7) shows that conservation of mass can be rewritten as

$$\frac{\partial \bar{\rho}}{\partial t} + \frac{\partial}{\partial x_i}\left(\bar{\rho}\tilde{u}_i\right) = 0 \qquad (5.11)$$

This is a remarkable simplification as Equation (5.11) looks just like the laminar mass-conservation equation. What we have done is treat the momentum per unit volume, $\rho u_i$, as the dependent variable rather than the velocity. This is a sensible thing to do from a physical point of view, especially when we focus upon the momentum equation in the next section. That is, the rate of change of momentum per unit volume, not velocity, is equal to the sum of the imposed forces per unit volume in a flow.

When we use Favre averaging, it is customary to decompose the instantaneous velocity into the mass-averaged part, $\tilde{u}_i$, and a fluctuating part, $u_i''$, wherefore

$$u_i = \tilde{u}_i + u_i'' \qquad (5.12)$$

Now, to form the Favre average, we simply multiply through by $\rho$ and do a time average in the manner established in Chapter 2. Hence, from Equation (5.12) we find

$$\overline{\rho u_i} = \bar{\rho}\tilde{u}_i + \overline{\rho u_i''} \qquad (5.13)$$

But, from the definition of the Favre average given in Equation (5.9), we see immediately that, as expected, the Favre average of the fluctuating velocity, $u_i''$, vanishes, i.e.,

$$\overline{\rho u_i''} = 0 \qquad (5.14)$$

By contrast, the conventional Reynolds average of $u_i''$ is not zero. To see this, note that

$$u_i'' = u_i - \tilde{u}_i \qquad (5.15)$$

Hence, using Equation (5.10) to eliminate $\tilde{u}_i$,

$$u_i'' = u_i - U_i - \frac{\overline{\rho' u_i'}}{\bar{\rho}} \qquad (5.16)$$

Therefore, performing the conventional Reynolds average, we find

$$\overline{u_i''} = -\frac{\overline{\rho' u_i'}}{\bar{\rho}} \neq 0 \qquad (5.17)$$

As a final comment, do not lose sight of the fact that while Favre averaging eliminates density fluctuations from the averaged equations, it does not remove the effect the density fluctuations have on the turbulence. Consequently, **Favre averaging is a mathematical simplification, not a physical one.**

## 5.3 Favre-Averaged Equations

For motion in a compressible medium, we must solve the equations governing conservation of mass, momentum and energy. The instantaneous equations are as follows:

$$\frac{\partial \rho}{\partial t} + \frac{\partial}{\partial x_i}(\rho u_i) = 0 \qquad (5.18)$$

$$\frac{\partial}{\partial t}(\rho u_i) + \frac{\partial}{\partial x_j}(\rho u_j u_i) = -\frac{\partial p}{\partial x_i} + \frac{\partial t_{ji}}{\partial x_j} \qquad (5.19)$$

$$\frac{\partial}{\partial t}\left[\rho\left(e + \frac{1}{2}u_i u_i\right)\right] + \frac{\partial}{\partial x_j}\left[\rho u_j\left(h + \frac{1}{2}u_i u_i\right)\right] = \frac{\partial}{\partial x_j}(u_i t_{ij}) - \frac{\partial q_j}{\partial x_j} \qquad (5.20)$$

where $e$ is specific internal energy and $h = e + p/\rho$ is specific enthalpy. For compressible flow, the viscous stress tensor, $t_{ij}$, involves the second viscosity, $\zeta$, as well as the conventional molecular viscosity, $\mu$. Although it is not necessary for our immediate purposes, we eventually must specify an equation of state. For gases, we use the perfect-gas law so that pressure, density and temperature are related by

$$p = \rho R T \qquad (5.21)$$

where $R$ is the perfect-gas constant. The constitutive relation between stress and strain rate for a Newtonian fluid is

$$t_{ij} = 2\mu s_{ij} + \zeta \frac{\partial u_k}{\partial x_k} \delta_{ij} \qquad (5.22)$$

where $s_{ij}$ is the instantaneous strain-rate tensor [Equation (2.19)] and $\delta_{ij}$ is the Kronecker delta. The heat-flux vector, $q_j$, is usually obtained from Fourier's law so that

$$q_j = -\kappa \frac{\partial T}{\partial x_j} \qquad (5.23)$$

where $\kappa$ is thermal conductivity. We can simplify our analysis somewhat by introducing two commonly used assumptions. First, we relate second viscosity to $\mu$ by assuming

$$\zeta = -\frac{2}{3} \mu \qquad (5.24)$$

This assumption is correct for a monatomic gas, and is generally used for all gases in standard CFD applications. Assuming Equation (5.24) holds in general guarantees $t_{ii} = 0$ so that viscous stresses do not contribute to the pressure, even when $s_{ii} = \partial u_i/\partial x_i \neq 0$. This is tidy, even if not necessarily true. Second, we assume the fluid is calorically perfect so that its specific-heat coefficients are constant, and thus the specific internal energy, $e$, and specific enthalpy, $h$, are

$$e = c_v T \qquad \text{and} \qquad h = c_p T \qquad (5.25)$$

where $c_v$ and $c_p$ are the specific-heat coefficients for constant volume and pressure processes, respectively. Then, we can say that

$$q_j = -\kappa \frac{\partial T}{\partial x_j} = -\frac{\mu}{Pr_L} \frac{\partial h}{\partial x_j} \qquad (5.26)$$

where $Pr_L$ is the **laminar Prandtl number** defined by

$$Pr_L = \frac{c_p \mu}{\kappa} \qquad (5.27)$$

In order to mass average the conservation equations, we now decompose the various flow properties as follows.

$$\left. \begin{array}{rcl} u_i &=& \tilde{u}_i + u_i'' \\ \rho &=& \bar{\rho} + \rho' \\ p &=& P + p' \\ h &=& \tilde{h} + h'' \\ e &=& \tilde{e} + e'' \\ T &=& \tilde{T} + T'' \\ q_j &=& q_{L_j} + q_j' \end{array} \right\} \qquad (5.28)$$

Note that we decompose $p$, $\rho$ and $q_j$ in terms of conventional mean and fluctuating parts. Substituting Equations (5.28) into Equations (5.18) – (5.21) and performing the mass-averaging operations, we arrive at what are generally referred to as the **Favre (mass) averaged mean conservation equations**.

$$\frac{\partial \bar{\rho}}{\partial t} + \frac{\partial}{\partial x_i} \left( \bar{\rho} \tilde{u}_i \right) = 0 \tag{5.29}$$

$$\frac{\partial}{\partial t} \left( \bar{\rho} \tilde{u}_i \right) + \frac{\partial}{\partial x_j} \left( \bar{\rho} \tilde{u}_j \tilde{u}_i \right) = -\frac{\partial P}{\partial x_i} + \frac{\partial}{\partial x_j} \left[ \bar{t}_{ji} - \overline{\rho u_j'' u_i''} \right] \tag{5.30}$$

$$\frac{\partial}{\partial t} \left[ \bar{\rho} \left( \tilde{e} + \frac{\tilde{u}_i \tilde{u}_i}{2} \right) + \frac{\overline{\rho u_i'' u_i''}}{2} \right] + \frac{\partial}{\partial x_j} \left[ \bar{\rho} \tilde{u}_j \left( \tilde{h} + \frac{\tilde{u}_i \tilde{u}_i}{2} \right) + \tilde{u}_j \frac{\overline{\rho u_i'' u_i''}}{2} \right]$$

$$= \frac{\partial}{\partial x_j} \left[ -q_{L_j} - \overline{\rho u_j'' h''} + \overline{t_{ji} u_i''} - \overline{\rho u_j'' \tfrac{1}{2} u_i'' u_i''} \right]$$

$$+ \frac{\partial}{\partial x_j} \left[ \tilde{u}_i \left( \bar{t}_{ij} - \overline{\rho u_i'' u_j''} \right) \right] \tag{5.31}$$

$$P = \bar{\rho} R \tilde{T} \tag{5.32}$$

Equations (5.29), (5.30) and (5.32) differ from their laminar counterparts only by the appearance of the Favre-averaged Reynolds-stress tensor, viz.,

$$\bar{\rho} \tau_{ij} = -\overline{\rho u_i'' u_j''} \tag{5.33}$$

As in the incompressible case, the Favre-averaged $\tau_{ij}$ is a symmetric tensor.

Equation (5.31), the Favre-averaged mean-energy equation for total energy, i.e., the sum of internal energy, mean-flow kinetic energy and turbulence kinetic energy has numerous additional terms, each of which represents an identifiable physical process or property. Consider first the double correlation between $u_i''$ and itself that appears in each of the two terms on the left-hand side. This is the kinetic energy per unit volume of the turbulent fluctuations, so that it makes sense to define

$$\bar{\rho} k = \frac{1}{2} \overline{\rho u_i'' u_i''} \tag{5.34}$$

Next, the correlation between $u_j''$ and $h''$ is the turbulent transport of heat. In analogy to the notation selected for the molecular transport of heat, we define

$$q_{T_j} = \overline{\rho u_j'' h''} \tag{5.35}$$

The two terms $\overline{t_{ji} u_i''}$ and $\overline{\rho u_j'' \tfrac{1}{2} u_i'' u_i''}$ on the right-hand side of Equation (5.31) correspond to molecular diffusion and turbulent transport of turbulence kinetic energy, respectively. These terms arise because the mass-averaged total enthalpy appearing in the convective term of Equation (5.31) is the sum of mass-averaged enthalpy, mean kinetic energy and turbulence kinetic energy. They represent transfers between mean energy and turbulence kinetic energy, that naturally arise when we derive the Favre-averaged turbulence kinetic energy equation. The simplest way to derive the equation for $k$ is to multiply the primitive-variable form of the instantaneous momentum equation by $u_i''$ and time average.

$$\overline{\rho u_i'' \frac{\partial u_i}{\partial t}} + \overline{\rho u_i'' u_j \frac{\partial u_i}{\partial x_j}} = -\overline{u_i'' \frac{\partial p}{\partial x_i}} + \overline{u_i'' \frac{\partial t_{ji}}{\partial x_j}} \qquad (5.36)$$

As in Chapter 2, the most illuminating way to carry out the indicated time-averaging operations is to proceed term by term, and to use tensor notation for all derivatives. Proceeding from left to right, we first consider the **unsteady term**.

$$
\begin{aligned}
\overline{\rho u_i'' u_{i,t}} &= \overline{\rho u_i'' (\tilde{u}_i + u_i'')_{,t}} \\
&= \overline{\rho u_i'' \tilde{u}_{i,t}} + \overline{\rho u_i'' u_{i,t}''} \\
&= \overline{\rho (\tfrac{1}{2} u_i'' u_i'')_{,t}} \\
&= \frac{\partial}{\partial t}(\bar{\rho} k) - \tfrac{1}{2} \overline{u_i'' u_i''} \frac{\partial \rho}{\partial t}
\end{aligned}
\qquad (5.37)
$$

Turning now to the **convective term**, we have the following.

$$
\begin{aligned}
\overline{\rho u_i'' u_j u_{i,j}} &= \overline{\rho u_i'' [(\tilde{u}_j + u_j'')\tilde{u}_{i,j} + u_j u_{i,j}'']} \\
&= \overline{\rho u_i'' \tilde{u}_j \tilde{u}_{i,j}} + \overline{\rho u_i'' u_j'' \, \tilde{u}_{i,j}} + \overline{\rho u_j u_i'' u_{i,j}''} \\
&= -\bar{\rho} \tau_{ij} \tilde{u}_{i,j} + \overline{\rho u_j (\tfrac{1}{2} u_i'' u_i'')_{,j}} \\
&= -\bar{\rho} \tau_{ij} \tilde{u}_{i,j} + \overline{(\rho u_j \tfrac{1}{2} u_i'' u_i'')_{,j}} - \tfrac{1}{2} \overline{u_i'' u_i'' (\rho u_j)_{,j}} \\
&= -\bar{\rho} \tau_{ij} \tilde{u}_{i,j} + \left( \overline{\rho \tilde{u}_j \tfrac{1}{2} u_i'' u_i''} + \overline{\rho u_j'' \tfrac{1}{2} u_i'' u_i''} \right)_{,j} - \tfrac{1}{2} \overline{u_i'' u_i'' (\rho u_j)_{,j}} \\
&= -\bar{\rho} \tau_{ij} \frac{\partial \tilde{u}_i}{\partial x_j} + \frac{\partial}{\partial x_j} \left( \overline{\rho \tilde{u}_j k} + \overline{\rho u_j'' \tfrac{1}{2} u_i'' u_i''} \right) - \tfrac{1}{2} \overline{u_i'' u_i'' \frac{\partial}{\partial x_j}(\rho u_j)}
\end{aligned}
$$
$$(5.38)$$

The **pressure-gradient term** simplifies immediately as follows.

$$\overline{u_i'' p_{,i}} = \overline{u_i'' P_{,i}} + \overline{u_i'' p_{,i}'} = \overline{u_i''} \frac{\partial P}{\partial x_i} + \frac{\partial}{\partial x_i} \left( \overline{p' u_i''} \right) - \overline{p' \frac{\partial u_i''}{\partial x_i}} \qquad (5.39)$$

Finally, the **viscous term** is simply rewritten as

$$\overline{u_i'' t_{ji,j}} = \frac{\partial}{\partial x_j}\left(\overline{t_{ji}u_i''}\right) - \overline{t_{ji}\frac{\partial u_i''}{\partial x_j}} \tag{5.40}$$

Thus, substituting Equations (5.37) through (5.40) into Equation (5.36), we arrive at the **Favre-averaged turbulence kinetic energy equation**. In arriving at the final result, we make use of the fact that the sum of the last terms on the right-hand sides of Equations (5.37) and (5.38) vanish since their sum is proportional to the two terms appearing in the instantaneous continuity equation. Additionally, to facilitate comparison with the incompressible turbulence kinetic energy equation [Equation (4.4)], we use the Favre-averaged continuity equation to rewrite the unsteady and convective terms in non-conservation form. The exact equation is as follows.

$$\begin{aligned}
\bar{\rho}\frac{\partial k}{\partial t} + \bar{\rho}\tilde{u}_j\frac{\partial k}{\partial x_j} &= \overline{\rho\tau_{ij}}\frac{\partial \tilde{u}_i}{\partial x_j} - \overline{t_{ji}\frac{\partial u_i''}{\partial x_j}} + \frac{\partial}{\partial x_j}\left[\overline{t_{ji}u_i''} - \overline{\rho u_j'' \tfrac{1}{2} u_i'' u_i''} - \overline{p' u_j''}\right] \\
&\quad - \underbrace{\overline{u_i''}\frac{\partial P}{\partial x_i}}_{\text{Pressure Work}} + \underbrace{\overline{p'\frac{\partial u_i''}{\partial x_i}}}_{\text{Pressure Dilatation}}
\end{aligned} \tag{5.41}$$

Comparing the mean energy Equation (5.31) with the turbulence kinetic energy Equation (5.41), we see that indeed the two terms $\overline{t_{ji}u_i''}$ and $\overline{\rho u_j'' \tfrac{1}{2} u_i'' u_i''}$ on the right-hand side of the mean-energy equation are **Molecular Diffusion** and **Turbulent Transport** of turbulence kinetic energy. Inspection of the turbulence kinetic energy equation also indicates that the **Favre-averaged dissipation rate** is given by

$$\bar{\rho}\epsilon = \overline{t_{ji}\frac{\partial u_i''}{\partial x_j}} = \frac{1}{2}\overline{t_{ji}\left(\frac{\partial u_i''}{\partial x_j} + \frac{\partial u_j''}{\partial x_i}\right)} = \overline{t_{ji}s_{ij}''} \tag{5.42}$$

where $s_{ij}''$ is the fluctuating strain-rate tensor. This is entirely consistent with the definition of dissipation for incompressible flows given in Equation (4.6).

Comparison of Equation (5.41) with the incompressible equation for $k$ [Equation (4.4)] shows that all except the last two terms, i.e., the **Pressure Work** and **Pressure-Dilatation** terms, have analogs in the incompressible equation. Both of these terms vanish in the limit of incompressible flow with zero density fluctuations. The **Pressure Work** vanishes because the time average of $u_i''$ is zero when density fluctuations are zero. The **Pressure-Dilatation** term vanishes because the fluctuating field has zero divergence for incompressible flow. Hence, Equation (5.41) simplifies to Equation (4.4) for incompressible flow with zero density fluctuations.

Note that the turbulence kinetic energy production, $\bar{\rho}\tau_{ij}\partial\tilde{u}_i/\partial x_j$, and pressure correlation terms represent a transfer from mean kinetic energy to turbulence kinetic energy. Also, dissipation is a transfer from turbulence kinetic energy to internal energy. Thus, since these transfers simply redistribute energy, they must cancel in the overall energy-conservation equation. Consequently, only the two terms involving spatial transport of turbulence kinetic energy appear in Equation (5.31).

Using a similar derivation (we omit the details here for the sake of brevity), the **Favre-averaged Reynolds-stress equation** assumes the following form.

$$\frac{\partial}{\partial t}\left(\bar{\rho}\tau_{ij}\right) + \frac{\partial}{\partial x_k}\left(\bar{\rho}\tilde{u}_k\tau_{ij}\right) = -\bar{\rho}\tau_{ik}\frac{\partial\tilde{u}_j}{\partial x_k} - \bar{\rho}\tau_{jk}\frac{\partial\tilde{u}_i}{\partial x_k} + \bar{\rho}\epsilon_{ij} - \bar{\rho}\Pi_{ij}$$

$$+ \frac{\partial}{\partial x_k}\left[-\overline{\left(t_{kj}u_i'' + t_{ki}u_j''\right)} + \bar{\rho}C_{ijk}\right]$$

$$+ \overline{u_i''}\frac{\partial P}{\partial x_j} + \overline{u_j''}\frac{\partial P}{\partial x_i} \tag{5.43}$$

where

$$\bar{\rho}\Pi_{ij} = \overline{p'\left(\frac{\partial u_i''}{\partial x_j} + \frac{\partial u_j''}{\partial x_i}\right)} \tag{5.44}$$

$$\bar{\rho}\epsilon_{ij} = \overline{t_{kj}\frac{\partial u_i''}{\partial x_k} + t_{ki}\frac{\partial u_j''}{\partial x_k}} \tag{5.45}$$

$$\bar{\rho}C_{ijk} = \overline{\rho u_i''u_j''u_k''} + \overline{p'u_i''}\delta_{jk} + \overline{p'u_j''}\delta_{ik} \tag{5.46}$$

Taking advantage of the definitions given in Equations (5.33), (5.34), (5.35) and (5.42), we can summarize the Favre-averaged mean equations and turbulence kinetic energy equation in conservation form.

$$\frac{\partial\bar{\rho}}{\partial t} + \frac{\partial}{\partial x_i}\left(\bar{\rho}\tilde{u}_i\right) = 0 \tag{5.47}$$

$$\frac{\partial}{\partial t}\left(\bar{\rho}\tilde{u}_i\right) + \frac{\partial}{\partial x_j}\left(\bar{\rho}\tilde{u}_j\tilde{u}_i\right) = -\frac{\partial P}{\partial x_i} + \frac{\partial}{\partial x_j}\left[\bar{t}_{ji} + \bar{\rho}\tau_{ji}\right] \tag{5.48}$$

$$\frac{\partial}{\partial t}(\bar{\rho}E) + \frac{\partial}{\partial x_j}(\bar{\rho}\tilde{u}_j H) = \frac{\partial}{\partial x_j}\left[-q_{L_j} - q_{T_j} + \overline{t_{ji}u_i''} - \overline{\rho u_j''\tfrac{1}{2}u_i''u_i''}\right]$$

$$+ \frac{\partial}{\partial x_j}\left[\tilde{u}_i(\bar{t}_{ij} + \bar{\rho}\tau_{ij})\right] \tag{5.49}$$

$$\frac{\partial}{\partial t}(\bar{\rho}k) + \frac{\partial}{\partial x_j}(\bar{\rho}\tilde{u}_j k) = \bar{\rho}\tau_{ij}\frac{\partial\tilde{u}_i}{\partial x_j} - \bar{\rho}\epsilon + \frac{\partial}{\partial x_j}\left[\overline{t_{ji}u_i''} - \overline{\rho u_j''\tfrac{1}{2}u_i''u_i''} - \overline{p'u_j''}\right]$$

$$- \overline{u_i''}\frac{\partial P}{\partial x_i} + \overline{p'\frac{\partial u_i''}{\partial x_i}} \tag{5.50}$$

$$P = \bar{\rho} R \tilde{T} \tag{5.51}$$

The quantities $E$ and $H$ are the **total energy** and **total enthalpy**, and include the kinetic energy of the fluctuating turbulent field, viz.,

$$E = \tilde{e} + \tfrac{1}{2}\tilde{u}_i\tilde{u}_i + k \quad \text{and} \quad H = \tilde{h} + \tfrac{1}{2}\tilde{u}_i\tilde{u}_i + k \tag{5.52}$$

## 5.4 Compressible-Flow Closure Approximations

As discussed in the preceding section, in addition to having variable mean density, $\bar{\rho}$, Equations (5.43) through (5.52) reflect effects of compressibility through various correlations that are affected by fluctuating density. For all but stress-transport models, diffusivity-type closure approximations must be postulated for the mass-averaged Reynolds-stress tensor and heat-flux vector. Depending on the type of turbulence model used, additional closure approximations may be needed to close the system of equations defining the model.

This section briefly reviews some of the most commonly used closure approximations for compressible flows. Because of the paucity of measurements compared to the incompressible case, and the additional complexities attending compressible flows, far less is available to guide development of closure approximations suitable for a wide range of applications. As a result, modeling of compressibility effects is in a continuing state of development as we approach the end of the twentieth century. The closure approximations discussed in this, and following, sections are those that have either stood the test of time or show the greatest promise.

Before focusing upon specific closure approximations, it is worthwhile to cite important guidelines that should be followed in devising compressible-flow closure approximations. Adhering to the following items will lead to the simplest and most elegant models.

1. All closure approximations should approach the proper limiting value for Mach number and density fluctuations tending to zero.

2. All closure terms should be written in proper tensor form, e.g., not dependent upon a specific geometrical configuration.

3. All closure approximations should be dimensionally consistent and invariant under a Galilean transformation.

It should be obvious that Items 2 and 3 apply for incompressible flows as well. In practice, Galilean invariance seems to be ignored more often than any other item listed, especially for compressible flows. Such models should be rejected as they violate a fundamental feature of the Navier-Stokes equation, and are thus physically unsound. We must be aware, for example, that total enthalpy, $H$, includes the kinetic energy and is not Galilean invariant, so its use as a dependent variable requires caution. For instance, it must not be used in diffusivity format in the manner that $\tilde{h}$ is used in Equation (5.54) below.

### 5.4.1  Reynolds-Stress Tensor

For zero-, one- and two-equation models, nearly all researchers use the Boussinesq approximation with suitable generalization for compressible flows. Specifically, denoting the eddy viscosity by $\mu_T$, the following form is generally assumed.

$$\bar{\rho}\tau_{ij} \equiv \overline{-\rho u_i'' u_j''} = 2\mu_T \left( S_{ij} - \frac{1}{3}\frac{\partial \tilde{u}_k}{\partial x_k}\delta_{ij} \right) - \frac{2}{3}\bar{\rho}k\delta_{ij} \qquad (5.53)$$

The most important consideration in postulating Equation (5.53) is guaranteeing that the trace of $\tau_{ij}$ is $-2k$. Note that this means the "second eddy viscosity" must be $-\frac{2}{3}\mu_T$ [recall Equation (5.24)].

### 5.4.2  Turbulent Heat-Flux Vector

The most commonly used closure approximation for the turbulent heat-flux vector, $q_{T_j}$, follows from appealing to the classical analogy [Reynolds (1874)] between momentum and heat transfer. It is thus assumed to be proportional to the mean temperature gradient, so that

$$q_{T_j} = \overline{\rho u_j'' h''} = -\frac{\mu_T c_p}{Pr_T}\frac{\partial \tilde{T}}{\partial x_j} = -\frac{\mu_T}{Pr_T}\frac{\partial \tilde{h}}{\partial x_j} \qquad (5.54)$$

where $Pr_T$ is the **turbulent Prandtl number**. A constant value for $Pr_T$ is often used and this is usually satisfactory for shock-free flows up to low supersonic speeds, provided the heat transfer rate is not too high. The most common values assumed for $Pr_T$ are 0.89 or 0.90, in the case of a boundary layer. Heat-transfer predictions can usually be improved somewhat by letting $Pr_T$ vary through the boundary layer. Near the edge of a boundary layer and throughout a free shear layer, a value of the order of 0.5 is more appropriate for $Pr_T$.

### 5.4.3  Molecular Diffusion and Turbulent Transport

If a zero-equation model is used, the $\frac{2}{3}\rho k\delta_{ij}$ contribution in Equation (5.53) is usually ignored as are the molecular diffusion, $\overline{t_{ji}u_i''}$, and turbulent transport, $\overline{\rho u_j'' \frac{1}{2}u_i'' u_i''}$, terms appearing in the mean-energy equation. Some researchers ignore these terms for higher-order models as well. This is usually a good approximation for flows with Mach numbers up to the supersonic range, which follows from the fact that $\bar{\rho}k \ll P$ (and hence $k \ll \tilde{h}$) in most flows of engineering interest. However, at hypersonic speeds, it is entirely possible to achieve conditions under which $\bar{\rho}k$ is a significant fraction of $P$. To ensure exact conservation of total energy (which includes turbulence kinetic energy), additional closure approximations are needed. The most straightforward procedure for one-equation, two-equation and stress-transport models is to generalize the low-speed closure approximations for the molecular diffusion and turbulent transport terms. The most commonly used approximation is:

$$\overline{t_{ji}u_i''} - \overline{\rho u_j'' \tfrac{1}{2} u_i'' u_i''} = \left(\mu + \frac{\mu_T}{\sigma_k}\right)\frac{\partial k}{\partial x_j} \tag{5.55}$$

## 5.4.4  Dilatation Dissipation

To understand what "dilatation dissipation" is, we must examine the turbulence-energy dissipation rate more closely. Recall from Equation (5.42) that

$$\bar{\rho}\epsilon = \overline{t_{ji}\frac{\partial u_i''}{\partial x_j}} \tag{5.56}$$

Hence, in terms of the instantaneous strain-rate tensor, $s_{ij}$, we have

$$\bar{\rho}\epsilon = \mu\left[\overline{2s_{ji}s_{ij}''} - \frac{2}{3}\overline{u_{k,k}u_{i,i}''}\right] \tag{5.57}$$

Assuming that the correlation between velocity-gradient fluctuations and kinematic viscosity fluctuations is negligible, we can rewrite this equation as

$$\bar{\rho}\epsilon = \bar{\nu}\left[2\overline{\rho s_{ji}'' s_{ij}''} - \frac{2}{3}\overline{\rho u_{k,k}'' u_{i,i}''}\right] \tag{5.58}$$

In terms of the fluctuating vorticity, $\omega_i''$, there follows

$$\bar{\rho}\epsilon = \bar{\nu}\left[\overline{\rho\omega_i''\omega_i''} + 2\overline{\rho u_{i,j}'' u_{j,i}''} - \frac{2}{3}\overline{\rho u_{i,i}'' u_{i,i}''}\right] \tag{5.59}$$

Finally, we can say $\overline{u_{i,j}'' u_{j,i}''} \approx \overline{(u_{i,i}'')^2}$, which is exactly true for homogeneous turbulence, and is a very good approximation for high-Reynolds-number, inhomogeneous turbulence [see, for example, Tennekes and Lumley (1983)]. Hence, we conclude that the dissipation can be written as

$$\bar{\rho}\epsilon = \bar{\rho}\epsilon_s + \bar{\rho}\epsilon_d \tag{5.60}$$

where

$$\bar{\rho}\epsilon_s = \bar{\nu}\,\overline{\rho\omega_i''\omega_i''} \quad \text{and} \quad \bar{\rho}\epsilon_d = \frac{4}{3}\bar{\nu}\,\overline{\rho u_{i,i}''u_{i,i}''} \tag{5.61}$$

Thus, we have shown that the compressible turbulence dissipation rate can logically be written in terms of the fluctuating vorticity and the divergence of the fluctuating velocity. Equivalently, we could have written the fluctuating velocity as the sum of a divergence-free and a curl-free component. At high Reynolds number, these components presumably are uncorrelated (again, an exact result for homogeneous turbulence), and Equation (5.59) would follow directly. The quantity $\epsilon_s$ is known as the **solenoidal dissipation**, while $\epsilon_d$ is known as the **dilatation dissipation**. Clearly, the latter contribution is present only for compressible flows.

Based on observations from some older Direct Numerical Simulations (DNS), Sarkar et al. (1989) and Zeman (1990) postulate that the dilatation dissipation should be a function of turbulence Mach number, $M_t$, defined by

$$M_t^2 = 2k/a^2 \tag{5.62}$$

where $a$ is the speed of sound. They further argue that the $k$ and $\epsilon$ equations should be replaced by

$$\bar{\rho}\frac{dk}{dt} = -\bar{\rho}\left(\epsilon_s + \epsilon_d\right) + \cdots \tag{5.63}$$

$$\bar{\rho}\frac{d\epsilon_s}{dt} = -C_{\epsilon 2}\bar{\rho}\epsilon_s^2/k + \cdots \tag{5.64}$$

where $C_{\epsilon 2}$ is a closure coefficient. Only the dissipation terms are shown explicitly in Equations (5.63) and (5.64) since no changes occur in any other terms. Particularly noteworthy, both Sarkar and Zeman postulate that the equation for $\epsilon_s$ is unaffected by compressibility. The dilatation dissipation is further assumed to be proportional to $\epsilon_s$ so that they say

$$\epsilon_d = \xi^* F(M_t)\,\epsilon_s \tag{5.65}$$

where $\xi^*$ is a closure coefficient and $F(M_t)$ is a prescribed function of $M_t$. The Sarkar and Zeman formulations differ in the value of $\xi^*$ and the functional form of $F(M_t)$, which we will discuss in Section 5.5.

Interestingly, while both Sarkar and Zeman arrive at similar formulations, their basic postulates are fundamentally different. Sarkar et al. postulate that $\epsilon_d$ "varies on a fast compressibility time scale relative to $\epsilon_s$." As a consequence, they conclude that dilatation dissipation increases with $M_t$ in a monotone manner. By contrast, Zeman postulates the existence of eddy shocklets, which are principally responsible for the dilatation dissipation. His analysis predicts that a threshold exists below which dilatation dissipation is negligible.

Although their arguments seem plausible when taken at face value, the premises are flawed. Most importantly, both draw from early DNS results for low-Reynolds-number, initially-isotropic turbulence subjected to strong compression or shear, where both dilatation dissipation and pressure-dilatation (see the next subsection) are significant. As pointed out by Ristorcelli et al. (1995), dilatation fluctuations occur mainly in the large eddies, where density fluctuations are large and viscous effects are small. That is, the mean-square dilatation fluctuation is virtually independent of Reynolds number, so $\epsilon_d$ varies as $1/Re$ and is therefore small at real-life Reynolds numbers.

DNS results for compressible thin shear layers [Coleman et al. (1995) and Huang et al. (1995) for channel flows, and Vreman et al. (1996) for mixing layers] show that dilatation dissipation is small or negligible, even in the presence of eddy shocklets and even at the fairly low Reynolds numbers of recent DNS studies. The channel results are consistent with the fact that compressibility corrections are not needed for boundary-layer flows. These DNS results also show insignificant pressure dilatation (see next subsection).

Nevertheless, the "dilatation-dissipation" corrections postulated by Zeman and Sarkar can, with adjustment of empirical coefficients, successfully correlate the decrease in mixing-layer growth rate with increasing Mach number. With care, they can also be arranged to have the desired lack of influence on non-hypersonic boundary layers, in which $M_t$ is generally lower at given $M_e$ because $k/U_e^2$ is smaller than in mixing layers. Evidently they should be regarded as completely empirical corrections rather than true models of dilatation dissipation. We return to the question of what these compressibility corrections really mean after discussing the other explicit compressibility terms in the turbulence kinetic energy equation, namely pressure diffusion and pressure dilatation.

## 5.4.5 Pressure Diffusion and Pressure-Dilatation

Section 4.1 discusses the lack of information regarding diffusion by pressure fluctuations in incompressible flows. So little is known that it is simply ignored; by implication, it is lumped in with triple-product turbulent transport. Even less is known for compressible flows. However, given the fundamentally different role that pressure plays in a compressible medium relative to its essentially passive role at low speeds, we might reasonably suspect that ignoring pressure diffusion and pressure dilatation might lead to significant error. However, DNS research shows that, as with dilatation dissipation, these terms are very small for both mixing layers and boundary layers. As in the case of dilatation dissipation, the early homogeneous-strain simulations were misleading. As Zeman (1993) shows, pressure-dilatation is large in flows with a large ratio of turbulence-energy production to dissipation – typical of strongly-strained initially-isotropic flows. In thin shear layers, production and dissipation are roughly the same and pressure-dilatation is small. (This is not a Reynolds-number effect: pressure-dilatation is determined by the large eddies, like mean-square dilatation, and does not involve viscosity.) Hence, models for these pressure terms, to the extent that they improve predictions, are ad hoc in nature and do not reflect the true physics of compressible turbulence.

New proposals, especially for the pressure-dilatation mean product, have been made by many authors [Sarkar et al. (1991,1992), Zeman (1991,1993) and Ristorcelli et al. (1993,1995)], but none has received general acceptance. For example, Sarkar (1992) proposes that the pressure dilatation can be approximated as

$$\overline{p'\frac{\partial u_i''}{\partial x_i}} = \alpha_2 \bar{\rho} \tau_{ij} \frac{\partial \tilde{u}_i}{\partial x_j} M_t + \alpha_3 \bar{\rho} \epsilon M_t^2 \tag{5.66}$$

where $M_t$ is the **turbulence Mach number** defined in Equation (5.62). The closure coefficients $\alpha_2$ and $\alpha_3$ are given by

$$\alpha_2 = 0.15 \quad \text{and} \quad \alpha_3 = 0.2 \tag{5.67}$$

The model has been calibrated for a range of compressible-flow applications including the mixing layer and attached boundary layers (but apparently not with respect to DNS results for these flows).

### 5.4.6  Pressure Work

The pressure work term, $\overline{u_i''} P_{,i}$ (or $\overline{u_i''} P_{,j} + \overline{u_j''} P_{,i}$ for stress-transport models), arises because the time average of $u_i''$ does not vanish. It is proportional to the density/velocity correlation $\overline{\rho' u_i'}$, and illustrates how Favre averaging does not completely eliminate the need to know how these fluctuating properties are correlated.

Wilcox and Alber (1972) postulate an empirical model for this term that improves two-equation model predictions for hypersonic base flows. Oh (1974) proposes a closure approximation postulating existence of "eddy shocks" and accurately simulates compressible mixing layers with a one-equation turbulence model. Neither model is entirely satisfactory however as they both involve the mean velocity in a manner that violates Galilean invariance of the Navier-Stokes equation.

More recently, Zeman (1993) and Ristorcelli (1993) have argued that the time average of $u_i''$ for boundary layers behaves as

$$\overline{u_i''} \sim \frac{M_t k}{\bar{\rho} \epsilon} \tau_{ij} \frac{\partial \bar{\rho}}{\partial x_j} \tag{5.68}$$

Although corroborating measurements to verify this model are essentially nonexistent, we can at least say that it is dimensionally correct and does not violate Galilean invariance.

# 5.5  Mixing-Layer Compressibility Corrections

The decrease in mixing-layer growth rate with increasing Mach number has been known for many years [e.g., Birch and Eggers (1972)]. This decrease is not likely to be the result of density changes across the layer. We know from the measurements of Brown and Roshko (1974) that low-speed mixing layers between flows of two different gases show only a moderate effect of density ratio. Most researchers believe no current turbulence model predicts the Mach-number dependence of spreading rate without an explicit compressibility correction.

We have seen above that the explicit compressibility terms in the $k$ equation are small in practical cases. Also, empirical functions of turbulence Mach number, $M_t$, calibrated to reproduce compressibility effects in mixing layers, are liable to have unwanted effects on boundary layers. From this we can deduce two things. First, compressibility effects result from Mach-number dependence of the main terms in the equations, i.e., those which are present even in incompressible flow. Second, these effects appear mainly in the mixing layer, but are not entirely attributable to the typically higher $M_t$ in mixing layers.

There is now fairly conclusive evidence, both from simulations [e.g., Vreman et al. (1996)] and experiment [e.g., Clemens and Mungal (1995)], that quasi-two-dimensional spanwise vortex rolls, which form the large-scale structure of low-speed mixing layers, become more three-dimensional as Mach number increases. This is in line with the Mach-number dependence of the most-unstable disturbances in laminar mixing layers, which are vortex rolls with gradually-increasing sweepback. This "inflection-point" instability is essentially inviscid, capable of growing in the presence of viscosity, and may therefore be at least qualitatively relevant to the behavior of large structure in the presence of small-scale turbulence.

It seems unlikely that laminar stability theory will lead directly to a quantitative correlation for turbulent flow, which must therefore rest on empiricism. An important question not yet settled by experiment is whether the spreading rate reaches an asymptotic value at high Mach number, or continues to decrease indefinitely. Acoustic radiation from the turbulence, which in the past was occasionally blamed entirely for the observed compressibility effects, will certainly become an important mechanism of energy loss at very high Mach number and may therefore prevent an asymptotic state from being reached.

The arguments above strongly suggest that compressibility effects manifest themselves in the pressure-strain "redistribution" term, $\Pi_{ij}$, defined in Equation (5.44) and a major term appearing in the Reynolds-stress transport equation for $\tau_{ij}$, Equation (5.43). Unless some of the smaller unknown terms on the right-hand side of Equation (5.43) increase very greatly with Mach number, the empirical compressibility correction terms which are *added* to the turbulence equations are a substitute for compressibility *factors* on $\Pi_{ij}$. To date, most compressibility corrections have been applied to the turbulence energy equation, as used in two-equation models. In these models, the $\tau_{ij}$ equation is not treated explicitly, and corrections to $\Pi_{ij}$ have not yet been explored.

## 5.5.1 The Sarkar/Zeman/Wilcox Compressibility Corrections

As noted in Subsection 5.4.4, focusing upon the $k$-$\epsilon$ model, Sarkar et al. (1989) and Zeman (1990) have devised particularly elegant models for the $k$ equation that correct the deficiency for the compressible mixing layer. Although their physical arguments have since been shown to apply at best only to low-Reynolds-number, strained homogeneous flows (the subjects of early DNS studies), their models are nevertheless quite useful. Building upon the Sarkar/Zeman formulations, and upon dimensional analysis, Wilcox (1992b) has postulated a similar model that enjoys an important advantage for wall-bounded flows.

To implement the Sarkar or Zeman modification in the $k$-$\omega$ model, we begin by making the formal change of variables given by $\epsilon_s = \beta^* \omega k$. If we treat $\beta^*$ as a pure constant for simplicity, this tells us immediately that

$$\bar{\rho} \frac{d\omega}{dt} = \frac{\bar{\rho}}{\beta^* k} \left[ \frac{d\epsilon_s}{dt} - \frac{\epsilon_s}{k} \frac{dk}{dt} \right] \tag{5.69}$$

Consequently, a compressibility term must appear in the $\omega$ equation as well as in the $k$ equation. Inspection of Equations (4.37) and (4.38) shows that the Sarkar/Zeman compressibility modifications correspond to letting closure coefficients $\beta$ and $\beta^*$ in the $k$-$\omega$ model vary with $M_t$. In terms of $\xi^*$ and the compressibility function $F(M_t)$, $\beta$ and $\beta^*$ become:

$$\beta^* = \beta_o^* f_{\beta^*} \left[ 1 + \xi^* F(M_t) \right], \qquad \beta = \beta_o f_\beta - \beta_o^* f_{\beta^*} \xi^* F(M_t) \tag{5.70}$$

where $\beta_o^* f_{\beta^*}$ and $\beta_o f_\beta$ are the corresponding incompressible values of $\beta^*$ and $\beta$. The values of $\xi^*$ and $F(M_t)$ for the Sarkar, Zeman and Wilcox models are:

**Sarkar's Model**[2]

$$\xi^* = 1, \qquad F(M_t) = M_t^2 \tag{5.71}$$

**Zeman's Model**

$$\xi^* = 3/4, \qquad F(M_t) = \left[ 1 - e^{-\frac{1}{2}(\gamma+1)(M_t - M_{t_o})^2 / \Lambda^2} \right] \mathcal{H}(M_t - M_{t_o}) \tag{5.72}$$

**Wilcox's Model**

$$\xi^* = 3/2, \qquad M_{t_o} = 1/4, \qquad F(M_t) = \left[ M_t^2 - M_{t_o}^2 \right] \mathcal{H}(M_t - M_{t_o}) \tag{5.73}$$

---

[2] When Sarkar's pressure-dilatation term, Equation (5.66), is used in combination with Equation (5.71), the coefficient $\xi^*$ should be reduced to 1/2.

where $\gamma$ is the specific-heat ratio and $\mathcal{H}(x)$ is the Heaviside step function. Zeman recommends using $\Lambda = 0.60$ and $M_{t_o} = 0.10\sqrt{2/(\gamma + 1)}$ for free shear flows. For boundary layers, their values must increase to $\Lambda = 0.66$ and $M_{t_o} = 0.25\sqrt{2/(\gamma + 1)}$. Zeman uses a different set of closure coefficients for boundary layers because he postulates that they depend upon the kurtosis, $\overline{u'^4}/(\overline{u'^2})^2$. The kurtosis is presumed to be different for free shear flows as compared to boundary layers. While this may be true, it is not much help for two-equation or stress-transport models since such models only compute double correlations and make closure approximations for triple correlations. Quadruple correlations such as $\overline{u'^4}$ are beyond the scope of these models.

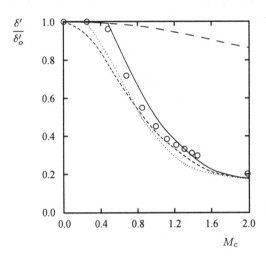

Figure 5.1: *Computed and measured spreading rate for a compressible mixing layer:* — — *Unmodified k-ω model;* —— *Wilcox,* $\xi^* = 3/2$; - - - *Sarkar,* $\xi^* = 1$; ··· *Zeman,* $\xi^* = 3/4$; ○ *Langley curve [Kline et al. (1981)].*

## 5.5.2 Applications

To illustrate how well these models perform, we consider mixing of a supersonic stream and a quiescent fluid with constant total temperature. For simplicity, we present results only for the $k$-$\omega$ model as $k$-$\epsilon$ results are nearly identical. The equations of motion have been transformed to similarity form for the far field and integrated using Program **MIXER** (see Appendix C). Figure 5.1 compares computed and measured [see Kline et al. (1981)] spreading rate, $\delta'$. As in the incompressible case, spreading rate is defined as the difference between the values of $y/x$ where $(U - U_2)^2/(U_1 - U_2)^2$ is 9/10 and 1/10. The quantity $\delta'_o$

denotes the incompressible spreading rate and $M_c$ is convective Mach number [Papamoschou and Roshko (1988)], viz.,

$$M_c = \frac{U_1 - U_2}{a_1 + a_2} \tag{5.74}$$

Since $U_2 = 0$, Equation (5.74) simplifies to

$$M_c = \frac{M_1}{1 + \left[1 + \frac{(\gamma-1)}{2} M_1^2\right]^{1/2}} \tag{5.75}$$

As shown, the unmodified $k$-$\omega$ model fails to predict a significant decrease in spreading rate as Mach number increases. By contrast, the Sarkar, Zeman and Wilcox modifications, all applied to the $k$-$\omega$ model, yield much closer agreement between computed and measured spreading rates.

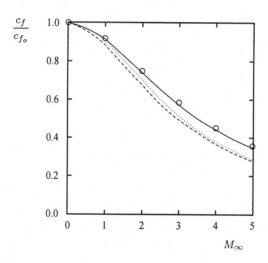

Figure 5.2: *Computed and measured k-ω model skin friction for a compressible flat-plate boundary layer:* —— *Wilcox with* $\xi^* = 0$ *and* $\xi^* = 3/2$; - - - *Sarkar,* $\xi^* = 1$; $\cdots$ *Zeman,* $\xi^* = 3/4$; ○ *Van Driest correlation.*

We turn now to the adiabatic-wall flat-plate boundary layer. The equations of motion for the $k$-$\omega$ model have been solved with Program **EDDYBL** (see Appendix D). Figure 5.2 compares computed skin friction, $c_f$, with a correlation of measured values for freestream Mach number between 0 and 5. As shown, the unmodified model virtually duplicates measured skin friction. By contrast, the Sarkar compressibility modification yields a value for $c_f$ at Mach 5 that is 18% lower than the value computed with $\xi^* = 0$. Using the Wilcox compressibility correction with $\xi^* = 3/2$ yields very little difference in skin friction.

Using $\Lambda = 0.60$ and $M_{t_o} = 0.10\sqrt{2/(\gamma + 1)}$ in Zeman's model, computed $c_f$ at Mach 5 is 15% smaller than the value obtained with the unmodified model. Increasing $\Lambda$ and $M_{t_o}$ to 0.66 and $0.25\sqrt{2/(\gamma + 1)}$, respectively, eliminates this discrepancy. However, using this large a value for $M_{t_o}$ for the mixing layer results in larger-than-measured spreading rates, with differences in excess of 100% between computed and measured spreading rate for $M_c$ in excess of 1.

These results make it clear that neither the Sarkar nor the Zeman compressibility term is completely satisfactory for both the mixing layer and boundary layers. The Wilcox model was formulated to resolve this dilemma. Making $\beta$ and $\beta^*$ functions of $M_t$ is a useful innovation, and is not the root cause of the problem. Rather, the postulated form of the function $F(M_t)$ is the weak link.

Inspection of turbulence Mach numbers in mixing layers and boundary layers shows that all we need is an alternative to the Sarkar and Zeman functional dependencies of $\epsilon_d$ upon $M_t$. Table 5.1 shows why the Sarkar term improves predictions for the mixing layer. The unmodified $k$-$\omega$ model predicts peak values of $M_t$ in the mixing layer that are more than twice the values in the boundary layer for the same freestream Mach number. The Sarkar compressibility term reduces $(M_t)_{max}$ by about one third for the mixing layer when $M_\infty \geq 2$. Even with this much reduction, $(M_t)_{max}$ for the mixing layer remains higher than the largest value of $(M_t)_{max}$ in the boundary layer all the way up to Mach 5.

Table 5.1: *Maximum Turbulence Mach Number,* $(M_t)_{max}$

| $M_\infty$ | Boundary Layer | | Mixing Layer | |
|---|---|---|---|---|
| | $\xi^* = 0$ | $\xi^* = 1$ | $\xi^* = 0$ | $\xi^* = 1$ |
| 0 | 0 | 0 | 0 | 0 |
| 1 | 0.061 | 0.061 | 0.180 | 0.159 |
| 2 | 0.114 | 0.107 | 0.309 | 0.227 |
| 3 | 0.149 | 0.135 | 0.384 | 0.245 |
| 4 | 0.174 | 0.154 | 0.424 | 0.254 |
| 5 | 0.191 | 0.171 | 0.453 | 0.266 |

For Mach 1, the Sarkar term reduces mixing-layer spreading rate below measured values (Figure 5.1). Zeman's term predicts a somewhat larger spreading rate at Mach 1, mainly because of the Mach number threshold in Zeman's model. That is, Zeman postulates that the compressibility effect is absent for $M_t < M_{t_o}$. Zeman's Mach number threshold also yields smaller differences between computed and measured boundary-layer skin friction at lower Mach numbers (see Figure 5.2). These observations show that an improved compressibility term can be devised by extending Zeman's threshold Mach number to a larger value of $M_t$. The Wilcox model simply combines the relative simplicity of Sarkar's functional form for $F(M_t)$ with Zeman's Mach number threshold to accomplish this end.

## 5.6   Compressible Law of the Wall

In this section, we use perturbation methods to examine $k$-$\omega$ and $k$-$\epsilon$ model predicted, compressible log-layer structure. The results are particularly illuminating
and clearly demonstrate why the Sarkar and Zeman compressibility terms ad-
versely affect boundary-layer predictions. Recall from Section 4.6.1 that the log
layer is the region sufficiently close to the solid boundary for neglect of convective terms and far enough distant for molecular diffusion terms to be dropped.
In the log layer, the equations for the $k$-$\omega$ model simplify to the following.

$$\mu_T \frac{d\tilde{u}}{dy} = \bar{\rho}_w u_\tau^2 \tag{5.76}$$

$$\mu_T \frac{d}{dy} \left[ \frac{c_p \tilde{T}}{Pr_T} + \frac{1}{2}\tilde{u}^2 + \sigma^* k \right] = -q_w \tag{5.77}$$

$$\sigma^* \frac{d}{dy} \left[ \mu_T \frac{dk}{dy} \right] + \mu_T \left( \frac{d\tilde{u}}{dy} \right)^2 - \beta^* \bar{\rho} \omega k = 0 \tag{5.78}$$

$$\sigma \frac{d}{dy} \left[ \mu_T \frac{d\omega}{dy} \right] + \alpha \bar{\rho} \left( \frac{d\tilde{u}}{dy} \right)^2 - \beta \bar{\rho} \omega^2 = 0 \tag{5.79}$$

$$\bar{\rho}\tilde{T} = \bar{\rho}_w \tilde{T}_w \tag{5.80}$$

The quantity $u_\tau$ is friction velocity defined as $\sqrt{\tau_w/\bar{\rho}_w}$ where $\tau_w$ is surface
shear stress and $\bar{\rho}_w$ is density at the surface. Also, $\tilde{T}_w$ is surface temperature,
$q_w$ is surface heat flux and $c_p$ is specific heat at constant pressure. Finally, $y$ is
distance from the surface.

At this point, we can simplify our analysis for the $k$-$\omega$ model by assuming
the cross-diffusion parameter, $\chi_k = \nabla k \cdot \nabla \omega / \omega^3$, is negligibly small in the
compressible log layer. This is indeed true and can be verified once the solution
has been obtained. This approximation permits us to approximate $f_{\beta^*} \approx 1$. Also,
since the flow is two dimensional, we have $f_\beta = 1$. Then, we introduce Sarkar's
compressibility modification [Equation (5.71)], so that Equation (5.70) for $\beta^*$
and $\beta$ simplifies to

$$\beta^* = \beta_o^* \left[ 1 + \xi^* M_t^2 \right] \quad \text{and} \quad \beta = \beta_o - \beta_o^* \xi^* M_t^2 \tag{5.81}$$

Following Saffman and Wilcox (1974), we change independent variables
from $y$ to $\tilde{u}$. Consequently, derivatives transform according to

$$\mu_T \frac{d}{dy} = \mu_T \frac{d\tilde{u}}{dy} \frac{d}{d\tilde{u}} = \bar{\rho}_w u_\tau^2 \frac{d}{d\tilde{u}} \tag{5.82}$$

With this change of variables, we replace Equations (5.77) – (5.79) by

$$\frac{d}{d\tilde{u}}\left[\frac{c_p\tilde{T}}{Pr_T} + \frac{1}{2}\tilde{u}^2 + \sigma^*k\right] = -\frac{q_w}{\bar{\rho}_w u_\tau^2} \tag{5.83}$$

$$\sigma^*\frac{d^2k}{d\tilde{u}^2} + 1 - \frac{\beta^*\bar{\rho}^2k^2}{\bar{\rho}_w^2 u_\tau^4} = 0 \tag{5.84}$$

$$\sigma\frac{d^2\omega}{d\tilde{u}^2} + \alpha\frac{\omega}{k} - \frac{\beta\bar{\rho}^2k\omega}{\bar{\rho}_w^2 u_\tau^4} = 0 \tag{5.85}$$

Integrating Equation (5.83) yields the temperature, and hence the density, as a function of velocity and Mach number based on friction velocity, $M_\tau \equiv u_\tau/a_w$.

$$\frac{\tilde{T}}{\tilde{T}_w} = \frac{\bar{\rho}_w}{\bar{\rho}} = 1 - (\gamma-1)Pr_T M_\tau^2\left[\frac{1}{2}\left(\frac{\tilde{u}}{u_\tau}\right)^2 + \frac{q_w}{\bar{\rho}_w u_\tau^3}\left(\frac{\tilde{u}}{u_\tau}\right) + \sigma^*\left(\frac{k}{u_\tau^2}\right)\right] \tag{5.86}$$

Next, we assume a solution of the form:

$$\bar{\rho}k = \Gamma\bar{\rho}_w u_\tau^2 \tag{5.87}$$

where $\Gamma$ is a constant to be determined. Substituting Equations (5.86) and (5.87) into Equation (5.84), and noting that $M_t^2 = 2\Gamma M_\tau^2$, leads to the following quartic equation for $\Gamma$.

$$\beta_o^*\left[1 + 2\xi^*M_\tau^2\Gamma\right]\left[1 + (\gamma-1)Pr_T\sigma^*M_\tau^2\Gamma\right]\Gamma^2 = 1 \tag{5.88}$$

As can easily be verified, when $M_\tau^2 \ll 1$ the asymptotic solution for $\Gamma$ is

$$\Gamma = \frac{1}{\sqrt{\beta_o^*}} - \left[\frac{\xi^* + \frac{(\gamma-1)}{2}Pr_T\sigma^*}{\beta_o^*}\right]M_\tau^2 + \cdots \tag{5.89}$$

Finally, in terms of $\Gamma$, Equation (5.85) simplifies to

$$\sigma\frac{d^2\omega}{d\tilde{u}^2} + \left[\alpha - \left(\beta_o - 2\beta_o^*\xi^*M_\tau^2\Gamma\right)\Gamma^2\right]\frac{\bar{\rho}\,\omega}{\bar{\rho}_w u_\tau^2\Gamma} = 0 \tag{5.90}$$

Combining Equations (5.86) and (5.87) yields the density as a function of velocity and $\Gamma$.

$$\frac{\bar{\rho}_w}{\bar{\rho}} = \frac{1 - \frac{(\gamma-1)}{2}Pr_T M_\tau^2\left[\left(\frac{\tilde{u}}{u_\tau}\right)^2 + \frac{2q_w}{\bar{\rho}_w u_\tau^3}\left(\frac{\tilde{u}}{u_\tau}\right)\right]}{1 + (\gamma-1)Pr_T\sigma^*\Gamma M_\tau^2} \tag{5.91}$$

Equation (5.91) assumes a more compact form if we introduce the freestream velocity, $U_\infty$. A bit more algebra yields

$$\frac{\bar{\rho}_w}{\bar{\rho}} = \frac{1 + Bv - A^2v^2}{1 + (\gamma - 1)Pr_T\sigma^*\Gamma M_\tau^2} \tag{5.92}$$

where

$$
\left.\begin{aligned}
v &= \tilde{u}/U_\infty \\[2mm]
A^2 &= \frac{(\gamma-1)}{2}Pr_T M_\infty^2(\tilde{T}_\infty/\tilde{T}_w) \\[2mm]
B &= -Pr_T q_w U_\infty/(c_p\tilde{T}_w\tau_w)
\end{aligned}\right\} \tag{5.93}
$$

Using Equations (5.89), (5.92) and (5.93), and retaining terms up to $O(M_\tau^2)$, Equation (5.90) assumes the following form,

$$\frac{d^2\omega}{dv^2} - \left[\frac{\kappa_\omega^2(U_\infty/u_\tau)^2}{1 + Bv - A^2v^2}\right]\omega = 0 \tag{5.94}$$

where the constant $\kappa_\omega$ is defined by

$$\kappa_\omega^2 = \kappa^2 - \left[\frac{(2 + \alpha + \beta_o/\beta_o^*)\xi^*}{\sigma} + \frac{(\gamma - 1)Pr_T(3\alpha - \beta_o/\beta_o^*)\sigma^*}{2\sigma}\right]M_\tau^2 + \cdots \tag{5.95}$$

and $\kappa$ is Kármán's constant. Because $U_\infty/u_\tau \gg 1$, we can use the WKB method [see Kevorkian and Cole (1981) or Wilcox (1995a)] to solve Equation (5.94). Noting that $\omega$ decreases as $\tilde{u}/U_\infty$ increases, the asymptotic solution for $\omega$ is

$$\omega \sim C_0\left[1 + Bv - A^2v^2\right]^{1/4}\exp\left[-\kappa_\omega u^*/u_\tau\right] \tag{5.96}$$

where $C_0$ is a constant of integration and $u^*$ is defined by

$$\frac{u^*}{U_\infty} = \frac{1}{A}\left[\sin^{-1}\left(\frac{2A^2v - B}{\sqrt{B^2 + 4A^2}}\right) + \sin^{-1}\left(\frac{B}{\sqrt{B^2 + 4A^2}}\right)\right] \tag{5.97}$$

The second $\sin^{-1}$ term is needed to ensure $u^* = 0$ when $v = 0$. Combining Equations (5.76), (5.87) and (5.96), we can relate velocity and distance, $y$.

$$\int\left[1 + Bv - A^2v^2\right]^{-1/4}\exp\left[\kappa_\omega u^*/u_\tau\right]dv \sim \frac{C_0 y}{\Gamma U_\infty} \tag{5.98}$$

We integrate by parts to generate the asymptotic expansion of the integral in Equation (5.98) as $U_\infty/u_\tau \to \infty$. Hence,

$$\left[1 + Bv - A^2v^2\right]^{1/4}\exp\left[\kappa_\omega u^*/u_\tau\right] \sim \frac{\kappa_\omega C_0 y}{\Gamma u_\tau} \tag{5.99}$$

Finally, we set the constant of integration $C_0 = \Gamma u_\tau^2/(\kappa_\omega \nu_w)$. Taking the natural log of Equation (5.99), we conclude that

$$\frac{u^*}{u_\tau} \sim \frac{1}{\kappa_\omega} \ell n \left(\frac{u_\tau y}{\nu_w}\right) + C_\omega \tag{5.100}$$

The quantity $C_\omega$ is the effective "constant" in the law of the wall defined by

$$C_\omega = C + \frac{1}{\kappa_\omega} \ell n \left(\frac{\bar{\rho}}{\bar{\rho}_w}\right)^{1/4} \tag{5.101}$$

where $C$ is a true constant.

Most of the analysis above holds for the $k$-$\epsilon$ model. The only significant difference is in the $\epsilon$ equation which is as follows.

$$\sigma_\epsilon^{-1} \frac{d}{dy}\left[\mu_T \frac{d\epsilon}{dy}\right] + C_\mu C_{\epsilon 1} \bar{\rho} k \left(\frac{d\tilde{u}}{dy}\right)^2 - C_{\epsilon 2} \frac{\bar{\rho}\epsilon^2}{k} = 0 \tag{5.102}$$

Equations (5.87), (5.89) and (5.92) are still valid for the turbulence kinetic energy and density, provided $\sigma^*$ is replaced by $\sigma_k^{-1}$. The transformed equation for $\epsilon$ is

$$\frac{d^2\epsilon}{dv^2} - \left[\frac{\kappa_\epsilon^2(U_\infty/u_\tau)^2}{1 + Bv - A^2v^2}\right]\epsilon = 0 \tag{5.103}$$

where the constant $\kappa_\epsilon$ is defined by

$$\kappa_\epsilon^2 = \kappa^2 - \left[(C_{\epsilon 1} + C_{\epsilon 2})\sigma_\epsilon \xi^* + \frac{(\gamma - 1)Pr_T(3C_{\epsilon 1} - C_{\epsilon 2})\sigma_\epsilon}{2\sigma_k}\right]M_\tau^2 + \cdots \tag{5.104}$$

In arriving at Equation (5.104), recall from Equation (4.134) that the $k$-$\epsilon$ model's closure coefficients are related by

$$\kappa^2 = \sqrt{C_\mu}\,(C_{\epsilon 2} - C_{\epsilon 1})\,\sigma_\epsilon \tag{5.105}$$

The asymptotic solution for $\epsilon$ is

$$\epsilon \sim C_1\left[1 + Bv - A^2v^2\right]^{1/4}\exp\left[-\kappa_\epsilon u^*/u_\tau\right] \tag{5.106}$$

where $C_1$ is a constant of integration. Velocity and distance from the surface are related by

$$\int\left[1 + Bv - A^2v^2\right]^{3/4}\exp\left[\kappa_\epsilon u^*/u_\tau\right]dv \sim C_1 y \tag{5.107}$$

Consequently, Equation (5.99) is replaced by

$$\left[1 + Bv - A^2v^2\right]^{5/4}\exp\left[\kappa_\epsilon u^*/u_\tau\right] \sim C_2 y \tag{5.108}$$

where $C_2$ is another constant of integration. Finally, the law of the wall for the $k$-$\epsilon$ model is

$$\frac{u^*}{u_\tau} \sim \frac{1}{\kappa_\epsilon} \ell n \left(\frac{u_\tau y}{\nu_w}\right) + C_\epsilon \qquad (5.109)$$

where $C_\epsilon$ is given by

$$C_\epsilon = C + \frac{1}{\kappa_\epsilon} \ell n \left(\frac{\bar{\rho}}{\bar{\rho}_w}\right)^{5/4} \qquad (5.110)$$

Equations (5.100) and (5.109) are very similar to the compressible law of the wall deduced by Van Driest (1951) [cf. Equation (5.4)]. There are two ways in which these equations differ from the Van Driest law.

**The first difference** is the effective Kármán constants, $\kappa_\omega$ and $\kappa_\epsilon$, which vary with $M_\tau$ according to Equation (5.95) for the $k$-$\omega$ model and according to Equation (5.104) for the $k$-$\epsilon$ model. In terms of each model's closure coefficients, $\kappa_\omega$ and $\kappa_\epsilon$ are given by (for $M_\tau \ll 1$):

$$\kappa_\omega^2 \sim \kappa^2 \left[1 - (39.52\xi^* + 0.80) M_\tau^2 + \cdots\right] \qquad (5.111)$$

and

$$\kappa_\epsilon^2 \sim \kappa^2 \left[1 - (23.92\xi^* + 3.07) M_\tau^2 + \cdots\right] \qquad (5.112)$$

Table 5.2 summarizes results obtained in the boundary-layer computations of Section 5.5 for the unmodified $k$-$\omega$ model ($\xi^* = 0$) and for the $k$-$\omega$ model with the Sarkar compressibility term ($\xi^* = 1$). The value of $\kappa_\omega$ for the unmodified model deviates from the Kármán constant, $\kappa = 0.41$, by less than 0.12% for freestream Mach numbers between 0 and 5. By contrast, when $\xi^* = 1$, the deviation is as much as 5.10%. This large a deviation in the effective Kármán constant is consistent with the observed differences between computed and measured skin friction. Similarly, with $M_\tau = .05$, $\kappa_\epsilon$ differs from $\kappa$ by 0.5% and 3.5% for $\xi^* = 0$ and 1, respectively. Thus the Sarkar compressibility term has a somewhat smaller effect on $\kappa$ for the $k$-$\epsilon$ model relative to the effect on $\kappa$ for the $k$-$\omega$ model.

Table 5.2: *Effective Kármán Constant for the $k$-$\omega$ Model*

| $M_\infty$ | $M_\tau\|_{\xi^*=0}$ | $\kappa_\omega$ | $M_\tau\|_{\xi^*=1}$ | $\kappa_\omega$ |
|---|---|---|---|---|
| 0 | 0 | 0.410 | 0 | 0.410 |
| 1 | 0.032 | 0.410 | 0.031 | 0.402 |
| 2 | 0.048 | 0.410 | 0.046 | 0.392 |
| 3 | 0.052 | 0.410 | 0.049 | 0.390 |
| 4 | 0.050 | 0.410 | 0.046 | 0.392 |
| 5 | 0.048 | 0.410 | 0.043 | 0.394 |

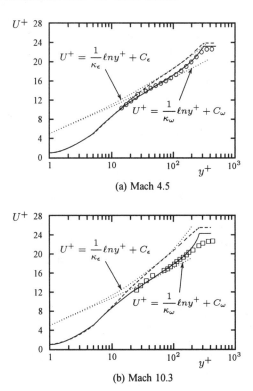

(a) Mach 4.5

(b) Mach 10.3

Figure 5.3: *Comparison of computed and measured velocity profiles for compressible flat-plate boundary layers:* —— $k$-$\omega$; - - - *Chien* $k$-$\epsilon$; ∘ *Coles*; □ *Watson*.

To see why a small perturbation in $\kappa$ corresponds to a larger perturbation in $c_f$, differentiate the law of the wall with respect to $\kappa$. Noting that $c_f = \frac{1}{2}u_\tau^2/U_\infty^2$, a little algebra shows that

$$\frac{dc_f}{d\kappa} \approx \frac{2}{\kappa}c_f \qquad (5.113)$$

Thus, we should expect $\Delta c_f/c_f$ to be double the value of $\Delta\kappa/\kappa$. The numerical results indicate somewhat larger differences in $c_f$, but the trend is clear.

**The second difference** between Equations (5.100) and (5.109) and the Van Driest compressible law of the wall is in the effective variation of the "constant" terms $C_\omega$ and $C_\epsilon$ with $(\bar{\rho}/\bar{\rho}_w)$. Because the exponent is only 1/4 for the $k$-$\omega$ model, the effect is minor. By contrast, the exponent is 5/4 for the $k$-$\epsilon$ model. This large an exponent has a much stronger effect on predicted boundary-layer properties. Figure 5.3 compares computed and measured [Fernholz and Finley (1981)] velocity profiles for adiabatic-wall boundary layers at Mach numbers 4.5 and 10.3. The computed results are for the $k$-$\omega$ model and for Chien's (1982) low-Reynolds-number $k$-$\epsilon$ model. Equations (5.100) and (5.109) are also shown (with $C = 5.0$) to underscore the importance of the models' variable "constant" in the compressible law of the wall.

These results are consistent with the analysis of Huang, Bradshaw and Coakley (1992), which shows how poorly the $k$-$\epsilon$ model performs for compressible boundary layers. Since $\bar{\rho}/\bar{\rho}_w > 1$ for all but strongly cooled walls, its effect is to increase the "constant" in the law of the wall with a corresponding decrease in $c_f$. The Sarkar and Zeman terms will thus amplify this inherent deficiency of the $k$-$\epsilon$ model.

To put these results in proper perspective, we must not lose sight of the fact that the $k$-$\epsilon$ model requires the use of either wall functions or viscous damping functions in order to calculate wall-bounded flows. If these functions have an effect that persists well into the log layer, it may be possible to suppress the $k$-$\epsilon$ model's inherent flaws at low Reynolds numbers. However, the perturbation analysis shows that such a model will not be asymptotically consistent with the compressible law of the wall in the limit of infinite Reynolds number. In effect, such a model would have compensating errors that may fortuitously yield reasonably close agreement with the law of the wall at low Reynolds numbers.

As a final comment, if we had used $\rho\epsilon$ as the dependent variable in Equation (5.102) in place of $\epsilon$, the exponent 5/4 in Equation (5.110) would be reduced to 1/4. Presumably, this change would improve $k$-$\epsilon$ model predictions for compressible boundary layers. The effect of this rescaling on the mixing layer is unclear.

## 5.7  Compressible Boundary Layers

Most turbulence models are capable of providing reasonably accurate predictions for constant-pressure, adiabatic-wall boundary layers provided the Mach number does not exceed about 5. Similar to the incompressible situation, adverse pressure gradients continue to be anathema to the $k$-$\epsilon$ model, while presenting no major problem for the $k$-$\omega$ model. When surface heat transfer is present, model predictions often show nontrivial discrepancies from measured values.

Algebraic models such as the Cebeci-Smith and Baldwin-Lomax models (see Subsections 3.4.1 and 3.4.2) require no special compressibility corrections. For the sake of clarity, recall that the Cebeci-Smith model uses the velocity thickness, $\delta_v^*$, defined in Equation (3.114) for both compressible and incompressible flow. The velocity thickness differs from the displacement thickness, $\delta^*$, which is defined for compressible flows by

$$\delta^* = \int_0^\infty \left(1 - \frac{\bar{\rho}}{\bar{\rho}_e}\frac{\tilde{u}}{\tilde{u}_e}\right)\,dy \qquad (5.114)$$

The primary reason algebraic models should fare well for compressible boundary layers without special compressibility modifications is illustrated by Maise and McDonald (1967). Using the best experimental data of the time for compressible boundary layers, they inferred the mixing-length variation. Their analysis shows that for Mach numbers up to 5:

- Velocity profiles for adiabatic walls correlate with the incompressible profile when the Van Driest (1951) scaling is used, i.e.,

$$\frac{u^*}{U_\infty} = \frac{1}{A}\sin^{-1}\left(A\frac{\tilde{u}}{U_\infty}\right), \qquad A^2 = \frac{(\gamma-1)}{2}M_\infty^2(\tilde{T}_\infty/\tilde{T}_w) \quad (5.115)$$

- The Van Driest scaling fails to correlate compressible velocity profiles when surface heat transfer is present, especially for surfaces that are very cold.

- The classical mixing length is independent of Mach number.

Using singular-perturbation methods, Barnwell (1992) shows that algebraic models are consistent with the Maise-McDonald observations. Many researchers have applied the Cebeci-Smith model to compressible boundary layers, showing excellent agreement with measurements for adiabatic walls and somewhat larger differences when surface heat transfer is present. The Baldwin-Lomax model yields similar predictions.

Because the length scale employed in most older $k$-equation oriented one-equation models is patterned after the mixing length, they should also be expected to apply to compressible flows without ad hoc compressibility modifications. This is indeed the case, especially for these and for newer models, which have been designed for compressible-flow applications.

As we have seen in the last subsection, the issue is more complicated for two-equation turbulence models. The log-layer solution indicates that the length scale for the $k$-$\omega$ and $k$-$\epsilon$ models varies linearly with distance from the surface, independent of Mach number. The models even predict the Van Driest velocity scaling. Thus, two-equation models are consistent with two of the most important observations made by Maise and McDonald, at least in the log layer.

However, we have also seen that the $\epsilon$ equation includes a nonphysical density effect that distorts the model's log-layer structure [see Equations (5.109) and (5.110)], and precludes a satisfactory solution. By contrast, the $\omega$ equation is entirely consistent with the Maise-McDonald observations. As shown in Figures 5.2 and 5.3, the $k$-$\omega$ model provides good quantitative agreement with measurements for Mach numbers up to about 5.

Turning to effects of pressure gradient, Figures 5.4 and 5.5 compare computed and measured skin friction and velocity profiles for two compressible boundary

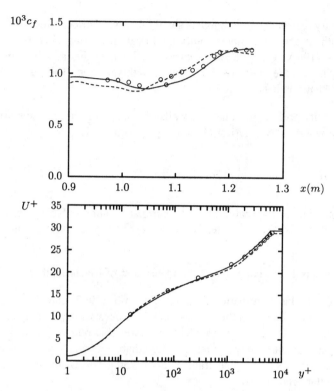

Figure 5.4: *Computed and measured skin friction and velocity profile (at x = 1.18 m) for a Mach 4, adiabatic-wall boundary layer with an adverse pressure gradient:* —— *k-ω model;* - - - *Chien k-ε model;* ○ *Zwarts.*

layers with adverse pressure gradient. Figure 5.4 corresponds to a Mach 4, adiabatic-wall experiment conducted by Zwarts [see Kline et al. (1981) — Flow 8411]. Computed results are shown for the Wilcox (1998) $k$-$\omega$ model without viscous corrections and for the Chien (1982) $k$-$\epsilon$ model. Although the effect is small for this flow, neither computation includes a compressibility correction. As shown, $k$-$\omega$ model predictions generally fall within the scatter of the experimental data. By contrast, the $k$-$\epsilon$ model skin friction is about 8% lower than measured at the beginning of the computation where the Mach number is 4. This is consistent with results shown in Figure 5.3(a). Because the flow is decelerating, the Mach number decreases with distance, and falls to 3 by the end of the run. As a result, $\bar{\rho}_e/\bar{\rho}_w$ is only half its upstream value, and the corresponding distortion of the $k$-$\epsilon$ model's log-layer velocity profile is greatly reduced. Consequently, the $k$-$\epsilon$ model's velocity profile is fortuitously in close agreement with the measured profile.

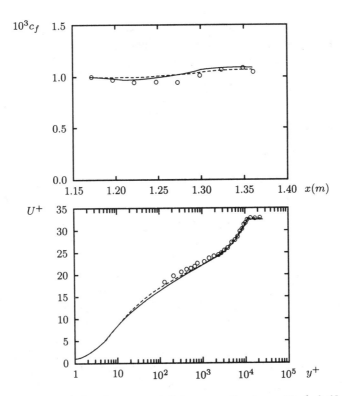

Figure 5.5: *Computed and measured flow properties for a Mach 2.65, heated-wall boundary layer with an adverse pressure gradient:* —— *k-ω model;* - - - *Chien k-ε model;* ○ *Fernando and Smits.*

Figure 5.5 presents a similar comparison for a Mach 2.65 boundary layer with adverse pressure gradient and mild surface heating. The ratio of wall temperature to the adiabatic-wall temperature, $T_w/T_{aw}$, varies between 1.07 and 1.13 for the flow. Again, because the Mach number is in the low supersonic range, the density term in the $k$-$\epsilon$ model's law of the wall is small. The value of $\kappa_\epsilon^{-1}\ell n(\bar{\rho}/\bar{\rho}_w)^{5/4}$ ranges between 0.50 at $y^+ = 100$ to 1.45 at $y^+ = 5000$. By comparison, the distortion in the $k$-$\omega$ model's law of the wall is just a fifth of these values.

While $k$-$\epsilon$ solutions for both of these adverse pressure gradient cases are nearly as close to measurements as $k$-$\omega$ solutions, similar results should not be expected for higher Mach numbers. Many compressible-flow experiments have been conducted for Mach numbers of 3 and less. Far fewer experiments have been done at higher Mach numbers. Hence, these results show how a turbulence model calibrated for the best data available may not apply at higher Mach numbers.

The $k$-$\epsilon$ model's near-wall behavior has a significant impact on model predictions, and Chien's model happens to be optimum for these two flows. The Jones-Launder (1972) and Launder-Sharma (1974) models, for example, predict skin friction values more than twice the measured values for these two flows. Zhang et al. (1993) have developed a low-Reynolds-number $k$-$\epsilon$ model that yields close agreement with constant-pressure boundary layer data for Mach numbers up to 10. Interestingly, they note from the work of Coleman and Mansour (1991) that the exact Favre-averaged equation for solenoidal dissipation, $\epsilon_s$, includes a term proportional to the rate of change of the kinematic viscosity, $\bar{\nu}$, as follows:

$$\bar{\rho}\frac{d\epsilon_s}{dt} = \frac{\bar{\rho}\epsilon_s}{\bar{\nu}}\frac{d\bar{\nu}}{dt} + \cdots \quad \Longrightarrow \quad \bar{\rho}\bar{\nu}\frac{d}{dt}\left(\frac{\epsilon_s}{\bar{\nu}}\right) = \cdots \qquad (5.116)$$

This corresponds to an effective change of dependent variable in the $\epsilon_s$ equation. Assuming a power-law for viscosity, i.e., $\bar{\mu} \propto \tilde{T}^n$, the effective rescaled dependent variable would be $\bar{\rho}^{(1+n)}\epsilon_s$. Correspondingly, the exponent 5/4 in Equation (5.110) would become $(n + 1/4)$. For a typical value $n = 7/10$, the new coefficient would be 0.95. Hence this term should yield only a slight improvement for the model's distorted law of the wall. Through a series of closure approximations, Zhang et al. combine this and other terms to arrive at a rescaling that effectively leads to using $\bar{\rho}^{-0.61}\epsilon_s$. This would correspond to replacing the exponent 5/4 by 1.86 which would yield even more distortion. It is unclear how Zhang et al. have circumvented the inherent flaw in the $k$-$\epsilon$ model for compressible flows. Since virtually all of their applications to date have been for low-Reynolds-number flows, it is possible that their damping functions penetrate far enough above the sublayer to offset the behavior indicated in Equation (5.110).

Turning to effects of surface heat transfer, Figure 5.6 compares computed skin friction with a correlation of measured values [see Kline et al. (1981) — Flow 8201]. The $k$-$\omega$ model virtually duplicates the Van Driest correlation in the absence of compressibility modifications. Using the Wilcox compressibility modification, Equation (5.73), reduces predicted $c_f/c_{f_o}$ by up to 15%. The $k$-$\epsilon$ model predictions of Zhang et al. (1993) show a similar trend, with differences from measured values of less than 10%.

As the final application, consider compressible flow over roughened flat plates. Note that this provides a test of the $k$-$\omega$ model rough-surface boundary condition on flows for which it has not been calibrated. Figure 5.7 compares computed skin friction with the data summarized by Reda, Ketter and Fan (1974). Computations have been done for Mach numbers of 0, 1 and 5 and dimensionless roughness height, $k_s^+$, ranging from 0 to 100. For each Mach number, the reference smooth-wall skin friction coefficient, $c_{f_o}$, corresponds to a momentum-thickness Reynolds number, $Re_\theta$, of 10,000. As shown, computed skin friction falls well within experimental data scatter for the entire range of roughness heights considered in the computations.

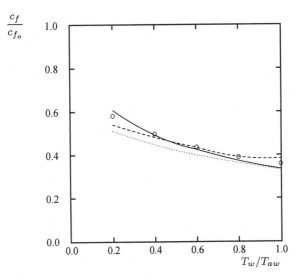

Figure 5.6: *Computed and measured effects of surface cooling on skin friction for a Mach 5 flat-plate boundary layer:* —— *$k$-$\omega$ model, $\xi^* = 0$; $\cdots$ $k$-$\omega$ model, $\xi^* = 3/2$; - - - Zhang et al. $k$-$\epsilon$ model; $\circ$ Van Driest correlation.*

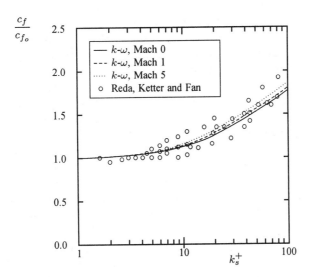

Figure 5.7: *Computed and measured effects of surface roughness on skin friction for compressible flat-plate boundary layers.*

The computations also demonstrate consistency with the observation originally made by Goddard (1959) that "the effect of surface roughness on skin-friction drag is localized deep within the boundary layer at the surface itself and is independent of the external flow, i.e., Mach number, per se, is eliminated as a variable." Consistent with Goddard's observation, Mach number has little effect on predicted $c_f/c_{f_o}$. Additionally, consistent with Reda's findings, computed skin friction departs noticeably from the smooth-wall value for $k_s^+$ values near 4 to 5 as opposed to Goddard's correlation which indicates no effect for $k_s^+ < 10$.

## 5.8 Shock-Induced Boundary-Layer Separation

One of the most interesting and challenging CFD problems is the interaction of a turbulent boundary layer with a shock wave. Many researchers have analyzed this problem since the 1960's, with varying degrees of success. The earliest efforts were confined to algebraic models, largely because of the long computing times required to solve the full Favre-averaged continuity, Navier-Stokes and energy-conservation equations. The fastest computer of the late 1960's and early 1970's was the CDC 7600, a machine comparable in speed to a 50 MHz 80486-based microcomputer. Additionally, the best compressible-flow numerical algorithms of that era were explicit time-marching methods that required tens of thousands of timesteps to achieve a solution.

Wilcox (1974) obtained the first solutions to the Favre-averaged Navier-Stokes equation, using an advanced turbulence model, for shock-induced separation of a turbulent boundary layer. This early CFD study included six computations, three for reflection of an oblique shock from a flat plate and three for flow into a compression corner. Results of the study indicate that, using a two-equation turbulence model, a reasonably accurate description of the flowfield can be obtained for reflection of an oblique shock from a flat plate. However, the numerical flowfields for compression corners differ significantly from the experimentally observed flowfields, even though Mach and Reynolds numbers and shock strength are identical to those of the flat-plate cases. Thus, a seemingly simple change in flow geometry causes a major difference in predictive accuracy.

To put these computations in proper perspective, note that the turbulence model used was the Saffman-Wilcox (1974) $k$-$\omega^2$ model with surface boundary conditions given by matching to the law of the wall. The numerical algorithm used was a first-order accurate explicit time-marching procedure. The computations took 40 to 50 hours of UNIVAC 1108 computer time.

Since that time, computational methods have improved dramatically thanks to the innovative work of many researchers such as Beam and Warming (1976), Steger and Warming (1979), Roe (1981), Van Leer (1982), MacCormack (1985), and Roache and Salari (1990), to name just a few. As a result of their innovations, converged solutions for separated flows can often be obtained in less than 200 timesteps or iterations. A two-equation turbulence model computation now takes about 10 minutes of 200 MHz Pentium-based microcomputer CPU time for a shock-separated flow.

While great advances have been made in developing accurate and efficient finite-difference algorithms, far less improvement has been made with turbulence models for such flows. A veritable plethora of CFD researchers including Viegas and Horstman (1979), Viegas, Rubesin and Horstman (1985), Champney (1989), Horstman (1992), Huang and Liou (1994), Liou and Huang (1996), and Knight (1997) provides clear substantiation of this claim. They have applied many turbulence models to shock-separated flows with almost universal results, viz.:

1. too little upstream influence, as shown by pressure starting to rise well downstream of the measured start of adverse pressure gradient;

2. surface pressure in excess of measured values in the separation bubble;

3. skin friction and heat transfer higher than measured downstream of reattachment;

4. velocity profiles downstream of reattachment that indicate flow deceleration within the boundary layer in excess of corresponding measurements.

On the one hand, using wall functions and the $k$-$\epsilon$ model, Viegas, Horstman and Rubesin (1985) are able to remove Item 3 from this list. On the other hand, they achieve only modest improvements in the other items. As we will see in this section, the $k$-$\omega$ model removes Item 2 and half of Item 3 (the skin friction) from the list. While its predictions are closer to measurements than most other models, it also displays the symptoms cited in Items 1 and 4, as well as excessive heat transfer downstream of reattachment. This lack of success on the compression-corner problem, which has persisted for more than two decades, is excellent testimony to the oft quoted statement that...

**Turbulence modeling is the pacing item in CFD.**

Most modern shock-separated computations are done without introducing wall functions. There is no evidence that the law of the wall holds in separated regions, and its use via wall functions is therefore a questionable approximation. The primary motivation for using wall functions in large scale computations that require substantial computer resources is in reducing CPU time.

Viegas, Horstman and Rubesin (1985), in effect, create a two-layer turbulence model where their wall functions apply in the sublayer, and the Standard $k$-$\epsilon$ model applies above the sublayer. While their procedure yields significant reduction in computing time, numerical results are sensitive to the location of the grid point closest to the surface, $y_2^+$. In fact, there is no obvious convergence to a well defined limiting value as $y_2^+ \to 0$. Consequently, the value of $y_2^+$ is effectively an adjustable parameter in their model equations, to be selected by the user. In practice, it is typical for the user to fix $y_2$ at each location, rather than modify it locally as the solution develops, which would be required to achieve a constant value of $y_2^+$. Thus, in practice, $y_2^+$ actually varies throughout the flow in a manner that cannot be determined a priori, so that the sensitivity to its value is a computational liability.

The sensitivity can be removed by using perturbation methods to devise suitable wall functions. Following Wilcox (1989), for example, we can deduce the following compressible-flow wall functions for the $k$-$\omega$ model as given in Equations (4.36) through (4.42):

$$\left.\begin{array}{l} u^* = u_\tau \left[\dfrac{1}{\kappa}\ell n\left(\dfrac{u_\tau y}{\nu_w}\right) + C - 1.13\dfrac{u_\tau y}{\nu_w}P^+ + O(P^+)^2\right] \\[3mm] k = \dfrac{\bar{\rho}_w}{\bar{\rho}}\dfrac{u_\tau^2}{\sqrt{\beta^*}}\left[1 + 1.16\dfrac{u_\tau y}{\nu_w}P^+ + O(P^+)^2\right] \\[3mm] \omega = \sqrt{\dfrac{\bar{\rho}_w}{\bar{\rho}}}\dfrac{u_\tau}{\sqrt{\beta^*}\kappa y}\left[1 - 0.30\dfrac{u_\tau y}{\nu_w}P^+ + O(P^+)^2\right] \end{array}\right\} \qquad (5.117)$$

where $P^+$ is the dimensionless pressure-gradient parameter defined by

$$P^+ = \frac{\nu_w}{\rho u_\tau^3}\frac{dP}{dx} \qquad (5.118)$$

As with the incompressible wall functions deduced for the $k$-$\omega$ model (see Subsection 4.7.1), the expansions in Equation (5.117) have been derived assuming $P^+$ is a small parameter. Using these wall functions, numerical solutions show very little sensitivity to placement of the grid point closest to the surface, provided it lies below $y^+ = 100$.

Figure 5.8 compares computed and measured [Settles, Vas and Bogdanoff (1976)] surface pressure for Mach 3 flow into a 24° compression corner using algebraic models, a one-equation model and several two-equation models. None of the algebraic, one-equation or two-equation models provides a satisfactory solution. In more recent computations, Huang and Liou (1994) show that the RNG $k$-$\epsilon$ model [Yakhot and Orszag (1986)] consistently predicts separation bubbles that are: (a) nearly double the length of those predicted by the standard version; and (b) much longer than measured.

Figure 5.9 compares computed and measured surface pressure and skin friction for the same flow using the $k$-$\omega$ model with ($\xi^* = 3/2$) and without ($\xi^* = 0$) the Wilcox compressibility term [Equation (5.73)]. The compressibility correction increases the length of the separation bubble by 56%. Observe that the

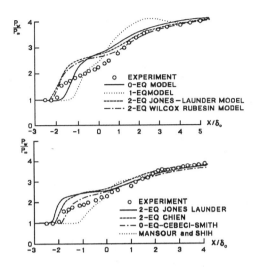

Figure 5.8: *Comparison of computed and measured surface pressure for Mach 3 flow into a 24° compression corner for several turbulence models. [From Marshall and Dolling (1992) — Copyright © AIAA 1992 — Used with permission.]*

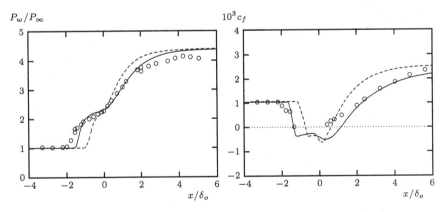

Figure 5.9: *Effect of the Wilcox compressibility correction on surface pressure for Mach 3 flow into a 24° compression corner for the $k$-$\omega$ model: —— $\xi^* = 3/2$; - - - $\xi^* = 0$; ∘ Settles, Vas and Bogdanoff (1976).*

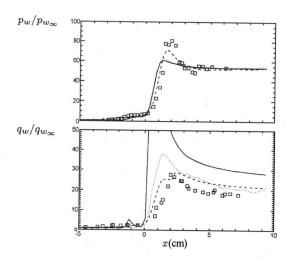

Figure 5.10: *Comparison of computed and measured surface pressure and heat transfer for Mach 9.2 flow past a 40° cylinder flare:* — k-ε *model;* - - - *Compressible model;* ···· *Two-layer model;* □ *Experiment.* *[From Horstman (1992) — Copyright © AIAA 1992 — Used with permission.]*

predicted pressure plateau in the separation bubble and skin friction downstream of reattachment are much closer to measurements than any of the results shown in Figure 5.8. By contrast, Viegas, Rubesin and Horstman (1985) predict pressure plateau values about 20% higher than measured for this flow, and are unable to simultaneously make accurate predictions for skin friction downstream of reattachment and the initial rise in surface pressure. That is, their solutions can match either skin friction or surface pressure, but not both.

The numerical separation points for this flow are further upstream than indicated by oil flow measurements. Marshall and Dolling (1992) indicate that the flow includes a low-frequency oscillation of the separation shock. This phenomenon is also observed in three-dimensional shock-separated flows [Brusniak and Dolling (1996)]. The time-mean pressure distribution upstream of the corner is affected by these oscillations, whose frequency content includes substantial energy at time scales of the mean motion. This unsteadiness is responsible for the apparent mismatch between the beginning of the pressure rise and the separation point. Since computations with none of the turbulence models considered display any low-frequency oscillation of the shock, more research is needed to arrive at a completely satisfactory solution.

Figure 5.10 illustrates a critical problem regarding prediction of surface heating rates. Results are shown for three k-ε models, viz., the Jones-Launder (1972) model, the same model with compressibility corrections devised by Rubesin

(1990), and a two-layer $k$-$\epsilon$ model developed by Rodi (1991) that uses a one-equation model rather than wall functions. As shown, the Jones-Launder model surface heat transfer, $q_w$, is off scale and is roughly triple the measured value. While the modified models predict peak heating rates closer to measured values, differences between computed and measured heat transfer are in excess of 25% throughout the flow.

Coakley and Huang (1992) propose and test numerous compressibility modifications, one of which is very effective in reducing predicted heating rates at the reattachment point for shock-separated flows. Specifically, they first define the so-called **von Kármán length scale**, $\ell_\mu$, as follows.

$$\ell_\mu = \begin{cases} \min\left(2.5y, k^{1/2}/\omega\right), & k - \omega \text{ model} \\ \min\left(2.5y, k^{3/2}/\epsilon\right), & k - \epsilon \text{ model} \end{cases} \qquad (5.119)$$

where $y$ is distance normal to the surface. Then, the value of $\omega$ or $\epsilon$ is recomputed according to

$$\omega = k^{1/2}/\ell_\mu, \qquad \epsilon = k^{3/2}/\ell_\mu \qquad (5.120)$$

This compressibility correction is very effective and yields realistic heating rates at reattachment for both $k$-$\omega$ and $k$-$\epsilon$ models.

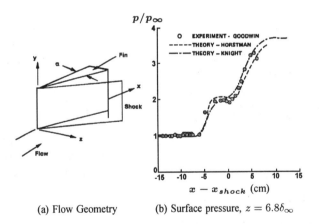

|  |  |
|---|---|
| (a) Flow Geometry | (b) Surface pressure, $z = 6.8\delta_\infty$ |

Figure 5.11: *Single-fin shock wave/boundary layer interaction at Mach 2.9. [Figure provided by D. D. Knight (1993).]*

There has been substantial progress in the capability for prediction of three-dimensional shock wave, turbulent boundary layer interactions. Recent reviews by Knight (1993,1997) describe the status of research for five basic geometries. Figure 5.11(a) illustrates the three-dimensional single fin, arguably the most extensively studied such interaction. The deflection of the fin surface by an angle

$\alpha$ generates an oblique shock that interacts with the boundary layer on the flat plate. This interaction is of some practical interest, as it represents a geometric abstraction of a fin-body juncture for a high-speed aircraft. Figure 5.11(b) compares computed and measured surface pressure for $M_\infty = 2.9$, $\alpha = 20°$, and $Re_{\delta_\infty} = 9 \cdot 10^5$, where $\delta_\infty$ is boundary-layer thickness upstream of the interaction. The comparison has been made at a spanwise distance, $z = 6.8\delta_\infty$ from the plane of symmetry. Computations using the Baldwin-Lomax (1978) model (labeled "Knight") and Rodi's (1991) $k$-$\epsilon$ model (labeled "Horstman") are in close agreement with measurements. Similar close agreement has been obtained with experimental data for pitot pressure and yaw angle [Knight et al. (1987)]. These results imply that the flowfield is predominantly rotational and inviscid, except within a thin region adjacent to the solid boundaries. This result is similar to the triple-deck theory developed for interacting boundary layers [e.g., Stewartson (1981)] and extended to non-separated three-dimensional shock wave, turbulent boundary layer interactions by Inger (1986). Consequently, the choice of turbulence model is unimportant for comparison with all but the inner (lower deck) provided the upstream boundary layer is correct. However, predicted skin friction and surface heat transfer are very sensitive to the turbulence model chosen, and can exhibit significant disagreement with experiment [Knight (1993)].

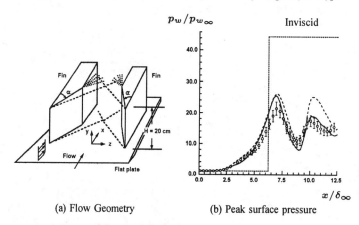

(a) Flow Geometry          (b) Peak surface pressure

Figure 5.12: *Double-fin shock wave/boundary layer interaction at Mach 8.3:* ∘ *Experiment;* —— *Baldwin-Lomax model;* - - - *Rodi k-$\epsilon$ model. [Figure provided by D. D. Knight.]*

Figure 5.12(a) shows the double-fin geometry. This geometry is of practical interest as it represents a geometric simplification of a hypersonic inlet using sidewall compression, or a sidewall interaction for a supersonic mixed compression inlet. The two fins generate opposing shocks that intersect on the centerline, and interact with the boundary layers on the flat plate and fin. Figure 5.12(b)

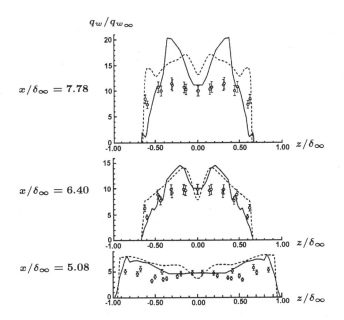

Figure 5.13: *Transverse profiles of flat plate surface heat transfer at three streamwise locations for a double-fin shock wave/boundary layer interaction at Mach 8.3:* —— *Baldwin-Lomax model;* - - - *Rodi k-ε model;* ∘ *Experiment.* *[Figure provided by D. D. Knight.]*

compares computed [Narayanswami, Horstman and Knight (1993)] and measured peak surface pressure (on the centerline) for $M_\infty = 8.3$, $\alpha = 15°$, and $Re_{\delta_\infty} = 1.7 \cdot 10^5$. The turbulence models are the Baldwin-Lomax (1978) model and the Rodi (1991) version of the $k$-$\epsilon$ model. The predictions are reasonably close except at the peak near $x/\delta_\infty = 10$. Baldwin-Lomax predictions are within about 20% of measurements, while $k$-$\epsilon$ predictions differ by as much as 45%. It is interesting to note that the peak pressure is approximately half the theoretical inviscid level because of the viscous-inviscid interaction. Reasonable agreement is obtained between computed and measured pitot pressure and yaw-angle profiles. Comparison of computed eddy viscosity shows significant differences, however. As a result, Knight concludes that, similar to the single-fin case, the flow is predominantly rotational and inviscid, except within a thin region near the surface. The turbulence model has a very significant effect on computed heat transfer, and neither model yields acceptable results (Figure 5.13). Although the $k$-$\omega$ model has not been applied to these flows, we can reasonably conclude that further research is needed in the development and application of turbulence models for three-dimensional shock wave, turbulent boundary layer interactions.

# Problems

**5.1** Derive the Reynolds-averaged momentum-conservation equation for compressible flow.

**5.2** Derive the Favre-averaged Reynolds-stress equation [Equation (5.43)].

**5.3** Verify that Equations (5.58) and (5.59) are equivalent.

**5.4** The classical Crocco temperature-velocity relationship for an adiabatic-wall boundary layer is

$$\frac{T}{T_w} = 1 - A^2 \left(\frac{\tilde{u}}{U_\infty}\right)^2$$

where $A$ is a constant. Use this approximation to evaluate the following integral.

$$u^* = \int_0^{\tilde{u}} \sqrt{\frac{\rho}{\rho_w}}\, du$$

Compare your result with Equation (5.115).

**5.5** To use the WKB method in solving an equation such as

$$\frac{d^2\omega}{dv^2} - \lambda^2 f(v)\omega = 0, \qquad \lambda \to \infty$$

we assume a solution of the form

$$\omega(v) \sim \exp\left[\lambda \sum_{n=0}^{\infty} S_n(v)\lambda^{-n}\right] \sim \exp\left[\lambda S_0(v) + S_1(v) + O(\lambda^{-1})\right]$$

(a)  Verify that $S_0(v)$ and $S_1(v)$ are given by

$$S_0(v) = \pm \int \sqrt{f(v)}\, dv + \text{constant}$$

$$S_1(v) = \ell n\, |f(v)|^{-1/4} + \text{constant}$$

(b)  Use the result of Part (a) to show that the leading-order solution to Equation (5.94) is given by Equations (5.96) and (5.97).

(c)  Now, complete the derivation of Equation (5.99).

**5.6** Derive the compressible law of the wall implied by the Cebeci-Smith model.

**5.7** Using the compressible log-layer solution, show that the turbulence length scale for the $k$-$\omega$ model defined by $\ell = k^{1/2}/\omega$ varies linearly with distance from the surface in the compressible log layer.

**5.8** Using the compressible log-layer solution, show that the turbulence length scale for the $k$-$\epsilon$ model defined by $\ell = k^{3/2}/\epsilon$ varies linearly with distance from the surface in the compressible log layer.

**5.9** Under rapid distortion due to sudden flow compression or expansion, the equations for $k$ and $\epsilon$ assume the form

$$\frac{dk}{dt} \approx -\frac{2}{3}\frac{\partial \tilde{u}_i}{\partial x_i}k \quad \text{and} \quad \frac{d\epsilon}{dt} \approx -c\frac{\partial \tilde{u}_i}{\partial x_i}\epsilon$$

where $c$ is a constant. If the turbulence length scale, $\ell = k^{3/2}/\epsilon$, is such that $\rho\ell$ remains constant under a sudden compression or expansion, what is the value of $c$?

**5.10** Using Program **MIXER** (Appendix C), compute $\delta'/\delta'_o$ at Mach 5 for the $k$-$\omega$, $k$-$\epsilon$ and RNG $k$-$\epsilon$ models. That is, let the Mach number of the upper stream be $M_1 = 5$, and let the lower stream be at rest. Do your computations using 101 grid points, and exercise the program for the Sarkar, Zeman and Wilcox compressibility corrections, Equations (5.71) through (5.73).

**5.11** The object of this problem is to compare predictions of modern turbulence models with measured properties of a Mach 2.65 turbulent boundary layer with adverse pressure gradient and surface heat transfer. The experiment to be simulated was conducted by Fernando and Smits [see Fernholz and Finley (1981)]. Use Program **EDDYBL** and its menu-driven setup utility, Program **SETEBL**, to do the computations (see Appendix D).

(a) Using **SETEBL**, change the appropriate input parameters to accomplish the following: use SI units (IUTYPE); set the freestream conditions to $p_{t_\infty} = 6.7221 \cdot 10^5$ N/m$^2$, $T_{t_\infty} = 270.74$ K, $M_\infty = 2.653$ (PT1, TT1, XMA); use an initial stepsize of $\Delta s = 0.02$ m (DS) and 151 grid points (IEDGE); set the initial boundary-layer properties so that $c_f = 0.000992$, $\delta = 0.02249$ m, $H = 5.3056$, $Re_\theta = 88483$, $s_i = 1.151$ m (CF, DELTA, H, RETHET, SI); set the maximum arc length to $s_f = 1.361$ m and maximum number of steps to 200 (SSTOP, IEND1); set up for $N = 12$ points to define the pressure and set up for prescribed surface temperature (NUMBER, KODWAL).

(b) Use the following data to define the pressure distribution in a file named **presur.dat**. The initial and final pressure gradients are both zero.

| $s$ (m) | $p_e$ (N/m$^2$) | $s$ (m) | $p_e$ (N/m$^2$) | $s$ (m) | $p_e$ (N/m$^2$) |
|---|---|---|---|---|---|
| 0.000 | $3.1039 \cdot 10^4$ | 1.197 | $3.5839 \cdot 10^4$ | 1.299 | $4.2510 \cdot 10^4$ |
| 1.000 | $3.1039 \cdot 10^4$ | 1.222 | $3.9890 \cdot 10^4$ | 1.324 | $4.1000 \cdot 10^4$ |
| 1.151 | $3.0490 \cdot 10^4$ | 1.248 | $4.2870 \cdot 10^4$ | 1.349 | $3.9630 \cdot 10^4$ |
| 1.172 | $3.1500 \cdot 10^4$ | 1.273 | $4.3650 \cdot 10^4$ | 1.361 | $4.0170 \cdot 10^4$ |

(c) Use the following data to define the surface-temperature distribution in a file named **heater.dat** along with zero heat flux. The initial and final temperature gradients are both zero.

| $s$ (m) | $T_w$ (K) | $s$ (m) | $T_w$ (K) | $s$ (m) | $T_w$ (K) | $s$ (m) | $T_w$ (K) |
|---|---|---|---|---|---|---|---|
| 0.000 | 281.56 | 1.172 | 285.76 | 1.248 | 276.94 | 1.324 | 281.49 |
| 1.000 | 281.56 | 1.197 | 281.97 | 1.273 | 279.32 | 1.349 | 284.48 |
| 1.151 | 281.56 | 1.222 | 279.99 | 1.299 | 280.53 | 1.361 | 270.92 |

(d) Do three computations using the $k$-$\omega$ model, one of the $k$-$\epsilon$ models and any other model.

(e) Compare computed skin friction with the following measured values.

| $s$ (m) | $c_f$ | $s$ (m) | $c_f$ | $s$ (m) | $c_f$ |
|---------|-------|---------|-------|---------|-------|
| 1.151 | $9.92 \cdot 10^{-4}$ | 1.248 | $9.46 \cdot 10^{-4}$ | 1.349 | $1.08 \cdot 10^{-3}$ |
| 1.172 | $9.96 \cdot 10^{-4}$ | 1.273 | $9.41 \cdot 10^{-4}$ | 1.361 | $1.04 \cdot 10^{-3}$ |
| 1.197 | $9.67 \cdot 10^{-4}$ | 1.299 | $1.01 \cdot 10^{-3}$ | | |
| 1.222 | $9.43 \cdot 10^{-4}$ | 1.324 | $1.07 \cdot 10^{-3}$ | | |

**5.12** The object of this problem is to compare predictions of modern turbulence models with measured properties of a Mach 2.2 flat-plate turbulent boundary layer. The experiment to be simulated was conducted by Shutts [see Fernholz and Finley (1981)]. Use Program **EDDYBL** and its menu-driven setup utility, Program **SETEBL**, to do the computations (see Appendix D).

(a) Using **SETEBL**, change the appropriate input parameters to accomplish the following: use USCS units (IUTYPE); set the freestream conditions to $p_{t_\infty} = 5297.382$ lb/ft$^2$, $T_{t_\infty} = 609.354°$ R, $M_\infty = 2.244$ (PT1, TT1, XMA); use an initial stepsize $\Delta s = 0.01$ ft (DS); start from laminar conditions by setting IBOUND = 0 and SI = 0 ft; set the maximum arc length to $s_f = 3.02$ ft and maximum number of steps to 1000 (SSTOP, IEND1).

(b) Prepare a file named **presur.dat** with constant freestream pressure, $p_e = 462.4495$ lb/ft$^2$, and a file **heater.dat** with constant wall temperature, $T_w = 575.7°$ R, and zero heat flux.

(c) Do three computations using the $k$-$\omega$ model, the Spalart-Allmaras model and any other model. (Avoid the $k$-$\epsilon$ models — they are very difficult to implement numerically for transitional flows.)

(d) Compare computed velocity profiles with the following measured values. Also, compare to the measured skin friction at $s = 3.02$ ft, which is $c_f = 0.00162$.

| $y^+$ | $u^*/u_\tau$ | $y^+$ | $u^*/u_\tau$ | $y^+$ | $u^*/u_\tau$ |
|-------|--------------|-------|--------------|-------|--------------|
| $6.1100 \cdot 10^1$ | 16.056 | $3.3197 \cdot 10^2$ | 19.064 | $1.5200 \cdot 10^3$ | 24.527 |
| $7.4670 \cdot 10^1$ | 16.069 | $4.0052 \cdot 10^2$ | 19.838 | $1.8607 \cdot 10^3$ | 25.544 |
| $8.7570 \cdot 10^1$ | 16.030 | $4.6841 \cdot 10^2$ | 20.580 | $2.1995 \cdot 10^3$ | 26.445 |
| $1.1540 \cdot 10^2$ | 16.030 | $5.7090 \cdot 10^2$ | 20.962 | $2.8776 \cdot 10^3$ | 27.749 |
| $1.4420 \cdot 10^2$ | 16.030 | $6.7206 \cdot 10^2$ | 21.360 | $3.5573 \cdot 10^3$ | 28.056 |
| $1.8261 \cdot 10^2$ | 16.961 | $8.4178 \cdot 10^2$ | 22.098 | $4.2367 \cdot 10^3$ | 28.081 |
| $2.2402 \cdot 10^2$ | 17.894 | $1.0115 \cdot 10^3$ | 22.764 | $4.9150 \cdot 10^3$ | 28.105 |
| $2.7900 \cdot 10^2$ | 19.218 | $1.1812 \cdot 10^3$ | 23.423 | | |

**5.13** The object of this problem is to compare predictions of modern turbulence models with measured properties of a Mach 4.5 flat-plate turbulent boundary layer. The experiment to be simulated was conducted by Coles [see Fernholz and Finley (1981)]. Use Program **EDDYBL** and its menu-driven setup utility, Program **SETEBL**, to do the computations (see Appendix D).

(a) Using **SETEBL**, change the appropriate input parameters to accomplish the following: use USCS units (IUTYPE); set the freestream conditions to $p_{t_\infty} = 8409.111$ lb/ft$^2$, $T_{t_\infty} = 563°$ R, $M_\infty = 4.544$ (PT1, TT1, XMA); use an initial stepsize $\Delta s = 0.001$ ft (DS); start from laminar conditions by setting IBOUND = 0 and SI = 0 ft; set the maximum arc length to $s_f = 1.90$ ft and maximum number of steps to 1000 (SSTOP, IEND1); and, set the freestream turbulence energy to $k_e = 5 \cdot 10^{-4}$ (ZIOTAE).

(b) Prepare a file named **presur.dat** with constant freestream pressure, $p_e = 27.50813$ lb/ft$^2$, and a file **heater.dat** with constant wall temperature, $T_w = 513.1°$ R, and zero heat flux.

(c) Do three computations using the $k$-$\omega$ model, the Baldwin-Barth model any other model. (Avoid the $k$-$\epsilon$ models — they are very difficult to implement numerically for transitional flows.)

(d) Compare computed velocity profiles with the following measured values. Also, compare to the measured skin friction at $s = 1.90$ ft, which is $c_f = 0.00126$.

| $y^+$ | $u^*/u_\tau$ | $y^+$ | $u^*/u_\tau$ | $y^+$ | $u^*/u_\tau$ |
|---|---|---|---|---|---|
| $1.4420 \cdot 10^1$ | 10.295 | $5.1570 \cdot 10^1$ | 14.990 | $1.9650 \cdot 10^2$ | 19.893 |
| $1.7100 \cdot 10^1$ | 10.972 | $6.2450 \cdot 10^1$ | 15.472 | $2.3909 \cdot 10^3$ | 20.951 |
| $2.0380 \cdot 10^1$ | 11.713 | $7.5510 \cdot 10^1$ | 15.968 | $2.8953 \cdot 10^3$ | 21.951 |
| $2.4440 \cdot 10^1$ | 12.456 | $9.1470 \cdot 10^1$ | 16.559 | $3.5196 \cdot 10^3$ | 22.523 |
| $2.6230 \cdot 10^1$ | 13.182 | $1.1099 \cdot 10^2$ | 17.258 | $4.2800 \cdot 10^3$ | 22.540 |
| $3.5590 \cdot 10^1$ | 13.848 | $1.3466 \cdot 10^2$ | 18.052 | | |
| $4.2930 \cdot 10^1$ | 14.465 | $1.6282 \cdot 10^2$ | 18.943 | | |

# Chapter 6

# Beyond the Boussinesq Approximation

The Boussinesq eddy-viscosity approximation assumes that the principal axes of the Reynolds-stress tensor, $\tau_{ij}$, are coincident with those of the mean strain-rate tensor, $S_{ij}$, at all points in a turbulent flow. This is the analog of Stokes' postulate for laminar flows. The coefficient of proportionality between $\tau_{ij}$ and $S_{ij}$ is the eddy viscosity, $\nu_T$. Unlike the molecular viscosity which is a property of the fluid, the eddy viscosity depends upon many details of the flow under consideration. It is affected by the shape and nature (e.g., roughness height) of any solid boundaries, freestream turbulence intensity, and, perhaps most significantly, flow-history effects. Flow-history effects on $\tau_{ij}$ often persist for long distances in a turbulent flow, thus casting doubt on the validity of a simple linear relationship between $\tau_{ij}$ and $S_{ij}$, even for the primary shear stress. In this chapter, we examine several flows for which the Boussinesq approximation yields a completely unsatisfactory description. We then examine some of the remedies that have been proposed to provide more accurate predictions for such flows. Although our excursion into the realm beyond the Boussinesq approximation is brief, we will see how useful the analytical tools developed in preceding chapters are for even the most complicated turbulence models.

## 6.1 Boussinesq-Approximation Deficiencies

While models based on the Boussinesq eddy-viscosity approximation provide excellent predictions for many flows of engineering interest, there are some applications for which predicted flow properties differ greatly from corresponding measurements. Generally speaking, such models are inaccurate for flows with sudden changes in mean strain rate and for flows with what Bradshaw (1973a) refers to as **extra rates of strain**. It is not surprising that flows with sudden changes in mean strain rate pose a problem. The Reynolds stresses adjust to such changes at a rate unrelated to mean-flow processes and time scales, so that the Boussinesq approximation must fail. Similarly, when a flow experiences extra rates of strain caused by rapid dilatation, out of plane straining, or significant streamline curvature, all of which give rise to unequal normal Reynolds stresses, the approximation again becomes suspect. Some of the most noteworthy types of applications for which models based on the Boussinesq approximation fail are:

1. flows with sudden changes in mean strain rate;

2. flow over curved surfaces;

3. flow in ducts with secondary motions;

4. flow in rotating fluids;

5. three-dimensional flows;

6. flows with boundary-layer separation.

As an example of a flow with a sudden change in strain rate, consider the experiment of Tucker and Reynolds (1968). In this experiment, a nearly isotropic turbulent flow is subjected to uniform mean normal strain rate produced by the following mean velocity field:

$$U = \text{constant}, \qquad V = -ay, \qquad W = az \qquad (6.1)$$

The coefficient $a$ is the constant strain rate. The strain rate is maintained for a finite distance in the $x$ direction in the experiment and then removed. The turbulence becomes anisotropic as a result of the uniform straining, and gradually approaches isotropy downstream of the point where the straining ceases. Wilcox and Rubesin (1980) have applied their $k$-$\omega^2$ eddy-viscosity model to this flow to demonstrate the deficiency of the Boussinesq approximation for flows in which mean strain rate abruptly changes. Figure 6.1 compares the computed and measured distortion parameter, $K$, defined by

$$K \equiv \frac{\overline{v'^2} - \overline{w'^2}}{\overline{v'^2} + \overline{w'^2}} \qquad (6.2)$$

As shown, when the strain rate is suddenly removed at $x \approx 2.3$ m, the model predicts an instantaneous return to isotropy, i.e., all normal Reynolds stresses become equal. By contrast, the turbulence approaches isotropy at a finite rate. Note also that the model predicts a discontinuous jump in $K$ when the straining

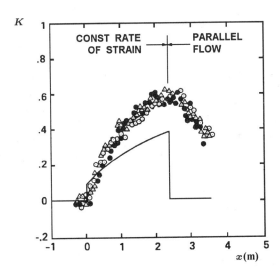

Figure 6.1: *Computed and measured distortion parameter for the Tucker-Reynolds plane-strain flow:* —— *$k$-$\omega^2$ model; $\circ$ $\bullet$ $\triangle$ Tucker-Reynolds. [From Wilcox and Rubesin (1980).]*

begins at $x = 0$. Interestingly, if the computation is extended downstream of $x = 2.3$ m without removing the strain rate, the model-predicted asymptotic value of $K$ matches the measured value at $x = 2.3$ m, but approaches this value at a slower-than-measured rate.

As an example of a flow with significant streamline curvature, consider flow over a curved surface. Meroney and Bradshaw (1975), and later investigators, find that for both convex and concave walls, when the radius of curvature, $\mathcal{R}$, is 100 times the local boundary-layer thickness, $\delta$, skin friction differs from its corresponding plane-wall value by as much as 10%. By contrast, laminar skin friction changes by about 1% for $\delta/\mathcal{R} = 0.01$. Similar results have been obtained by Thomann (1968) for supersonic boundary layers; for constant-pressure flow over surfaces with $\delta/\mathcal{R} \sim 0.02$, heat transfer changes by nearly 20%. Clearly, many practical aerodynamic surfaces are sufficiently curved to produce significant curvature effects. For such flows, a reliable turbulence model must be capable of predicting effects of curvature on the turbulence.

Standard two-equation turbulence models fail to predict any significant effect of streamline curvature. For an incompressible boundary layer on a surface with radius of curvature $\mathcal{R}$, the $k$ equation is

$$U\frac{\partial k}{\partial x} + V\frac{\partial k}{\partial y} = \nu_T\left(\frac{\partial U}{\partial y} - \frac{U}{\mathcal{R}}\right)^2 - \epsilon + \frac{\partial}{\partial y}\left[(\nu + \sigma^*\nu_T)\frac{\partial k}{\partial y}\right] \qquad (6.3)$$

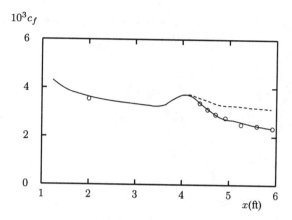

Figure 6.2: *Computed and measured skin friction for flow over a convex surface with constant pressure: - - - k-ω model without curvature correction; —— k-ω model with curvature correction; ○ So and Mellor.*

The effects of curvature appear only in the production term, and have a negligible impact on model predictions, since $(U/\mathcal{R})/(\partial U/\partial y)$ is somewhat less than $\delta/\mathcal{R}$ over most of the boundary layer.

For example, Figure 6.2 compares computed and measured skin friction for flow over a convex wall. The flow, experimentally investigated by So and Mellor (1972), has nearly constant pressure. The wall is planar up to $x = 4.375$ ft and has $\delta/\mathcal{R} \sim 0.075$ beyond that location. As shown, computed skin friction for the k-ω model (the dashed curve) is as much as 40% higher than measured.

Wilcox and Chambers (1977) propose a curvature correction to the turbulence kinetic energy equation that provides an accurate prediction for flow over curved surfaces. Appealing to the classical stability arguments for flow over a curved wall advanced by von Kármán (1934), they postulate that the equation for $k$ should more appropriately be thought of as the equation for $\overline{v'^2}$. For flow over a curved surface, again with radius of curvature $\mathcal{R}$, the equation for $\overline{v'^2}$ is

$$U\frac{\partial \overline{v'^2}}{\partial x} + V\frac{\partial \overline{v'^2}}{\partial y} - 2\frac{U}{\mathcal{R}}\overline{u'v'} = \cdots \qquad (6.4)$$

The last term on the left-hand side of Equation (6.4) results from transforming to surface-aligned coordinates.[1] Approximating $\overline{v'^2} \approx \frac{4}{9}k$ and $-\overline{u'v'} \approx \nu_T\partial U/\partial y$, Wilcox and Chambers model this effect by adding a term to the k-ω model's

---

[1]The equation for $\overline{u'^2}$ has the same term with the opposite sign and the $\overline{w'^2}$ equation has no additional curvature-related term. Thus, when we contract the Reynolds-stress equation to form the $k$ equation, no curvature-related term appears.

*k* equation. The boundary-layer form of the equations for flow over a curved surface is as follows.

$$U\frac{\partial U}{\partial x} + V\frac{\partial U}{\partial y} = -\frac{1}{\rho}\frac{dP}{dx} + \frac{\partial}{\partial y}\left[(\nu + \nu_T)\left(\frac{\partial U}{\partial y} - \frac{U}{\mathcal{R}}\right)\right], \qquad \nu_T = \frac{k}{\omega} \quad (6.5)$$

$$U\frac{\partial k}{\partial x} + V\frac{\partial k}{\partial y} + \frac{9}{2}\nu_T\frac{U}{\mathcal{R}}\frac{\partial U}{\partial y} = \nu_T\left(\frac{\partial U}{\partial y} - \frac{U}{\mathcal{R}}\right)^2 - \beta^*\omega k + \frac{\partial}{\partial y}\left[(\nu + \sigma^*\nu_T)\frac{\partial k}{\partial y}\right] \quad (6.6)$$

$$U\frac{\partial \omega}{\partial x} + V\frac{\partial \omega}{\partial y} = \alpha\left(\frac{\partial U}{\partial y} - \frac{U}{\mathcal{R}}\right)^2 - \beta\omega^2 + \frac{\partial}{\partial y}\left[(\nu + \sigma\nu_T)\frac{\partial \omega}{\partial y}\right] \quad (6.7)$$

The last term on the left-hand side of Equation (6.6) is the Wilcox-Chambers curvature-correction term. As shown in Figure 6.2, including the curvature term brings model predictions into much closer agreement with measurements. A perturbation analysis of Equations (6.5) to (6.7) for the log layer (see problems section) shows that the model predicts a modified law of the wall given by

$$\left[1 - \beta_R\frac{y}{\mathcal{R}}\right]\frac{U}{u_\tau} = \frac{1}{\kappa}\ell n\left(\frac{u_\tau y}{\nu}\right) + \text{constant} \quad (6.8)$$

with $\beta_R \approx 8.5$. This is very similar to the modified law of the wall deduced by Meroney and Bradshaw (1975), who conclude from correlation of measurements that $\beta_R \approx 12.0$.

Other curvature corrections have been proposed for two-equation models. Lakshminarayana (1986) and Patel and Sotiropoulos (1997) present comprehensive overviews. Often, in the context of the $k$-$\epsilon$ model, a correction term is added to the $\epsilon$ equation. Launder, Priddin and Sharma (1977), for example, replace the coefficient $C_{\epsilon 2}$ [see Equation (4.48)] by

$$C_{\epsilon 2} \rightarrow C_{\epsilon 2}\left(1 - 0.2Ri_T\right) \quad (6.9)$$

where $Ri_T$ is the turbulence Richardson number defined by

$$Ri_T = \frac{2U}{\mathcal{R}\partial U/\partial y} \quad (6.10)$$

This type of correction yields improved accuracy comparable to that obtained with the Wilcox-Chambers curvature correction.

While two-equation model curvature-correction terms greatly improve predictive accuracy for flow over curved walls, they are *ad hoc* modifications that cannot be generalized for arbitrary flows. The Wilcox-Chambers curvature term is introduced by making analogy to the full Reynolds-stress equation and by assuming that $k$ behaves more like $\overline{v'^2}$ than the turbulence kinetic energy for such flows. This implicitly assumes that a stress-transport model will naturally predict effects of streamline curvature. We will see in Section 6.3 that this can indeed be the case, at least for convex curvature.

These two applications alone are sufficient to serve as a warning that models based on the Boussinesq approximation will fail under some frequently encountered flow conditions. We have also seen in preceding chapters that such models are unreliable for separated flows, especially when the flow is compressible. Such models also fail to predict secondary motions that commonly occur in straight, non-circular ducts, and in the absence of ad hoc corrections, fail to predict salient features of rotating and stratified flows. While these are more subtle and specialized applications, each failure underscores the fact that models based on the Boussinesq approximation are not universal. The following sections explore some of the proposals made to remove many of these deficiencies in a less ad hoc fashion.

## 6.2  Nonlinear Constitutive Relations

One approach to achieving a more appropriate description of the Reynolds-stress tensor without introducing any additional differential equations is to assume the Boussinesq approximation is simply the leading term in a series expansion of functionals. Proceeding with this premise, Lumley (1970) and Saffman (1976) show that for incompressible flow the expansion must proceed through second order according to

$$
\tau_{ij} = -\frac{2}{3}k\delta_{ij} + 2\nu_T S_{ij} - B\frac{k}{\omega^2}S_{mn}S_{nm}\delta_{ij} - C\frac{k}{\omega^2}S_{ik}S_{kj}
$$
$$
-D\frac{k}{\omega^2}\left(S_{ik}\Omega_{kj} + S_{jk}\Omega_{ki}\right) - F\frac{k}{\omega^2}\Omega_{mn}\Omega_{nm}\delta_{ij} - G\frac{k}{\omega^2}\Omega_{ik}\Omega_{kj} \qquad (6.11)
$$

where $B$, $C$, $D$, $F$ and $G$ are closure coefficients, and $k/\omega^2$ may be equivalently written as $k^3/\epsilon^2$. Also, $S_{ij}$ and $\Omega_{ij}$ are the mean strain-rate and rotation tensors, viz.,

$$
S_{ij} = \frac{1}{2}\left(\frac{\partial U_i}{\partial x_j} + \frac{\partial U_j}{\partial x_i}\right) \quad \text{and} \quad \Omega_{ij} = \frac{1}{2}\left(\frac{\partial U_i}{\partial x_j} - \frac{\partial U_j}{\partial x_i}\right) \qquad (6.12)
$$

In order to guarantee that the trace of $\tau_{ij}$ is $-2k$, we must have $B = -C/3$ and $F = -G/3$. Equation (6.11) can be simplified by requiring it to conform with certain fundamental experimental observations. In the experiment of Tucker and Reynolds (1968), for example, the normal Reynolds stresses are related approximately by

$$\tau_{xx} \approx \frac{1}{2}(\tau_{yy} + \tau_{zz}) \tag{6.13}$$

Substituting Equations (6.1) and (6.13) into Equation (6.11) shows that necessarily $C = 0$. In addition, Ibbetson and Tritton (1975) show that homogeneous turbulence in rigid body rotation decays without developing anisotropy. This observation requires $G = 0$. Finally, if Equation (6.11) with $C = G = 0$ is applied to a classical shear layer where the only significant velocity gradient is $\partial U/\partial y$, Equation (6.13) again applies with $\tau_{xx}$ and $\tau_{zz}$ interchanged, independent of the value of $D$. Thus, Saffman's general expansion simplifies to:

$$\tau_{ij} = -\frac{2}{3}k\delta_{ij} + 2\nu_T S_{ij} - D\frac{k}{\omega^2}\left(S_{ik}\Omega_{kj} + S_{jk}\Omega_{ki}\right) \tag{6.14}$$

In analogy to this result, Wilcox and Rubesin (1980) propose the following simplified **nonlinear constitutive relation** for their $k$-$\omega^2$ model.

$$\tau_{ij} = -\frac{2}{3}k\delta_{ij} + 2\nu_T\left(S_{ij} - \frac{1}{3}\frac{\partial U_k}{\partial x_k}\delta_{ij}\right) + \frac{8}{9}\frac{k(S_{ik}\Omega_{kj} + S_{jk}\Omega_{ki})}{(\beta^*\omega^2 + 2S_{mn}S_{nm})} \tag{6.15}$$

The term $2S_{mn}S_{nm}$ in the denominator of the last term is needed to guarantee that $\overline{u'^2}$, $\overline{v'^2}$ and $\overline{w'^2}$ are always positive. The primary usefulness of this prescription for the Reynolds-stress tensor is in predicting the normal stresses. The coefficient 8/9 is selected to guarantee

$$\overline{u'^2} : \overline{v'^2} : \overline{w'^2} = 4 : 2 : 3 \tag{6.16}$$

for the flat-plate boundary layer. Equation (6.16) is a good approximation throughout the log layer and much of the defect layer. The model faithfully predicts the ratio of the normal Reynolds stresses for boundary layers with adverse pressure gradient where the ratios are quite different from those given in Equation (6.16). Bardina, Ferziger and Reynolds (1983) have used an analog of this stress/strain-rate relationship in their Large Eddy Simulation studies. However, the model provides no improvement for flows over curved surfaces.

Speziale (1987b) proposes a nonlinear constitutive relation for the $k$-$\epsilon$ model as follows (for incompressible flow):

$$\tau_{ij} = -\frac{2}{3}k\delta_{ij} + 2\nu_T S_{ij} + 4C_D C_\mu^2\frac{k^3}{\epsilon^2}\left(S_{ik}S_{kj} - \frac{1}{3}S_{mn}S_{nm}\delta_{ij}\right)$$
$$+ 4C_E C_\mu^2\frac{k^3}{\epsilon^2}\left(\overset{\circ}{S}_{ij} - \frac{1}{3}\overset{\circ}{S}_{mm}\delta_{ij}\right) \tag{6.17}$$

where $\overset{\circ}{S}_{ij}$ is the frame-indifferent Oldroyd derivative of $S_{ij}$ defined by

$$\overset{\circ}{S}_{ij} = \frac{\partial S_{ij}}{\partial t} + U_k\frac{\partial S_{ij}}{\partial x_k} - \frac{\partial U_i}{\partial x_k}S_{kj} - \frac{\partial U_j}{\partial x_k}S_{ki} \tag{6.18}$$

The closure coefficients $C_D$ and $C_E$ are given by

$$C_D = C_E = 1.68 \tag{6.19}$$

This model satisfies three key criteria that assure consistency with properties of the exact Navier-Stokes equation. **First**, like the Saffman and Wilcox-Rubesin models, it satisfies general coordinate and dimensional invariance. **Second**, it satisfies a limited form of the Lumley (1978) realizability constraints (i.e., positiveness of $k \equiv -\frac{1}{2}\tau_{ii}$). **Third**, it satisfies material-frame indifference in the limit of two-dimensional turbulence. The latter consideration leads to introduction of the Oldroyd derivative of $S_{ij}$.

The appearance of the rate of change of $S_{ij}$ in the constitutive relation is appropriate for a viscoelastic-like medium. While, to some degree, the Speziale constitutive relation includes rate effects, it still fails to describe the gradual adjustment of the Reynolds stresses following a sudden change in strain rate. For example, consider the Tucker-Reynolds flow discussed above. The Oldroyd derivative of $S_{ij}$ is given by

$$\overset{\circ}{S}_{yy} = \overset{\circ}{S}_{zz} = -2a^2; \qquad \text{all other } \overset{\circ}{S}_{ij} = 0 \tag{6.20}$$

Clearly, when the strain rate is abruptly removed, the Speziale model predicts that the normal Reynolds stresses instantaneously return to isotropy. Hence, the model is no more realistic than other eddy-viscosity models for such flows.

For flow over a curved surface, the contribution of the nonlinear terms in the Speziale model to the shear stress is negligible. Consequently, this model, like the Wilcox-Rubesin model, offers no improvement over the Boussinesq approximation for curved-wall flows.

While the Speziale model fails to improve model predictions for flows with sudden changes in strain rate and flows with curved streamlines, it does make a dramatic difference for flow through a rectangular duct [see Figure 6.3(a)]. For such a flow, the difference between $\tau_{zz}$ and $\tau_{yy}$ according to Speziale's relation is, to leading order,

$$\tau_{zz} - \tau_{yy} = C_D C_\mu^2 \frac{k^3}{\epsilon^2}\left[\left(\frac{\partial U}{\partial z}\right)^2 - \left(\frac{\partial U}{\partial y}\right)^2\right] \tag{6.21}$$

while, to the same order, the shear stresses are

$$\tau_{xy} = \nu_T \frac{\partial U}{\partial y}, \qquad \tau_{xz} = \nu_T \frac{\partial U}{\partial z}, \qquad \tau_{yz} = C_D C_\mu^2 \frac{k^3}{\epsilon^2}\frac{\partial U}{\partial y}\frac{\partial U}{\partial z} \tag{6.22}$$

Having a difference between $\tau_{zz}$ and $\tau_{yy}$ is critical in accurately simulating secondary motions of the second kind, i.e., stress-induced motions.[2] Using his

---

[2]By contrast, secondary motions of the first kind, by definition, are pressure driven, and can be predicted by eddy-viscosity models.

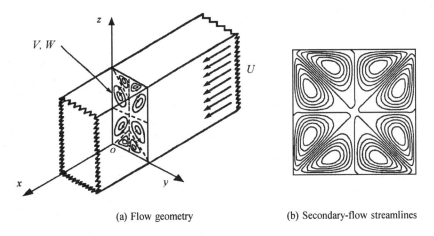

(a) Flow geometry      (b) Secondary-flow streamlines

Figure 6.3: *Fully developed turbulent flow in a rectangular duct. [From Speziale (1991) — Published with the author's permission.]*

model, Speziale (1987b) has computed flow through a rectangular duct. Figure 6.3(b) shows computed secondary-flow streamlines, which clearly illustrates that there is an eight-vortex secondary-flow structure as seen in experiments. Using the Boussinesq approximation, no secondary flow develops, so that the Speziale model obviously does a better job of capturing this missing feature. Although Speziale presents no comparison of computed and measured results, the net effect of the nonlinear terms is very dramatic.

Speziale's nonlinear constitutive relation also improves $k$-$\epsilon$ model predictions for the backward-facing step. Focusing on the experiment of Kim, Kline and Johnston (1980), Thangam and Speziale (1992) have shown that using the nonlinear model with a low-Reynolds-number $k$-$\epsilon$ model increases predicted reattachment length for this flow from 6.3 step heights to 6.9 step heights. The measured length is 7.0 step heights.

Rodi (1976) deduces a nonlinear constitutive equation by working with a model for the full Reynolds-stress equation [Equation (2.34)]. Rodi begins by approximating the convective and turbulent transport terms for incompressible flow as proportional to the Reynolds-stress component considered, i.e.,

$$\frac{\partial \tau_{ij}}{\partial t} + U_k \frac{\partial \tau_{ij}}{\partial x_k} - \frac{\partial}{\partial x_k}\left(\nu \frac{\partial \tau_{ij}}{\partial x_k} + C_{ijk}\right)$$
$$\approx \frac{\tau_{ij}}{k}\left[\frac{\partial k}{\partial t} + U_k \frac{\partial k}{\partial x_k} - \frac{\partial}{\partial x_k}\left(\nu \frac{\partial k}{\partial x_k} + \frac{1}{2}C_{jjk}\right)\right] \quad (6.23)$$

This approximation yields a nonlinear algebraic equation that can be used to determine the Reynolds-stress tensor, viz.,

$$\frac{\tau_{ij}}{k}\left\{\tau_{mn}\frac{\partial U_m}{\partial x_n}-\epsilon\right\}=-\tau_{ik}\frac{\partial U_j}{\partial x_k}-\tau_{jk}\frac{\partial U_i}{\partial x_k}+\epsilon_{ij}-\Pi_{ij} \qquad (6.24)$$

With suitable closure approximations for the dissipation tensor, $\epsilon_{ij}$, and the pressure-strain correlation tensor, $\Pi_{ij}$, Equation (6.24) defines a nonlinear constitutive relation. A model derived in this manner is known as an **Algebraic Stress Model** or, in abbreviated form, as an **ASM**.

Gatski and Speziale (1992) regard such models as strain-dependent generalizations of nonlinear constitutive relations, which can be solved explicitly to yield anisotropic eddy-viscosity models. That is, an ASM can be written in a form similar to Saffman's expansion [Equation (6.11)]. The various closure coefficients then become functions of certain Reynolds-stress tensor invariants. The complexity of the constitutive relation depends on the closure approximations, and alternative approximations have been tried by many researchers [see Lakshminarayana (1986) or Speziale (1997a)].

One of the most inconvenient features of the traditional ASM is the fact that it provides **implicit** equations for the several Reynolds stresses. However, experience has shown that such models also have unpleasant mathematical behavior. For example, Speziale (1997a) explains how such models can have either multiple solutions or singularities, defects that can wreak havoc with any numerical solver. To remedy this situation, Speziale describes a methodology that can be used to deduce algebraic stress models which are **explicit** functions of $S_{ij}$, $\Omega_{ij}$, etc. For example, building on the procedures pioneered by Pope (1975), Gatski and Speziale (1992) argue that the Reynolds-stress tensor can be approximated by

$$\tau_{ij}=-\frac{2}{3}k\delta_{ij}+\frac{3}{3-2\eta^2+6\xi^2}\left[\alpha_1\frac{k^2}{\epsilon}S_{ij}+\alpha_2\frac{k^3}{\epsilon^2}\left(S_{ik}\Omega_{kj}+S_{jk}\Omega_{ki}\right)\right.$$
$$\left.-\alpha_3\frac{k^3}{\epsilon^2}\left(S_{ik}S_{kj}-\frac{1}{3}S_{mn}S_{nm}\delta_{ij}\right)\right] \qquad (6.25)$$

where $\alpha_1$, $\alpha_2$ and $\alpha_3$ are constants that depend upon the stress-transport model used. The quantities $\xi$ and $\eta$ are

$$\xi^2=C_\xi\frac{k}{\epsilon}\sqrt{\Omega_{ij}\Omega_{ij}} \qquad \eta=C_\eta\frac{k}{\epsilon}\sqrt{S_{ij}S_{ij}} \qquad (6.26)$$

with the coefficients $C_\xi$ and $C_\eta$ depending upon the ratio of production to dissipation.

As it turns out, this *explicit* model, like *implicit* models, can exhibit singular behavior. Specifically, the Reynolds stresses can become infinite when

$$3-2\eta^2+6\xi^2\to 0 \qquad (6.27)$$

To remove this shortcoming, Gatski and Speziale (1992) **regularize** the relationship by using a Padé approximation [cf. Bender and Orszag (1978)] whereby

$$\frac{3}{3 - 2\eta^2 + 6\xi^2} \approx \frac{3\left(1 + \eta^2\right)}{3 + \eta^2 + 6\xi^2\eta^2 + 6\xi^2} \tag{6.28}$$

These two algebraic relations are virtually identical for turbulent flows that are close to equilibrium, i.e., for $\xi$ and $\eta$ less than 1. However, the expression on the right-hand side of Equation (6.28) remains finite for all values of $\xi$ and $\eta$, which correspond to strongly nonequilibrium flows. Subsequently, Speziale and Xu (1996) regularize the relationship for consistency with *Rapid Distortion Theory*.

When an ASM is used for a flow with zero mean strain rate, Equation (6.24) simplifies to

$$\tau_{ij} = \frac{k}{\epsilon}\left(\Pi_{ij} - \epsilon_{ij}\right) \tag{6.29}$$

As we will discuss in Subsection 6.3.1, in the limit of vanishing mean strain rate, the most common closure approximations for $\epsilon_{ij}$ and $\Pi_{ij}$ simplify to

$$\Pi_{ij} \rightarrow C_1\frac{\epsilon}{k}\left(\tau_{ij} + \frac{2}{3}k\delta_{ij}\right) \quad \text{and} \quad \epsilon_{ij} \rightarrow \frac{2}{3}\epsilon\delta_{ij} \tag{6.30}$$

where $C_1$ is a closure coefficient. Hence, when the mean strain rate vanishes, the algebraic stress model simplifies to

$$\tau_{ij} = -\frac{2}{3}k\delta_{ij} \tag{6.31}$$

This shows that the ASM predicts an instantaneous return to isotropy in the Tucker-Reynolds flow discussed above. Hence, like the Wilcox-Rubesin and Speziale nonlinear constitutive relations, the ASM fails to properly account for sudden changes in the mean strain rate. The ASM does provide significant improvement for flows with streamline curvature however. So and Mellor (1978), for example, show that excellent agreement between computed and measured flow properties is possible using an ASM for boundary layers on curved surfaces. The model predicts most qualitative features and provides fair quantitative agreement for flows with secondary motions as shown, for example, by Demuren (1991).

In summary, the primary advantage of nonlinear constitutive relations appears to be in predicting the anisotropy of the normal Reynolds stresses. The most important application for which this is of interest is for flow in ducts with secondary motions of the second kind. In the case of algebraic stress models, greatly improved predictions can be obtained for flows with nontrivial streamline curvature. It is doubtful that the nonlinear stress models discussed in this section yield any significant improvement for separating and reattaching flows. While the $k$-$\epsilon$ model's predicted reattachment length is closer to the measured length when the Speziale, or any other, nonlinear model is used, it is not clear that a better description of the physics of this flow has been provided. The excellent solutions obtained with the $k$-$\omega$ model [see Section 4.10] strongly suggest that the $k$-$\epsilon$ model's inaccuracy for such flows has nothing to do with the basic eddy-viscosity assumption. While the improvements attending use of a nonlinear constitutive relation with two-equation models are nontrivial, the models still retain many of their deficiencies.

## 6.3  Stress-Transport Models

Although posing a more formidable task with regard to establishing suitable closure approximations, there are potential gains in universality that can be realized by devising a stress-transport model. In general turbulence-modeling literature, such models are often referred to as *second-order closure* or *second-moment closure* models. As we will see, stress-transport models naturally include effects of streamline curvature, sudden changes in strain rate, secondary motions, etc. We will also see that there is a significant price to be paid in complexity and computational difficulty for these gains.

Virtually all researchers use the same starting point for developing such a model, viz., the exact differential "transport" equation describing the behavior of the specific Reynolds-stress tensor, $\tau_{ij} \equiv -\overline{u_i' u_j'}$. Note that, as we have done throughout this book and consistent with common practice, we usually drop the term "specific" in referring to $\tau_{ij}$. As shown in Chapter 2, the incompressible form of the exact equation is

$$\frac{\partial \tau_{ij}}{\partial t} + U_k \frac{\partial \tau_{ij}}{\partial x_k} = -\tau_{ik}\frac{\partial U_j}{\partial x_k} - \tau_{jk}\frac{\partial U_i}{\partial x_k} + \epsilon_{ij} - \Pi_{ij} + \frac{\partial}{\partial x_k}\left[\nu\frac{\partial \tau_{ij}}{\partial x_k} + C_{ijk}\right] \quad (6.32)$$

where

$$\Pi_{ij} = \overline{\frac{p'}{\rho}\left(\frac{\partial u_i'}{\partial x_j} + \frac{\partial u_j'}{\partial x_i}\right)} \quad (6.33)$$

$$\epsilon_{ij} = 2\nu \overline{\frac{\partial u_i'}{\partial x_k}\frac{\partial u_j'}{\partial x_k}} \quad (6.34)$$

$$\rho C_{ijk} = \rho\overline{u_i' u_j' u_k'} + \overline{p' u_i'}\delta_{jk} + \overline{p' u_j'}\delta_{ik} \quad (6.35)$$

Inspection of Equation (6.32) shows why we can expect a stress-transport model to correct some of the Boussinesq approximation's shortcomings. **First,** since the equation automatically accounts for the convection and diffusion of $\tau_{ij}$, a stress-transport model will include effects of flow history. The dissipation and turbulent-transport terms indicate the presence of time scales unrelated to mean-flow time scales, so history effects should be more realistically represented than with a two-equation model. **Second**, Equation (6.32) contains convection, production and (optionally) body-force terms that respond automatically to effects such as streamline curvature, system rotation and stratification, at least qualitatively. Thus, there is potential for naturally representing such effects with a well-formulated stress-transport model. **Third**, Equation (6.32) gives no a priori reason for the normal stresses to be equal even when the mean strain rate vanishes. Rather, their values will depend upon initial conditions and other flow processes, so that the model should behave properly for flows with sudden changes in strain rate.

Chou (1945) and Rotta (1951) were the first to accomplish closure of the Reynolds-stress equation, although they did not carry out numerical computations. Many researchers have made important contributions since their pioneering efforts. Two of the most important conceptual contributions have been made by Donaldson and Lumley. Donaldson [cf. Donaldson and Rosenbaum (1968)] was the first to advocate the concept of **invariant modeling**, i.e., establishing closure approximations that rigorously satisfy coordinate invariance. Lumley (1978) has tried to develop a systematic procedure for representing closure approximations that guarantees **realizability**, i.e., that all physically positive-definite turbulence properties be computationally positive definite and that all computed correlation coefficients lie between $\pm 1$. However, Speziale, Abid and Durbin (1994) have recently cast doubt on some aspects of the Lumley approach.

## 6.3.1 Closure Approximations

To close Equation (6.32), we must model the dissipation tensor, $\epsilon_{ij}$, the turbulent-transport tensor, $C_{ijk}$, and the pressure-strain correlation tensor, $\Pi_{ij}$. Because each of these terms is a tensor, the approximations required for closure may be much more elaborate than the approximations used for the simpler scalar and vector terms in the $k$ equation. In this subsection, we will discuss some of the most commonly used closure approximations.

**Dissipation:** Because dissipation occurs at the smallest scales, most modelers use the Kolmogorov (1941) hypothesis of local isotropy, which implies

$$\epsilon_{ij} = \frac{2}{3}\epsilon\delta_{ij} \qquad (6.36)$$

where

$$\epsilon = \nu\overline{\frac{\partial u_i'}{\partial x_k}\frac{\partial u_i'}{\partial x_k}} \qquad (6.37)$$

The scalar quantity $\epsilon$ is the dissipation rate appearing in the turbulence kinetic energy equation of standard two-equation models. This becomes evident upon contracting Equation (6.32) to form an equation for $k = -\frac{1}{2}\tau_{ii}$. As with simpler models, we must establish a procedure for determining $\epsilon$. In most of his work, for example, Donaldson specified $\epsilon$ algebraically, similar to what is done with a one-equation model. Most researchers use the $\epsilon$ equation as formulated for the $k$-$\epsilon$ model. Wilcox and Rubesin (1980) and Wilcox (1988b) compute $\epsilon$ by using an equation for the specific dissipation rate.

Since dissipation is in reality anisotropic, particularly close to solid boundaries, efforts have been made to model this effect. Generalizing a low-Reynolds-number proposal of Rotta (1951), Hanjalić and Launder (1976) write[3]

$$\epsilon_{ij} = \frac{2}{3}\epsilon\delta_{ij} + 2f_s\epsilon b_{ij} \tag{6.38}$$

where $b_{ij}$ is the dimensionless **Reynolds-stress anisotropy tensor**, viz.,

$$b_{ij} = \frac{\overline{u_i'u_j'} - \frac{2}{3}k\delta_{ij}}{2k} \tag{6.39}$$

Also, $f_s$ is a low-Reynolds-number damping function, which they choose empirically to vary with turbulence Reynolds number, $Re_T \equiv k^2/(\epsilon\nu)$, as

$$f_s = \left(1 + \frac{1}{10}Re_T\right)^{-1} \tag{6.40}$$

**Turbulent Transport:** As with the turbulence energy equation, pressure fluctuations, as well as triple products of velocity fluctuations, appear in the tensor $C_{ijk}$. Pressure fluctuations within the fluid cannot be measured with any assurance of accuracy, so there are no experimental data to provide any guidance for modeling the pressure-correlation terms. Currently-available DNS data seem to support neglecting pressure fluctuations. Traditionally, they are effectively ignored. The most common approach used in modeling $C_{ijk}$ is to assume a gradient transport process. Daly and Harlow (1970), argue that the simplest tensor of rank three that can be obtained from the second-order correlation $\tau_{ij}$ is $\partial\tau_{ij}/\partial x_k$, and make the following approximation.

$$C_{ijk} \propto \frac{\partial\tau_{ij}}{\partial x_k} \tag{6.41}$$

This form, although mathematically simple, is inconsistent with the fact that $C_{ijk}$ is symmetric in all three of its indices, i.e., it is **rotationally invariant** [provided pressure fluctuations are neglected – see Equation (6.35)].

To properly reproduce the symmetry of $C_{ijk}$, Donaldson (1972) postulates

$$C_{ijk} \propto \frac{\partial\tau_{jk}}{\partial x_i} + \frac{\partial\tau_{ik}}{\partial x_j} + \frac{\partial\tau_{ij}}{\partial x_k} \tag{6.42}$$

---

[3]Note that $b_{ii} = 0$ and $\delta_{ii} = 3$ so that Equation (6.38), like Equation (6.36), gives $\epsilon_{ii} = 2\epsilon$.

This tensor has the proper symmetry, but is not dimensionally correct. We require a factor whose dimensions are length$^2$/time — a gradient diffusivity — and the ratio of $k^2/\epsilon$ has been employed by Mellor and Herring (1973) and Launder, Reece and Rodi (1975). Using the notation of Launder et al., the final form of the closure approximation is

$$C_{ijk} = \frac{2}{3}C_s\frac{k^2}{\epsilon}\left[\frac{\partial\tau_{jk}}{\partial x_i} + \frac{\partial\tau_{ik}}{\partial x_j} + \frac{\partial\tau_{ij}}{\partial x_k}\right] \tag{6.43}$$

where $C_s \approx 0.11$ is a scalar closure coefficient.

Launder, Reece and Rodi also postulate a more general form based on analysis of the transport equation for $C_{ijk}$. Through a series of heuristic arguments, they infer the following alternative closure approximation:

$$C_{ijk} = -C_s'\frac{k}{\epsilon}\left[\tau_{im}\frac{\partial\tau_{jk}}{\partial x_m} + \tau_{jm}\frac{\partial\tau_{ik}}{\partial x_m} + \tau_{km}\frac{\partial\tau_{ij}}{\partial x_m}\right] \tag{6.44}$$

where $C_s' \approx 0.25$ is also a scalar closure coefficient. Note that optimizing $C_s$ and $C_s'$ implies that any pressure diffusion is combined with triple-product diffusion.

**Pressure-Strain Correlation:** The tensor $\Pi_{ij}$ defined in Equation (6.33), which is often referred to as the **pressure-strain redistribution** term, has received the greatest amount of attention from turbulence modelers. The reason for this interest is twofold. First, being of the same order as production, the term plays a critical role in most flows of engineering interest. Second, because it involves essentially unmeasurable correlations, a great degree of ingenuity is required to establish a rational closure approximation.

To determine pressure fluctuations in an incompressible flow we must, in principle, solve the following Poisson equation for $p'$.

$$\frac{1}{\rho}\nabla^2 p' = -2\frac{\partial U_i}{\partial x_j}\frac{\partial u_j'}{\partial x_i} - \frac{\partial^2}{\partial x_i\partial x_j}\left(u_i'u_j' - \overline{u_i'u_j'}\right) \tag{6.45}$$

This equation follows from taking the divergence of the Navier-Stokes equation, using the continuity equation and subtracting the time-averaged equation from the instantaneous equation.

As an aside, note that in a compressible flow, these operations lead to a transport equation for the divergence of $u_i'$, viz., $\partial u_i'/\partial x_i$. Equation (6.45), depending upon $\partial u_i'/\partial x_i = 0$, is a degenerate case. This is consistent with the fact that pressure signals travel through a fluid at the speed of sound, which is infinite for Mach number approaching zero. Hence, we should expect the fluctuating pressure to be governed by an elliptic equation such as Equation (6.45) for incompressible flow. By contrast, pressure signals travel at finite speed in a compressible flow. Thus, we should expect the pressure fluctuations to satisfy a transport equation, which is typically hyperbolic in nature, for compressible flows.

The classical approach to solving Equation (6.45) is to write $p'$ as the sum of two contributions, viz.,

$$p' = p'_{slow} + p'_{rapid} \qquad (6.46)$$

By construction, the **slow** and **rapid** pressure fluctuations satisfy the following equations.

$$\frac{1}{\rho}\nabla^2 p'_{slow} = -\frac{\partial^2}{\partial x_i \partial x_j}\left(u'_i u'_j - \overline{u'_i u'_j}\right) \qquad (6.47)$$

$$\frac{1}{\rho}\nabla^2 p'_{rapid} = -2\frac{\partial U_i}{\partial x_j}\frac{\partial u'_j}{\partial x_i} \qquad (6.48)$$

The general notion implied by the nomenclature is that changes in the mean strain rate contribute immediately to $p'_{rapid}$ because the mean velocity gradient appears explicitly in Equation (6.48). By contrast, such effects are implicitly represented in Equation (6.47). The terminology **slow** and **rapid** should not be taken too literally, however, since in real-life flows the mean strain rate does not necessarily change more rapidly than $u'_i u'_j$.

For homogeneous turbulence, these equations can be solved in terms of appropriate Green's functions, and the resulting form of $\Pi_{ij}$ is

$$\Pi_{ij} = A_{ij} + M_{ijkl}\frac{\partial U_k}{\partial x_l} \qquad (6.49)$$

where $A_{ij}$ is the **slow pressure strain** and the tensor $M_{ijkl}\partial U_k/\partial x_l$ is the **rapid pressure strain**. The tensors $A_{ij}$ and $M_{ijkl}$ are given by the following.

$$A_{ij} = \frac{1}{4\pi}\iiint_V \overline{\left(\frac{\partial u'_i}{\partial x_j} + \frac{\partial u'_j}{\partial x_i}\right)\frac{\partial^2 (u'_k u'_l)}{\partial y_k \partial y_l}}\frac{d^3 y}{|\mathbf{x} - \mathbf{y}|} \qquad (6.50)$$

$$M_{ijkl} = \frac{1}{2\pi}\iiint_V \overline{\left(\frac{\partial u'_i}{\partial x_j} + \frac{\partial u'_j}{\partial x_i}\right)\frac{\partial u'_l}{\partial y_k}}\frac{d^3 y}{|\mathbf{x} - \mathbf{y}|} \qquad (6.51)$$

The integration range for Equations (6.50) and (6.51) is the entire flowfield. For inhomogeneous turbulence, the second term in Equation (6.49) becomes an integral with the mean velocity gradient inside the integrand. This emphasizes a shortcoming of single-point closure schemes that has not been as obvious in any of the closure approximations we have discussed thus far. That is, we are postulating that we can accomplish closure based on correlations of fluctuating quantities at the same physical location. The pressure-strain correlation very clearly is not a localized process, but rather, involves contributions from every point in the flow. This would suggest that two-point correlations, i.e., products of fluctuating properties at two separate physical locations, are more appropriate. Nevertheless, we expect contributions from more than one or two large eddy sizes away to be negligible, and this would effectively define what is usually referred to as the **locally-homogeneous approximation**. Virtually all modelers assume that turbulent flows behave as though they are locally homogeneous, and use Equation (6.49).

The forms of the tensors $A_{ij}$ and $M_{ijkl}$ must adhere to a variety of constraints resulting from the symmetry of indices, mass conservation and other kinematic constraints. We know, for example, that the trace of $\Pi_{ij}$ must vanish and this is true for the slow and rapid parts individually. Rotta (1951) postulates that the **slow pressure-strain** term, often referred to as the **return-to-isotropy** term, is given by

$$A_{ij} = C_1 \frac{\epsilon}{k} \left( \tau_{ij} + \frac{2}{3} k \delta_{ij} \right) \tag{6.52}$$

where $C_1$ is a closure coefficient whose value can be inferred from measurements [Uberoi (1956)] to lie in the range

$$1.4 < C_1 < 1.8 \tag{6.53}$$

Turning now to the **rapid pressure strain**, early research efforts of Donaldson [Donaldson and Rosenbaum (1968)], Daly and Harlow (1970), and Lumley (1972) assume that the rapid pressure strain is negligible compared to the slow pressure strain. However, Crow (1968) and Reynolds (1970) provide simple examples of turbulent flows for which the effect of the rapid pressure strain far outweighs the slow pressure strain.

Launder, Reece and Rodi (1975) have devised a particularly elegant closure approximation based almost entirely on kinematical considerations. Building upon the analysis of Rotta (1951), they write $M_{ijkl}$ in terms of a tensor $a_{ijkl}$ as follows.

$$M_{ijkl} = a_{ijkl} + a_{jikl} \tag{6.54}$$

This relation is strictly valid only for homogeneous turbulence. Rotta demonstrated that the tensor $a_{ijkl}$ must satisfy the following *symmetry* and *normalization* constraints:

$$a_{ijkl} = a_{ljki} = a_{lkji} \qquad \text{(symmetry)} \tag{6.55}$$

$$a_{iikl} = 0, \qquad a_{ijjl} = -2\tau_{il} \qquad \text{(normalization)} \tag{6.56}$$

Launder et al. propose that the fourth-rank tensor $a_{ijkl}$ can be expressed as a linear function of the Reynolds-stress tensor. The most general tensor, linear in $\tau_{ij}$, satisfying the symmetry constraints of Equation (6.55) is

$$a_{ijkl} = -\alpha \delta_{kj} \tau_{li} - \beta (\delta_{lk} \tau_{ij} + \delta_{lj} \tau_{ik} + \delta_{ik} \tau_{lj} + \delta_{ij} \tau_{lk})$$
$$-C_2 \delta_{li} \tau_{kj} + [\eta \delta_{li} \delta_{kj} + \upsilon (\delta_{lk} \delta_{ij} + \delta_{lj} \delta_{ik})] k \tag{6.57}$$

where $\alpha$, $\beta$, $C_2$, $\eta$ and $\upsilon$ are closure coefficients. Invoking the conditions of Equation (6.56), all of the coefficients can be expressed in terms of $C_2$, viz.,

$$\alpha = \frac{4C_2 + 10}{11}, \quad \beta = -\frac{3C_2 + 2}{11}, \quad \eta = -\frac{50C_2 + 4}{55}, \quad \upsilon = \frac{20C_2 + 6}{55} \quad (6.58)$$

Finally, combining Equations (6.54) through (6.58), we arrive at the well-known LRR model for the rapid pressure strain.

**LRR Rapid Pressure-Strain Model:**

$$M_{ijkl}\frac{\partial U_k}{\partial x_l} = -\hat{\alpha}\left(P_{ij} - \frac{1}{3}P_{kk}\delta_{ij}\right) - \hat{\beta}\left(D_{ij} - \frac{1}{3}D_{kk}\delta_{ij}\right) - \hat{\gamma}kS_{ij} \quad (6.59)$$

$$P_{ij} = \tau_{im}\frac{\partial U_j}{\partial x_m} + \tau_{jm}\frac{\partial U_i}{\partial x_m} \quad \text{and} \quad D_{ij} = \tau_{im}\frac{\partial U_m}{\partial x_j} + \tau_{jm}\frac{\partial U_m}{\partial x_i} \quad (6.60)$$

$$\hat{\alpha} = \frac{8 + C_2}{11}, \quad \hat{\beta} = \frac{8C_2 - 2}{11}, \quad \hat{\gamma} = \frac{60C_2 - 4}{55}, \quad 0.4 < C_2 < 0.6 \quad (6.61)$$

Note that for compressible flows, the mean strain-rate tensor, $S_{ij}$, is usually replaced by $S_{ij} - \frac{1}{3}S_{kk}\delta_{ij}$ in Equation (6.59).

One of the most remarkable features of this closure approximation is the presence of just one undetermined closure coefficient, namely, $C_2$. The value of $C_2$ has been established by comparison of model predictions with measured properties of homogeneous turbulent flows. Launder, Reece and Rodi (1975) suggested using $C_2 = 0.40$. Morris (1984) revised its value upward to $C_2 = 0.50$, while Launder (1992) currently recommends $C_2 = 0.60$. Section 6.4 discusses the kind of flows used to calibrate this model.

Bradshaw (1973b) has shown that there is an additional contribution to Equations (6.50) and (6.51) that has a nontrivial effect close to a solid boundary. It is attributed to a surface integral that appears in the Green's function for Equation (6.45), equivalent to a volume integral over an identical "image" flowfield below the solid surface. This has come to be known as the **pressure-echo effect** or **wall-reflection effect**. Launder, Reece and Rodi (1975), and most others until recently, propose a near-wall correction to their model for $\Pi_{ij}$ that explicitly involves distance from the surface. Gibson and Launder (1978), Craft and Launder (1992) and Launder and Li (1994) propose alternative models to account for the pressure-echo effect. For example, the LRR wall-reflection term, $\Pi_{ij}^{(w)}$, is

$$\Pi_{ij}^{(w)} = \left[0.125\frac{\epsilon}{k}(\tau_{ij} + \frac{2}{3}k\delta_{ij}) - 0.015(P_{ij} - D_{ij})\right]\frac{k^{3/2}}{\epsilon n} \quad (6.62)$$

where $n$ is distance normal to the surface.

More recent efforts at devising a suitable closure approximation for $\Pi_{ij}$ have focused on developing a nonlinear expansion in terms of the anisotropy tensor,

$b_{ij}$, defined in Equation (6.39). Lumley (1978) has systematically developed a general representation for $\Pi_{ij}$ based on Equations (6.45) through (6.51). In addition to insisting upon coordinate invariance and other required symmetries, Lumley insists upon **realizability**. As noted earlier, this means that all quantities known to be strictly positive must be guaranteed to be positive by the closure model. Additionally, all computed correlation coefficients must lie between $\pm 1$. This limits the possible form of the functional expansion for $\Pi_{ij}$. Lumley shows that the most general form of the complete tensor $\Pi_{ij}$ for incompressible flow is as follows.

**Lumley Pressure-Strain Model:**

$$
\begin{aligned}
\Pi_{ij} = a_0 \epsilon b_{ij} + a_1 \epsilon \left( b_{ik} b_{jk} - \frac{1}{3} II \delta_{ij} \right) + a_2 k S_{ij} \\
+ k \left( a_3 b_{kl} S_{lk} + a_4 b_{kl} b_{lm} S_{mk} \right) b_{ij} \\
+ k \left( a_5 b_{kl} S_{lk} + a_6 b_{kl} b_{lm} S_{mk} \right) \left( b_{ik} b_{kj} - \frac{1}{3} II \delta_{ij} \right) \\
+ a_7 k \left( b_{ik} S_{jk} + b_{jk} S_{ik} - \frac{2}{3} b_{kl} S_{lk} \delta_{ij} \right) \\
+ a_8 k \left( b_{ik} b_{kl} S_{jl} + b_{jk} b_{kl} S_{il} - \frac{2}{3} b_{kl} b_{lm} S_{mk} \delta_{ij} \right) \\
+ a_9 k \left( b_{ik} \Omega_{jk} + b_{jk} \Omega_{ik} \right) + a_{10} k \left( b_{ik} b_{kl} \Omega_{jl} + b_{jk} b_{kl} \Omega_{il} \right)
\end{aligned}
\tag{6.63}
$$

The eleven closure coefficients are assumed to be functions of the **tensor invariants** $II$ and $III$, i.e.,

$$
a_i = a_i(II, III), \qquad II = b_{ij} b_{ij}, \qquad III = b_{ik} b_{kl} b_{li}
\tag{6.64}
$$

The tensor $\Omega_{ij}$ is the mean rotation tensor defined in Equation (6.12). The LRR model can be shown to follow from Lumley's general expression when nonlinear terms in $b_{ij}$ are neglected, i.e., when all coefficients except $a_0$, $a_2$, $a_7$ and $a_9$ are zero.

A similar, but simpler, nonlinear model has been postulated by Speziale, Sarkar and Gatski (1991). For incompressible flows, this model, known as the **SSG** model, is as follows.

**SSG Pressure-Strain Model:**

$$
\begin{aligned}
\Pi_{ij} = - \left( C_1 \epsilon + C_1^* \tau_{mn} \frac{\partial U_m}{\partial x_n} \right) b_{ij} + C_2 \epsilon \left( b_{ik} b_{kj} - \frac{1}{3} b_{mn} b_{nm} \delta_{ij} \right) \\
+ \left( C_3 - C_3^* \sqrt{II} \right) k S_{ij} + C_4 k \left( b_{ik} S_{jk} + b_{jk} S_{ik} - \frac{2}{3} b_{mn} S_{mn} \delta_{ij} \right) \\
+ C_5 k ( b_{ik} \Omega_{jk} + b_{jk} \Omega_{ik} )
\end{aligned}
\tag{6.65}
$$

$$\left. \begin{array}{llll} C_1 = 3.4, & C_1^* = 1.8, & C_2 = 4.2, & C_3 = 0.8 \\ C_3^* = 1.3, & C_4 = 1.25, & C_5 = 0.4 \end{array} \right\} \tag{6.66}$$

Interestingly, the SSG model does not appear to require a correction for the pressure-echo effect in order to obtain a satisfactory log-layer solution.

Many other proposals have been made for closing the Reynolds-stress equation, with most of the attention on $\Pi_{ij}$. Weinstock (1981), Shih and Lumley (1985), Haworth and Pope (1986), Reynolds (1987), Shih, Mansour and Chen (1987), Fu, Launder and Tselepidakis (1987) and Craft et al. (1989) have formulated nonlinear pressure-strain correlation models.

As with the $k$-$\epsilon$ model, low-Reynolds-number damping functions are needed to integrate through the sublayer when the $\epsilon$ equation is used. Damping functions appear in the pressure-strain correlation tensor as well as in the dissipation. So et al. (1991) give an excellent review of stress-transport models including low-Reynolds-number corrections. Compressibility, of course, introduces an extra complication, and a variety of new proposals are being developed.

While the discussion in this subsection is by design brief, it illustrates the nature of the closure problem for stress-transport models. Although dimensional analysis combined with physical insight still plays a role, there is a greater dependence upon the formalism of tensor calculus. To some extent, this approach focuses more on the differential equations than on the physics of turbulence. This appears to be necessary because the increased complexity mandated by having to model second and higher rank tensors makes it difficult to intuit the proper forms solely on the strength of physical reasoning. Fortunately, the arguments developed during the past decade have a stronger degree of rigor than the drastic surgery approach to modeling terms in the dissipation-rate equation discussed in Subsection 4.3.2.

Increasingly, stress-transport models are being tested for nontrivial flows. The paper by Schwarz and Bradshaw (1994), for example, illustrates the actual performance of some of these models in three-dimensional boundary layers. So and Yuan (1998) test 8 two-equation models and 3 stress-transport models for flow past a backward-facing step. The study by Parneix, Laurence and Durbin (1998) also provides useful assessments of modern stress-transport models.

## 6.3.2  Launder-Reece-Rodi Model

The model devised by Launder, Reece and Rodi (1975) is the best known and most thoroughly tested stress-transport model based on the $\epsilon$ equation. Most newer stress-transport models are based on the LRR model and differ primarily in the closure approximation chosen for $\Pi_{ij}$. Combining the closure approximations discussed in the preceding subsection, we have the following high-Reynolds-number, compressible form of the model.

**Reynolds-Stress Tensor:**

$$\rho \frac{\partial \tau_{ij}}{\partial t} + \rho U_k \frac{\partial \tau_{ij}}{\partial x_k} = -\rho P_{ij} + \frac{2}{3}\rho\epsilon\delta_{ij} - \rho\Pi_{ij}$$

$$-C_s \frac{\partial}{\partial x_k}\left[\frac{\rho k}{\epsilon}\left(\tau_{im}\frac{\partial \tau_{jk}}{\partial x_m} + \tau_{jm}\frac{\partial \tau_{ik}}{\partial x_m} + \tau_{km}\frac{\partial \tau_{ij}}{\partial x_m}\right)\right] \quad (6.67)$$

**Dissipation Rate:**

$$\rho \frac{\partial \epsilon}{\partial t} + \rho U_j \frac{\partial \epsilon}{\partial x_j} = C_{\epsilon 1}\frac{\rho\epsilon}{k}\tau_{ij}\frac{\partial U_i}{\partial x_j} - C_{\epsilon 2}\frac{\rho\epsilon^2}{k} - C_\epsilon \frac{\partial}{\partial x_k}\left[\frac{\rho k}{\epsilon}\tau_{km}\frac{\partial \epsilon}{\partial x_m}\right] \quad (6.68)$$

**Pressure-Strain Correlation:**

$$\Pi_{ij} = C_1 \frac{\epsilon}{k}\left(\tau_{ij} + \frac{2}{3}k\delta_{ij}\right) - \hat{\alpha}\left(P_{ij} - \frac{2}{3}P\delta_{ij}\right)$$

$$-\hat{\beta}\left(D_{ij} - \frac{2}{3}P\delta_{ij}\right) - \hat{\gamma}k\left(S_{ij} - \frac{1}{3}S_{kk}\delta_{ij}\right)$$

$$+\left[0.125\frac{\epsilon}{k}(\tau_{ij} + \frac{2}{3}k\delta_{ij}) - 0.015(P_{ij} - D_{ij})\right]\frac{k^{3/2}}{\epsilon n} \quad (6.69)$$

**Auxiliary Relations:**

$$P_{ij} = \tau_{im}\frac{\partial U_j}{\partial x_m} + \tau_{jm}\frac{\partial U_i}{\partial x_m}, \quad D_{ij} = \tau_{im}\frac{\partial U_m}{\partial x_j} + \tau_{jm}\frac{\partial U_m}{\partial x_i}, \quad P = \frac{1}{2}P_{kk} \quad (6.70)$$

**Closure Coefficients [Launder (1992)]:**

$$\left.\begin{array}{lll}\hat{\alpha} = (8 + C_2)/11, & \hat{\beta} = (8C_2 - 2)/11, & \hat{\gamma} = (60C_2 - 4)/55 \\ C_1 = 1.8, & C_2 = 0.60, & C_s = 0.11 \\ C_\epsilon = 0.18, & C_{\epsilon 1} = 1.44, & C_{\epsilon 2} = 1.92\end{array}\right\} \quad (6.71)$$

Note that Equation (6.68) differs from the $\epsilon$ equation used with the Standard $k$-$\epsilon$ model [Equation (4.48)] in the form of the diffusion term. Rather than introduce an isotropic eddy viscosity, Launder, Reece and Rodi opt to use the analog of the turbulent transport term, $C_{ijk}$. The values of the closure coefficients in Equation (6.71) are specific to the LRR model of course, and their values are influenced by the specific form assumed for $\Pi_{ij}$. In their original paper, Launder, Reece and Rodi recommend $C_1 = 1.5$, $C_2 = 0.4$, $C_s = 0.11$, $C_\epsilon = 0.15$, $C_{\epsilon 1} = 1.44$ and $C_{\epsilon 2} = 1.90$. The values quoted in Equation (6.71) are those currently recommended by Launder (1992).

### 6.3.3  Wilcox Stress-$\omega$ Model

Not all stress-transport models use the $\epsilon$ equation to compute the dissipation. Wilcox and Rubesin (1980) postulate a stress-transport model based on their $\omega^2$ equation and the LRR model for $\Pi_{ij}$, with $\epsilon = \beta^*\omega k$. Although the model showed some promise for flows over curved surfaces and for swirling flows, its applications were very limited. More recently, Wilcox (1988b) has proposed a stress-transport model, known as the **multiscale model**, that has had a wider range of application. While the multiscale model has proven to be as accurate as the $k$-$\omega$ model for wall-bounded flows, including separation, its equations are ill conditioned for free shear flows. In this section, we present a new Reynolds-stress model that we will refer to as the **Stress-$\omega$ model**. The high-Reynolds-number, compressible version of the Stress-$\omega$ model is as follows.

**Reynolds-Stress Tensor:**

$$\rho\frac{\partial\tau_{ij}}{\partial t} + \rho U_k\frac{\partial\tau_{ij}}{\partial x_k} = -\rho P_{ij} + \frac{2}{3}\beta^*\rho\omega k\delta_{ij} - \rho\Pi_{ij} + \frac{\partial}{\partial x_k}\left[(\mu + \sigma^*\mu_T)\frac{\partial\tau_{ij}}{\partial x_k}\right]$$

(6.72)

**Specific Dissipation Rate:**

$$\rho\frac{\partial\omega}{\partial t} + \rho U_j\frac{\partial\omega}{\partial x_j} = \alpha\frac{\rho\omega}{k}\tau_{ij}\frac{\partial U_i}{\partial x_j} - \beta\rho\omega^2 + \frac{\partial}{\partial x_k}\left[(\mu + \sigma\mu_T)\frac{\partial\omega}{\partial x_k}\right] \quad (6.73)$$

**Pressure-Strain Correlation:**

$$\Pi_{ij} = \beta^*C_1\omega\left(\tau_{ij} + \frac{2}{3}k\delta_{ij}\right) - \hat{\alpha}\left(P_{ij} - \frac{2}{3}P\delta_{ij}\right)$$
$$-\hat{\beta}\left(D_{ij} - \frac{2}{3}P\delta_{ij}\right) - \hat{\gamma}k\left(S_{ij} - \frac{1}{3}S_{kk}\delta_{ij}\right) \quad (6.74)$$

**Auxiliary Relations:**

$$\mu_T = \rho k/\omega \quad (6.75)$$

$$P_{ij} = \tau_{im}\frac{\partial U_j}{\partial x_m} + \tau_{jm}\frac{\partial U_i}{\partial x_m}, \quad D_{ij} = \tau_{im}\frac{\partial U_m}{\partial x_j} + \tau_{jm}\frac{\partial U_m}{\partial x_i}, \quad P = \frac{1}{2}P_{kk}$$

(6.76)

**Closure Coefficients:**

$$\hat{\alpha} = (8 + C_2)/11, \quad \hat{\beta} = (8C_2 - 2)/11, \quad \hat{\gamma} = (60C_2 - 4)/55 \quad (6.77)$$

$$C_1 = \frac{9}{5}, \quad C_2 = \frac{13}{25} \quad (6.78)$$

$$\alpha = \frac{13}{25}, \quad \beta = \beta_o f_\beta, \quad \beta^* = \beta_o^* f_{\beta^*}, \quad \sigma = \frac{1}{2}, \quad \sigma^* = \frac{1}{2} \quad (6.79)$$

$$\beta_o = \frac{9}{125}, \quad f_\beta = \frac{1 + 70\chi_\omega}{1 + 80\chi_\omega}, \quad \chi_\omega \equiv \left| \frac{\Omega_{ij}\Omega_{jk}S_{ki}}{(\beta_o^*\omega)^3} \right| \tag{6.80}$$

$$\beta_o^* = \frac{9}{100}, \quad f_{\beta^*} = \begin{cases} 1, & \chi_k \leq 0 \\ \dfrac{1 + 640\chi_k^2}{1 + 400\chi_k^2}, & \chi_k > 0 \end{cases}, \quad \chi_k \equiv \frac{1}{\omega^3}\frac{\partial k}{\partial x_j}\frac{\partial \omega}{\partial x_j} \tag{6.81}$$

With the exception of the coefficient 640 appearing in the function $f_{\beta^*}$ [Equation (6.81)], all closure coefficients shared by the $k$-$\omega$ and Stress-$\omega$ models have the same values. The $k$-$\omega$ model uses the same functional form with 680 instead of 640. The latter value yields somewhat closer agreement between Stress-$\omega$ model predictions and measurements for both free shear flows and attached boundary layers. The values chosen for $\hat{\alpha}$, $\hat{\beta}$ and $\hat{\gamma}$ are those used in the original Launder, Reece and Rodi (1975) pressure-strain correlation model. This means there are two new closure coefficients to be determined, namely, $C_1$ and $C_2$.

In analyzing the Stress-$\omega$ model's sublayer predictions (see Subsection 6.6.1), we find that the constant in the law of the wall, $C$, is sensitive to the values of $\sigma$ and $C_2$. Retaining $\sigma = 1/2$ from the $k$-$\omega$ model, selecting $C_2 = 13/25$ yields $C \approx 5.0$ with no viscous damping functions. The traditional procedure for determining $C_1$ and $C_2$ is to appeal to measurements of homogeneous turbulent flows to determine these coefficients, which we do in the next section. Because we have selected $C_2$ to optimize sublayer predictions, in addressing homogeneous turbulence, we effectively seek the optimum value of $C_1$ that is compatible with $C_2 = 13/25$.

Unlike the LRR model, the Stress-$\omega$ model does not require a wall-reflection term such as $\Pi_{ij}^{(w)}$ defined in Equation (6.62). By design, the most significant difference between the LRR and Stress-$\omega$ models is in the scale-determining equation. The LRR model uses the $\epsilon$ equation while the Stress-$\omega$ model uses the $\omega$ equation. All other differences are minor by comparison. This strongly suggests that the end accomplished by the LRR wall-reflection term may be to mitigate a shortcoming of the model equation for $\epsilon$ rather than to correctly represent the physics of the pressure-echo process [see Parneix et al. (1998)].

Before proceeding to applications, it is worthwhile to pause and discuss two guidelines followed in formulating the Stress-$\omega$ model. **First**, a key objective is to create as simple and elegant a stress-transport model as possible. This dictates use of the LRR model for $\Pi_{ij}$, for example, but certainly does not preclude the use of a nonlinear model such as that developed by Speziale, Sarkar and Gatski (1991). Similarly, the Daly-Harlow (1970) approximation for $C_{ijk}$ could be replaced by a rotationally-invariant form with little additional effort. **Second**, because of the $k$-$\omega$ model's good predictions for a wide range of turbulent flows, the Stress-$\omega$ model is designed to resemble the $k$-$\omega$ model to as great an extent as possible for the flows to which both models apply. As we will see in the following sections, its predictions for free shear flows and attached boundary layers are usually within less than 5% of $k$-$\omega$ model predictions. Also, low-Reynolds-number modifications and surface boundary conditions for rough surfaces and for mass injection are very similar to those used with the $k$-$\omega$ model.

## 6.4  Application to Homogeneous Turbulent Flows

Homogeneous turbulent flows are useful for establishing the new closure coefficients introduced in modeling the pressure-strain correlation tensor, $\Pi_{ij}$. This is the primary type of flow normally used to calibrate a stress-transport model. Recall that homogeneous turbulence is defined as a turbulent flow that is statistically uniform in all directions. This means that the diffusion terms in all of the equations of motion are identically zero, as is the pressure-echo correction. Hence, the only difference between the $\epsilon$-based LRR model and the $\omega$-based Stress-$\omega$ model when applied to homogeneous turbulent flows is in the scale-determining equation. That is, both models use the LRR pressure-strain model and the Kolmogorov isotropy hypothesis for $\epsilon_{ij}$, so that the equations for the Reynolds stresses are nearly identical, with $C_2$ lying within the range of values recommended for the LRR model, viz., between 0.4 and 0.6.

Additionally, since the diffusion terms vanish, the equations simplify to first-order, ordinary differential equations, which can sometimes be solved in closed form. At most, a simple Runge-Kutta integration is required. Such flows are ideal for helping establish values of closure coefficients such as $C_1$ and $C_2$ in the LRR model, provided of course that we believe the same values apply to all turbulent flows. As noted in the preceding section, we have already selected $C_2 = 13/25$ for the Stress-$\omega$ model. Hence, we seek the optimum value for $C_1$ compatible with this value of $C_2$.

The simplest of all homogeneous flows is the decay of isotropic turbulence, which we discussed in Section 4.4 and used to set the ratio of $\beta_o^*$ to $\beta_o$ for the $k$-$\omega$ model. The Stress-$\omega$ model equations for $k$ and $\omega$ simplify to

$$\frac{dk}{dt} = -\beta_o^* \omega k \quad \text{and} \quad \frac{d\omega}{dt} = -\beta_o \omega^2 \qquad (6.82)$$

For large time, the asymptotic solution for $k$ is given by

$$k \sim t^{-\beta_o^*/\beta_o} \qquad (6.83)$$

Similarly, for the LRR model, $k$ varies with $t$ according to

$$k \sim t^{-1/(C_{\epsilon 2}-1)} \qquad (6.84)$$

Experimental observations summarized by Townsend (1976) indicate that turbulence energy varies according to $k \sim t^{-n}$ where $n = 1.25 \pm 0.06$ for decaying homogeneous, isotropic turbulence. Hence, we can conclude that our closure coefficients must lie in the following ranges.

$$1.19 < \beta_o^*/\beta_o < 1.31, \qquad 1.76 < C_{\epsilon 2} < 1.84 \qquad (6.85)$$

The Stress-$\omega$ model's chosen values for $\beta_o$ and $\beta_o^*$ [Equations (6.80) and (6.81)] give $\beta_o^*/\beta_o = 1.25$, which satisfies Equation (6.85). However, the value chosen for $C_{\epsilon 2}$ in the LRR model is 1.92, which lies outside the range indicated in Equation (6.85).

Figures 6.4(a) and (b) compare computed and measured $k$ for decaying homogeneous, isotropic turbulence as predicted by the Stress-$\omega$ model. The experimental data in (a) and (b) are those of Comte-Bellot and Corrsin (1971) and Wigeland and Nagib (1978), respectively.

Because the equations we solve for homogeneous shear flows are initial-value problems, the entire solution is affected by the assumed initial conditions, especially the initial value of $\epsilon$ or $\omega$. Estimates of the initial dissipation rate, $\epsilon_o$, are often quoted for homogeneous turbulence experiments. However, any errors in these estimates can have a large effect on the solution at all subsequent times. An alternative method for setting initial conditions is to estimate $\epsilon_o$ from the differential equation for $k$ at the initial station. In the case of homogenous isotropic turbulence, this means

$$\epsilon_o = -\left(\frac{dk}{dt}\right)_o \qquad \text{or} \qquad \omega_o = -\frac{1}{\beta_o^*}\left(\frac{1}{k}\frac{dk}{dt}\right)_o \qquad (6.86)$$

The initial value of $\omega$ has been selected to match the initial shape of the measured curves for the two cases, and the inferred values are quoted in Figure 6.4. Computed and measured values of $k$ are within 5% for both cases.

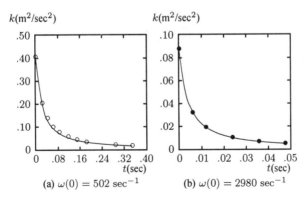

(a) $\omega(0) = 502 \text{ sec}^{-1}$        (b) $\omega(0) = 2980 \text{ sec}^{-1}$

Figure 6.4: *Computed and measured decay of turbulence energy for homogeneous, isotropic turbulence:* —— *Stress-omega model;* ○ *Comte-Bellot and Corrsin (1971);* • *Wigeland and Nagib (1978).*

The second type of homogeneous turbulent flow that is useful for establishing the value of pressure-strain correlation closure coefficients is decaying, anisotropic turbulence.. Such flows are created in the laboratory, for example, by subjecting turbulence to uniform strain-rate, which yields unequal normal Reynolds stresses. The turbulence then enters a region free of strain, and gradually returns to isotropy. The Tucker-Reynolds (1968) experiment that we discussed in Section 6.1 is an example of this type of flow (see Figure 6.1).

Because the mean strain rate is zero, the rapid pressure-strain term vanishes. Then, assuming dissipation follows the Kolmogorov (1941) isotropy hypothesis [Equation (6.36)], and using Rotta's (1951) slow pressure-strain term [Equation (6.52)], the Reynolds-stress equation written in terms of $\epsilon$ is

$$\frac{d\tau_{ij}}{dt} = \frac{2}{3}\epsilon\delta_{ij} - C_1\frac{\epsilon}{k}\left(\tau_{ij} + \frac{2}{3}k\delta_{ij}\right) \tag{6.87}$$

If the scale-determining equation is for $\omega$ rather than for $\epsilon$, we simply replace $C_1\epsilon/k$ by $C_1\beta^*\omega$. The solutions according to the LRR and Stress-$\omega$ models are

$$\frac{\tau_{ij} + \frac{2}{3}\rho k\delta_{ij}}{\left(\tau_{ij} + \frac{2}{3}\rho k\delta_{ij}\right)_o} = \left(\frac{k_o\epsilon}{k\epsilon_o}\right)^{C_1/(C_{\epsilon 2}-1)} = \left(\frac{\omega}{\omega_o}\right)^{C_1\beta_o^*/\beta_o} \tag{6.88}$$

where subscript $o$ denotes initial value. Measurements of decaying anisotropic turbulence have been used to determine the closure coefficient $C_1$. The data of Uberoi (1956), for example, indicate that $C_1$ lies between 1.4 and 1.8 [see Equation (6.53)]. More recent experiments such as those of Le Penven et al. (1984) further confirm that $C_1$ lies in this range.

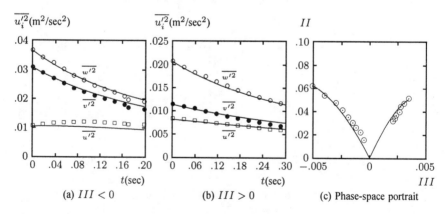

Figure 6.5: *Computed and measured decay of Reynolds stresses for homogeneous, anisotropic turbulence:* —— *Stress-omega model;* □, ●, ○, ⊙ *Le Penven et al. data* $(\overline{u'^2},\ \overline{v'^2},\ \overline{w'^2},\ II)$.

Figure 6.5 compares computed normal Reynolds stresses with the measurements of Le Penven et al. for decaying homogeneous, anisotropic turbulence as predicted by the Stress-$\omega$ model. The figure presents results in terms of the Reynolds-stress anisotropy second and third tensor invariants, $II$ and $III$, defined by [see Equations (6.39) and (6.64)]

$$II = b_{ij}b_{ij} \quad \text{and} \quad III = b_{ik}b_{kl}b_{li} \quad \text{where} \quad b_{ij} \equiv \frac{\overline{u'_i u'_j} - \frac{2}{3}k\delta_{ij}}{2k} \quad (6.89)$$

Parts (a) and (b) of the figure correspond to $III$ assuming negative and positive values, respectively. Part (c) displays $II$ as a function of $III$, which is generally referred to as the **phase-space portrait** for this type of flow. As shown, differences between theory and experiment are small, indicating that $C_1 = 1.8$ is optimum for the Stress-$\omega$ model.

Figure 6.6(a) compares computed and measured [Choi and Lumley (1984)] normal components of the Reynolds-stress anisotropy tensor, $b_{ij}$. This experiment is similar to that of Tucker and Reynolds, with turbulence initially subjected to plain strain and then returning to isotropy after the strain is removed. Figure 6.6(b) shows the phase-space portrait of the return-to-isotropy problem, plotted as $II^{1/2}$ versus $III^{1/3}$. As shown, $II^{1/2}$ is essentially a linear function of $III^{1/3}$ for this flow according to the LRR model. This computation has been done with the original LRR coefficients, i.e., with $C_1 = 1.5$.

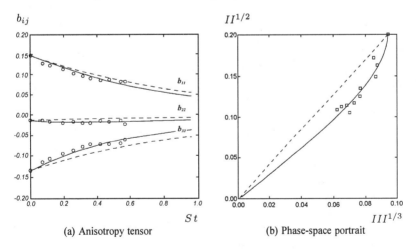

(a) Anisotropy tensor        (b) Phase-space portrait

Figure 6.6: *Comparison of computed and measured anisotropy tensor and phase-space portrait for homogeneous, anisotropic turbulence: - - - LRR model; —— Sarkar-Speziale model;* ○, □ *Choi and Lumley. [From Speziale and So (1996) — Published with the authors' permission.]*

While discrepancies between computed and measured stresses are satisfactory, even closer agreement between theory and experiment can be obtained with a nonlinear model for the slow pressure-strain model. Sarkar and Speziale (1990), for example, propose a simple quadratic model for the slow pressure-strain, i.e.,

$$A_{ij} = -\hat{C}_1 \epsilon b_{ij} + \hat{C}_2 \epsilon \left( b_{ik} b_{kj} - \frac{1}{3} b_{mn} b_{nm} \delta_{ij} \right) \qquad (6.90)$$

where $\hat{C}_1 = 3.4$ and $\hat{C}_2 = 4.2$ [see Equation (6.65)]. Figures 6.6(a) and (b) compare computed and measured anisotropy tensor components and phase-space portraits. The nonlinear model clearly falls within the scatter of the experimental data, while the LRR model prediction provides a less satisfactory description. The phase-space portrait is especially revealing, with the nonlinear model faithfully reflecting the nonlinear variation of $II^{1/2}$ with $III^{1/3}$.

Homogeneous turbulence experiments have also been performed that include irrotational plane strain [Townsend (1956) and Tucker and Reynolds (1968)] and uniform shear [Champagne, Harris and Corrsin (1970), Harris, Graham and Corrsin (1977), Tavoularis and Corrsin (1981), and Tavoularis and Karnik (1989)]. These flows can be used to establish closure coefficients such as $C_2$ in the LRR pressure-strain model. The velocity gradient tensor for these flows is

$$\frac{\partial U_i}{\partial x_j} = \begin{bmatrix} 0 & S & 0 \\ 0 & -a & 0 \\ 0 & 0 & a \end{bmatrix} \qquad (6.91)$$

where $a$ is the constant strain rate and $S$ is the constant rate of mean shear.

While closed form solutions generally do not exist when mean strain rate and/or shear are present, analytical progress can be made for the asymptotic forms in the limit $t \to \infty$. In general, the specific dissipation rate, $\omega \sim \epsilon/k$, approaches a constant limiting value while $k$ and the Reynolds stresses grow exponentially. Assuming solutions of this form yields closed-form expressions for the Reynolds stresses.

Using such analysis for uniform shear ($a = 0, S \neq 0$), Abid and Speziale (1993) have analyzed the LRR and SSG pressure-strain models and two nonlinear pressure-strain models developed by Shih and Lumley (1985) [SL model] and by Fu, Launder and Tselepidakis (1987) [FLT model]. Table 6.1 summarizes their results, along with results for the Stress-$\omega$ model and asymptotic values determined experimentally by Tavoularis and Karnik (1989). Inspection of the table shows that the SSG model most faithfully reproduces measured asymptotic values of the Reynolds stresses.

The parameter $Sk/\epsilon$ is the ratio of the turbulence time scale, $\epsilon/k$, to the mean-flow time scale as represented by the reciprocal of $S$. Inspection of the table shows that the Stress-$\omega$ model predicts a value for $Sk/\epsilon$ that is within less

Table 6.1: *Anisotropy-Tensor Limiting Values for Uniform Shear*

| Property | Stress-$\omega$ | LRR | SL | FLT | SSG | Measured |
|---|---|---|---|---|---|---|
| $b_{xx}$ | .142 | .152 | .120 | .196 | .218 | .210 |
| $b_{xy}$ | -.157 | -.186 | -.121 | -.151 | -.164 | -.160 |
| $b_{yy}$ | -.136 | -.119 | -.122 | -.136 | -.145 | -.140 |
| $b_{zz}$ | -.006 | -.033 | .002 | -.060 | -.073 | -.070 |
| $Sk/\epsilon$ | 4.904 | 4.830 | 7.440 | 5.950 | 5.500 | 5.000 |

than 2% of the measured value, closest of the models tested. While the LRR model is also very close with the predicted and measured values differing by 3.4%, the SL, FLT and SSG models all differ by at least 10% and by as much as 49%. This is important because all of the other models use the $\epsilon$ equation, which plays a role in determining the time scale of the various physical processes represented in the Reynolds-stress equation. Errors associated with the $\epsilon$ equation clearly have an adverse impact on the balance of these physical processes.

Note that the anisotropy tensor is proportional to the difference between the Reynolds-stress tensor and $\frac{2}{3}k$. Hence, percentage differences between computed and measured values of $b_{ij}$ present an exaggerated estimate of the differences between computed and measured Reynolds stresses. For example, the SL model-predicted value of $b_{xx}$ is 43% smaller than the measured value. However, this corresponds to a difference between computed and measured $\overline{u'^2}$ of only 17%.

Figure 6.7 compares Stress-$\omega$ model Reynolds stresses and corresponding measured values for the Tavoularis and Karnik (1989) uniform-shear experiments with $S = 29.0$ sec$^{-1}$, $39.9$ sec$^{-1}$ and $84.0$ sec$^{-1}$. The initial values used for $\omega$

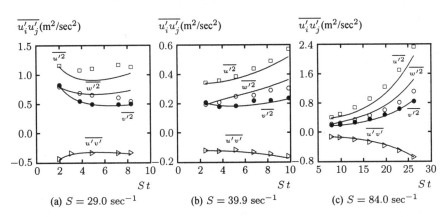

Figure 6.7: *Computed and measured Reynolds stresses for homogeneous, plane shear: —— Stress-$\omega$ model; $\square$, $\bullet$, $\circ$, $\triangleright$ Tavoularis and Karnik ($\overline{u'^2}$, $\overline{v'^2}$, $\overline{w'^2}$, $\overline{u'v'}$).*

correspond to having $Sk_o/\epsilon_o = S/(\beta_o^* \omega_o)$ equal to 1.86, 3.10 and 3.81, respectively. Consistent with the asymptotic results summarized in Table 6.1, computed values of $\overline{uv}$ and $\overline{v'^2}$ are very close to measurements, while computed $\overline{u'^2}$ and $\overline{w'^2}$ values are generally 10% below and above measurements, respectively.

Turning to flows with irrotational strain rate ($a \neq 0, S = 0$), Figures 6.8(a) and (b) compare Stress-$\omega$ model and measured [Townsend (1956) and Tucker and Reynolds (1968), respectively] Reynolds stresses. The strain rate for the Townsend case is $a = 9.44 \text{ sec}^{-1}$ (with $ak_o/\epsilon_o = 0.57$), while that of the Tucker-Reynolds case is $a = 4.45 \text{ sec}^{-1}$ (with $ak_o/\epsilon_o = 0.49$). Launder, Reece and Rodi (1975) report very similar results for the Tucker-Reynolds case.

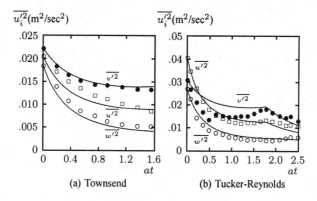

(a) Townsend                    (b) Tucker-Reynolds

Figure 6.8: *Computed and measured Reynolds stresses for homogeneous, plane strain:* —— *Stress-$\omega$ model;* □, •, ○ *Experiment* ($\overline{u'^2}$, $\overline{v'^2}$, $\overline{w'^2}$).

To illustrate how much of an improvement stress-transport models make for flows with sudden changes in mean strain rate, Figure 6.9 compares measured distortion parameter, $K$, for the Tucker-Reynolds experiment with computed results obtained using the Stress-$\omega$ model and the Wilcox-Rubesin (1980) $k$-$\omega^2$ model. As shown, the Stress-$\omega$ model predicts a gradual approach to isotropy and the computed $K$ more closely matches the experimental data.

## 6.5  Application to Free Shear Flows

While stress-transport models eliminate many of the shortcomings of the Boussinesq eddy-viscosity approximation, they are not necessarily more accurate than two-equation models for free shear flows. This is true because the scale determining equation ($\omega$, $\epsilon$, $\ell$, etc.) used by a stress-transport model plays a key role. For example, the Wilcox (1988b) multiscale model uses the $\omega$ equation of the Wilcox (1988a) $k$-$\omega$ model. Just as the spreading rates of this $k$-$\omega$ model are

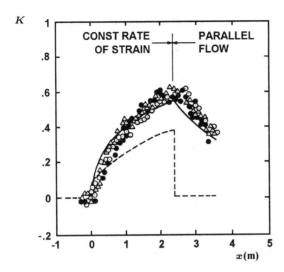

Figure 6.9: *Computed and measured distortion parameter for the Tucker-Reynolds plane-strain flow: —— Stress-$\omega$ model; - - - Wilcox-Rubesin $k$-$\omega^2$ model; $\circ$ $\bullet$ $\triangle$ Tucker-Reynolds.*

significantly larger than measured (see Table 4.5), so are those predicted by the multiscale model. Other shortcomings, such as the round-jet/plane-jet anomaly, also carry through from two-equation models to stress-transport models.

Table 6.2 summarizes computed and measured spreading rates for the LRR model and the Stress-$\omega$ model. Comparison with Table 4.4 shows that the Stress-$\omega$ model predicts spreading rates similar to those of the Wilcox (1998) $k$-$\omega$ model. The average difference between computed and measured spreading rates is 4%. Thus, like the Wilcox (1998) $k$-$\omega$ model, the Stress-$\omega$ model provides credible solutions for plane, round and radial jets. The LRR model's spreading rates are roughly 10% larger than those of the Standard $k$-$\epsilon$ model

Table 6.2: *Free Shear Flow Spreading Rates for the Stress-$\omega$ and LRR Models*

| Flow | Stress-$\omega$ Model | LRR Model | Measured |
|---|---|---|---|
| Far Wake | .355 | — | .365 |
| Mixing Layer | .097 | .104 | .115 |
| Plane Jet | .105 | .123 | .100-.110 |
| Round Jet | .092 | .135 | .086-.096 |
| Radial Jet | .101 | — | .096-.110 |

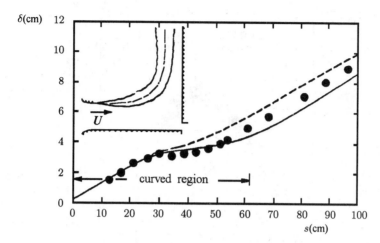

Figure 6.10: *Comparison of computed and measured width for a curved mixing layer:* —— *LRR model;* - - - *Standard k-ε model;* • *Castro and Bradshaw. [From Rodi (1981) — Copyright © AIAA 1981 — Used with permission.]*

[cf. Table 4.4]. As noted by Launder and Morse (1979), because the predicted round-jet spreading rate exceeds the predicted plane-jet spreading rate, the LRR model fails to resolve the round-jet/plane-jet anomaly.

Figure 6.10 compares computed and measured width of a curved mixing layer. The computation was done using the LRR model [Rodi (1981)], and the measurements correspond to an experiment of Castro and Bradshaw (1976) with stabilizing curvature. The LRR model predicts a greater reduction in width than the Standard k-ε model. However, the LRR model's predicted width lies as far below the measured width as the k-ε model's prediction lies above. Although not shown in the figure, Rodi's (1976) Algebraic Stress Model predicts a width that is about midway between, and thus in close agreement with measured values.

As a final comment, with all of the additional new closure coefficients attending *nonlinear* pressure-strain models, it is very likely that such models can be fine tuned to correct the round-jet/plane-jet anomaly. However, we should keep in mind that the anomaly underscores a deficiency in our physical description and understanding of jets. Such fine tuning reveals nothing regarding the nature of these flows, and thus amounts to little more than a curve-fitting exercise.

By contrast, the physically-plausible argument presented by Pope (1978) regarding the role of vortex stretching (Subsection 4.5.5) offers a more credible solution. While the modification to the ε equation fails to rectify the k-ε model's deficiencies for jets, the modification to the coefficient β appearing in the Stress-ω model [Equation (6.80)] implements Pope's ideas quite effectively: recall that this model uses the *linear* LRR pressure-strain model.

# 6.6 Application to Wall-Bounded Flows

This section focuses upon wall-bounded flows, including channel and pipe flow, and boundary layers with a variety of complicating effects. Before addressing such flows, however, we discuss surface boundary conditions. As with two-equation models, we have the option of using wall functions or integrating through the viscous sublayer.

## 6.6.1 Surface Boundary Conditions

Wall-bounded flows require boundary conditions appropriate to a solid boundary for the mean velocity and the scale-determining parameter, e.g., $\epsilon$ or $\omega$. Additionally, surface boundary conditions are needed for each component of the Reynolds-stress tensor (implying a boundary condition for $k$). The exact surface boundary conditions follow from the no-slip condition:

$$\tau_{ij} = 0 \quad \text{at} \quad y = 0 \tag{6.92}$$

Stress-transport models, like two-equation models, may or may not predict a satisfactory value of the constant $C$ in the law of the wall when the equations are integrated through the viscous sublayer. If the model fails to predict a satisfactory value for $C$, we have the choice of either introducing viscous damping factors or using wall functions to obviate integration through the sublayer. The near-wall behavior of stress-transport models is strongly influenced by the scale-determining equation. Models based on the $\epsilon$ equation fail to predict an acceptable value of $C$ unless damping factors are applied. When damping factors are used, the equations become very stiff and are very difficult to integrate through the sublayer [see Durbin (1991) and Laurence and Durbin (1994)]. By contrast, models based on the $\omega$ equation often predict an acceptable value of $C$ and are generally quite easy to integrate through the sublayer.

The most rational procedure for devising wall functions is to analyze the log layer with perturbation methods. As with the $k$-$\epsilon$ model, the velocity, $k$ and either $\epsilon$ or $\omega$ are given by

$$U = u_\tau \left[ \frac{1}{\kappa} \ell n \left( \frac{u_\tau y}{\nu} \right) + C \right] \tag{6.93}$$

$$k = \frac{u_\tau^2}{\sqrt{\beta_o^*}}, \quad \omega = \frac{k^{1/2}}{(\beta_o^*)^{1/4} \kappa y}, \quad \epsilon = (\beta_o^*)^{3/4} \frac{k^{3/2}}{\kappa y} \tag{6.94}$$

Similar relations are needed for the Reynolds stresses, and the precise forms depend upon the approximations used to close the Reynolds-stress equation. Regardless of the model, the general form of the Reynolds-stress tensor is

$$\tau_{ij} = W_{ij} k \quad \text{as} \quad y \to 0 \tag{6.95}$$

where $W_{ij}$ is a constant tensor whose components depend upon the model's closure coefficients. The problems section examines log-layer structure for the LRR and Stress-$\omega$ models. The tensor $W_{ij}$ for these two models is

$$W_{ij} = \begin{bmatrix} -0.906 & 0.306 & 0 \\ 0.306 & -0.438 & 0 \\ 0 & 0 & -0.656 \end{bmatrix} \qquad \text{(Stress-}\omega \text{ model)} \qquad (6.96)$$

$$W_{ij} = \begin{bmatrix} -0.852 & 0.301 & 0 \\ 0.301 & -0.469 & 0 \\ 0 & 0 & -0.679 \end{bmatrix} \qquad \text{(LRR model)} \qquad (6.97)$$

So, Lai, Zhang, and Hwang (1991) review low-Reynolds-number corrections for stress-transport models based on the $\epsilon$ equation. The damping functions generally introduced are similar to those proposed for the $k$-$\epsilon$ model (see Section 4.9). As with the $k$-$\epsilon$ model, many authors have postulated low-Reynolds-number damping functions, and the topic remains in a continuing state of development.

As with the $k$-$\omega$ model, the surface value of specific dissipation rate, $\omega_w$, determines the value of the constant $C$ in the law of the wall for the Stress-$\omega$ model. Perturbation analysis of the sublayer shows that the limit $\omega_w \to \infty$ corresponds to a perfectly-smooth wall and, without low-Reynolds-number corrections, the asymptotic behavior of $\omega$ approaching the surface for both the $k$-$\omega$ and Stress-$\omega$ models is

$$\omega \to \frac{6\nu_w}{\beta_o y^2} \qquad \text{as} \qquad y \to 0 \qquad \text{(Smooth Wall)} \qquad (6.98)$$

Using Program **SUBLAY** (Appendix C), the Stress-$\omega$ model's sublayer behavior can be readily determined. Most importantly, the constant, $C$, in the law of the wall is predicted to be

$$C = 5.04 \qquad (6.99)$$

This is close enough to 5.0 to justify integrating the Stress-$\omega$ model equations through the sublayer without the aid of viscous damping functions. Figure 6.11 compares Stress-$\omega$ model smooth-wall velocity profiles with corresponding measurements of Laufer (1952), Andersen, Kays and Moffat (1972), and Wieghardt [as tabulated by Coles and Hirst (1969)]. Figure 6.12 compares computed turbulence production and dissipation terms with Laufer's (1952) near-wall pipe-flow measurements. Aside from the erroneous dissipation data for $y^+ < 10$, predictions are within experimental error bounds.

As with the $k$-$\omega$ model, the value of $C$ is sensitive to the value of $\sigma$. Its value is also affected by the value chosen for $C_2$. For consistency with the $k$-$\omega$ model, the value of $\sigma$ has been chosen to be $1/2$. Then, selecting $C_2 = 13/25 = 0.52$ gives the value quoted in Equation (6.99). To illustrate the sensitivity of $C$ to $C_2$, note that choosing $C_2 = 0.546$ gives $C = 5.50$.

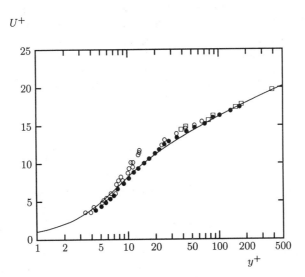

Figure 6.11: *Computed and measured sublayer velocity:* ∘ *Laufer;* • *Andersen et al.;* □ *Wieghardt;* —— *Stress-ω model.*

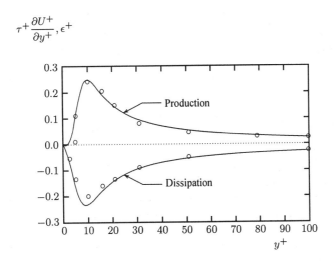

Figure 6.12: *Computed and measured production and dissipation:* ∘ *Laufer;* —— *Stress-ω model.*

As noted above, the Stress-$\omega$ model has the property that the constant $C$ varies with the surface value of $\omega$. We can thus correlate $\omega_w$ with surface roughness height, $k_s$, and surface mass-injection velocity, $v_w$. The resulting correlations are a little different from those appropriate for the $k$-$\omega$ model (see Subsections 4.7.2 and 4.7.3). The surface boundary conditions based on these correlations are as follows.

**For rough surfaces:**

$$\omega = \frac{u_\tau^2 S_R}{\nu_w} \quad \text{at} \quad y = 0 \quad \text{(Rough Wall)} \quad (6.100)$$

where the dimensionless coefficient $S_R$ is defined in terms of the dimensionless roughness height, $k_s^+ = u_\tau k_s / \nu_w$, by

$$S_R = \begin{cases} (50/k_s^+)^2, & k_s^+ \le 25 \\ \\ 500/(k_s^+)^{3/2} & k_s^+ > 25 \end{cases} \quad (6.101)$$

**For surfaces with mass injection:**

$$\omega = \frac{u_\tau^2 S_B}{\nu_w} \quad \text{at} \quad y = 0 \quad \text{(Mass Injection)} \quad (6.102)$$

where the dimensionless coefficient $S_B$ is defined in terms of the dimensionless injection velocity, $v_w^+ = v_w / u_\tau$, by

$$S_B = \frac{18}{v_w^+(1 + 4v_w^+)} \quad (6.103)$$

As with the $k$-$\omega$ model, for flows with suction ($v_w < 0$), either the **smooth-surface** [Equation (6.98)] or **slightly-rough-surface** [Equation (4.205) with $k_s^+ < 5$] boundary condition for $\omega$ is appropriate.

While the Stress-$\omega$ model does not require viscous damping functions to achieve a satisfactory sublayer solution, introducing low-Reynolds-number corrections can improve model predictions for a variety of flows. Most importantly, with straightforward viscous damping functions very similar to those introduced for the $k$-$\omega$ model (see Subsection 4.9.2), the model's ability to predict transition can be greatly improved. As with the $k$-$\omega$ model, we let

$$\nu_T = \alpha^* \frac{k}{\omega} \quad \text{and} \quad Re_T = \frac{k}{\omega \nu} \quad (6.104)$$

and the closure coefficients in Equations (6.77) – (6.81) are replaced by:

$$\alpha^* = \frac{\alpha_o^* + Re_T/R_k}{1 + Re_T/R_k}$$

$$\alpha = \frac{13}{25} \cdot \frac{\alpha_o + Re_T/R_\omega}{1 + Re_T/R_\omega} \cdot \frac{3 + Re_T/R_\omega}{3\alpha_o^* + Re_T/R_\omega}$$

$$\beta^* = \frac{9}{100} \cdot \frac{4/15 + Re_T/R_\beta}{1 + Re_T/R_\beta} \cdot f_{\beta^*}$$

$$\hat{\alpha} = \frac{1 + \hat{\alpha}_\infty Re_T/R_k}{1 + Re_T/R_k}$$

$$\hat{\beta} = \hat{\beta}_\infty \cdot \frac{Re_T/R_k}{1 + Re_T/R_k}$$  (6.105)

$$\hat{\gamma} = \hat{\gamma}_\infty \cdot \frac{\hat{\gamma}_o + Re_T/R_k}{1 + Re_T/R_k}$$

$$C_1 = \frac{9}{5} \cdot \frac{5/3 + Re_T/R_k}{1 + Re_T/R_k}$$

$$\hat{\alpha}_\infty = (8 + C_2)/11, \quad \hat{\beta}_\infty = (8C_2 - 2)/11, \quad \hat{\gamma}_\infty = (60C_2 - 4)/55 \quad (6.106)$$

$$\beta = \frac{9}{125} f_\beta, \quad \sigma^* = \sigma = \frac{1}{2}, \quad C_2 = \frac{13}{25} \quad (6.107)$$

$$\alpha_o^* = \frac{1}{3}\beta_o, \quad \alpha_o = \frac{21}{200}, \quad \hat{\gamma}_o = \frac{7}{1000} \quad (6.108)$$

$$R_\beta = 12, \quad R_k = 12, \quad R_\omega = 6.20 \quad (6.109)$$

With these viscous corrections, the Stress-$\omega$ model reproduces all of the low-Reynolds-number $k$-$\omega$ model transition-predictions discussed in Subsection 4.9.2, and other subtle features such as asymptotic consistency. The modification to the coefficient $C_1$ guarantees that the Reynolds shear stress goes to zero as $y^3$ as $y \to 0$. The values chosen for $R_\beta$, $R_k$ and $R_\omega$ yield $C = 5.0$.

## 6.6.2 Channel and Pipe Flow

Figure 6.13 compares computed and measured velocity and Reynolds-stress profiles for the original Launder-Reece-Rodi model. The computation was done using wall functions. Velocity profile data shown are those of Laufer (1951) and Hanjalić (1970), while the Reynolds-stress data are those of Comte-Bellot (1965). As shown, with the exception of $\overline{u'^2}$, computed and measured profiles differ by less than 5%. The computed and measured $\overline{u'^2}$ profiles differ by no more than 20%. Although not shown, even closer agreement between computed and measured Reynolds stresses can be obtained with low-Reynolds-number versions of the LRR model [see So et al. (1991)].

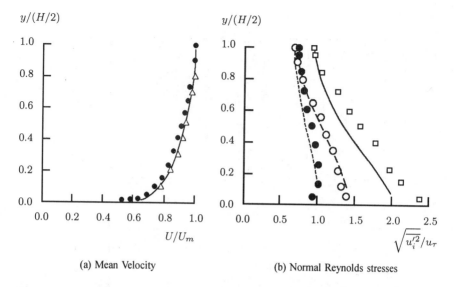

(a) Mean Velocity                    (b) Normal Reynolds stresses

Figure 6.13:  *Computed and measured flow properties for channel flow:  (a)* —— *LRR model,* △ *Laufer,* • *Hanjalić; (b)* ——, - - -, − − *LRR model* ($\sqrt{\overline{u'^2}}/u_\tau$, $\sqrt{\overline{v'^2}}/u_\tau$, $\sqrt{\overline{w'^2}}/u_\tau$), □, •, ∘ *Comte-Bellot* ($\sqrt{\overline{u'^2}}/u_\tau$, $\sqrt{\overline{v'^2}}/u_\tau$, $\sqrt{\overline{w'^2}}/u_\tau$).

One of the most controversial features of the LRR-model solution for channel flow is the importance of the pressure-echo term throughout the flow.  The pressure-echo contribution on the centerline is approximately 15% of its peak value.  It is unclear that a supposed near-wall effect should have this large an impact at the channel centerline.  On the one hand, some researchers argue that the echo effect scales with maximum eddy size which, for channel flow, would be about half the channel height.  What matters is the ratio of eddy size to $y$. This is (nominally) constant through the log layer and doesn't fall much in the defect layer.

On the other hand, the Stress-$\omega$ model — despite all its similarity to the LRR model aside from its use of the $\omega$ equation in place of the $\epsilon$ equation — does not require a pressure-echo contribution to achieve a satisfactory channel-flow solution.  As noted earlier, this strongly suggests that the unreasonably-large pressure-echo term used in the LRR and other $\epsilon$-equation-based stress-transport models is needed to accommodate a deficiency of the modeled $\epsilon$ equation, most likely its ill-conditioned near-wall behavior.

Figures 6.14 and 6.15 compare computed and measured channel-flow and pipe-flow properties for the Stress-$\omega$ model with and without viscous corrections. As shown, the computed skin friction is generally within 6% of the Halleen and Johnston (1967) correlation [see Equation (3.138)] for channel flow.

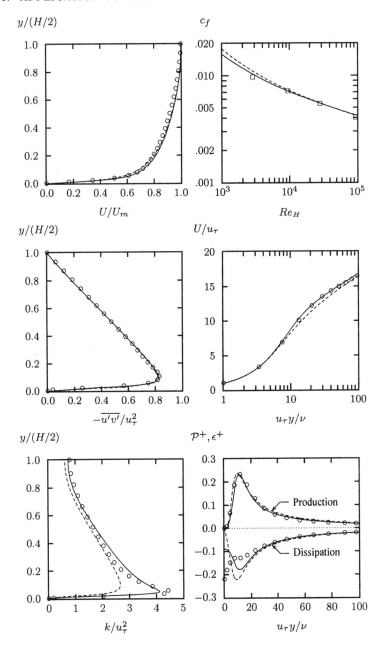

Figure 6.14: *Comparison of computed and measured channel-flow properties,*
$Re_H = 13,750.$ —— *Low-$Re$ Stress-$\omega$ model;* - - - *High-$Re$ Stress-$\omega$ model;*
$\circ$ *Mansour et al. (DNS);* $\square$ *Halleen-Johnston correlation.*

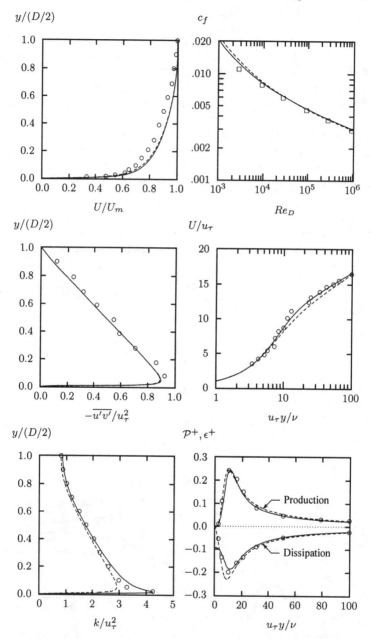

Figure 6.15: *Comparison of computed and measured pipe-flow properties,* $Re_D = 40,000.$ —— *Low Re Stress-ω model;* - - - *High Re Stress-ω model;* ○ *Laufer;* □ *Prandtl correlation.*

Similarly, computed $c_f$ differs from Prandtl's universal law of friction [see Equation (3.139)] by less than 7% except at the lowest Reynolds numbers, where the formula is known to be inaccurate. For both channel and pipe flow, the velocity, Reynolds shear stress, and turbulence kinetic energy profiles differ by less than 6%. Most notably, the low-Reynolds-number model predicts the peak value of $k$ near the wall to within 7% of the DNS value for channel flow and 1% of the measured value for pipe flow. For both cases, the turbulence-energy production, $\mathcal{P}^+ = \nu \tau_{xy}(\partial U/\partial y)/u_\tau^4$, is within 5% of the DNS and measured results. The dissipation rate, $\epsilon^+ = \nu \epsilon/u_\tau^4$, is within 10% of the DNS and measured results except very close to the surface.

The most noticeable difference between computed and measured flow properties occurs for the dissipation when $y^+ < 20$. The DNS channel-flow data show that dissipation achieves its maximum value at the surface, a feature that is not captured by the low-Reynolds-number version of the Stress-$\omega$ model. Several low-Reynolds-number versions of the LRR model have been developed that closely mimic the near-wall behavior of the dissipation. This is accomplished with viscous damping functions that are much more complicated than the simple bilinear forms used for the Stress-$\omega$ model [see Equations (6.105)]. The excellent overall agreement between theory and experiment for all other features of the channel- and pipe-flow solutions casts doubt on the importance of duplicating this subtle feature of the solution, with the attendant complication (and the potential source of numerical mischief) that would be involved in forcing the model to duplicate the measured surface value of $\epsilon$.

Capturing other subtle details such as the sharp peak in $k$ near the surface, and achieving asymptotic consistency (e.g., $k \sim y^2$ and $\tau_{xy} \sim y^3$) has been done with virtually no change in skin friction and in mean-flow and turbulence-property profiles above $y^+ \approx 10$. Similarly, low-Reynolds-number versions of the LRR model have their most significant changes in turbulence-property profiles confined to the portion of the channel below $y^+ \approx 20$.

Unlike the Stress-$\omega$ model however, some low-Reynolds-number variants of the LRR model provide accurate descriptions of near-wall dissipation while simultaneously giving nontrivial discrepancies between computed and measured skin friction for typical wall-bounded flows. By contrast, the low-Reynolds-number corrections have virtually no effect on the Stress-$\omega$ model's predicted skin friction.

Rotating channel flow is an interesting application of stress-transport models. As with flow over a curved surface, two-equation models require ad hoc corrections for rotating channel flow in order to make realistic predictions [e.g., Launder, Priddin and Sharma (1977) and Wilcox and Chambers (1977)]. To understand the problem, note that in a rotating coordinate frame, the Coriolis acceleration yields additional inertial terms in the Reynolds-stress equation. Specifically, in a coordinate system that is rotating with angular velocity, $\Omega$, the

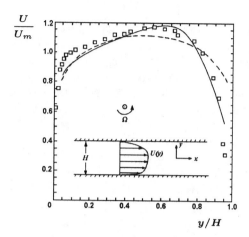

Figure 6.16: *Computed and measured velocity profiles for rotating channel flow with $\Omega H/U_m = 0.21$:* —— *Gibson-Launder model;* - - - *$k$-$\epsilon$ model;* □ *Johnston et al. [From Speziale (1991) — Published with the author's permission.]*

Reynolds-stress equation is

$$\frac{\partial \tau_{ij}}{\partial t} + U_k \frac{\partial \tau_{ij}}{\partial x_k} + 2 \left( \epsilon_{jkm} \Omega_k \tau_{im} + \epsilon_{ikm} \Omega_k \tau_{jm} \right)$$

$$= -\tau_{ik} \frac{\partial U_j}{\partial x_k} - \tau_{jk} \frac{\partial U_i}{\partial x_k} + \epsilon_{ij} - \Pi_{ij} + \frac{\partial}{\partial x_k} \left[ \nu \frac{\partial \tau_{ij}}{\partial x_k} + C_{ijk} \right] \quad (6.110)$$

where $\epsilon_{jkm}$ is the permutation tensor. Note that if the rotation tensor, $\Omega_{ij}$, appears in any of the closure approximations for $\epsilon_{ij}$, $\Pi_{ij}$ or $C_{ijk}$, it must be replaced by $\Omega_{ij} + \epsilon_{ikj} \Omega_k$. Contracting Equation (6.110) yields the turbulence kinetic energy equation. Because the trace of the Coriolis term is zero, there is no explicit effect of rotation appearing in the equation for $k$. Since rotation has a strong effect on turbulence, this shows why ad hoc modifications are needed for a two-equation model.

Figure 6.16 compares a computed and measured velocity profile for a channel with a constant angular velocity about the spanwise ($z$) direction. The computations have been done using the Gibson-Launder (1978) stress-transport model and the Standard $k$-$\epsilon$ model. The experimental data are those of Johnston et al. (1972), and correspond to an inverse Rossby number, $\Omega H/U_m = 0.21$, where $H$ is the height of the channel and $U_m$ is the average velocity. As shown, the $k$-$\epsilon$ model predicts a velocity profile that is symmetric about the center line. Consistent with measurements, the Gibson-Launder model predicts an asymmetric profile. However, as clearly shown in the figure, the velocity and shear stress on the "stable" side near $y = 0$ are underestimated.

## 6.6.3  Boundary Layers

Figure 6.17 compares computed and measured skin friction for the 16 incompressible boundary layers considered in Chapters 3 and 4 (see Figures 3.16, 3.18, 4.2 and 4.24). The figure includes numerical results for the Stress-$\omega$ model with and without low-Reynolds-number corrections.

As shown, both versions of the Stress-$\omega$ model provide acceptable predictions for all ranges of pressure gradients, from favorable to strong adverse. Table 6.3 summarizes differences between computed and measured skin friction at the final station for the various cases. The overall average difference for 15 cases (Flow 5300 has been excluded) is approximately 6% for both low- and high-Reynolds-number versions of the Stress-$\omega$ model.

Table 6.3: *Differences Between Computed and Measured Skin Friction.*

| Pressure Gradient | Flows | Low $Re$ Stress-$\omega$ | High $Re$ Stress-$\omega$ |
|---|---|---|---|
| Favorable | 1400, 1300, 2700, 6300 | 3.8% | 4.3% |
| Mild Adverse | 1100, 2100, 2500, 4800 | 6.0% | 4.5% |
| Moderate Adverse | 2400, 2600, 3300, 4500 | 10.0% | 10.4% |
| Strong Adverse | 0141, 1200, 4400 | 4.1% | 3.4% |
| All | – | 6.0% | 5.7% |

Although no results are included for $\epsilon$-based stress-transport models, older versions are generally only a bit closer to measurements than the $k$-$\epsilon$ model (cf. Figure 4.2). By contrast, newer versions, especially those with nonlinear pressure-strain terms, appear to more faithfully reproduce experimental data. For example, Hanjalić et al. (1997) have developed an $\epsilon$-based stress-transport model (with a linear pressure-strain term) that accurately predicts effects of adverse pressure gradient. Using perturbation methods, Henkes (1998a) has shown that this model is as close to measurements as the $k$-$\omega$ model for equilibrium boundary layers (i.e., for constant $\beta_T$ – see Section 4.6), strongly suggesting that it will perform well in general boundary-layer applications.

Although the Hanjalić et al. model has a large number of empirical functions designed to permit the model to achieve asymptotic consistency and the ability to predict transition and relaminarization, the improved accuracy for effects of pressure gradient appears to result from a single modification to the $\epsilon$ equation. Specifically, Hanjalić et al. add a term of the form

$$S_\ell \propto \max\left[\left(\frac{1}{C_\ell^2}\frac{\partial \ell}{\partial x_j}\frac{\partial \ell}{\partial x_j} - 1\right)\frac{1}{C_\ell^2}\frac{\partial \ell}{\partial x_j}\frac{\partial \ell}{\partial x_j}; \, 0\right]\frac{\epsilon^2}{k} \qquad (6.111)$$

where $C_\ell = 2.5$ is a closure coefficient and $\ell = k^{3/2}/\epsilon$ is the turbulence length scale. This term limits the growth of $\ell$ in the log layer, and cancels the undesirable

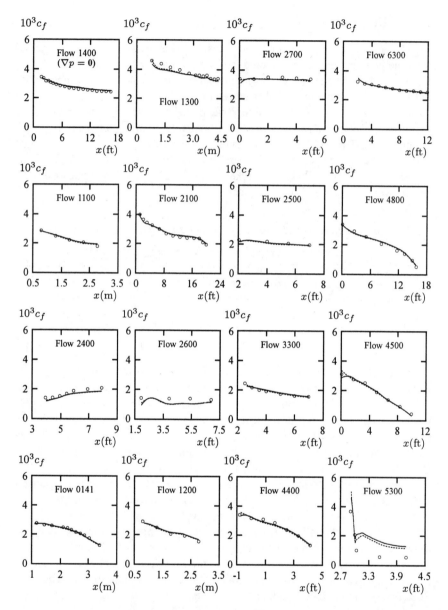

Figure 6.17: *Computed and measured skin friction for boundary layers subjected to a pressure gradient. Top row - favorable $\nabla p$; next to top row - mild adverse $\nabla p$; next to bottom row - moderate adverse $\nabla p$; bottom row - strong adverse $\nabla p$. —— Low $Re$ Stress-$\omega$ model; - - - High $Re$ Stress-$\omega$ model; ∘ measured.*

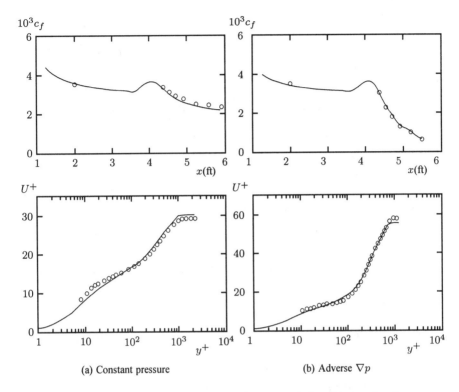

Figure 6.18: *Comparison of computed and measured skin friction for flow over a convex wall: —— Stress-$\omega$ model; o So and Mellor.*

effects of cross diffusion (relative to the $k$-$\omega$ model) that plague the $k$-$\epsilon$ model [see the discussion at the end of Subsection 4.6.2].

Surface curvature, like system rotation, has a significant effect on structural features of the turbulent boundary layer. As discussed in Section 6.1, in the absence of ad hoc modifications, such effects cannot be accurately predicted with a two-equation model, as curvature has a trivial effect on the turbulence kinetic energy equation. In principle, stress-transport models display none of these shortcomings. Thus, computing curved-wall boundary layers poses an interesting test of stress-transport models.

Figure 6.18 presents results of two computations done with the Stress-$\omega$ model for flow over a convex surface. The two cases are the constant-pressure and adverse-pressure-gradient flows that So and Mellor (1972) have investigated experimentally. To insure accurate starting conditions, the measured momentum and displacement thickness at $x = 2$ ft. have been matched to within 1% for both cases, a point well upstream of the beginning of the curved-wall portion of the flow at $x$ = 4.375 ft. For both cases, computed and measured flow properties differ by less than 6%.

The LRR model also offers important improvement in predictive accuracy relative to the $k$-$\epsilon$ model for flows with secondary motions. Lai et al. (1991), for example, have successfully applied three variants of the LRR model with wall functions to flow in a curved pipe. Consistent with measurements, their computations predict existence of secondary flows.

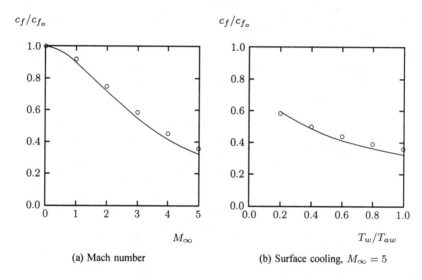

(a) Mach number          (b) Surface cooling, $M_\infty = 5$

Figure 6.19: *Computed and measured effects of freestream Mach number and surface cooling on flat-plate boundary-layer skin friction:* —— *Stress-$\omega$ model;* o *Van Driest correlation.*

Turning to effects of compressibility, a stress-transport model's performance is intimately tied to the scale-determining equation. Models based on the $\epsilon$ equation will share the $k$-$\epsilon$ model's incorrect density scaling (see Section 5.6). By contrast, models based on the $\omega$ equation should share the $k$-$\omega$ model's ability to accurately predict the compressible law of the wall. Figure 6.19 confirms this point for the Stress-$\omega$ model. The figure compares computed effects of Mach number and surface cooling on flat-plate boundary layer skin friction. The turbulent heat-flux vector has been computed according to Equation (5.54) with constant turbulent Prandtl number. Figure 6.19(a) compares computed ratio of skin friction to the incompressible value, $c_{f_o}$, as a function of Mach number with the Van Driest correlation. Figure 6.19(b) focuses upon effects of surface temperature on flat-plate skin friction at Mach 5. Inspection of the figure shows that differences between the predicted values and the correlated values nowhere exceed 9% in both cases.

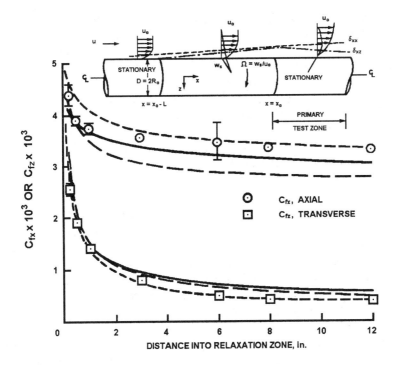

Figure 6.20: *Skin friction on a segmented spinning cylinder;* —— *Cebeci-Smith model;* – – *Wilcox-Rubesin $k$-$\omega^2$ model;* - - - *Wilcox-Rubesin stress-transport model;* ○□ *Higuchi and Rubesin. [From Rubesin (1989) — Copyright ©AIAA — Used with permission.]*

Stress-transport models hold promise of more accurate predictions for three-dimensional flows. The primary reason two-equation models are inaccurate for three-dimensional boundary layers, for example, lies in their use of an isotropic eddy viscosity. However, the eddy viscosities in the streamwise and crossflow directions of a typical three-dimensional boundary layer can differ significantly. Figure 6.20 compares computed and measured skin friction for such a flow, a boundary layer on a segmented cylinder, part of which rotates about its axis. The experiment was performed by Higuchi and Rubesin (1978). As shown, the Wilcox-Rubesin (1980) stress-transport model most accurately describes both the axial ($c_{f_x}$) and transverse ($c_{f_z}$) skin friction components in the relaxation zone, i.e., the region downstream of the spinning segment. The Cebeci-Smith algebraic model and the Wilcox-Rubesin (1980) two-equation model yield skin friction components that differ from measured values by as much as 20% and 10%, respectively.

The final round of applications is for incompressible, unsteady turbulent boundary layers. These flows pose a difficult challenge to a turbulence model because many complicated frequency-dependent phenomena are generally present, including periodic separation and reattachment.

Using a simplified stress-transport model (viz. the multiscale model), Wilcox (1988b) has simulated the experiments performed by Jayaraman, Parikh and Reynolds (1982). In these experiments, a well developed steady turbulent boundary layer enters a test section which has been designed to have freestream velocity that varies according to:

$$U_e = U_o \left\{ 1 - ax'[1 - cos(2\pi ft)] \right\}, \qquad x' = (x - x_0)/(x_1 - x_0) \quad (6.112)$$

The quantity $x'$ is fractional distance through the test section where $x_0$ and $x_1$ are the values of streamwise distance, $x$, at the beginning and end of the test section, respectively. Thus, an initially steady turbulent boundary layer is subjected to a sinusoidally varying adverse pressure gradient. The experiments were performed for low- and high-amplitude unsteadiness characterized by having $a \approx 0.05$ and 0.25, respectively. For both amplitudes, experiments were conducted for five frequencies, $f$, ranging from 0.1 Hz to 2.0 Hz. Wilcox simulates nine of the experiments, including all of the low-amplitude cases and all four of the high-amplitude cases.

In order to compare computed and measured flow properties, we must decompose any flow property $y(\mathbf{x}, t)$ in terms of three components, viz.,

$$y(\mathbf{x}, t) = \bar{y}(\mathbf{x}) + \tilde{y}(\mathbf{x}, t) + y'(\mathbf{x}, t) \quad (6.113)$$

where $\bar{y}(\mathbf{x})$ is the long-time averaged value of $y(\mathbf{x}, t)$, $\tilde{y}(\mathbf{x}, t)$ is the organized response component due to the imposed unsteadiness, and $y'(\mathbf{x}, t)$ is the turbulent fluctuation. Using an unsteady boundary-layer program, Wilcox computes the phase-averaged component, $< y(\mathbf{x}, t) >$, defined by

$$< y(\mathbf{x}, t) >= \bar{y}(\mathbf{x}) + \tilde{y}(\mathbf{x}, t) \quad (6.114)$$

Jayaraman et al. expand $< y(\mathbf{x}, t) >$ in a Fourier series according to

$$< y(\mathbf{x}, t) >= \bar{y}(\mathbf{x}) + \sum_{n=1}^{\infty} A_{n,y}(\mathbf{x}) \cos \left[ 2n\pi ft + \phi_{n,y}(\mathbf{x}) \right] \quad (6.115)$$

Velocity profile data, for example, are presented by Jayaraman et al. in terms of $\bar{u}(\mathbf{x})$, $A_{1,u}(\mathbf{x})$ and $\phi_{1,u}(\mathbf{x})$. These quantities can be extracted from the boundary-layer solution by the normal Fourier decomposition, viz., by computing the following integrals.

$$\bar{u}(\mathbf{x}) = f \int_0^{1/f} < u(\mathbf{x}, t) > dt \quad (6.116)$$

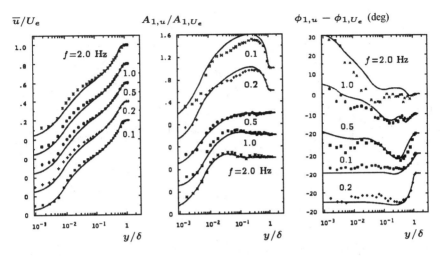

Figure 6.21: *Comparison of computed and measured mean velocity, $A_{1,u}$ and phase profiles at $x' = 0.88$ for low amplitude: —— multiscale model; • Jayaraman et al. [From Wilcox (1988b) — Copyright © AIAA 1988 — Used with permission.]*

$$A_{1,u}(\mathbf{x}) \cos \phi_{1,u} = f \int_0^{1/f} < u(\mathbf{x}, t) > \cos (2\pi f t) \, dt \qquad (6.117)$$

$$A_{1,u}(\mathbf{x}) \sin \phi_{1,u} = -f \int_0^{1/f} < u(\mathbf{x}, t) > \sin (2\pi f t) \, dt \qquad (6.118)$$

Figure 6.21 compares computed and measured velocity profiles at $x' = 0.88$ for the five low-amplitude cases. As shown, computed mean velocity profiles differ from corresponding measured profiles by no more than 5% of scale. Comparison of computed and measured $A_{1,u}$ profiles shows that, consistent with measurements, unsteady effects are confined to the near-wall Stokes layer at the higher frequencies ($f > .5$ Hz). By contrast, at the two lowest frequencies, the entire boundary layer is affected, with significant amplification of the organized component occurring away from the surface. Differences between the numerical and experimental $A_{1,u}$ profiles are less than 10%. Computed and measured phase, $\phi_{1,u}$, profiles are very similar with differences nowhere exceeding 5°.

Figure 6.22 compares computed and measured velocity profiles at $x' = 0.94$ for the high-amplitude cases. As for the low amplitude cases, computed and measured $\bar{u}(\mathbf{x})$ profiles lie within 5% of scale of each other. Similarly, computed $A_{1,u}$ and $\phi_{1,u}$ profiles differ from corresponding measurements by less than 10%. To provide a measure of how accurately temporal variations have been

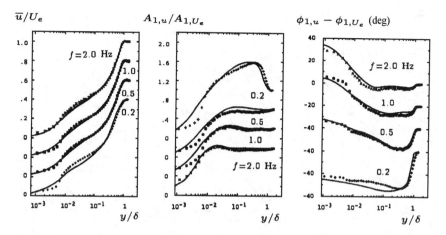

Figure 6.22: *Comparison of computed and measured mean velocity, $A_{1,u}$ and phase profiles at $x' = 0.94$ for high amplitude: —— multiscale model; • Jayaraman et al. [From Wilcox (1988b) — Copyright © AIAA 1988 — Used with permission.]*

predicted, Figure 6.23 compares computed and measured shape factor through a complete cycle for all four frequencies. Differences between computed and measured shape factors are less than 5%.

The four high-amplitude cases have also been computed using the Wilcox (1988a) $k$-$\omega$ model. Figure 6.23 shows that $k$-$\omega$ and multiscale-model predictions differ by only a few percent. Although it is possible the test cases are not as difficult as might be expected, this seems unlikely in view of the wide Strouhal number range and the fact that periodic separation and reattachment are present. More likely, the $k$-$\omega$ model fares well because all of the cases have attached boundary layers through most of each cycle and in the mean.

As a closing comment, many recent turbulence modeling efforts focusing on unsteady boundary layers mistakenly credit their success (or lack of it) to achieving asymptotic consistency as $y \to 0$ with the $k$-$\epsilon$ model or with a stress-transport model based on the $\epsilon$ equation. The computations described above were done using the high-Reynolds-number versions of the $k$-$\omega$ and multiscale models, neither of which is asymptotically consistent. All that appears to be necessary is to achieve a satisfactory value for the constant $C$ in the law of the wall. This makes sense physically as the dissipation time scale is so short in the sublayer that the sublayer responds to changes in the mean flow almost instantaneously and thus behaves as a quasi-steady region. Consequently, achieving asymptotically consistent behavior in the sublayer is neither more nor less important for unsteady flows than it is for steady flows.

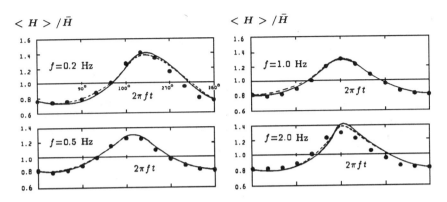

Figure 6.23: *Comparison of computed and measured temporal variation of shape factor for the high-amplitude cases: - - - k-ω model; —— multiscale model; • Jayaraman et al. [From Wilcox (1988b) — Copyright © AIAA 1988 — Used with permission.]*

## 6.7 Application to Separated Flows

As we have seen in preceding chapters, turbulence models that use the Boussinesq approximation are generally unreliable for separated flows, especially shock-induced separation. Figure 5.8, for example, illustrates how poorly such models perform for Mach 3 flow into a compression corner. In this section, we will take a close look at how well stress-transport models perform for several separated flows.

Because stress-transport models require more computer resources than algebraic and two-equation models, applications to separated flows have been rare until recently. Consequently, only preliminary conclusions can be drawn. Incompressible applications have generally been limited to the backward-facing step, while compressible-flow applications have been done for compression corners for a limited range of Mach numbers.

Focusing first on the backward-facing step, So et al. (1988) and So and Yuan (1998) have done interesting studies using a variety of turbulence models and closure approximations. The 1988 computations assess the effect of various models for the pressure-strain correlation, while the 1998 study focuses on low-Reynolds-number $k$-$\epsilon$ and $\epsilon$-equation based stress-transport models.

The So et al. (1988) computations use Chien's (1982) low-Reynolds number version of the $\epsilon$ equation. Most importantly, they have used three different models for the pressure-strain correlation, viz., the models of Rotta (1951) [Model A1], Launder, Reece and Rodi (1975) [Model A2], and Gibson and Younis (1986) [Model A4]. Using the Rotta model, computations have been done with wall

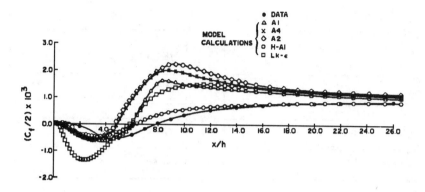

Figure 6.24: *Computed and measured skin friction for flow past a backward-facing step: A1=Rotta model; A2=LRR model; A4=Gibson-Younis model; H-A1=Rotta model with wall functions; Lk-ε=Chien k-ε model; •=Eaton and Johnston. [From So et al. (1988) — Published with permission.]*

functions as well [Model H-A1]. For reference, their computations also include the Chien (1982) low-Reynolds-number $k$-$\epsilon$ model [Model Lk-$\epsilon$]. These models differ mainly in their representation of the fast pressure-strain term, with the Rotta model ignoring it altogether. The computations simulate the experiments of Eaton and Johnston (1980) in a duct with a large expansion ratio, for which the measured reattachment length is 8 step heights.

As shown in Figure 6.24, computed reattachment length for all of the computations lies between 5 and 6 step heights, so that the result closest to measurements differs from the measured value by 25%. All of the models show large discrepancies between computed and measured wall pressure, while peak skin friction values are as much as 3 times measured values downstream of reattachment for the low-Reynolds-number models. In general, the stress-transport model skin friction results are as far from measurements as those of the low-Reynolds-number $k$-$\epsilon$ model. Only when wall functions are used with the stress-transport model does the computed skin friction lie reasonably close to measured values. So et al. note that the smallest discrepancies between computed and measured flow properties are obtained with the Rotta pressure-strain model, which omits the rapid pressure-strain correlation. That is, the LRR and Gibson-Younis models for the rapid pressure strain appear to yield larger discrepancies between computed and measured values.

Recalling how close to measurements $k$-$\omega$ model predictions are for flow past a backward-facing step (Section 4.10), the So et al. computations suggest that their poor predictions are caused by the $\epsilon$ equation. On the one hand, comparison of Figures 4.37 and 6.24 shows that for stress-transport model H-A1, $c_f$ is very similar to $k$-$\epsilon$ model results when wall functions are used. Although the flows are a little different, the reattachment length is 25% smaller than measured for both cases. On the other hand, using the same low-Reynolds-number $\epsilon$ equation, $c_f$ for stress-transport model A1 is very similar to the low-Reynolds-number $k$-$\epsilon$ model's skin friction, except in the reverse-flow region. Despite the latter difference, the reattachment length is the same in this case also. Thus, as with two-equation models, a stress-transport model's performance for the backward-facing step is intimately linked to the scale-determining equation. This strongly suggests that much closer agreement between computed and measured flow properties would be obtained with a stress-transport model based on the $\omega$ equation, such as the Stress-$\omega$ model. Unfortunately, the this model has not yet been applied to the backward-facing step, so this must remain a point of conjecture until such a computation is done.

In the more recent study, So and Yuan (1998) compute backstep flow with seven low-Reynolds-number $k$-$\epsilon$ models, the four-equation model of Durbin (1991) and three stress-transport models. The flow considered is a relatively low-Reynolds-number (5100 based on step height) case studied experimentally by Jovic and Driver (1994) and computed with DNS by Le, Moin and Kim (1997). Overall, the computations are in closer agreement with measurements for this flow than those of the high-Reynolds-number Eaton and Johnston (1980) case discussed above. The So-Yuan computations show the following:

- The average difference between the predicted and measured reattachment length for the seven $k$-$\epsilon$ models is 9%. For Chien's (1982) model, the reattachment length is within 7% of the measured value. Note that this model's reattachment length is 25% shorter than measured for the high-Reynolds-number Eaton and Johnson (1980) flow.

- Durbin's four-equation model gives one of the best overall solutions, with a reattachment length within 4% of the measured value. As shown by Durbin (1995), this model also provides a credible solution for the high-Reynolds-number backstep experiment of Driver and Seegmiller (1985).

- The three stress-transport models considered predict reattachment lengths within 3%, 7% and 10% of the measured value.

It is difficult to draw any firm conclusions from the So-Yuan study as it concentrates on just one flow. Furthermore, the flow chosen is one that $k$-$\epsilon$ models predict reasonably well, and that doesn't reveal their inherent weakness for this type of application. Given their inaccurate predictions for backstep flows at higher Reynolds numbers, the close agreement is probably a lucky coincidence. This suggests that either: (a) the same assessment is warranted for the stress-transport models considered, or (b) additional computations at higher Reynolds numbers are needed.

We now consider three shock-separated turbulent boundary-layer computations using both the high-Reynolds-number $k$-$\omega$ and Stress-$\omega$ models. The flows include two planar compression-corner flows and an axisymmetric compression-corner flow. Computations with both models have been done using the Wilcox compressibility modification defined in Equation (5.73). For the $k$-$\omega$ model, we use $\xi^* = 3/2$, while a smaller value of $\xi^* = 1/2$ proves to be more appropriate for the Stress-$\omega$ model. In all cases, the equations are integrated through the viscous sublayer, using the high-Reynolds-number versions of the models.

The first of the three applications is for Mach 2.79 flow into a 20° compression corner. This flow has been experimentally investigated by Settles, Vas and Bogdonoff (1976) and includes a small region over which separation of the incident turbulent boundary layer occurs. Figure 6.25(a) compares computed and measured surface pressure, $p_w/p_\infty$, and skin friction, $c_f$. The Stress-$\omega$ model predicts more upstream influence, a lower pressure plateau at separation, and a more gradual increase in skin friction downstream of reattachment relative to the $k$-$\omega$ results. All of these features represent significant improvement in predictive accuracy. Using the $k$-$\epsilon$ model and specially devised wall functions, Viegas, Rubesin and Horstman (1985) are able to achieve similar accuracy for this flow.

The second of the three applications is for Mach 2.84 flow into a 24° compression corner. This flow has also been experimentally investigated by Settles, Vas and Bogdonoff (1976) and includes a larger region over which separation of the incident turbulent boundary layer occurs than in the 20° case. Figure 6.25(b) compares computed and measured surface pressure and skin friction. As in the 20° compression-corner computation, the Stress-$\omega$ model predicts more upstream influence. Interestingly, the $k$-$\omega$ predicted pressure plateau at separation is very close to the measured level, and there is little difference between $k$-$\omega$ and Stress-$\omega$ predicted increase in skin friction downstream of reattachment. Note that, for this flow, Viegas, Rubesin and Horstman (1985) predict pressure plateau values about 20% higher than measured, and are unable to simultaneously make accurate predictions for skin friction downstream of reattachment and the initial rise in surface pressure. That is, their solutions can match either skin friction or surface pressure, but not both.

The third application is for Mach 2.85 flow into a 30° axisymmetric compression corner. This flow has been experimentally investigated by Brown (1986) and includes a separation bubble of length comparable to the 24° planar compression corner. Figure 6.25(c) compares computed and measured surface pressure. Computed skin friction is also shown. Once again the Stress-$\omega$ model predicts more upstream influence, and the pressure plateau at separation is close to measurements. The predicted pressure plateau for the $k$-$\omega$ model is about 10% higher the measured level. There is a 10% difference between $k$-$\omega$ and Stress-$\omega$ predicted increase in skin friction downstream of reattachment. The overall pressure rise is predicted by both models to be 4.5, while the measurements indicate a value

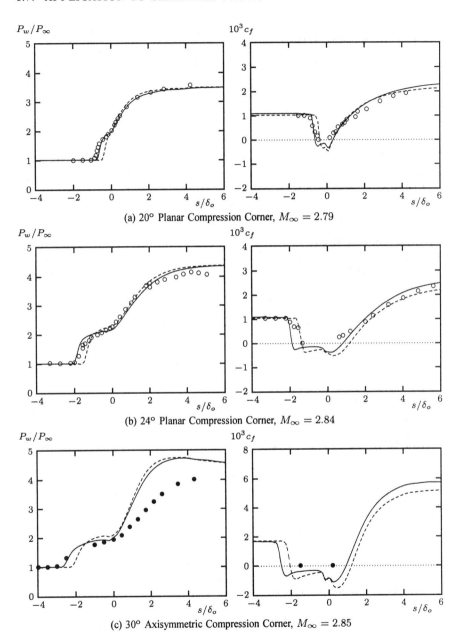

Figure 6.25: *Comparison of computed and measured properties for supersonic compression corners (s is tangential distance from corner):* —— *Stress-ω model;* - - - *k-ω model;* ○ *Settles et al.;* • *Brown.*

of 4.0. The inviscid pressure rise for a 30° axisymmetric compression corner is 4.4, so that the experiment appears to be inconsistent with the physics of this flow. However, the experiment is known to have some unsteadiness, and this may account for the difference.

Clearly, for the three compression-corner cases considered, while the predictions are not in complete agreement with measurements, the Stress-$\omega$ model provides a flowfield more consistent with experimental observations than does the $k$-$\omega$ model. The primary reason for the difference in the two models' predictions can be found by examining predicted behavior of the Reynolds shear stress near the separation point. Figure 6.26 shows the maximum Reynolds shear stress, $\tau_{max}$, throughout the interaction region for the three compression-corner computations. As shown, the $k$-$\omega$ model predicts a more abrupt increase in $\tau_{max}$ at separation and a much larger peak value than predicted by the Stress-$\omega$ model. For the axisymmetric case, the figure includes experimental data for points ahead of the measured separation point. The Stress-$\omega$ model-predicted values of $\tau_{max}$ are closer to measurements than those of the $k$-$\omega$ model.

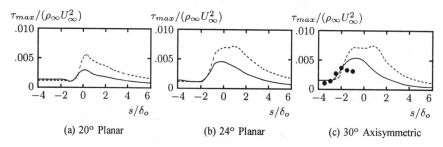

(a) 20° Planar          (b) 24° Planar          (c) 30° Axisymmetric

Figure 6.26: *Computed and measured maximum Reynolds shear stress for supersonic compression-corner flows:* —— *Stress-$\omega$ model;* - - - *$k$-$\omega$ model;* • *Brown.*

The physical implication of the pronounced difference in the rate of amplification of the Reynolds shear stress is clear. Using the Boussinesq approximation, the $k$-$\omega$ model makes a far more rapid adjustment to the rotation of the mean strain rate tensor's principal axes than the Stress-$\omega$ model. Consequently, the predicted separation point and initial pressure rise lie closer to the corner with the $k$-$\omega$ model than measured. Predicting more physically realistic growth of the Reynolds stresses, the Stress-$\omega$ model predicts overall flow properties which are in closer agreement with measurements.

Note that for the Stress-$\omega$ model, although the pressure is in such close agreement with measurements, the numerical separation points are further upstream than indicated by oil-flow measurements for all three compression corner cases. Marshall and Dolling (1992) indicate that these flows include a low-frequency oscillation of the separation shock. The time-mean pressure distribution upstream of the corner is affected by these oscillations whose frequency content includes substantial energy at time scales of the mean motion. This unsteadiness is responsible for the apparent mismatch between the beginning of the pressure rise and the separation point. Since computations with the Stress-$\omega$ model and the $k$-$\omega$ model fail to display any low-frequency oscillation of the shock, more research is needed to arrive at a completely satisfactory solution.

# 6.8  Range of Applicability

The two primary approaches to removing the limitations of the Boussinesq approximation are to use either a **nonlinear constitutive relation** or a **stress-transport model**. As discussed in Section 6.2, nonlinear constitutive relations offer some advantage over the Boussinesq approximation, most notably for flows in which anisotropy of the normal Reynolds stresses is important. **Algebraic Stress Models** provide a straightforward method for accurately predicting effects of streamline curvature and system rotation, although ad hoc corrections to standard two-equation models are just as effective. However, nonlinear constitutive relations offer no improvement over the Boussinesq approximation for flows with sudden changes in mean strain rate.

Despite their complexity, stress-transport models have great potential for removing shortcomings of the Boussinesq approximation in a natural way. Without ad hoc corrections, stress-transport models provide physically realistic predictions for flows with curved streamlines, system rotation, sudden changes in mean strain rate and secondary motions of the second kind. While more research is needed for separated flows, these models may also improve predictions for shock-separated flows. However, to be completely objective in our assessment, we must also note that in many such applications only qualitative agreement between theory and experiment has been obtained.

Just as older $k$-equation oriented, one-equation turbulence models share the shortcomings and successes of the mixing-length model, stress-transport models reflect the strengths and weaknesses of the scale-determining equation used with the model. There is an increasing pool of evidence that many of the shortcomings of stress-transport models are caused by the scale-determining equation. Results obtained for the backward-facing step (Section 6.7), for example, strongly suggest that predictions of standard stress-transport models can be improved by using the $\omega$ equation in place of the $\epsilon$ equation. This is not to say all of the ills of stress-transport models are caused by their use of the $\epsilon$ equation. Based on DNS results for backstep flows, Parneix et al. (1998) show that even when the dissipation rate is accurately predicted, current models are capable of predicting large discrepancies from measurements. We can reasonably infer that this points to deficiencies in modeling of the pressure-strain correlation, $\Pi_{ij}$.

From a numerical point of view, stress-transport models are at least as difficult to solve as the corresponding two-equation model. Models based on the $\epsilon$ equation fail to predict a satisfactory law of the wall and require complicated viscous damping functions. Correspondingly, such models are generally very difficult to integrate. By contrast, models based on the $\omega$ equation require no special viscous corrections, and are much easier to integrate. In particular, the Stress-$\omega$ model usually requires only about 25% to 40% more computing time relative to the $k$-$\omega$ model. Hence, the scale-determining equation may be even more important for stress-transport models than for two-equation models.

# Problems

**6.1** The objective of this problem is to derive the modified law of the wall for flow over a curved wall according to the $k$-$\omega$ model.

(a) Verify that the dimensionless form of Equations (6.5) to (6.7) in the log layer is [with $\epsilon \equiv \nu/(u_\tau \mathcal{R})$ and assuming $\beta^* \approx \beta_o^*$ and $\beta \approx \beta_o$]:

$$\nu_T^+ \left( \frac{dU^+}{dy^+} - \epsilon U^+ \right) = 1, \qquad \nu_T^+ = \frac{k^+}{\omega^+}$$

$$\sigma^* \nu_T^+ \frac{d}{dy^+} \left[ \nu_T^+ \frac{dk^+}{dy^+} \right] = \beta_o^* (k^+)^2 - 1 + \frac{9}{2} \epsilon (\nu_T^+)^2 U^+ \frac{dU^+}{dy^+}$$

$$\sigma \nu_T^+ \frac{d}{dy^+} \left[ \nu_T^+ \frac{d\omega^+}{dy^+} \right] = \beta_o k^+ \omega^+ - \alpha \frac{\omega^+}{k^+}$$

(b) Assume a solution of the form

$$\frac{dU^+}{dy^+} \sim \frac{1}{\kappa y^+} \left[ 1 + \epsilon a y^+ \ell n y^+ + O(\epsilon^2) \right]$$

$$k^+ \sim \frac{1}{\sqrt{\beta_o^*}} \left[ 1 + \epsilon b y^+ \ell n y^+ + O(\epsilon^2) \right]$$

$$\omega^+ \sim \frac{1}{\sqrt{\beta_o^*} \kappa y^+} \left[ 1 + \epsilon c y^+ \ell n y^+ + O(\epsilon^2) \right]$$

with $\epsilon \ll 1$. Substitute into the equations for $k^+$ and $\omega^+$ and verify that the coefficients $b$ and $c$ are given by

$$b = -\frac{9/2}{2 - \sigma^* \kappa^2 / \sqrt{\beta_o^*}} \qquad \text{and} \qquad c = \frac{\alpha}{\alpha - \beta_o / \beta_o^*} b$$

**NOTE:** Use the fact that for the $k$-$\omega$ model $\sigma \kappa^2 = (\beta_o / \beta_o^* - \alpha) \sqrt{\beta_o^*}$, and ignore terms proportional to $y^+$ relative to terms proportional to $y^+ \ell n y^+$.

(c) Substitute into the momentum equation and verify that

$$a + b - c = 1$$

(d) Using $\alpha = 13/25$, $\beta_o = 9/125$, $\beta_o^* = 9/100$, $\sigma = 1/2$ and $\sigma^* = 1/2$, determine the numerical values of $a$, $b$ and $c$, and show that the modified law of the wall is of the form

$$\left[ 1 - \beta_R \frac{y}{\mathcal{R}} \right] \frac{U}{u_\tau} \sim \frac{1}{\kappa} \ell n \left( \frac{u_\tau y}{\nu} \right) + \cdots$$

where $\beta_R \approx 8.5$.

**6.2** For incompressible flow, we wish to use Speziale's nonlinear constitutive relation with the $k$-$\omega$ model. In terms of $k$-$\omega$ model parameters, the relation can be written as

$$\tau_{ij} = -\frac{2}{3}k\delta_{ij} + 2\nu_T S_{ij} + C_D \frac{k}{\beta_o^* \omega^2}\left(S_{ik}S_{kj} - \frac{1}{3}S_{mn}S_{nm}\delta_{ij}\right)$$
$$+C_E \frac{k}{\beta_o^* \omega^2}\left(\overset{\circ}{S}_{ij} - \frac{1}{3}\overset{\circ}{S}_{mm}\delta_{ij}\right)$$

where $C_D$ and $C_E$ are closure coefficients whose values are to be determined.

(a) Verify for incompressible boundary layers that

$$S_{xy} = S_{yx} \approx \frac{1}{2}\frac{\partial U}{\partial y}, \qquad \text{all other } S_{ij} \approx 0$$

$$\overset{\circ}{S}_{xx} \approx -\left(\frac{\partial U}{\partial y}\right)^2, \qquad \text{all other } \overset{\circ}{S}_{ij} \approx 0$$

(b) Express the Reynolds-stress components $\tau_{xy}$, $\tau_{xx}$, $\tau_{yy}$ and $\tau_{zz}$ in terms of $k$, $\nu_T$, $\beta_o^*$, $\omega$ and $\partial U/\partial y$ for incompressible boundary layers.

(c) Using the stresses derived in part (b), write the log-layer form of the mean-momentum, $k$ and $\omega$ equations. Assume that $\beta^* \approx \beta_o^*$ and $\beta \approx \beta_o$.

(d) Assuming a solution of the form $\partial U/\partial y = u_\tau/(\kappa y)$ and $k$ = constant, verify that

$$\left(\frac{\partial U}{\partial y}\right)^2 = \beta_o^* \omega^2$$

(e) Verify that

$$\overline{u'^2}/k = (8 - C_D + 8C_E)/12$$
$$\overline{v'^2}/k = (8 - C_D - 4C_E)/12$$
$$\overline{w'^2}/k = (8 + 2C_D - 4C_E)/12$$

(f) Determine the values of $C_D$ and $C_E$ that are consistent with the normal Reynolds stresses standing in the ratio

$$\overline{u'^2} : \overline{v'^2} : \overline{w'^2} = 4 : 2 : 3$$

**6.3** Verify that in the log layer of an incompressible flat-plate boundary layer, the Wilcox-Rubesin nonlinear constitutive relation [Equation (6.15)] predicts that the normal Reynolds stresses stand in the ratio
$$\overline{u'^2} : \overline{v'^2} : \overline{w'^2} = 4 : 2 : 3$$
**HINT:** Recall that in the log layer, $\partial U/\partial y \approx \sqrt{\beta_o^*}\,\omega$.

**6.4** Check the accuracy of Speziale's regularization approximation as quoted in Equation (6.28). To do so, let $\eta$ vary from 0 to 1 and compare the right- and left-hand sides of the equation for $\xi = \frac{1}{4}, \frac{1}{2}$ and 1.

**6.5** For incompressible flow in a rectangular duct, the strain rate and rotation tensors are approximately

$$
S_{ij} = \begin{bmatrix} 0 & \frac{1}{2}\frac{\partial U}{\partial y} & \frac{1}{2}\frac{\partial U}{\partial z} \\ \frac{1}{2}\frac{\partial U}{\partial y} & 0 & 0 \\ \frac{1}{2}\frac{\partial U}{\partial x} & 0 & 0 \end{bmatrix} \quad \text{and} \quad \Omega_{ij} = \begin{bmatrix} 0 & \frac{1}{2}\frac{\partial U}{\partial y} & \frac{1}{2}\frac{\partial U}{\partial z} \\ -\frac{1}{2}\frac{\partial U}{\partial y} & 0 & 0 \\ -\frac{1}{2}\frac{\partial U}{\partial x} & 0 & 0 \end{bmatrix}
$$

Determine $\tau_{xy}$, $\tau_{xz}$, $\tau_{yz}$ and $(\tau_{zz} - \tau_{yy})$ according to the Wilcox-Rubesin nonlinear constitutive relation [Equation (6.15)].

**6.6** Derive the Poisson equation [Equation(6.45)] for the fluctuating pressure.

**6.7** Consider the Launder-Reece-Rodi (LRR) rapid-pressure-strain closure approximation, Equation (6.57).

(a) Verify that $a_{ijkl}$ satisfies the symmetry constraints in Equation (6.55).

(b) Invoke the constraints of Equation (6.56) and verify that $\alpha$, $\beta$, $\eta$ and $\upsilon$ are given by Equation (6.58).

(c) Form the tensor product

$$
M_{ijkl}\frac{\partial U_k}{\partial x_l} = (a_{ijkl} + a_{jikl})\frac{\partial U_k}{\partial x_l}
$$

and verify Equations (6.59) through (6.61).

**6.8** Consider Lumley's general representation for $\Pi_{ij}$ in Equation (6.63). Show that the LRR pressure-strain model [including $A_{ij}$ as defined in Equation (6.52)] is the limiting case where all coefficients other than $a_0$, $a_2$, $a_7$ and $a_9$ equal to zero. Also, assuming $C_1 = 1.8$, determine the values of $a_0$, $a_2$, $a_7$ and $a_9$ that correspond to $C_2 = 0.4$, 0.5 and 0.6. Assume the flow is incompressible.

**6.9** Suppose we have flow in a coordinate frame rotating with angular velocity $\mathbf{\Omega} = \Omega\mathbf{k}$, where $\mathbf{k}$ is a unit vector in the $z$ direction. The incompressible Navier-Stokes equation is

$$
\rho\frac{d\mathbf{u}}{dt} + 2\rho\mathbf{\Omega} \times \mathbf{u} = -\nabla p - \rho\mathbf{\Omega} \times \mathbf{\Omega} \times \mathbf{x} + \mu\nabla^2\mathbf{u}
$$

where $\mathbf{x}$ is position vector and $d/dt$ is the Eulerian derivative. Verify that the Reynolds-stress equation's inertial terms in a two-dimensional flow are as follows:

$$
\frac{d}{dt}\begin{bmatrix} \tau_{xx} & \tau_{xy} & 0 \\ \tau_{xy} & \tau_{yy} & 0 \\ 0 & 0 & \tau_{zz} \end{bmatrix} + \begin{bmatrix} -4\Omega\tau_{xy} & 2\Omega(\tau_{xx} - \tau_{yy}) & 0 \\ 2\Omega(\tau_{xx} - \tau_{yy}) & 4\Omega\tau_{xy} & 0 \\ 0 & 0 & 0 \end{bmatrix} = \cdots
$$

**6.10** Consider the Launder, Reece and Rodi stress-transport model, Equations (6.67) through (6.71). This problem analyzes the model's predicted asymptotic solution for homogeneous plane shear, in which

$$\frac{\partial U_i}{\partial x_j} = \begin{bmatrix} 0 & S & 0 \\ 0 & 0 & 0 \\ 0 & 0 & 0 \end{bmatrix}$$

(a) Assuming that $\epsilon/k \to$ constant as $t \to \infty$, verify that

$$\frac{P}{\epsilon} \to \frac{C_{\epsilon 2} - 1}{C_{\epsilon 1} - 1}$$

where $P = S\tau_{xy}$.

(b) Neglecting the pressure-echo effect, verify that

$$P_{ij} = \begin{bmatrix} 2S\tau_{xy} & S\tau_{yy} & S\tau_{yz} \\ S\tau_{yy} & 0 & 0 \\ S\tau_{yz} & 0 & 0 \end{bmatrix}, \quad D_{ij} = \begin{bmatrix} 0 & S\tau_{xx} & 0 \\ S\tau_{xx} & 2S\tau_{xy} & S\tau_{xz} \\ 0 & S\tau_{xz} & 0 \end{bmatrix}$$

(c) Assuming a solution of the form $\tau_{ij} = C_{ij}e^{\lambda t}$ where $C_{ij}$ is independent of time and $\lambda$ is a constant, verify that if $\tau_{xz}$ and $\tau_{yz}$ are initially zero, they are always zero, provided $\hat{\beta}(1 - \hat{\alpha}) > 0$.

(d) Determine $\epsilon/k$ and $P/k$ as functions of $C_{\epsilon 1}$, $C_{\epsilon 2}$ and $\lambda$ under the assumption that $\tau_{ij} = C_{ij}e^{\lambda t}$.

(e) Using results of Parts (a) – (d), determine $\overline{u'^2}/k$, $\overline{v'^2}/k$ and $\overline{w'^2}/k$ as algebraic functions of the closure coefficients. **HINT:** You can simplify your computations somewhat by first writing the equation for $\tau_{ij}$ as an equation for $\tau_{ij} + \frac{2}{3}k\delta_{ij}$.

(f) Using the following two sets of closure coefficient values, compute the numerical values of $\overline{u'^2}/k$, $\overline{v'^2}/k$ and $\overline{w'^2}/k$.

1. Original LRR: $C_1 = 1.5$, $C_2 = 0.4$, $C_{\epsilon 1} = 1.44$, $C_{\epsilon 2} = 1.90$
2. Revised LRR: $C_1 = 1.8$, $C_2 = 0.6$, $C_{\epsilon 1} = 1.44$, $C_{\epsilon 2} = 1.92$

**6.11** Consider the Stress-$\omega$ model, Equations (6.72) through (6.81). In the following computations, you can assume $\beta^* = \beta_o^*$ and $\beta = \beta_o$.

(a) State the limiting form of the equations for the incompressible, two-dimensional log layer.

(b) Assuming a solution of the form

$$\frac{dU}{dy} \sim \frac{u_\tau}{\kappa y}, \quad k \sim \frac{u_\tau^2}{\sqrt{\beta_o^*}}, \quad \omega \sim \frac{u_\tau}{\sqrt{\beta_o^*}\kappa y}$$

determine $\kappa$, $-\overline{u'v'}/k$, $\overline{u'^2}/k$, $\overline{v'^2}/k$ and $\overline{w'^2}/k$ as algebraic functions of the closure coefficients. **HINT:** All are constant.

(c) Using the closure coefficient values in Equations (6.77) through (6.81), verify that $\kappa \approx 0.41$, $-\overline{u'v'}/k \approx 0.31$ and $\overline{u'^2} : \overline{v'^2} : \overline{w'^2} \approx 4.0 : 1.9 : 2.9$.

**6.12** Consider the Launder, Reece and Rodi stress-transport model, Equations (6.67) through (6.71).

(a) State the limiting form of the equations for the incompressible, two-dimensional log layer.

(b) Assuming a solution of the form

$$\frac{dU}{dy} \sim \frac{u_\tau}{\kappa y}, \qquad k \sim \frac{u_\tau^2}{\sqrt{C_\mu}}, \qquad \epsilon \sim \frac{u_\tau^3}{\kappa y}$$

determine $\kappa$, $-\overline{u'v'}/k$, $\overline{u'^2}/k$, $\overline{v'^2}/k$ and $\overline{w'^2}/k$ as algebraic functions of the closure coefficients. **HINTS:** All are constant. Also, the $\epsilon$ equation yields $\kappa$ as a function of the closure coefficients and $\overline{v'^2}/k$. You needn't simplify further.

(c) Using the closure coefficient values in Equation (6.71), verify that $\kappa \approx 0.39$, $-\overline{u'v'}/k \approx 0.30$, and $\overline{u'^2} : \overline{v'^2} : \overline{w'^2} \approx 4 : 2.2 : 3.2$. **HINT:** Combining the simplified $\epsilon$ and $\tau_{yy}$ equations yields a cubic equation for $\kappa$. It can be solved in closed form by assuming $\kappa = 0.4(1 + \delta)$, linearizing and solving for $\delta$.

**6.13** Consider the Stress-$\omega$ model, Equations (6.72) through (6.81). In the following computations, you can assume $\beta^* = \beta_o^*$ and $\beta = \beta_o$.

(a) State the limiting form of the equations for the incompressible, two-dimensional log layer.

(b) Assuming a solution of the form

$$\frac{dU}{dy} \sim \frac{u_\tau}{\kappa y}, \qquad k \sim \frac{u_\tau^2}{\sqrt{\beta_o^*}}, \qquad \omega \sim \frac{u_\tau}{\sqrt{\beta_o^*}\kappa y}$$

determine $\kappa$ and verify that the Reynolds stress components according to the Stress-$\omega$ model are:

$$\frac{\tau_{xx}}{k} \sim \frac{1}{C_1}\left[\frac{2}{3}(1 - C_1) - \left(2 - \frac{4}{3}\hat{\alpha} + \frac{2}{3}\hat{\beta}\right)\right]$$

$$\frac{\tau_{yy}}{k} \sim \frac{1}{C_1}\left[\frac{2}{3}(1 - C_1) - \left(\frac{2}{3}\hat{\alpha} - \frac{4}{3}\hat{\beta}\right)\right]$$

$$\frac{\tau_{zz}}{k} \sim \frac{1}{C_1}\left[\frac{2}{3}(1 - C_1) - \left(\frac{2}{3}\hat{\alpha} + \frac{2}{3}\hat{\beta}\right)\right]$$

$$\frac{\tau_{xy}}{k} \sim \frac{1}{C_1}\left[(\hat{\alpha} - 1)\frac{\tau_{yy}}{\tau_{xy}} + \hat{\beta}\frac{\tau_{xx}}{\tau_{xy}} + \frac{1}{2}\hat{\gamma}\frac{k}{\tau_{xy}}\right]$$

(c) Substituting for $\hat{\alpha}$, $\hat{\beta}$ and $\hat{\gamma}$ in terms of $C_1$ and $C_2$ from Equation (6.77), show that the normal stresses are

$$\frac{\tau_{xx}}{k} \sim -\frac{22C_1 + 12C_2 + 8}{33C_1}$$

$$\frac{\tau_{yy}}{k} \sim -\frac{22C_1 - 30C_2 + 2}{33C_1}$$

$$\frac{\tau_{zz}}{k} \sim -\frac{22C_1 + 18C_2 - 10}{33C_1}$$

(d) Substituting the results of Part (c) into the last of the equations developed in Part (b), show that the shear stress satisfies the following equation.

$$\left(\frac{\tau_{xy}}{k}\right)^2 = \frac{2}{15(11C_1)^2}\left(55 + 242C_1 - 330C_2 - 165C_2^2\right) \qquad (6.119)$$

(e) Verify that if we insist upon $\tau_{xy}/k$ being equal to Bradshaw's constant so that $\sqrt{\beta_o^*} = \beta_r = 0.3$, necessarily $C_1$ and $C_2$ are related by a simple quadratic. Make a table of values of $C_2$ as a function for $C_1$ for the range of physically-realistic values based on measurements [cf. Equation (6.53)].

**6.14** Consider the low-Reynolds-number version of the Stress-$\omega$ model, Equations (6.72) through (6.74), (6.76) and (6.104) through (6.109). Modify Program **SUBLAY** as needed to permit specifying the values of $C_2$ and $R_\omega$ (see Subroutine START).

(a) With $R_\omega = 6.20$, compute the value of the constant in the law of the wall, $C$, for $C_2 = 0.40, 0.45, 0.50$ and $0.55$.

(b) Leaving all other values unchanged, determine the value of $R_\omega$ that gives $C = 5.00$ for $C_2 = 0.40, 0.45, 0.50$ and $0.55$.

**6.15** The object of this problem is to compare predictions of the Stress-$\omega$ model with measured properties of a turbulent boundary layer with surface mass injection. The experiment to be simulated was conducted by Andersen et al. (1972). Use Program **EDDYBL** and its menu-driven setup utility, Program **SETEBL**, to do the computations (see Appendix D).

(a) Using **SETEBL**, change appropriate input parameters to accomplish the following: use USCS units (IUTYPE); set freestream conditions to $p_{t_\infty} = 2131.1$ lb/ft$^2$, $T_{t_\infty} = 536.1°$ R, $M_\infty = 0.02776$ (PT1, TT1, XMA); use an initial stepsize given by $\Delta s = 0.005$ ft (DS); set initial boundary-layer properties so that $\Lambda = 0.5$, $c_f = 0.00266$, $\delta = 0.0296$ ft, $H = 1.689$, $Re_\theta = 619$, $s_i = 0.22667$ ft (ALAMM, CF, DELTA, H, RETHET, SI); set the maximum arc length to $s_f = 7.5$ ft (SSTOP); set up to use surface mass injection (NFLAG); and, set up to integrate the $\omega$ equation to $y = 0$ (USTOP).

(b) The edge pressure is constant and given by $p_e = 2130$ lb/ft$^2$. Use this information to define the pressure distribution in a file named **presur.dat**. Prepare a file **heater.dat** with constant wall temperature, $T_w = 536.1°$ R, and zero heat flux. Finally. prepare a file **blocrv.dat** with constant surface mass flux which is equal to $\rho v_w = 2.735540 \cdot 10^{-4}$ slug/(ft$^2$·sec). **NOTE:** When preparing **blocrv.dat**, be sure to set the body radius equal to 1.

(c) Do computations using the high-$Re$ and low-$Re$ versions of the Stress-$\omega$ model.

(d) Compare computed skin friction with the following measured values.

| $s$ (ft) | $c_f$ | $s$ (ft) | $c_f$ | $s$ (ft) | $c_f$ |
|---|---|---|---|---|---|
| 0.1312 | $2.65 \cdot 10^{-3}$ | 2.8208 | $1.31 \cdot 10^{-3}$ | 5.8384 | $9.70 \cdot 10^{-4}$ |
| 0.8462 | $1.92 \cdot 10^{-3}$ | 3.8376 | $1.16 \cdot 10^{-3}$ | 6.8224 | $9.00 \cdot 10^{-4}$ |
| 1.8368 | $1.55 \cdot 10^{-3}$ | 4.8216 | $1.04 \cdot 10^{-3}$ | 7.5112 | $8.50 \cdot 10^{-4}$ |

**6.16** The object of this problem is to compare predictions of the Stress-$\omega$ model with measured properties of a turbulent boundary layer with adverse pressure gradient. The experiment to be simulated was conducted by Bradshaw [see Coles and Hirst (1969) – Flow 3300]. Use Program **EDDYBL** and its menu-driven setup utility, Program **SETEBL**, to do the computations (see Appendix D).

(a) Using **SETEBL**, change appropriate input parameters to accomplish the following: use USCS units (IUTYPE); set freestream conditions to $p_{t_\infty} = 2148$ lb/ft$^2$, $T_{t_\infty} = 537.6°$ R, $M_\infty = 0.106$ (PT1, TT1, XMA); use an initial stepsize given by $\Delta s = 0.05$ ft and a geometric-progression ratio $k_g = 1.08$ (DS, XK); set initial boundary-layer properties so that $c_f = 0.00225$, $\delta = 0.125$ ft, $H = 1.40$, $Re_\theta = 9216$, $s_i = 2.55$ ft (CF, DELTA, H, RETHET, SI); set the maximum arc length to $s_f = 7.0$ ft (SSTOP); and, set up for $N = 11$ points to define the pressure (NUMBER).

(b) Use the following data to define the pressure distribution in a file named **presur.dat**. The initial and final pressure gradients are given by $(dp_e/dx)_i = 3.410939$ lb/ft$^3$ and $(dp_e/dx)_f = 0.64929$ lb/ft$^3$, respectively. Also, prepare a file **heater.dat** with constant wall temperature, $T_w = 537.6°$ R, and zero heat flux.

| $s$ (ft) | $p_e$ (lb/ft$^2$) | $s$ (ft) | $p_e$ (lb/ft$^2$) | $s$ (ft) | $p_e$ (lb/ft$^2$) |
|---|---|---|---|---|---|
| 2.5 | $2.13123 \cdot 10^3$ | 4.5 | $2.13556 \cdot 10^3$ | 6.5 | $2.13768 \cdot 10^3$ |
| 3.0 | $2.13272 \cdot 10^3$ | 5.0 | $2.13621 \cdot 10^3$ | 7.0 | $2.13806 \cdot 10^3$ |
| 3.5 | $2.13387 \cdot 10^3$ | 5.5 | $2.13677 \cdot 10^3$ | 7.5 | $2.13841 \cdot 10^3$ |
| 4.0 | $2.13480 \cdot 10^3$ | 6.0 | $2.13726 \cdot 10^3$ | | |

(c) Do computations using the high-$Re$ and low-$Re$ versions of the Stress-$\omega$ model.

(d) Compare computed skin friction with the following measured values.

| $s$ (ft) | $c_f$ | $s$ (ft) | $c_f$ | $s$ (ft) | $c_f$ |
|---|---|---|---|---|---|
| 2.5 | $2.45 \cdot 10^{-3}$ | 4.00 | $1.91 \cdot 10^{-3}$ | 7.00 | $1.56 \cdot 10^{-3}$ |
| 3.0 | $2.17 \cdot 10^{-3}$ | 5.00 | $1.74 \cdot 10^{-3}$ | | |
| 3.5 | $2.00 \cdot 10^{-3}$ | 6.00 | $1.61 \cdot 10^{-3}$ | | |

**6.17** The object of this problem is to compare predictions of the Stress-$\omega$ model with measured properties of a Mach 2.65 turbulent boundary layer with adverse pressure gradient and surface heat transfer. The experiment to be simulated was conducted by Fernando and Smits [see Fernholz and Finley (1981)]. Use Program **EDDYBL** and its menu-driven setup utility, Program **SETEBL**, to do the computations (see Appendix D).

(a) Using **SETEBL**, change the appropriate input parameters to accomplish the following: use SI units (IUTYPE); set the freestream conditions to $p_{t_\infty} = 6.7221 \cdot 10^5$ N/m$^2$, $T_{t_\infty} = 270.74$ K, $M_\infty = 2.653$ (PT1, TT1, XMA); use an initial stepsize of $\Delta s = 0.02$ m (DS) and 151 grid points (IEDGE); set the initial boundary-layer properties so that $c_f = 0.000992$, $\delta = 0.02249$ m, $H = 5.3056$, $Re_\theta = 88483$, $s_i = 1.151$ m (CF, DELTA, H, RETHET, SI); set the maximum arc length to $s_f = 1.361$ m and maximum number of steps to 200 (SSTOP, IEND1); set up for $N = 12$ points to define the pressure and set up for prescribed surface temperature (NUMBER, KODWAL).

(b) Use the following data to define the pressure distribution in a file named **presur.dat**. The initial and final pressure gradients are both zero.

| $s$ (m) | $p_e$ (N/m$^2$) | $s$ (m) | $p_e$ (N/m$^2$) | $s$ (m) | $p_e$ (N/m$^2$) |
|---------|-----------------|---------|-----------------|---------|-----------------|
| 0.000 | $3.1039 \cdot 10^4$ | 1.197 | $3.5839 \cdot 10^4$ | 1.299 | $4.2510 \cdot 10^4$ |
| 1.000 | $3.1039 \cdot 10^4$ | 1.222 | $3.9890 \cdot 10^4$ | 1.324 | $4.1000 \cdot 10^4$ |
| 1.151 | $3.0490 \cdot 10^4$ | 1.248 | $4.2870 \cdot 10^4$ | 1.349 | $3.9630 \cdot 10^4$ |
| 1.172 | $3.1500 \cdot 10^4$ | 1.273 | $4.3650 \cdot 10^4$ | 1.361 | $4.0170 \cdot 10^4$ |

(c) Use the following data to define the surface-temperature distribution in a file named **heater.dat** along with zero heat flux. The initial and final temperature gradients are both zero.

| $s$ (m) | $T_w$ (K) | $s$ (m) | $T_w$ (K) | $s$ (m) | $T_w$ (K) | $s$ (m) | $T_w$ (K) |
|---------|-----------|---------|-----------|---------|-----------|---------|-----------|
| 0.000 | 281.56 | 1.172 | 285.76 | 1.248 | 276.94 | 1.324 | 281.49 |
| 1.000 | 281.56 | 1.197 | 281.97 | 1.273 | 279.32 | 1.349 | 284.48 |
| 1.151 | 281.56 | 1.222 | 279.99 | 1.299 | 280.53 | 1.361 | 270.92 |

(d) Do computations using the high-$Re$ and low-$Re$ versions of the Stress-$\omega$ model.

(e) Compare computed skin friction with the following measured values.

| $s$ (m) | $c_f$ | $s$ (m) | $c_f$ | $s$ (m) | $c_f$ |
|---------|-------|---------|-------|---------|-------|
| 1.151 | $9.92 \cdot 10^{-4}$ | 1.248 | $9.46 \cdot 10^{-4}$ | 1.349 | $1.08 \cdot 10^{-3}$ |
| 1.172 | $9.96 \cdot 10^{-4}$ | 1.273 | $9.41 \cdot 10^{-4}$ | 1.361 | $1.04 \cdot 10^{-3}$ |
| 1.197 | $9.67 \cdot 10^{-4}$ | 1.299 | $1.01 \cdot 10^{-3}$ | | |
| 1.222 | $9.43 \cdot 10^{-4}$ | 1.324 | $1.07 \cdot 10^{-3}$ | | |

# Chapter 7

# Numerical Considerations

Modern turbulence model equations pose special numerical difficulties that must be understood in order to obtain reliable numerical solutions, even for boundary-layer flows where the equations are parabolic. For one-equation, two-equation and stress-transport models, these difficulties can include stiffness caused by the presence of an additional time scale, singular behavior near solid boundaries, and non-analytical behavior at sharp turbulent/nonturbulent interfaces. This chapter focuses on these difficulties and on the solution methods for turbulence-model equations that have evolved.

## 7.1 Multiple Time Scales and Stiffness

One key issue that must be addressed in developing a numerical algorithm for fluid-flow problems is that of the physically relevant time scales. Taking proper account of these time scales is a necessary condition for numerical accuracy. For example, when we deal with non-chemically-reacting laminar flow, there are two distinct time scales corresponding to different physical processes. If $L$ and $U$ denote characteristic length and velocity for the flowfield, $a$ is sound speed and $\nu$ is kinematic viscosity, the time scales are:

- Wave propagation, $t_{wave} \sim L/|U \pm a|$

- Molecular diffusion, $t_{diff} \sim L^2/\nu$

When we use turbulence transport equations, we have yet another time scale corresponding to the rate of decay of turbulence properties. In terms of the specific dissipation rate, $\omega \sim \epsilon/k$, this time scale is:

- Dissipation, $t_{diss} \sim 1/\omega \sim k/\epsilon$

Any numerical algorithm designed for use with turbulence transport equations should take account of all three of these time scales.

In terms of the Reynolds number, $Re_L = UL/\nu$, and the Mach number, $M = U/a$, the ratio of $t_{diff}$ to $t_{wave}$ is given by

$$\frac{t_{diff}}{t_{wave}} \sim \frac{|M \pm 1|Re_L}{M} \tag{7.1}$$

Clearly, for high Reynolds number flows the diffusion time scale is much longer than the wave-propagation time scale regardless of Mach number. Diffusion will generally be important over very short distances such as the thickness of a boundary layer, $\delta$, i.e., when $L \sim \delta$. For specified freestream Mach and Reynolds numbers, the relative magnitudes of the diffusion and wave-propagation time scales are more-or-less confined to a limited range. This is not the case for the dissipation time scale.

The specific dissipation rate, $\omega$, can vary by many orders of magnitude across a turbulent boundary layer. Consequently, in the same flow, $t_{diss}$ can range from values much smaller than the other time scales to much larger. This is a crude reminder of the physical nature of turbulence, which consists of a wide range of frequencies. Thus, regardless of the flow speed, we should expect the dissipation time to have a nontrivial impact on numerical algorithms.

Because of the multiplicity of time scales attending use of turbulence transport equations, especially two-equation models and stress-transport models, we must contend with an unpleasant feature known as **stiffness**. An equation, or system of equations, is said to be stiff when there are two or more very different scales of the independent variable on which the dependent variables are changing. For example, consider the equation

$$\frac{d^2y}{dt^2} = 100y \tag{7.2}$$

The general solution to this equation is

$$y(t) = Ae^{-10t} + Be^{10t} \tag{7.3}$$

If we impose the initial conditions

$$y(0) = 1 \quad \text{and} \quad \dot{y}(0) = -10 \tag{7.4}$$

the exact solution becomes

$$y_{exact}(t) = e^{-10t} \tag{7.5}$$

Unfortunately, any roundoff or truncation error in a numerical solution can excite the $e^{10t}$ factor, viz., we can inadvertently wind up with

$$y_{numerical}(t) = e^{-10t} + \epsilon e^{10t}, \quad |\epsilon| \ll 1 \tag{7.6}$$

No matter how small $\epsilon$ is, the second term will eventually dominate the solution. The equivalent situation for a system of equations is to have eigenvalues of the characteristic equation of very different magnitudes.

It is easy to see that most turbulence transport equations hold potential for being stiff. The $k$-$\epsilon$ model is notoriously stiff when some of the commonly used viscous damping functions are introduced. Stress-transport models that use the $\epsilon$ equation are often so stiff as to almost preclude stable numerical solution. Some of the difficulty with the $\epsilon$ equation occurs because the dissipation time scale is a function of both $k$ and $\epsilon$. Transient solution errors in both parameters can yield large variations in $k/\epsilon$, so that the dissipation time scale can assume an unrealistic range of values. By contrast, near-wall solutions to models based on the $\omega$ equation have well-defined algebraic solutions approaching a solid boundary, and are thus much easier to integrate.

## 7.2  Numerical Accuracy Near Boundaries

Proper treatment of boundary conditions is necessary for all numerical solutions, regardless of the equations being solved. Because of the special nature of turbulence transport equations, there are two types of boundary behavior that require careful treatment. Specifically, quantities such as dissipation rate, $\epsilon$, and specific dissipation rate, $\omega$, grow so rapidly approaching a solid boundary that they appear to be singular. In fact, $\omega$ is singular for a perfectly-smooth wall. Also, at interfaces between turbulent and nonturbulent regions, velocity and other properties have nearly discontinuous slopes approaching the interface. Because wall-bounded flows typically involve both types of boundaries, accurate numerical solutions must account for the special problems presented by this unusual solution behavior.

### 7.2.1  Solid Surfaces

We know that for a perfectly-smooth wall, the specific dissipation rate varies in the sublayer as $y^{-2}$ approaching the surface (see Subsection 4.6.3). Even if we choose to use wall functions to obviate integration through the viscous sublayer, analysis of the log layer (see Subsection 4.6.1) shows that both $\epsilon$ and $\omega$ are inversely proportional to distance from the surface. In either case, care must be taken to accurately compute derivatives of such functions.

To illustrate the difficulty imposed by singular behavior approaching a solid boundary, consider the function $\phi$ defined by

$$\phi = \frac{1}{y^n}, \qquad n = 1 \text{ or } 2 \tag{7.7}$$

The exact first and second derivatives are

$$\frac{d\phi}{dy} = -\frac{n}{y^{n+1}} \quad \text{and} \quad \frac{d^2\phi}{dy^2} = \frac{n(n+1)}{y^{n+2}} \tag{7.8}$$

Using central differences on a uniform grid with $y_j = j\Delta y$, a straightforward calculation shows that

$$\left(\frac{d\phi}{dy}\right)_j \approx \frac{\phi_{j+1} - \phi_{j-1}}{2\Delta y} = \left[\frac{j^2}{j^2 - 1}\right]^n \left(\frac{d\phi}{dy}\right)_{exact} \tag{7.9}$$

and

$$\left(\frac{d^2\phi}{dy^2}\right)_j \approx \frac{\phi_{j+1} - 2\phi_j + \phi_{j-1}}{(\Delta y)^2} \approx \left[\frac{j^2}{j^2 - 1}\right]^n \left(\frac{d^2\phi}{dy^2}\right)_{exact} \tag{7.10}$$

where subscript $j$ denotes the value at $y = y_j$. Table 7.1 lists the errors attending use of central differences as a function of $\Delta y/y_j$ for $n = 1$ and $n = 2$.

Table 7.1: *Central Difference Errors for $\phi = y^{-n}$*

| $j$ | $\Delta y/y_j$ | (% Error)$_{n=1}$ | (% Error)$_{n=2}$ |
|-----|------|------|------|
| 2   | .50  | 33   | 78   |
| 3   | .33  | 13   | 27   |
| 5   | .20  | 4    | 9    |
| 7   | .14  | 2    | 4    |
| 10  | .10  | 1    | 2    |

Clearly, significant numerical errors are introduced if the ratio $\Delta y/y_j$ is not small. If wall functions are used (corresponding to $n = 1$), regardless of how close the grid point nearest the surface lies, nontrivial numerical errors in derivatives result for $j < 5$. Consequently, simply using wall functions as effective boundary conditions applied at the first grid point above the surface is unsatisfactory. Rather, the value for $\omega$ or $\epsilon$ should be specified for all points below $j = 4$ (at a minimum) to insure numerical accuracy. This is undoubtedly the primary reason why most researchers find their numerical solutions to be sensitive to near-wall grid-point spacing when they use wall functions. As an alternative, a relatively large cell can be used next to the surface, so that for example, $y_1 = 0$, $y_2 = \Delta y$, $y_3 = 1.2\Delta y$, etc. By using the Rubel-Melnik (1984) transformation, Program **DEFECT** (Appendix C) automatically generates such a grid.

When the $k$-$\omega$ or Stress-$\omega$ model is integrated through the viscous sublayer for a perfectly-smooth surface (corresponding to $n = 2$), there is no practical way to avoid having $\Delta y / y_2 \sim 1$. The exact solution to the $\omega$ equation in the viscous sublayer is

$$\omega \sim \frac{6 \nu_w}{\beta_o y^2}, \qquad y^+ < 2.5 \qquad (7.11)$$

If we simply use the value of $\omega$ according to Equation (7.11) at the first grid point above the surface, Table 7.1 shows that the molecular diffusion term will be in error by 78%. This, in turn, will increase values of $\omega$ at larger values of $y$. Recall that the surface value of $\omega$ has a strong effect on the additive constant, $C$, in the law of the wall (see Subsection 4.7.2). Thus, computing too large a value of $\omega$ near the surface will distort the velocity profile throughout the sublayer and into the log layer. That is, numerically inaccurate near-wall $\omega$ values can distort the entire boundary-layer solution.

The remedy that has proven very effective for eliminating this numerical error is to use Equation (7.11) for the first 7 to 10 grid points above the surface. Of course, these grid points must lie below $y^+ = 2.5$ since Equation (7.11) is not valid above this point. This procedure has been used in Programs **PIPE** and **SUBLAY** (Appendix C) and Program **EDDYBL** (Appendix D).

An alternative procedure for accurately computing near-surface behavior of $\omega$ is to use the rough-wall boundary condition. As shown in Subsection 4.7.2 for the $k$-$\omega$ model and Subsection 6.6.1 for the Stress-$\omega$ model,

$$\omega = \frac{u_\tau^2}{\nu_w} S_R \qquad \text{at} \qquad y = 0 \qquad (7.12)$$

where

$$S_R = (50/k_s^+)^2, \qquad k_s^+ < 25 \qquad (7.13)$$

The quantity $k_s^+ = u_\tau k_s / \nu_w$ is the scaled surface-roughness height.

In order to simulate a smooth surface, we simply require that $k_s^+$ be smaller than 5. Then, combining Equations (7.12) and (7.13), we arrive at the **slightly-rough-surface boundary condition** on $\omega$, viz.,

$$\omega = \frac{2500 \nu_w}{k_s^2} \qquad \text{at} \qquad y = 0 \qquad (7.14)$$

It is important to select a small enough value of $k_s$ to insure that $k_s^+ < 5$. If too large a value is selected, the skin friction values will be larger than smooth-wall values.

As a final comment, the near-wall solution to the $\omega$ equation for a rough wall is given by

$$\omega = \frac{\omega_w}{\left( 1 + \sqrt{\frac{\beta_o \omega_w}{6 \nu_w}} \, y \right)^2}, \qquad y^+ < 2.5 \qquad (7.15)$$

where $\omega_w$ is the surface value of $\omega$. An important test for numerical accuracy of any finite-difference program implementing the $\omega$ equation is to verify that solutions match either Equation (7.11) or (7.15). If the program fails to accurately reproduce the near-wall $\omega$ variation, the program is unlikely to yield accurate results.

For smooth-surface applications, Menter (1992c) proposes an alternative to the slightly-rough-surface boundary condition. In Menter's approach, the surface value of $\omega$ depends upon the distance of the first grid point above the surface, $\Delta y_2$, according to

$$\omega = \frac{N \nu_w}{(\Delta y_2)^2} \quad \text{at} \quad y = 0 \qquad (7.16)$$

where $N$ is a constant. Comparison with Equation (7.14) shows that this corresponds to setting the surface-roughness height according to

$$\frac{2500 \nu_w}{k_s^2} = \frac{N \nu_w}{(\Delta y_2)^2} \quad \Longrightarrow \quad k_s^+ = \frac{50 \Delta y_2^+}{\sqrt{N}} \qquad (7.17)$$

Choosing $N = 100$, for example, means that whenever the grid is such that $\Delta y_2 < 1$, the effective surface-roughness height will be less than 5. This, of course, corresponds to a hydraulically-smooth surface.

The advantage of Menter's method for smooth surfaces is simple. The solution is guaranteed to have sufficiently small $k_s$ to achieve hydraulic smoothness. The only disadvantage is that the boundary condition for $\omega$ is grid dependent, which complicates the task of determining grid independence of the solution. However, since the turbulence-model solution is more-or-less unaffected by decreasing $k_s$ below 5, the problem is minor.

Rapid variation of the dependent variable is not the only potential source of numerical error near solid boundaries. Another serious consideration is round-off error resulting from the relatively small difference between two numbers of comparable magnitude. This problem is frequently encountered with low-Reynolds-number $k$-$\epsilon$ models. For example, damping functions such as

$$f_2 = 1 - e^{-Re_T^2} \quad \text{and} \quad f_\mu = 1 - e^{-0.0115y^+} \qquad (7.18)$$

appear in the Lam-Bremhorst (1981) and Chien (1982) models. Approaching the surface, desired asymptotic behavior depends upon accurate values of these damping functions. If single-precision accuracy is used, it is advisable to use Taylor-series expansions for the damping functions close to the surface. For example, Chien's $f_\mu$ can be computed according to

$$f_\mu = \begin{cases} 1 - e^{-0.0115y^+}, & y^+ > 0.01 \\ 0.0115y^+, & y^+ \leq 0.01 \end{cases} \qquad (7.19)$$

This procedure is used in Program **EDDYBL** (Appendix D) to insure numerically accurate solutions.

## 7.2.2  Turbulent/Nonturbulent Interfaces

More often than not, turbulence-model equations that are in general usage appear to predict sharp interfaces between turbulent and nonturbulent regions, i.e., interfaces where discontinuities in derivatives of flow properties occur at the edge of the shear layer. As noted in earlier chapters, these interfaces bear no relation to the physical turbulent/nonturbulent interfaces that actually fluctuate in time and have smooth Reynolds-averaged properties. The mixing-length model, for example, exhibits a sharp interface for the far wake (see Subsection 3.3.1). That is, the predicted velocity profile is

$$U(x,y) = \begin{cases} U_\infty - 1.38\sqrt{\frac{D}{\rho x}}\left[1 - (y/\delta)^{3/2}\right]^2, & y < \delta \\ U_\infty, & y \geq \delta \end{cases} \qquad (7.20)$$

where $U_\infty$ is freestream velocity, $D$ is drag per unit width, $\rho$ is density, $y$ is distance from the centerline and $\delta$ is the half-width of the wake. Clearly, all derivatives of U above $\partial^2 U/\partial y^2$ are discontinuous at $y = \delta$. Such a solution is called a **weak solution** to the differential equation.

By definition [see Courant and Hilbert (1966)], a weak solution to a partial differential equation

$$\mathcal{L}[u] = \frac{\partial}{\partial x}P(x,y,u) + \frac{\partial}{\partial y}Q(x,y,u) + S(x,y,u) = 0 \qquad (7.21)$$

satisfies the following conditions.

1. $u(x,y)$ is piecewise continuous and has piecewise continuous first derivatives in two adjacent domains, $R_1$ and $R_2$.

2. $\mathcal{L}[u] = 0$ in $R_1$ and $R_2$.

3. For any test function $\phi(x,y)$ that is differentiable to all orders and that is identically zero outside of $R_1$ and $R_2$, the following integral over the combined region $R = R_1 \cup R_2$ must be satisfied.

$$\iint_R \left[P\frac{\partial\phi}{\partial x} + Q\frac{\partial\phi}{\partial y} - S\phi\right] dxdy = 0 \qquad (7.22)$$

A similar result holds for a system of equations. Clearly, Equation (7.22) can be rewritten as

$$\iint_R \left[\frac{\partial(\phi P)}{\partial x} + \frac{\partial(\phi Q)}{\partial y}\right] dxdy - \iint_R \phi\left[\frac{\partial P}{\partial x} + \frac{\partial Q}{\partial y} + S\right] dxdy = 0 \quad (7.23)$$

The second integral vanishes since $P$, $Q$ and $S$ satisfy the differential equation in both $R_1$ and $R_2$. Then, using Gauss' theorem, if $\Gamma$ is the curve of discontinuity that divides $R_1$ and $R_2$ and $\mathbf{n} = (n_x, n_y)$ is the unit normal to $\Gamma$, there follows:

$$\int_\Gamma \phi \left( [P]n_x + [Q]n_y \right) ds = 0 \tag{7.24}$$

The symbols $[P]$ and $[Q]$ denote the jumps in $P$ and $Q$ across $\Gamma$. Since the function $\phi$ is arbitrary, we can thus conclude that the **jump condition** across the surface of discontinuity is given by

$$[P]n_x + [Q]n_y = 0 \tag{7.25}$$

For example, in the case of the far-wake solution given by the mixing-length model, we have $P = U_\infty U$, $Q = -(\alpha \delta \partial U / \partial y)^2$ and $S = 0$. Inspection of Equation (7.20) shows that the jumps in $P$ and $Q$ are both zero, corresponding to the fact that the discontinuity appears in the second derivative rather than the first.

The occurrence of weak solutions causes problems on at least two counts. First, the jump condition is not unique. For example, if $Q$ can be written as a function of $P$, we can always multiply Equation (7.21) by an arbitrary function $\psi(P)$, and rearrange as follows:

$$\frac{\partial F}{\partial x} + \frac{\partial G}{\partial y} + S\psi = 0 \tag{7.26}$$

where

$$F = \int \psi(P)\, dP \quad \text{and} \quad G = \int \psi(P)Q'(P)\, dP \tag{7.27}$$

The jump condition then becomes

$$[F]n_x + [G]n_y = 0 \tag{7.28}$$

In other words, we can have any jump condition we want (and don't want!). This means we have no guarantee that our solution is unique.

The second difficulty posed by the presence of weak solutions has an adverse effect on accuracy and convergence of numerical-solution methods. For example, a central-difference approximation for a first derivative is second-order accurate provided the function of interest is twice differentiable. However, if the function has discontinuous first or second derivative, the accuracy of the central-difference approximation becomes indeterminate. Maintaining second-order accuracy is then possible only if we know the location of the curve of discontinuity in advance. For a hyperbolic equation, this curve is a characteristic curve so that the method of characteristics, for example, can provide a high degree of accuracy in the vicinity of such discontinuities. Since we don't know the location of the characteristics a priori in standard finite-difference computations, accuracy is suspect when the equations have weak solutions.

One-equation models have problems similar to the mixing-length model near turbulent/nonturbulent interfaces. Spalart and Allmaras (1992), for example, demonstrate existence of weak solutions to their one-equation model at such interfaces. Saffman (1970) was the first to illustrate weak solutions for a two-equation model. He discusses the nature of solutions to his $k$-$\omega^2$ model approaching a turbulent/nonturbulent interface. In fact, he builds in weak-solution behavior by choosing his closure coefficients to insure that approaching the interface from within the turbulent region, the streamwise velocity and turbulence length scale vary as

$$U_e - U \propto (\delta - y) \quad \text{and} \quad \ell = k^{1/2}/\omega \propto \text{constant} \quad \text{as} \quad y \to \delta \quad (7.29)$$

where the interface lies at $y = \delta$. Vollmers and Rotta (1977) discuss solution behavior near a turbulent/nonturbulent interface for their $k$-$k\ell$ model, while Rubel and Melnik (1984) perform a similar analysis for the $k$-$\epsilon$ model. Cazalbou, Spalart and Bradshaw (1994) confirm existence of weak solutions for most $k$-$\epsilon$, $k$-$k\ell$ and $k$-$\omega$ models (while demonstrating that there are parametric ranges of the closure coefficients where regular solutions exist). Finally, inspection of the $k$-$\epsilon$ model free shear flow velocity profiles [Figures 4.6 – 4.9] illustrates the nonanalytic behavior at the edge of the shear layer.

Rubel and Melnik (1984) offer an interesting solution for thin shear layers that effectively maps the turbulent/nonturbulent interface to infinity and implicitly clusters grid points near the interface. Their transformation consists of introducing a new independent variable, $\xi$, defined in terms of the normal distance, $y$, by

$$d\xi = \frac{dy}{\nu_T} \qquad \text{or} \qquad \frac{d}{d\xi} = \nu_T \frac{d}{dy} \qquad (7.30)$$

where $\nu_T$ is kinematic eddy viscosity. The Rubel-Melnik transformation, which is useful primarily for self-similar flows, improves numerical accuracy because the edge of the shear layer that occurs at a finite value of $y$ moves to infinity in terms of the transformed independent variable $\xi$ (provided $\nu_T = 0$ in the freestream). Since $\nu_T \to 0$, the transformation produces fine resolution near the interface. For example, if the freestream velocity, $U_e$, is constant, close to the shear-layer edge, convection balances turbulent diffusion in the streamwise momentum equation. Hence,

$$V \frac{dU}{dy} = \frac{d}{dy}\left(\nu_T \frac{dU}{dy}\right) \qquad (7.31)$$

where $V$ is the entrainment velocity, which must also be constant in order to satisfy continuity. Since shear layers grow in thickness, necessarily $V < 0$.

Multiplying both sides of Equation (7.31) by $\nu_T$ and using Equation (7.30), we arrive at

$$V \frac{dU}{d\xi} = \frac{d^2 U}{d\xi^2} \tag{7.32}$$

for which the solution is

$$U = U_e - \mathcal{U}e^{V\xi} \tag{7.33}$$

where $\mathcal{U}$ is a constant of integration.

Using the Rubel-Melnik transformation, it is a straightforward matter to determine the nature of solutions to turbulence-model equations approaching a turbulent/nonturbulent interface. Applying the transformation to the $k$-$\epsilon$ model, for example, we find

$$V \frac{dk}{d\xi} = \left(\frac{dU}{d\xi}\right)^2 - C_\mu k^2 + \frac{1}{\sigma_k} \frac{d^2 k}{d\xi^2} \tag{7.34}$$

$$V \frac{d\epsilon}{d\xi} = C_{\epsilon 1} \frac{\epsilon}{k} \left(\frac{dU}{d\xi}\right)^2 - C_{\epsilon 2} C_\mu k\epsilon + \frac{1}{\sigma_\epsilon} \frac{d^2 \epsilon}{d\xi^2} \tag{7.35}$$

Provided the closure coefficients $\sigma_k$ and $\sigma_\epsilon$ are both less than 2, the production and dissipation terms are negligible in both equations. The solution approaching the interface is

$$k \sim \mathcal{K}e^{\sigma_k V\xi}, \qquad \epsilon \sim \mathcal{E}e^{\sigma_\epsilon V\xi} \qquad (\sigma_k < 2, \quad \sigma_\epsilon < 2) \tag{7.36}$$

where $\mathcal{K}$ and $\mathcal{E}$ are integration constants. Thus, the eddy viscosity is

$$\nu_T \sim C_\mu \frac{\mathcal{K}^2}{\mathcal{E}} e^{(2\sigma_k - \sigma_\epsilon)V\xi} \tag{7.37}$$

Finally, substituting Equation (7.37) into Equation (7.30) and integrating yields

$$e^{V\xi} \propto (1 - y/\delta)^{(2\sigma_k - \sigma_\epsilon)^{-1}} \tag{7.38}$$

So, the solution to the $k$-$\epsilon$ model equations approaching a turbulent/nonturbulent interface from the turbulent side behaves according to

$$\left.\begin{array}{rcl} U_e - U & \sim & \mathcal{U}(1 - y/\delta)^{(2\sigma_k - \sigma_\epsilon)^{-1}} \\[2mm] k & \sim & \mathcal{K}(1 - y/\delta)^{\sigma_k(2\sigma_k - \sigma_\epsilon)^{-1}} \\[2mm] \epsilon & \sim & \mathcal{E}(1 - y/\delta)^{\sigma_\epsilon(2\sigma_k - \sigma_\epsilon)^{-1}} \end{array}\right\} \quad \text{as} \quad y \to \delta \tag{7.39}$$

Using the standard values $\sigma_k = 1.0$ and $\sigma_\epsilon = 1.3$, the $k$-$\epsilon$ model predicts

$$\left.\begin{array}{rcl} U_e - U & \sim & \mathcal{U}(1 - y/\delta)^{10/7} \\[2mm] k & \sim & \mathcal{K}(1 - y/\delta)^{10/7} \\[2mm] \epsilon & \sim & \mathcal{E}(1 - y/\delta)^{13/7} \end{array}\right\} \quad \text{as} \quad y \to \delta \tag{7.40}$$

The solution for the $k$-$\omega$ model is a bit more complicated when $\sigma$ and $\sigma^*$ are both equal to 1/2. As a result, only the dissipation terms are negligible, and the production term in the transformed $k$ equation yields a secular term, which complicates the solution. That is, the approximate transformed equations for $k$ and $\omega$ are as follows.

$$\frac{d^2 k}{d\xi^2} - 2V\frac{dk}{d\xi} = 2V^2 \mathcal{U}^2 e^{2V\xi} \tag{7.41}$$

$$\frac{d^2 \omega}{d\xi^2} - 2V\frac{d\omega}{d\xi} = 2\alpha V^2 \mathcal{U}^2 \frac{\omega}{k} e^{2V\xi} \tag{7.42}$$

The solution for $k$ and $\omega$ is

$$k \sim \mathcal{U}^2 V\xi\, e^{2V\xi}, \qquad \omega \sim \mathcal{W}\xi^{-\alpha} \tag{7.43}$$

where $\mathcal{W}$ is an integration constant. Computing the eddy viscosity and substituting into Equation (7.30), we arrive at

$$y \sim \delta + \frac{\mathcal{U}^2 V}{\mathcal{W}} \int_\xi^\infty \xi^{1+\alpha} e^{2V\xi} d\xi \tag{7.44}$$

Integrating by parts, we can approximate the limiting form of the integral for $\xi \to \infty$ as follows.

$$\delta - y \sim \frac{\mathcal{U}^2}{2\mathcal{W}}\xi^{1+\alpha} e^{2V\xi} \tag{7.45}$$

Now, we must solve this equation for $\xi$ as a function of $\delta - y$. To do this, let

$$\eta = \frac{2\mathcal{W}}{\mathcal{U}^2}(\delta - y) \tag{7.46}$$

Then, Equation (7.45) simplifies to

$$\eta \sim \xi^{1+\alpha} e^{2V\xi} \tag{7.47}$$

This equation can be solved for $\xi$ as a function of $\eta$ by assuming

$$2V\xi \sim \ell n\eta + \phi(\eta) \tag{7.48}$$

where $\phi(\eta)$ is a function to be determined. In the limit $\xi \to \infty$, which corresponds to $\eta \to 0$, the approximate solution for $\phi(\eta)$ is

$$\phi(\eta) \sim -(1+\alpha)\ell n\left(\frac{\ell n\eta}{2V}\right) \tag{7.49}$$

With a bit more algebra, there follows

$$e^{V\xi} \propto \eta^{1/2}\left(\frac{2V}{\ell n\eta}\right)^{(1+\alpha)/2} \tag{7.50}$$

Thus, for the $k$-$\omega$ model approaching a turbulent/nonturbulent interface from within the turbulent region, we have

$$
\left.
\begin{aligned}
U_e - U &\sim \mathcal{U}\sqrt{\lambda} \\
k &\sim -\mathcal{K}\lambda\ell n\lambda \\
\omega &\sim \mathcal{W}(-\ell n\lambda)^{-\alpha} \\
\lambda &\sim \frac{(1 - y/\delta)}{[-\ell n(1 - y/\delta)]^{1+\alpha}}
\end{aligned}
\right\}
\qquad \text{as} \quad y \to \delta
\qquad (7.51)
$$

Clearly, $\omega$ approaches zero very slowly from the turbulent side as compared to the variation of $\epsilon/k \sim (\delta - y)^{3/7}$ predicted by the $k$-$\epsilon$ model. Also, the velocity profile has discontinuous first derivative at the shear-layer edge, or more generally, at any turbulent/nonturbulent interface.

To verify that the asymptotic behavior predicted in Equations (7.51) is consistent with results of numerical computations, we can examine boundary-layer solutions from Program **EDDYBL** (Appendix D). Figure 7.1 compares numerical solutions with Equation (7.51). The computations are for a Mach 0.17 flat-plate boundary layer, and have been done using **EDDYBL** with two finite-difference grids. The first grid has 140 points normal to the surface, while the second grid has 289 points. As shown, the 289-point solution matches the closed-form solution to within 3% of scale for $U$, $k$ and $\omega$. The largest discrepancies are present for points very close to the interface. This is true because the computation has a nonzero value for $\omega$ in the freestream, while the closed-form solution is strictly valid for $\omega = 0$ in the freestream. Because of the coarser resolution, the 140-point solution shows slightly larger differences, again mainly for points closest to the interface. Results shown clearly indicate that the numerical solution is consistent with the weak-solution.

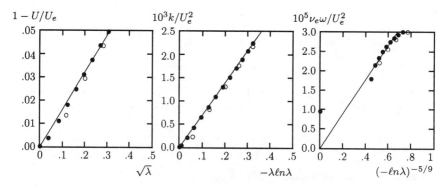

Figure 7.1: *Velocity, $k$ and $\omega$ profiles near a turbulent/nonturbulent interface:* o *140 points;* • *289 points;* —— *Equation (7.51).*

Usually it is more convenient to assign small nonzero values to $k$ and other turbulence parameters in the freestream, especially when the parameter appears in the denominator of the eddy viscosity. Cazalbou, Spalart and Bradshaw (1994) show that when this is done in boundary-layer computations with the $k$-$\epsilon$ model, the weak solution prevails below the interface. Small gradients in $k$ and $\epsilon$ appear above the interface that yield an asymptotic approach to the prescribed freestream values. There is "no significant influence on the predicted flow."

By contrast, Menter (1992a) shows that for the far wake, in which the entrainment velocity increases in magnitude linearly with distance from the centerline, the Wilcox (1988a) $k$-$\omega$ model predicts that $k$ and $\omega$ decay exponentially with distance squared. However, they decay at the same rate so that the eddy viscosity remains constant. As a consequence, consistent with results presented in Section 4.5, the freestream value of $\omega$ has a nontrivial effect on the solution, even for the Wilcox (1998) version. Menter indicates a smaller effect on boundary layers, primarily because of the large values of $\omega$ prevailing near the surface. The behavior of $\omega$ in Equation (7.51) is consistent with Menter's observation that the $k$-$\omega$ model solutions appear to have discontinuous derivatives at the shear layer edge. However, the discontinuity in $d\omega/dy$ would probably be difficult to detect.

In principle, solutions with discontinuous derivatives will not occur if molecular viscosity is included in the diffusion terms of the equations of motion. As shown by Saffman (1970), there is a thin **viscous-interface layer** of thickness

$$\delta_{vi} \sim \nu/|V| \qquad (7.52)$$

in which the discontinuities are resolved. This is a singular-perturbation problem in the limit $|V|\delta_{vi}/\nu \to \infty$, and the weak solution discussed above is the outer solution. The inner solution holds in the viscous-interface layer. For example, in the interface layer, Saffman's equations simplify to

$$\left.\begin{aligned}
V\frac{dU}{dy} &= \frac{d}{dy}\left[\left(\nu+\frac{k}{\omega}\right)\frac{dU}{dy}\right] \\[2mm]
V\frac{dk}{dy} &= \frac{d}{dy}\left[\left(\nu+\sigma^*\frac{k}{\omega}\right)\frac{dk}{dy}\right] \\[2mm]
V\frac{d\omega^2}{dy} &= \frac{d}{dy}\left[\left(\nu+\sigma\frac{k}{\omega}\right)\frac{d\omega^2}{dy}\right]
\end{aligned}\right\} \qquad (7.53)$$

These equations must be solved subject to the following boundary conditions, which correspond to formal matching of the solutions that hold on each side of the turbulent/nonturbulent interface:

$$U_e - U \to \mathcal{U}(\delta-y), \quad k \to \mathcal{K}(\delta-y)^2, \quad \omega \to \frac{\mathcal{K}}{|V|}(\delta-y) \quad \text{as} \quad \frac{|V|(\delta-y)}{\nu} \to \infty \qquad (7.54)$$

and

$$U_e - U \to 0, \quad k \to 0, \quad \omega \to 0 \quad \text{as} \quad \frac{|V|(\delta - y)}{\nu} \to -\infty \qquad (7.55)$$

As can be easily verified, for $\sigma = \sigma^* = 1/2$, the solution is given by

$$\left.\begin{aligned}
U_e - U &= \frac{\mathcal{U}|V|^3}{\mathcal{K}^2 \nu} \left( \frac{\omega^2}{1 + V^2\omega/\mathcal{K}\nu} \right) \\[2mm]
k &= \frac{V^2\omega^2}{\mathcal{K}} \\[2mm]
y - \delta &= \frac{|V|\omega}{\mathcal{K}} + \frac{2\nu}{|V|} \ell n \left( \frac{V^2\omega}{\mathcal{K}\nu} \right)
\end{aligned}\right\} \qquad (7.56)$$

In practice, finite-difference grids are never sufficiently fine to resolve the viscous-interface layer. Generally, grid points are packed close to the surface to permit accurate resolution of the sublayer. Hence, even when molecular viscosity is included in a typical finite-difference computation, turbulent/nonturbulent interfaces are not sufficiently resolved. As a consequence, the interfaces are sharp, and the weak solutions generally prevail. However, truncation error, numerical diffusion and dissipation will generally yield diffused solutions close to the interfaces. The most significant numerical problem typically encountered is the appearance of nonphysical negative values of $k$ and/or other normally positive turbulence parameters such as $\omega$, $\epsilon$ and $\ell$.

For self-similar flows such as the far wake, mixing layer, jet and defect layer, the Rubel-Melnik transformation cures the problem by mapping the interface to $\infty$. Programs **WAKE**, **MIXER**, **JET** and **DEFECT** described in Appendix C all use this transformation. In addition to eliminating difficulties associated with the turbulent/nonturbulent interface, the transformation linearizes the first and second derivative terms in the equations. This linearization tends to improve the rate of convergence of most numerical methods. The only shortcoming of the method is its sensitivity to the location of "$\infty$." Using too large or too small a value of $\xi_{max}$ (the farfield value of $\xi$) sometimes impedes convergence of the numerical solution.

In general finite-difference computations, the correct jump condition will be obtained provided the diffusion terms in all equations are differenced in a conservative manner. For the same reasons, we use conservative differencing for the Navier-Stokes equation to guarantee that the exact shock relations are satisfied across a shock wave in a finite-difference computation. Program **EDDYBL** (Appendix D), for example, uses conservative differencing for diffusion terms and rarely ever encounters numerical difficulties attending the presence of sharp turbulent/nonturbulent interfaces.

For nonzero freestream values of $k$, etc., some researchers prefer zero-gradient boundary conditions at a boundary-layer edge. While such conditions are clean

from a theoretical point of view, they are undesirable from a numerical point of
view. Almost universally, convergence of iterative schemes is much slower with
zero-gradient (Neumann-type) conditions than with directly-specified (Dirichlet-
type) conditions.

In order to resolve this apparent dilemma, we can appeal directly to the equations of motion. Beyond the boundary-layer edge, we expect to have vanishing
normal gradients so that the equations for $k$ and $\omega$ simplify to the following:

$$U_e \frac{dk_e}{dx} = -\beta_o^* \omega_e k_e \tag{7.57}$$

$$U_e \frac{d\omega_e}{dx} = -\beta_o \omega_e^2 \tag{7.58}$$

where subscript $e$ denotes the value at the boundary-layer edge. The solution to
Equations (7.57) and (7.58) can be obtained by simple quadrature, independent
of integrating the equations of motion through the boundary layer. Once $k_e$ and
$\omega_e$ are determined from Equations (7.57) and (7.58), it is then possible to specify
Dirichlet-type boundary conditions that guarantee zero normal gradients. Clearly,
the same procedure can be used for any turbulence model. Program **EDDYBL**
(Appendix D) uses this procedure.

# 7.3 Parabolic Marching Methods

In general, numerical methods for solving parabolic systems of equations such as
the boundary-layer equations are unconditionally stable. A second-order accurate
scheme like the Blottner (1974) variable-grid method, for example, involves inversion of a tridiagonal matrix. If the matrix is diagonally dominant, the scheme
will run stably with arbitrarily large streamwise stepsize, $\Delta x$. Turbulent boundary layer computations using algebraic models often run with $\Delta x/\delta$ between
1 and 10, where $\delta$ is boundary-layer thickness. By contrast, early experience
with two-equation models indicated that much smaller steps must be taken. Rastogi and Rodi (1978) found that their three-dimensional boundary-layer program
based on the Jones-Launder (1972) $k$-$\epsilon$ model required initial steps of about
$\delta/100$, and that ultimately $\Delta x$ could not exceed $\delta/2$. Similar results hold for
models based on the $\omega$ equation.

Wilcox (1981b) found that the problem stems from a loss of diagonal dominance caused by the production terms in the turbulence-model equations. To
illustrate the essence of the problem, consider the $k$-$\omega$ model's turbulence energy
equation for an incompressible two-dimensional boundary layer, viz.,

$$U \frac{\partial k}{\partial x} + V \frac{\partial k}{\partial y} = \left[ \frac{(\partial U/\partial y)^2}{\omega} - \beta^* \omega \right] k + \frac{\partial}{\partial y} \left[ (\nu + \sigma^* \nu_T) \frac{\partial k}{\partial y} \right] \tag{7.59}$$

The following analysis is based on the Blottner variable-grid method, which is the scheme implemented in Program **EDDYBL** (Appendix D). This algorithm uses a three-point forward difference formula [Adams-Bashforth — see Roache (1972)] in the streamwise direction, central differencing for the normal convection term, and conservative differencing for the diffusion terms. Hence, discretization approximations for all except the source terms are as follows:

$$U \frac{\partial k}{\partial x} \doteq \frac{U}{\Delta x} \left( 3k_{m+1,n} - 4k_{m,n} + k_{m-1,n} \right) \tag{7.60}$$

$$V \frac{\partial k}{\partial y} \doteq \frac{V}{2\Delta y} \left( k_{m+1,n+1} - k_{m+1,n-1} \right) \tag{7.61}$$

$$\frac{\partial}{\partial y} \left[ (\nu + \sigma^* \nu_T) \frac{\partial k}{\partial y} \right] \doteq \frac{\nu^+ (k_{m+1,n+1} - k_{m+1,n}) - \nu^- (k_{m+1,n} - k_{m+1,n-1})}{(\Delta y)^2}$$
$$\tag{7.62}$$

where $k_{m,n}$ denotes the value of $k$ at $x = x_m$ and $y = y_n$, and $\Delta y$ denotes the vertical distance between grid points. Unsubscripted quantities are assumed known during the typically iterative solution procedure. Also, the quantity $\nu^-$ denotes the value of $(\nu + \sigma^* \nu_T)$ midway between $y_{n-1}$ and $y_n$, while $\nu^+$ denotes the value midway between $y_n$ and $y_{n+1}$. For simplicity, we assume points are equally spaced in both the $x$ and $y$ directions, so that the grid consists of rectangular cells. Figure 7.2 shows the finite-difference molecule.

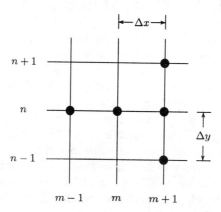

Figure 7.2: *Finite-difference molecule for Blottner's variable-grid method.*

Turning to the source terms, the simplest second-order accurate discretization approximation is

$$\left[ \frac{(\partial U/\partial y)^2}{\omega} - \beta^* \omega \right] k \doteq \left[ \frac{(\partial U/\partial y)^2}{\omega} - \beta^* \omega \right] k_{m+1,n} \tag{7.63}$$

where the quantity in brackets is also evaluated at $(m + 1, n)$ using values extrapolated from $(m, n)$ and $(m - 1, n)$. Substituting Equations (7.60) – (7.63) into Equation (7.59) and regrouping terms leads to a tridiagonal matrix system as follows:

$$A_n k_{m+1,n-1} + B_n k_{m+1,n} + C_n k_{m+1,n+1} = D_n \qquad (7.64)$$

where $A_n$, $B_n$, $C_n$ and $D_n$ are defined by

$$A_n = -\left[\frac{V}{2\Delta y} + \frac{\nu^-}{(\Delta y)^2}\right] \qquad (7.65)$$

$$B_n = 3\frac{U}{\Delta x} + \frac{\nu^- + \nu^+}{(\Delta y)^2} - \frac{(\partial U/\partial y)^2}{\omega} + \beta^* \omega \qquad (7.66)$$

$$C_n = \left[\frac{V}{2\Delta y} - \frac{\nu^+}{(\Delta y)^2}\right] \qquad (7.67)$$

$$D_n = \frac{U}{\Delta x}[4k_{m,n} - k_{m-1,n}] \qquad (7.68)$$

Now, in order to have a diagonally dominant system, the condition

$$B_n \geq -(A_n + C_n) \qquad (7.69)$$

must be satisfied. Substituting Equations (7.65) – (7.67) into Equation (7.69) yields the following condition.

$$3\frac{U}{\Delta x} - \frac{(\partial U/\partial y)^2}{\omega} + \beta^* \omega \geq 0 \qquad (7.70)$$

If dissipation exceeds production, so that $\beta^* \omega > (\partial U/\partial y)^2/\omega$, Equation (7.70) is satisfied so long as we march in the direction of flow (i.e., so long as $U$ and $\Delta x$ are of the same sign). The system is then said to be **unconditionally stable**. However, when production exceeds dissipation, we have the following limit on stepsize.

$$\Delta x \leq (\Delta x)_{theory} \equiv \frac{3\omega U}{(\partial U/\partial y)^2 - \beta^* \omega^2} \qquad (7.71)$$

Hence, the scheme is **conditionally stable**, the condition being that of Equation (7.71).

To demonstrate the validity of Equation (7.71), Wilcox (1981b) presents computed results for an incompressible flat-plate boundary layer using the Wilcox-Rubesin (1980) $k$-$\omega^2$ model. At a plate-length Reynolds number, $Re_x$, of $1.2 \cdot 10^6$, stable computation is found empirically to be possible provided the Reynolds number based on $\Delta x$ satisfies $Re_{\Delta x} < 2.2 \cdot 10^4$, which corresponds

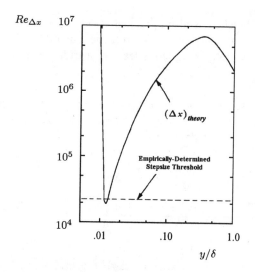

Figure 7.3: *Theoretical and empirically determined stepsize threshold for a flatplate boundary layer. [From Wilcox (1981b) — Copyright © AIAA 1981 — Used with permission.]*

to $\Delta x/\delta = 1.15$. Figure 7.3 shows $Re_{\Delta x}$ as predicted by Equation (7.71) throughout the boundary layer. The minimum value of $Re_{\Delta x}$ according to Equation (7.71) is $1.9 \cdot 10^4$ and occurs at $y/\delta \doteq 0.012$. This close agreement verifies that the source of instability is lack of diagonal dominance in the tridiagonal matrix system defined in Equations (7.64) – (7.68).

To remedy this situation, note that because of nonlinearity, Equation (7.64) always requires an iterative solution. Letting superscript $i$ denote iteration number, we replace $B_n$ and $D_n$ by the following revised discretization approximations:

$$B_n = 3\frac{U}{\Delta x} + \frac{\nu^- + \nu^+}{(\Delta y)^2} - \frac{(\partial U/\partial y)^2}{\omega} + (1 + \psi_k)\beta^*\omega \qquad (7.72)$$

$$D_n = \frac{U}{\Delta x}\left[4k_{m,n} - k_{m-1,n}\right] + \psi_k\beta^*\omega k_{m+1,n}^{i-1} \qquad (7.73)$$

where $\psi_k$ will be defined below. Then, Equation (7.64) is replaced by

$$A_n k_{m+1,n-1}^i + B_n k_{m+1,n}^i + C_n k_{m+1,n+1}^i = D_n \qquad (7.74)$$

Inspection of Equations (7.72) – (7.74) shows that when convergence has been achieved (i.e., when $k_{m+1,n}^i$ and $k_{m+1,n}^{i-1}$ differ by a negligible amount), terms on the right- and left-hand sides of Equation (7.74) proportional to $\psi_k$ cancel identically. Hence, $k_{m+1,n}^i$ satisfies the correct equation. The advantage of this procedure becomes obvious upon inspection of the stability condition, which now becomes

$$3\frac{U}{\Delta x} - \frac{(\partial U/\partial y)^2}{\omega} + (1 + \psi_k)\beta^*\omega \geq 0 \qquad (7.75)$$

Clearly, $\psi_k$ can be chosen to insure that this inequality is always satisfied, regardless of the value of $\Delta x$. This corresponds to unconditional stability.

Numerical experimentation shows that the best results are obtained when $(1 + \psi_k)\beta^*\omega$ exceeds $(\partial U/\partial y)^2/\omega$ by about 30%, a condition that is insured by defining $\psi_k$ as follows.

$$\psi_k = \begin{cases} \dfrac{3}{10}, & (\partial U/\partial y)^2 \leq \beta^*\omega^2 \\[2ex] \dfrac{(\partial U/\partial y)^2}{\beta^*\omega^2} - \dfrac{7}{10}, & (\partial U/\partial y)^2 > \beta^*\omega^2 \end{cases} \qquad (7.76)$$

A similar factor, $\psi_\omega$, must be introduced for the $\omega$ equation, and experience has shown that selecting

$$\psi_\omega = \psi_k \qquad (7.77)$$

is satisfactory to achieve both unconditional stability and rapid convergence.

The prescription for $\psi_k$ and $\psi_\omega$ given in Equations (7.76) and (7.77) permits stepsizes comparable to those used with algebraic models. While the numerical procedure is unconditionally stable for other values of $\psi_k$, using these values for $\psi_k$ and $\psi_\omega$ optimizes $k$-$\omega^2$ and $k$-$\omega$ model computations with respect to the number of iterations required for the solution to converge. Interestingly, if $\psi_k$ is too large, say $\psi_k = 2$, stable integration is inhibited. The value of $\psi_\omega$ cannot be too large either, although the upper bound appears to be dependent upon details of the specific model.

The same analysis applies to the $k$-$\epsilon$ model. For the $k$ equation, writing Equation (7.76) in terms of the model's variables leads to the following entirely equivalent form.

$$\psi_k = \begin{cases} \dfrac{3}{10}, & \nu_T(\partial U/\partial y)^2 \leq \epsilon \\[2ex] \dfrac{\nu_T(\partial U/\partial y)^2}{\epsilon} - \dfrac{7}{10}, & \nu_T(\partial U/\partial y)^2 > \epsilon \end{cases} \qquad (7.78)$$

By contrast, the value of the corresponding factor for the $\epsilon$ equation, $\psi_\epsilon$, is very much dependent upon details of the model. Low-Reynolds-number viscous damping functions have a pronounced effect on the most appropriate value. Table 7.2 lists the values of $\psi_\epsilon$ used in Program **EDDYBL** (Appendix D) for six different low-Reynolds-number $k$-$\epsilon$ models. The values listed have been found empirically to yield optimum convergence rates for incompressible boundary layers.

Table 7.2: *Values of $\psi_\epsilon$ for low-Reynolds-number $k$-$\epsilon$ models*

| Model | $\psi_\epsilon$ |
|---|---|
| Jones-Launder (1972) | 0.50 |
| Launder-Sharma (1974) | 0.50 |
| Lam-Bremhorst (1981) | 0.50 |
| Chien (1982) | -0.25 |
| Yang-Shih (1993) | -0.25 |
| Fan-Lakshminarayana-Barnett (1993) | -0.25 |

# 7.4   Elementary Time-Marching Methods

One of the most effective procedures for solving complex flowfields is the use of time-marching methods. If the desired solution is unsteady, time-marching solutions yield a true time history. Time-marching methods can also be used for steady-flow problems by letting the solution evolve in time until temporal variations become negligibly small. That is, we begin with an initial approximation and update the solution at each timestep until the solution differs between timesteps by less than a prescribed tolerance level. Prior to discussing the impact of turbulence-model source terms on explicit and implicit methods, this section presents a brief overview of these methods. For more complete details see a general text on Computational Fluid Dynamics such as Roache (1972), Peyret and Taylor (1983), Anderson et al. (1984), Minkowycz et al. (1988) or Ferziger and Perić (1996).

The simplest time-marching schemes are **explicit methods**, such as the DuFort-Frankel (1953), Godunov (1959), Lax-Wendroff (1960) and MacCormack (1969) methods. Most explicit schemes were developed prior to 1970. In an explicit scheme, the solution at time $t^{n+1}$ depends only on the past history, i.e., the solution at time $t^n$. For example, consider the one-dimensional wave equation:

$$\frac{\partial k}{\partial t} + U\frac{\partial k}{\partial x} = 0, \qquad U > 0 \tag{7.79}$$

where $k$ is a flow property, $U$ is velocity, $t$ is time and $x$ is streamwise direction. Letting $k_j^n$ denote $k(x_j, t^n)$, we approximate $\partial k/\partial t$ with a **forward-difference approximation** so that

$$\frac{\partial k}{\partial t} \doteq \frac{k_j^{n+1} - k_j^n}{\Delta t} + O(\Delta t) \tag{7.80}$$

where $\Delta t = t^{n+1} - t^n$. For simplicity, consider simple **upwind differencing** in which we approximate $\partial k / \partial x$ according to

$$\frac{\partial k}{\partial x} \doteq \frac{k_j^n - k_{j-1}^n}{\Delta x} + O(\Delta x) \tag{7.81}$$

Using these discretization approximations, we arrive at the following first-order accurate difference equation that approximates Equation (7.79).

$$k_j^{n+1} = k_j^n - \frac{U \Delta t}{\Delta x} \left( k_j^n - k_{j-1}^n \right) \tag{7.82}$$

This is not a particularly accurate method, but nevertheless illustrates the general nature of explicit schemes. Note that all terms on the right-hand side of Equation (7.82) are known from time $t^n$. Hence, $k_j^{n+1}$ is obtained from simple algebraic operations. Because only algebraic operations are needed (as opposed to inversion of a large matrix), explicit methods are easy to implement.

The primary shortcoming of explicit schemes is a limit on the timestep that can be used. For too large a timestep, solution errors will grow with increasing iterations and the computation becomes unstable. The most commonly used method for determining the stability properties of a time-marching finite-difference scheme is von Neumann stability analysis [see Roache (1972) or Anderson et al. (1984)]. In this method, we introduce a discrete Fourier series solution to the finite-difference equation under study, and determine the growth rate of each mode. If all Fourier modes decay as we march in time, the scheme is stable. However, if even a single mode grows, the scheme is unstable. We write each Fourier component as

$$k_j^n = G^n e^{i(j \kappa \Delta x)} \tag{7.83}$$

where $G$ is called the **amplitude factor**, $i = \sqrt{-1}$ and $\kappa$ is wavenumber. The stability of a scheme is determined as follows:

$$\left. \begin{array}{ll} |G| < 1, & \text{Stable} \\ |G| = 1, & \text{Neutrally Stable} \\ |G| > 1, & \text{Unstable} \end{array} \right\} \tag{7.84}$$

In general, $G$ is complex, and the notation $G^n$ means $G$ raised to the power $n$. The amplitude factor for Equation (7.82) is

$$G = 1 - \frac{U \Delta t}{\Delta x} \left( 1 - e^{-i\theta} \right), \qquad \text{where} \qquad \theta = \kappa \Delta x \tag{7.85}$$

Thus,

$$|G|^2 = 1 + 2(1 - \cos \theta) \frac{U \Delta t}{\Delta x} \left( \frac{U \Delta t}{\Delta x} - 1 \right) \tag{7.86}$$

In order to have a stable scheme, $|G|$ must be less than or equal to 1 for all possible values of $\theta$. Clearly, for the upwind-difference scheme, errors will not grow provided the condition

$$\Delta t < \frac{\Delta x}{U} \quad \text{or} \quad N_{CFL} = \frac{U \Delta t}{\Delta x} < 1 \tag{7.87}$$

is satisfied. This is the famous Courant-Friedrichs-Lewy (1967), or CFL condition. It arises because a disturbance traveling at speed $U$ cannot propagate a distance exceeding $\Delta x$ in a time equal to $\Delta t$. $N_{CFL}$ is known as the **CFL Number**.

Explicit methods are of interest in modern CFD applications mainly for time-dependent flows. Their algebraic simplicity makes them especially easy to implement on any computer. Their primary drawback is their conditional stability, and thousands of timesteps are often needed to achieve steady-flow conditions. There has been renewed interest in these methods because of their suitability for massively-parallel computers, where they may actually be more efficient than implicit schemes which can run with larger timesteps but are trickier to program.

**Implicit methods** date back to 1947 when the Crank-Nicolson (1947) method first appeared. Other methods such as the Euler [Lilly (1965)] and Alternating Direction Implicit (ADI) schemes [Peaceman and Rachford (1955)] are implicit. The solution at time $t^{n+1}$ and location $x_j$ in this type of scheme depends not only upon the solution at the earlier timestep, but upon the solution at other spatial locations at time $t^{n+1}$ as well. For example, the Crank-Nicolson method uses

$$\frac{\partial k}{\partial x} \doteq \frac{1}{2} \left( \frac{k_{j+1}^n - k_{j-1}^n}{2\Delta x} + \frac{k_{j+1}^{n+1} - k_{j-1}^{n+1}}{2\Delta x} \right) + O\left[(\Delta x)^2\right] \tag{7.88}$$

Thus, Equation (7.79) is approximated by the following second-order accurate difference equation:

$$-\lambda k_{j-1}^{n+1} + k_j^{n+1} + \lambda k_{j+1}^{n+1} = k_j^n - \lambda \left( k_{j+1}^n - k_{j-1}^n \right) \tag{7.89}$$

where

$$\lambda = \frac{U \Delta t}{4\Delta x} \tag{7.90}$$

Hence, as with the Blottner method discussed in the preceding section, a tridiagonal matrix system of equations must be solved. Although inverting any matrix is more time consuming than solving a simple algebraic equation, the increased complexity is attended by a significant increase in the maximum permissible timestep. That is, stability analysis shows that the scheme defined in Equation (7.89) is unconditionally stable.

Implicit schemes have proven to be especially useful for steady-flow computations where the CFL limit can be exceeded by factors as large as 5. While these schemes will run at a larger CFL number, using larger values of $\Delta t$ sometimes introduces significant truncation errors if convective effects have a significant effect on the physics of the flow. The number of timesteps required, relative to explicit methods, to achieve steady-flow conditions typically is reduced, although the factor is $N_{CFL}^{-n}$ where $n < 1$.

Recall from Section 7.1 that there are three physically relevant time scales when turbulence-model equations are used. If we use an explicit finite-difference scheme to approximate the Favre-averaged Navier-Stokes equation, stability analysis shows that the wave speed is $|\tilde{u}| + a$, where $\tilde{u}$ is mass averaged velocity and $a$ is sound speed. If $\nu$ denotes kinematic viscosity, the wave-propagation and diffusion timestep limitations are as follows.

$$\Delta t \leq \frac{\Delta x}{|\tilde{u}| + a} \quad \text{and} \quad \Delta t \leq \frac{(\Delta x)^2}{2\nu} \quad (7.91)$$

We might also anticipate that including source terms in the stability analysis would lead to an additional timestep constraint such as $\Delta t \leq t_{diss}$. This is indeed the case, and this timestep limitation is sometimes more restrictive than either condition in Equation (7.91).

To illustrate the problem, we add a source term, $Sk$, to Equation (7.79), giving

$$\frac{\partial k}{\partial t} + U \frac{\partial k}{\partial x} = Sk \quad (7.92)$$

If we suppose that $k$ denotes turbulence kinetic energy, the condition $S > 0$ corresponds to production exceeding dissipation, and vice versa for $S < 0$. To cast this equation in discretized form, we use Crank-Nicolson differencing and we approximate the source term as follows:

$$Sk \doteq S\left[\psi k_j^n + (1 - \psi)k_j^{n+1})\right] + O\left[(\psi - \tfrac{1}{2})\Delta t, (\Delta t)^2\right] \quad (7.93)$$

where $\psi$ lies between 0 and 1. Hence, our finite-difference approximation to Equation (7.92) is

$$k_j^{n+1} = k_j^n - \lambda\left(k_{j+1}^{n+1} + k_{j+1}^n - k_{j-1}^{n+1} - k_{j-1}^n\right) + S\Delta t\left[\psi k_j^n + (1 - \psi)k_j^{n+1})\right] \quad (7.94)$$

The complex amplification factor for this scheme is

$$G = \frac{1 + \psi S\Delta t - 2i\lambda \sin\theta}{1 - (1 - \psi)S\Delta t + 2i\lambda \sin\theta} \quad (7.95)$$

Hence, in order for this scheme to be stable, we must require

$$|G|^2 = \frac{[1 + \psi S\Delta t]^2 + 4\lambda^2 \sin^2\theta}{[1 - (1 - \psi)S\Delta t]^2 + 4\lambda^2 \sin^2\theta} \leq 1 \quad (7.96)$$

After a little algebra, the stability condition simplifies to

$$S\Delta t \left[1 + (\psi - \tfrac{1}{2})S\Delta t\right] \leq 0 \tag{7.97}$$

When $S < 0$, we find

$$\begin{cases} \Delta t \leq \dfrac{1}{(\psi - \tfrac{1}{2})|S|}; & \psi > \tfrac{1}{2}, \quad S < 0 \\[2ex] \text{Unconditionally Stable;} & \psi \leq \tfrac{1}{2}, \quad S < 0 \end{cases} \tag{7.98}$$

When $S > 0$, upon first inspection, von Neumann stability analysis indicates this scheme is unstable when $\psi \geq \tfrac{1}{2}$ and that $\Delta t$ must have a lower bound (as opposed to an upper bound) to insure stable computation when $\psi < \tfrac{1}{2}$. However, these results are irrelevant. This is true because the exact solution to Equation (7.92) is proportional to $e^{St}$, and is thus unbounded as $t \to \infty$. When this occurs, even if the error is a small fraction of the exact solution, it will also be unbounded. The requirement $|G| \leq 1$ is thus too stringent for an unbounded function. According to von Neumann, the condition for stability when the exact solution is unbounded is:

$$|G| \leq 1 + O(\Delta t) \tag{7.99}$$

With a little rearrangement of terms, Equation (7.96) can be written as

$$|G|^2 = 1 + \left(\frac{2[1 + (\psi - \tfrac{1}{2})S\Delta t]}{[1 - (1 - \psi)S\Delta t]^2 + 4\lambda^2 \sin^2\theta}\right) S\Delta t \tag{7.100}$$

Since the factor proportional to $\sin^2\theta$ serves only to increase the denominator, we can omit it and say that

$$|G|^2 \leq 1 + \left(\frac{2[1 + (\psi - \tfrac{1}{2})S\Delta t]}{[1 - (1 - \psi)S\Delta t]^2}\right) S\Delta t \tag{7.101}$$

Clearly, the function in parentheses is bounded as $\Delta t \to 0$ as long as the denominator doesn't vanish, so that Equation (7.99) is satisfied provided:

$$\Delta t \leq \frac{1}{(1 - \psi)S}, \quad S > 0 \tag{7.102}$$

Although this analysis has been done for implicit Crank-Nicolson differencing of the convective term, the same result holds for explicit methods. While Equation (7.93) involves $k_j^{n+1}$, the terms in an explicit scheme can be rearranged to preserve its explicit nature. For example, if we use upwind differencing for the convective term in Equation (7.92), the discretized equation becomes

$$k_j^{n+1} = \frac{\left[1 + \psi S\Delta t - \frac{U\Delta t}{\Delta x}\right] k_j^n + \frac{U\Delta t}{\Delta x}k_{j-1}^n}{1 - (1 - \psi)S\Delta t} \tag{7.103}$$

We now have sufficient information to discuss the most suitable discretization approximations for source terms in both explicit and implicit methods. If second-order accuracy is required, as it would be for numerical simulation of an unsteady flow, $\psi$ must be 1/2. On the other hand, if only steady-state solutions are needed, we can take advantage of the fact that using $\psi = 0$ when $S < 0$ and $\psi = 1$ when $S > 0$ yields an unconditionally stable (albeit first-order accurate in time) scheme. In summary, the following has proven satisfactory for turbulence-model equations.

**Second-Order Time Accuracy — Conditional Stability**

$$Sk \doteq \frac{1}{2} S \left( k_j^n + k_j^{n+1} \right), \qquad \Delta t \leq \frac{2}{|S|} \qquad (7.104)$$

**First-Order Time Accuracy — Unconditional Stability**

$$Sk \doteq \begin{cases} Sk_j^{n+1} & \text{for} \quad S < 0 \\ \\ Sk_j^n & \text{for} \quad S > 0 \end{cases} \qquad (7.105)$$

All of the programs in Appendix C use Equation (7.105).

# 7.5 Block-Implicit Methods

The most efficient numerical methods currently available for complex flowfields are **block-implicit** methods. They differ from elementary implicit methods in one very important respect. Specifically, when an elementary implicit scheme is applied to a coupled set of equations, each equation is solved in sequence. In the context of a system of equations, this is usually referred to as a **sequentially-implicit** method. By contrast, a block-implicit scheme solves all of the equations simultaneously at each grid point. The block-implicit formulation, generally requiring inversion of block-tridiagonal matrices, entails more computational effort than a sequentially-implicit method. The additional computation at each grid point and timestep is usually compensated for by a dramatically improved convergence rate. Block-implicit solvers can achieve CFL numbers in excess of 100, and often converge in 100 to 200 timesteps for flows including boundary-layer separation. For example, using a block-implicit method, a supersonic two-dimensional shock-separated turbulent flow can be simulated on a 133 MHz Pentium-based microcomputer in about 15 minutes of CPU time. On the same computer, a similar computation would take about 2 hours using a sequentially-implicit method [Wilcox (1990)] and 6 hours using an explicit method [Wilcox (1974)].

As in the preceding section, we begin with a brief overview of block-implicit methods. For simplicity, we focus on a well-known one-dimensional system. The primary concern in this section is, of course, upon how turbulence-model source terms impact such methods.

Consider the one-dimensional conservation equations for flow of a viscous, perfect gas, written in vector form, viz.,

$$\frac{\partial \mathbf{Q}}{\partial t} + \frac{\partial}{\partial x}(\mathbf{F} - \mathbf{F}_v) = 0 \tag{7.106}$$

where

$$\mathbf{Q} = \left\{ \begin{array}{c} \bar{\rho} \\ \bar{\rho}\tilde{u} \\ \bar{\rho}E \end{array} \right\}, \quad \mathbf{F} = \left\{ \begin{array}{c} \bar{\rho}\tilde{u} \\ \bar{\rho}\tilde{u}^2 + P \\ (\bar{\rho}E + P)\tilde{u} \end{array} \right\}, \quad \mathbf{F}_v = \left\{ \begin{array}{c} 0 \\ \hat{\tau}_{xx} \\ \tilde{u}\hat{\tau}_{xx} - \hat{q}_x \end{array} \right\} \tag{7.107}$$

The quantities $\hat{\tau}_{xx}$ and $\hat{q}_x$ denote total stress and heat flux, respectively. Also, the total energy for one-dimensional flow is $E = \tilde{e} + \frac{1}{2}\tilde{u}^2$ and the pressure is given by $P = (\gamma - 1)\bar{\rho}\tilde{e}$.

The first step often taken in establishing a block-implicit scheme for this system of equations is to introduce a first-order backward-difference (implicit backward-Euler) scheme, which can be written symbolically as follows.

$$\frac{\mathbf{Q}^{n+1} - \mathbf{Q}^n}{\Delta t} + \left[ \frac{\partial}{\partial x}(\mathbf{F} - \mathbf{F}_v) \right]^{n+1} = 0 \tag{7.108}$$

Now, we expand the flux vectors $\mathbf{F}$ and $\mathbf{F}_v$ in a Taylor series about time level $n$, wherefore

$$\mathbf{F}^{n+1} \doteq \mathbf{F}^n + \frac{\partial \mathbf{F}}{\partial t}\Delta t + O\left[(\Delta t)^2\right] \tag{7.109}$$

and similarly for $\mathbf{F}_v$. Then, using the chain rule of calculus, we have

$$\frac{\partial \mathbf{F}}{\partial t} = \frac{\partial \mathbf{F}}{\partial \mathbf{Q}}\frac{\partial \mathbf{Q}}{\partial t} \tag{7.110}$$

where $\partial \mathbf{F}/\partial \mathbf{Q}$ is the **inviscid-flux Jacobian matrix**. The incremental change in the dependent-variable vector, $\Delta \mathbf{Q}$, is defined by

$$\Delta \mathbf{Q} = \mathbf{Q}^{n+1} - \mathbf{Q}^n \tag{7.111}$$

Since we approximate the unsteady term according to $\partial \mathbf{Q}/\partial t \doteq \Delta \mathbf{Q}/\Delta t$, we can rewrite Equation (7.109) as

$$\mathbf{F}^{n+1} \doteq \mathbf{F}^n + \frac{\partial \mathbf{F}}{\partial \mathbf{Q}}\Delta \mathbf{Q} + O\left[(\Delta t)^2\right] \tag{7.112}$$

Because of the prominent role played by $\Delta\mathbf{Q}$, this approach is usually referred to as the **delta formulation**.

Finally, we must introduce a discretization approximation for the spatial derivatives of the vectors $\mathbf{F}$ and $\mathbf{F}_v$. In general, this means forming a matrix that multiplies $(\mathbf{F} - \mathbf{F}_v)$, and yields a desired degree of accuracy. Details of this matrix are unimportant for our discussion, and it is sufficient to introduce symbolic notation with the understanding that an approximation to spatial differentiation is implied. Thus, we introduce a finite-difference matrix operator, $\delta_x$, so that

$$\left[\frac{\partial}{\partial x}(\mathbf{F} - \mathbf{F}_v)\right]^{n+1} \doteq \delta_x(\mathbf{F}^n - \mathbf{F}_v^n) + \delta_x\left(\frac{\partial\mathbf{F}}{\partial\mathbf{Q}} - \frac{\partial\mathbf{F}_v}{\partial\mathbf{Q}}\right)\Delta\mathbf{Q} \qquad (7.113)$$

where $\partial\mathbf{F}_v/\partial\mathbf{Q}$ is the **viscous-flux Jacobian matrix**. Collecting all of this, we arrive at the symbolic form of a typical block-implicit method:

$$\left[\frac{I}{\Delta t} + \delta_x\left(\frac{\partial\mathbf{F}}{\partial\mathbf{Q}} - \frac{\partial\mathbf{F}_v}{\partial\mathbf{Q}}\right)\right]\Delta\mathbf{Q} = -\delta_x(\mathbf{F}^n - \mathbf{F}_v^n) \qquad (7.114)$$

where $I$ is the unit (identity) matrix. The matrix multiplying $\Delta\mathbf{Q}$ in Equation (7.114) is of block-tridiagonal form. In the present example, the blocks are 3 by 3, corresponding to the three equations being solved simultaneously at each mesh point.

Now, suppose we choose to use a two-equation turbulence model to determine the Reynolds stress, still considering one-dimensional flow for simplicity. The following three points that must be considered in modifying a block-implicit solution scheme.

1. Decide whether to solve all equations simultaneously or to solve the model equations and mean-flow equations sequentially.

2. If the preferred option is to solve all equations simultaneously, determine the changes to the flux-Jacobian matrices.

3. Make provision for handling source terms.

In principle, solving all equations simultaneously will yield the most rapidly convergent scheme in the number of iterations, but not necessarily in CPU time. However, the coupling between the turbulence-model equations and the mean-flow equations appears to be relatively weak. The primary coupling from the turbulence-model equations to the mean-flow equations is through the diffusion terms in the mean-momentum and mean-energy equations, and the eddy viscosity is usually treated as a constant in forming the viscous-flux Jacobian matrix. Limited experience to date seems to indicate there is little advantage to solving all equations simultaneously as opposed to solving the model equations and mean-flow equations sequentially.

If all equations are solved simultaneously, the basic system of equations for the $k$-$\omega$ model would be as follows:

$$\frac{\partial \mathbf{Q}}{\partial t} + \frac{\partial}{\partial x}(\mathbf{F} - \mathbf{F}_v) = \mathbf{S} \tag{7.115}$$

where the dependent-variable and inviscid-flux vectors are

$$\mathbf{Q} = \left\{ \begin{array}{c} \bar{\rho} \\ \bar{\rho}\tilde{u} \\ \bar{\rho}E \\ \bar{\rho}k \\ \bar{\rho}\omega \end{array} \right\}, \qquad \mathbf{F} = \left\{ \begin{array}{c} \bar{\rho}\tilde{u} \\ \bar{\rho}\tilde{u}^2 + P \\ (\bar{\rho}E + P)\tilde{u} \\ \bar{\rho}\tilde{u}k \\ \bar{\rho}\tilde{u}\omega \end{array} \right\} \tag{7.116}$$

The viscous-flux and source-term vectors are given by

$$\mathbf{F}_v = \left\{ \begin{array}{c} 0 \\ \frac{4}{3}\mu\frac{\partial\tilde{u}}{\partial x} + \bar{\rho}\tau_{xx} \\ \tilde{u}(\frac{4}{3}\mu\frac{\partial\tilde{u}}{\partial x} + \bar{\rho}\tau_{xx}) - \hat{q}_x \\ (\mu+\sigma^*\mu_T)\frac{\partial k}{\partial x} \\ (\mu+\sigma\mu_T)\frac{\partial\omega}{\partial x} \end{array} \right\}, \qquad \mathbf{S} = \left\{ \begin{array}{c} 0 \\ 0 \\ 0 \\ \bar{\rho}\tau_{xx}\frac{\partial\tilde{u}}{\partial x} - \beta^*\bar{\rho}\omega k \\ \alpha\left(\frac{\omega}{k}\right)\bar{\rho}\tau_{xx}\frac{\partial\tilde{u}}{\partial x} - \beta\bar{\rho}\omega^2 \end{array} \right\} \tag{7.117}$$

There are two places where the turbulence kinetic energy appears that have an impact on the flux-Jacobian matrices. Specifically, the specific total energy, $E$, should be written as

$$E = \tilde{e} + \frac{1}{2}\tilde{u}^2 + k \tag{7.118}$$

and the Reynolds-stress tensor is

$$\bar{\rho}\tau_{xx} = \frac{4}{3}\mu_T\frac{\partial\tilde{u}}{\partial x} - \frac{2}{3}\bar{\rho}k \tag{7.119}$$

Hence, since the vector $\mathbf{Q}$ contains $\bar{\rho}k$ as one of its elements, the inviscid- and viscous-flux Jacobian matrices must be evaluated from scratch. Some of the original 9 elements appropriate for laminar flow or an algebraic model will be affected by the appearance of $k$ in $E$ and $\hat{\tau}_{xx}$. For this system, the inviscid-flux Jacobian matrix assumes the following form:

$$\frac{\partial\mathbf{F}}{\partial\mathbf{Q}} = \begin{bmatrix} 0 & 1 & 0 & 0 & 0 \\ \left(\frac{\gamma-3}{2}\right)\tilde{u}^2 & (3-\gamma)\tilde{u} & (\gamma-1) & -(\gamma-1) & 0 \\ -\left[H - \frac{\gamma-1}{2}\tilde{u}^2\right]\tilde{u} & \left[H - (\gamma-1)\tilde{u}^2\right] & \gamma\tilde{u} & -(\gamma-1)\tilde{u} & 0 \\ -\tilde{u}k & k & 0 & \tilde{u} & 0 \\ -\tilde{u}\omega & \omega & 0 & 0 & \tilde{u} \end{bmatrix} \tag{7.120}$$

where $H$ is the specific total enthalpy defined by

$$H = \tilde{h} + \frac{1}{2}\tilde{u}^2 + k \tag{7.121}$$

As shown in Equation (7.120), the first two components on row 3 involve $H$, and are thus affected by $k$. In modifying an existing computer program based on this block-implicit scheme, all that would be required to modify the inviscid-flux Jacobian matrix components would be to have $H$ appear as indicated, and to include $k$ in the computation of $H$.

By contrast, if we solve the mean-flow and turbulence-model equations sequentially, we retain the original conservation equations [Equation (7.106)]. All of the flux-Jacobian matrices and, in fact, the entire algorithm remain the same. To determine $k$ and $\omega$, we then consider the following vector equation:

$$\frac{\partial \mathbf{q}}{\partial t} + \frac{\partial}{\partial x}(\mathbf{f} - \mathbf{f}_v) = \mathbf{s} \tag{7.122}$$

where

$$\mathbf{q} = \left\{ \begin{array}{c} \bar{\rho}k \\ \bar{\rho}\omega \end{array} \right\}, \quad \mathbf{f} = \left\{ \begin{array}{c} \bar{\rho}\tilde{u}k \\ \bar{\rho}\tilde{u}\omega \end{array} \right\}, \quad \mathbf{f}_v = \left\{ \begin{array}{c} (\mu + \sigma^* \mu_T)\frac{\partial k}{\partial x} \\ (\mu + \sigma \mu_T)\frac{\partial \omega}{\partial x} \end{array} \right\} \tag{7.123}$$

$$\mathbf{s} = \left\{ \begin{array}{c} \bar{\rho}\tau_{xx}\frac{\partial \tilde{u}}{\partial x} - \beta^* \bar{\rho}\omega k \\ \alpha\left(\frac{\omega}{k}\right)\bar{\rho}\tau_{xx}\frac{\partial \tilde{u}}{\partial x} - \beta\bar{\rho}\omega^2 \end{array} \right\} \tag{7.124}$$

Consistent with the block-implicit approach, we linearize the flux and source vectors according to

$$(\mathbf{f} - \mathbf{f}_v)^{n+1} \doteq (\mathbf{f} - \mathbf{f}_v)^n + \left(\frac{\partial \mathbf{f}}{\partial \mathbf{q}} - \frac{\partial \mathbf{f}_v}{\partial \mathbf{q}}\right)\Delta\mathbf{q} \tag{7.125}$$

$$\mathbf{s} \doteq \mathbf{s}^n + \frac{\partial \mathbf{s}}{\partial \mathbf{q}}\Delta\mathbf{q} \tag{7.126}$$

where $\partial \mathbf{s}/\partial \mathbf{q}$ is the **source-Jacobian matrix**. The flux-Jacobian matrices are generally much simpler than their counterparts in the mean-flow equations. For example, the inviscid-flux Jacobian matrix is

$$\frac{\partial \mathbf{f}}{\partial \mathbf{q}} = \left[ \begin{array}{cc} \tilde{u} & 0 \\ 0 & \tilde{u} \end{array} \right] \tag{7.127}$$

This brings us to the all important question of how to handle the source-term vector **s**. Several prescriptions are possible, and the primary considerations are to: maintain numerical stability; achieve rapid convergence rate; and guarantee

that $k$ and $\omega$ are positive definite. Wilcox (1991) has found the following linearization of the source terms to be quite satisfactory for the $k$-$\omega$ model, within
the framework of MacCormack's (1985) block-implicit method. Specifically, the
source-term vector is rearranged as follows.

$$\mathbf{s} = \left\{ \begin{array}{c} \bar{\rho}\tau_{xx}\frac{\partial\tilde{u}}{\partial x} - \beta^*\left(\frac{\omega}{k}\right)\frac{(\bar{\rho}k)^2}{\bar{\rho}} \\[2mm] \alpha\left(\frac{\omega}{k}\right)\bar{\rho}\tau_{xx}\frac{\partial\tilde{u}}{\partial x} - \beta\frac{(\bar{\rho}\omega)^2}{\bar{\rho}} \end{array} \right\} \tag{7.128}$$

Then, treating both $\bar{\rho}\tau_{xx}\partial\tilde{u}/\partial x$ and $\omega/k$ as constants in computing the source-
Jacobian matrix, we arrive at

$$\mathbf{s}^n = \left\{ \begin{array}{c} \bar{\rho}\tau_{xx}\frac{\partial\tilde{u}}{\partial x} - \beta^*\bar{\rho}\omega k \\[2mm] \alpha\left(\frac{\omega}{k}\right)\bar{\rho}\tau_{xx}\frac{\partial\tilde{u}}{\partial x} - \beta\bar{\rho}\omega^2 \end{array} \right\}, \quad \frac{\partial\mathbf{s}}{\partial\mathbf{q}} = \left[ \begin{array}{cc} -2\beta^*\omega & 0 \\ 0 & -2\beta\omega \end{array} \right] \tag{7.129}$$

In this treatment of the source-term vector the production terms are evaluated
explicitly (i.e., computed at time level $n$), and the dissipation terms are treated
implicitly (computed at time level $n + 1$). The block-tridiagonal scheme for the
turbulence-model equations becomes

$$\left[ \frac{I}{\Delta t} + \delta_x\left(\frac{\partial\mathbf{f}}{\partial\mathbf{q}} - \frac{\partial\mathbf{f}_v}{\partial\mathbf{q}}\right) - \frac{\partial\mathbf{s}}{\partial\mathbf{q}} \right]\Delta\mathbf{q} = -\delta_x\left(\mathbf{f}^n - \mathbf{f}_v^n\right) + \mathbf{s}^n \tag{7.130}$$

Since $\partial\mathbf{s}/\partial\mathbf{q}$ is a diagonal matrix and its diagonal elements are always negative, its contribution is guaranteed to enhance diagonal dominance of the matrix
multiplying $\Delta\mathbf{q}$. Additionally, Spalart and Allmaras (1992) show that this form
guarantees that $k$ and $\omega$ (or $\epsilon$ for a $k$-$\epsilon$ model) will always be positive.

However, Spalart and Allmaras also point out that in regions where production
and dissipation are both large and dominate the overall balance of terms in
the equation, this form can result in slow convergence. This appears to be a
more serious problem for the $k$-$\epsilon$ model than it is for the $k$-$\omega$ model. Wilcox
(1991), for example, has shown that the scheme described above yields very
rapid convergence in flows with attached equilibrium boundary layers and in
flows with large regions of separation. The procedure recommended by Spalart
and Allmaras is similar to the procedure recommended for elementary implicit
methods in Equation (7.105). That is, they recommend linearizing the source
term according to

$$\mathbf{s} \doteq \mathbf{s}^n + \text{neg}\left(\frac{\partial\mathbf{s}}{\partial\mathbf{q}}\right)\Delta\mathbf{q} \tag{7.131}$$

where the function $\text{neg}(x)$ is defined as

$$\text{neg}(x) = \left\{ \begin{array}{ll} x, & x < 0 \\ 0, & x \geq 0 \end{array} \right. \tag{7.132}$$

The production terms are then included in computing the source-Jacobian matrix. The neg operator is understood to apply to each element of the resulting (diagonal) matrix. Thus, as long as dissipation exceeds production, both production and dissipation are treated implicitly, and explicitly when production exceeds dissipation. Huang and Coakley (1992) have successfully applied a linearization similar to that recommended by Spalart and Allmaras. Gerolymos (1990) and Shih et al. (1993) also offer interesting information regarding stiffness and numerical issues resulting from source terms in turbulence-model equations.

# 7.6 Solution Convergence and Grid Sensitivity

Regardless of the application, there is a need for control of numerical accuracy in CFD [Roache (1990)]. This need is just as critical in CFD work as it is in experiments where the experimenter is expected to provide estimates for the accuracy of his or her measurements. All CFD texts of any value stress this need.

One key issue determining numerical accuracy is **iteration convergence**. Most numerical methods used in CFD applications require many iterations to converge. The iteration convergence error is defined as the difference between the current iterate and the exact solution to the difference equations. Often, the difference between successive iterates is used as a measure of the error in the converged solution, although this in itself is inadequate. A small relaxation factor can always give a false indication of convergence [Anderson et al. (1984)]. Whatever the algorithm is, you should always be careful to check that a converged solution has been obtained. This can be done by trying a stricter than usual convergence criterion, and demonstrating that there is a negligible effect on the solution. Most reputable engineering journals require demonstration of iteration convergence and grid independence as a condition for publication. This is not specific to turbulence-model applications — all of the usual criteria for standard CFD applications apply.

Specific to turbulence-model computations, the approach to iteration convergence often is more erratic, and typically much slower, than for laminar-flow computations. A variety of factors including stiffness and nonlinearity of the equations, as well as the severely stretched finite-difference grids needed to resolve thin viscous layers, yield less rapid and less monotone convergence. Ferziger (1989) explains the slow convergence often observed in terms of the eigenvalues of the matrix system corresponding to the discretized equations. He notes that any iteration scheme for a linear system can be written as

$$\phi^{n+1} = A\phi^n + S \qquad (7.133)$$

where $\phi^n$ is the solution after the $n^{th}$ iteration, $A$ is a matrix, and $S$ is a source term. He then shows that the actual solution error is given by

$$\phi_{exact} - \phi^n \approx \frac{\phi^{n+1} - \phi^n}{1 - \lambda_{max}} \tag{7.134}$$

where $\phi_{exact}$ denotes the **exact solution to the discretized equations** and $\lambda_{max}$ is the largest eigenvalue of the matrix $A$. Of course, all eigenvalues of $A$ must be less than 1 for the solution to converge. This result shows that the solution error is larger than the difference between iterates. Furthermore, the closer $\lambda_{max}$ is to 1, the larger the ratio of solution error to the difference between iterates. In other words, the slower the rate of convergence of the method, the smaller the difference between iterates must be to guarantee iteration convergence.

A second key issue is **grid convergence** or **grid independence**. Because of the finite size of finite-difference cells, discretization errors exist that represent the difference between the solution to the difference equations and the **exact (continuum) solution to the differential equations**. It is important to know the magnitude of these discretization errors and to insure that a fine enough grid has been used to reduce the error to an acceptable level.

As with iteration convergence, all CFD work should demonstrate grid convergence, regardless of what equations are being solved. In most engineering journals, it is no longer sufficient to publish results performed on a single fixed grid. While grid sensitivity studies should be done for all CFD work, they are even more crucial for turbulence-model computations because of the need to separate numerical error from turbulence-model error. This issue came into sharp focus at the 1980-81 AFOSR-HTTM-Stanford Conference on Complex Turbulent Flows [see Kline, Cantwell, and Lilley (1981)]. Clearly, no objective evaluation of the merits of different turbulence models can be made unless the discretization error of the numerical algorithm is known.

The most common way to demonstrate grid convergence is to repeat a computation on a grid with twice as many grid points, and compare the two solutions. If computer resources are unavailable to facilitate a grid doubling, a grid halving is also appropriate, although the error bounds will not be as sharp. Using results for two different grids, techniques such as **Richardson extrapolation** [see Roache (1972)] can be used to determine discretization error. This method is very simple to implement, and should be used whenever possible. For a second-order accurate method with central differences, Richardson extrapolation assumes the error, $E_h \equiv \phi_{exact} - \phi_h$, where $\phi_h$ denotes the solution when the grid-point spacing is $h$, can be expanded as a Taylor series in $h$, wherefore

$$E_h = e_2 h^2 + e_4 h^4 + e_6 h^6 + \cdots \tag{7.135}$$

Note that for three-point upwind differences the leading term is still $e_2 h^2$, but the next term is $e_3 h^2$, and Richardson extrapolation is only $O(h^3)$ rather than

$O(h^4)$. By hypothesis, the $e_i$ are, at worst, functions of the coordinates, but are nevertheless independent of $h$. Now, if we halve the number of grid points so that $h$ is doubled, the error is given by

$$E_{2h} = 4e_2h^2 + 16e_4h^4 + 64e_6h^6 + \cdots \qquad (7.136)$$

For small values of $h$, we can drop all but the leading terms, whence the discretization error is given by

$$E_h \approx \frac{1}{3}(\phi_h - \phi_{2h}) \qquad (7.137)$$

As a final comment, Richardson extrapolation has limitations. First, if it is applied to primitive variables such as velocity and internal energy, its implications regarding momentum and energy conservation may be inaccurate. Second, the method implicitly assumes the solution has continuous derivatives to all orders. Hence, its results are not meaningful near shock waves or turbulent/nonturbulent interfaces of the type discussed in Subsection 7.2.2.

There is another grid-related factor affecting solution accuracy. In order to resolve thin viscous layers, for example, highly stretched grids are normally used. Conventional central-difference approximations are only first-order accurate on such a grid, and care must be taken to account for this. Also, the location of the grid point nearest the surface has a nontrivial effect on the accuracy of skin friction and surface heat flux. Wilcox (1989), for example, has found that grid-insensitive computations using wall functions that account for pressure gradient [e.g., Equation (5.117)] can be obtained with block-implicit methods provided:

$$10 < y_2^+ < 100, \quad \text{(wall functions)} \qquad (7.138)$$

where $y_2^+$ is the sublayer-scaled value of the first grid point above the surface. This range appears to hold for boundary-layer computations as well [Chambers and Wilcox (1977)], again provided pressure gradient is accounted for. When turbulence-model equations are integrated through the viscous sublayer, many researchers have shown that it is imperative to require:

$$y_2^+ < 1, \quad \text{(integration through the sublayer)} \qquad (7.139)$$

When these limits are not adhered to, consistent with the discussion in Subsection 7.2.1, solution errors throughout the boundary layer generally are large.

# Problems

**7.1** For a Mach 3 turbulent flat-plate boundary layer, it is a fact that $M c_f Re_L \approx Re_{\delta*}$.

(a) In the viscous sublayer, the appropriate scaling for the specific dissipation rate is $\omega \sim u_\tau^2/\nu$. Noting that $u_\tau \approx U\sqrt{c_f}$, express the ratio of $t_{diss}$ to $t_{wave}$ as a function of $Re_{\delta*}$ in the sublayer.

(b) In the defect layer, the appropriate scaling for the specific dissipation rate is given by $\omega \sim u_\tau/\Delta$ where $\Delta = U\delta^*/u_\tau$. Express the ratio of $t_{diss}$ to $t_{wave}$ as a function of $Re_{\delta*}$ in the defect layer.

(c) Comment on the implications of your estimates in Parts (a) and (b).

**7.2** Determine whether or not the following systems of equations are stiff with regard to the specified initial conditions.

(a)

$$\frac{d}{dt}\left\{\begin{array}{c} x \\ y \end{array}\right\} = \left[\begin{array}{cc} -3 & 4 \\ 4 & 3 \end{array}\right]\left\{\begin{array}{c} x \\ y \end{array}\right\}, \qquad \left\{\begin{array}{c} \dot{x}(0) \\ \dot{y}(0) \end{array}\right\} = -5\left\{\begin{array}{c} x(0) \\ y(0) \end{array}\right\}$$

(b)

$$\frac{d}{dt}\left\{\begin{array}{c} x \\ y \end{array}\right\} = \left[\begin{array}{cc} -3 & 1 \\ 4 & -3 \end{array}\right]\left\{\begin{array}{c} x \\ y \end{array}\right\}, \qquad \left\{\begin{array}{c} \dot{x}(0) \\ \dot{y}(0) \end{array}\right\} = -\left\{\begin{array}{c} x(0) \\ y(0) \end{array}\right\}$$

**7.3** Consider the high-Reynolds-number $k$-$\omega$ model's near-wall variation of specific dissipation rate, $\omega$, for a rough wall, i.e.,

$$\omega = \frac{\omega_w}{[1 + Ay]^2}, \qquad A = \sqrt{\frac{\beta_o \omega_w}{6\nu_w}}$$

(a) Assuming equally-spaced grid points, show that the central-difference approximation to $d^2\omega/dy^2$ at the first grid point above the surface (i.e., at $y = \Delta y$) is given by

$$\left(\frac{d^2\omega}{dy^2}\right)_2 \approx \Phi(\Delta y)\left(\frac{d^2\omega}{dy^2}\right)_{exact}$$

where

$$\Phi(\Delta y) = \frac{[1 + A\Delta y]^2[1 + 2A\Delta y + \frac{2}{3}(A\Delta y)^2]}{[1 + 2A\Delta y]^2}$$

(b) Assuming a slightly-rough wall so that $\omega_w = 2500\nu_w/k_s^2$ and using $\beta_o = 9/125$, show that

$$A\Delta y = \sqrt{30}\frac{\Delta y^+}{k_s^+}$$

(c) Determine the percentage error introduced by the central-difference approximation in computing $d^2\omega/dy^2$ when we assume a hydraulically-smooth wall with $k_s^+ = 5$, and set $\Delta y^+ = 1/2$.

**7.4** This problem shows that while trapezoidal-rule integration is second-order accurate for a piecewise continuous function with a discontinuous first derivative, the truncation error depends upon placement of the nodes. Using the trapezoidal rule, the integral of a function $f(x)$ is

$$\int_a^b f(x)\,dx \approx \sum_{k=1}^N f(x_k)\Delta x + \frac{1}{2}[f(a) - f(b)]\Delta x$$

where

$$x_k = k\Delta x \quad \text{and} \quad \Delta x = \frac{b-a}{N}$$

Consider the following piecewise continuous function $f(x)$:

$$f(x) = \begin{cases} x^2, & 0 \le x \le 1 \\ 1, & 1 < x \le 2 \end{cases}$$

Note that a node lies at $x = 1$ only for even values of $N$.

(a) Verify that the exact integral of $f(x)$ for $x$ ranging from 0 to 2 is

$$I \equiv \int_0^2 f(x)\,dx = \frac{4}{3}$$

(b) Assuming $N$ is odd show that the trapezoidal rule yields

$$I \approx \frac{4}{3}\left[1 - \frac{1}{16}(\Delta x)^2\right]$$

(c) Assuming $N$ is even show that the trapezoidal rule yields

$$I \approx \frac{4}{3}\left[1 + \frac{1}{8}(\Delta x)^2\right]$$

**HINT:** Use the fact that $\displaystyle\sum_{k=1}^M k^2 = \frac{1}{6}M(M+1)(2M+1)$.

**7.5** Consider the mixing-length model with $\ell_{mix} = \alpha\delta$, where $\alpha$ is a constant and $\delta$ is shear layer thickness.

(a) Assuming that $dU/dy > 0$, verify that according to the Rubel-Melnik transformation,

$$\nu_T = \ell_{mix}\sqrt{\frac{dU}{d\xi}}$$

(b) For flow near a turbulent/nonturbulent interface with constant entrainment velocity, $V < 0$, determine the velocity difference, $U_e - U$, as $y \to \delta$. Express your answer as a function of $|V|$, $\alpha$ and $y/\delta$.

**7.6** The object of this problem is to verify that Equation (7.56) is the solution to Equations (7.53) – (7.55).

(a) Integrate Equations (7.53) once and impose the freestream boundary condition [Equation (7.55)].

(b) Observing that $\sigma = \sigma^* = 1/2$ for Saffman's model, combine the $k$ and $\omega^2$ equations to show that

$$\frac{dk}{d\omega^2} = \frac{k}{\omega^2}$$

Solve this equation subject to the boundary conditions.

(c) Introduce the dimensionless variables

$$\bar{y} = \frac{|V|(\delta - y)}{\nu} \quad \text{and} \quad \bar{\omega} = \frac{V^2\omega}{K\nu}$$

and substitute the solution for $k$ into the equation for $\omega$. Set any arbitrary constant of integration equal to zero, and verify the solution for $y - \delta$.

(d) Letting $\bar{U} = U/|V|$, rewrite the momentum equation. Using the dimensionless equation for $\bar{\omega}$ derived in Part (c), verify that

$$\left(\frac{1+\bar{\omega}}{2+\bar{\omega}}\right)\frac{d\bar{U}}{d\bar{\omega}} = \frac{\bar{U} - \bar{U}_e}{\bar{\omega}}$$

and verify the solution for $U_e - U$.

**7.7** This problem illustrates how nonlinear terms affect numerical stability for parabolic marching methods. Consider the following limiting form of the $k$-$\omega$ model.

$$U\frac{\partial\omega}{\partial x} = \alpha\left(\frac{\partial U}{\partial y}\right)^2 - \beta_o\omega^2$$

We wish to analyze the stability of the solution to this equation under the following discretization approximations.

$$U\frac{\partial\omega}{\partial x} \doteq \frac{U}{\Delta x}\left[3\omega_{m+1}^i - 4\omega_m + \omega_{m-1}\right]$$

$$\alpha\left(\frac{\partial U}{\partial y}\right)^2 \doteq \frac{\alpha(\partial U/\partial y)^2}{\omega_{m+1}^{i-1}}\omega_{m+1}^i$$

$$\beta_o\omega^2 \doteq (1 + \psi_\omega)\beta_o\omega_{m+1}^{i-1}\omega_{m+1}^i - \psi_\omega\beta_o(\omega_{m+1}^{i-1})^2$$

(a) Assuming that $\omega_{m+1}^i$ is the sum of the exact solution to the discretized equation, $\omega_{m+1}$, and an error term, $\delta\omega^i$, viz.,

$$\omega_{m+1}^i = \omega_{m+1} + \delta\omega^i$$

linearize the discretized equation for $\omega$ and verify that

$$\frac{\delta\omega^i}{\delta\omega^{i-1}} = \frac{(\psi_\omega - 1)\beta_o\omega_{m+1}^2 - \alpha(\partial U/\partial y)^2}{3U\omega_{m+1}/\Delta x - \alpha(\partial U/\partial y)^2 + (\psi_\omega + 1)\beta_o\omega_{m+1}^2}$$

(b) Using the fact that $w_{m+1}$ satisfies the exact discretized equation, simplify the denominator and show that

$$\frac{\delta w^i}{\delta w^{i-1}} = \frac{(\psi_w - 1) - \alpha(\partial U/\partial y)^2/(\beta_o w^2_{m+1})}{\psi_w + U(4w_m - w_{m-1})/(\beta_o w^2_{m+1}\Delta x)}$$

(c) Assuming the term proportional to $U$ is negligible, determine the condition that $\psi_w$ must satisfy in order to insure that $|\delta w^i/\delta w^{i-1}| < 1$.

**7.8** Using von Neumann stability analysis, determine $G$ and any condition required for stability of the following finite-difference schemes. Assume $U > 0$, $\nu > 0$ and $S < 0$.

(a) Euler's method:

$$k_j^{n+1} = k_j^n - \frac{U\Delta t}{2\Delta x}\left(k_{j+1}^{n+1} - k_{j-1}^{n+1}\right)$$

(b) Richardson's method:

$$k_j^{n+1} = k_j^{n-1} + \frac{2\nu\Delta t}{(\Delta x)^2}\left(k_{j+1}^n - 2k_j^n + k_{j-1}^n\right)$$

(c) Crank and Nicolson's method:

$$k_j^{n+1} = k_j^n - \frac{U\Delta t}{4\Delta x}\left(k_{j+1}^{n+1} + k_{j+1}^n - k_{j-1}^{n+1} - k_{j-1}^n\right) + \frac{1}{2}S\Delta t\left(k_j^{n+1} + k_j^n\right)$$

**7.9** Using von Neumann stability analysis, determine $G$ and any condition required for stability of the following first-order accurate scheme applied to the inviscid Burgers' equation, $u_t + Uu_x = 0$:

$$u_j^{n+1} = u_j^n - \frac{U\Delta t}{2\Delta x}\left(u_{j+1}^{n+1} - u_{j-1}^n\right), \qquad U > 0$$

**7.10** Consider the following one-dimensional wave equation with source and diffusion terms.

$$\frac{\partial k}{\partial t} + U\frac{\partial k}{\partial x} = Sk + \nu\frac{\partial^2 k}{\partial x^2}$$

where $U > 0$, $\nu > 0$ and $S$ can be either positive or negative.

(a) Cast this equation in finite-difference form using Crank-Nicolson differencing and the following approximation for the source term.

$$Sk \doteq S\left[\psi k_j^n + (1 - \psi)k_j^{n+1}\right], \qquad 0 \le \psi \le 1$$

(b) Using von Neumann stability analysis, determine $G$ and any condition required for stability of this finite-difference scheme. How do your results compare to the analysis of Equation (7.92) in Section 7.4?

**7.11** Using von Neumann stability analysis, determine $G$ and any condition required for stability of *Lax's method* applied to the inviscid Burgers' equation, $u_t + Uu_x = 0$:

$$\frac{u_j^{n+1} - \frac{1}{2}\left(u_{j+1}^n + u_{j-1}^n\right)}{\Delta t} + U\frac{u_{j+1}^n - u_{j-1}^n}{2\Delta x} = 0, \qquad U > 0$$

**7.12** Verify that the dependent-variable and inviscid-flux vectors in Equation (7.116) can be written as

$$\mathbf{Q} = \left\{ \begin{array}{c} Q_1 \\ Q_2 \\ Q_3 \\ Q_4 \\ Q_5 \end{array} \right\}, \quad \mathbf{F} = \left\{ \begin{array}{c} Q_2 \\ (\frac{3-\gamma}{2})Q_2^2/Q_1 + (\gamma - 1)Q_3 - (\gamma - 1)Q_4 \\ \gamma Q_2 Q_3/Q_1 - (\frac{\gamma-1}{2})Q_2^3/Q_1^2 - (\gamma - 1)Q_2 Q_4/Q_1 \\ Q_2 Q_4/Q_1 \\ Q_2 Q_5/Q_1 \end{array} \right\}$$

and show that the flux-Jacobian matrix is given by Equation (7.120).

**7.13** Suppose a finite-difference method is only first-order accurate. When this is true, Richardson's estimate of the error must be revised. Assuming

$$E_h = e_1 h + e_2 h^2 + \cdots$$

propose an alternative to Equation (7.137).

**7.14** The following table represents partial results for one-dimensional finite-difference computations using a second-order accurate, time-marching method. The computations have been done on grids with 50, 100 and 200 points. Use Richardson extrapolation to estimate the discretization error at each point for the two finest grids. Based on your results, make a table of the results below and add a column with your best estimate of the continuum solution (grid-point spacing → 0) to the differential equation.

| $j$ | $\phi_{50}$ | $j$ | $\phi_{100}$ | $j$ | $\phi_{200}$ |
|---|---|---|---|---|---|
| 1 | 0.5592 | 1 | 0.5628 | 1 | 0.5607 |
| 2 | 0.5700 | 3 | 0.5740 | 5 | 0.5726 |
| 3 | 0.5737 | 5 | 0.5748 | 9 | 0.5745 |
| 4 | 0.5615 | 7 | 0.5557 | 13 | 0.5573 |

**7.15** The following table represents partial results for one-dimensional finite-difference computations using a second-order accurate, time-marching method. The computations have been done on grids with 50, 100 and 200 points. Use Richardson extrapolation to estimate the discretization error at each point for the two finest grids. Based on your results, make a table of the results below and add a column with your best estimate of the continuum solution (grid-point spacing → 0) to the differential equation.

| $j$ | $\phi_{50}$ | $j$ | $\phi_{100}$ | $j$ | $\phi_{200}$ |
|---|---|---|---|---|---|
| 1 | 3.00361 | 1 | 2.96624 | 1 | 2.95443 |
| 2 | 3.07446 | 3 | 3.06157 | 5 | 3.05965 |
| 3 | 3.09224 | 5 | 3.06523 | 9 | 3.07557 |
| 4 | 3.54523 | 7 | 3.53756 | 13 | 3.52365 |

# Chapter 8

# New Horizons

The focus of the previous chapters has been on approximate, Reynolds-averaged, models for use in general engineering applications. Throughout this text, we have stressed the virtue of using the minimum amount of complexity while capturing the essence of the relevant physics. This is the same notion that G. I. Taylor described as the "simple model/simple experiment" approach.

Nevertheless, no pretense has been made that any of the models devised in this spirit applies to all turbulent flows: such a "universal" model may not exist. We must always proceed with some degree of caution since there is no guarantee that Reynolds-averaged models are accurate beyond their established data base. Thus, while simplicity has its virtues for many practical engineering applications, there is a danger that must not be overlooked. Specifically, as quipped by H. L. Mencken...

> *"There is always an easy solution to every human problem — neat, plausible and wrong."*

This chapter discusses modern efforts that more directly address the physics of turbulence without introducing Reynolds-closure approximations. We begin by discussing Direct Numerical Simulation (DNS) in which the exact Navier-Stokes and continuity equations are solved, though currently at relatively low Reynolds numbers because of the limitations of present-day computers. Next, we turn to Large Eddy Simulation (LES) in which the largest eddies are computed exactly and the smallest, "subgrid-scale" (SGS) eddies are modeled, hopefully with a less critical impact on the simulation than in Reynolds-stress modeling. Finally, we discuss current efforts in chaos studies, and their possible relevance to turbulence.

# 8.1  Background Information

Before plunging into these topics, it is worthwhile to pause and review the key aspects of turbulence that we discussed in Chapters 1 and 2. It may even be helpful for the unhurried reader to revisit Sections 1.3 and 2.5 before proceeding. Note that in pursuing a more fundamental approach to turbulence in DNS and LES studies, we still have a need to understand important aspects of turbulence such as the roles played by the largest and smallest eddies and the cascade process. The reason for this need changes however. In developing a turbulence model, we are trying to mimic the physics in our mathematical formulation. As our understanding of turbulent-flow physics improves, so the quality of our approximations improves (assuming we make intelligent use of the improved understanding). Even in DNS we need some knowledge of turbulence physics to check for the physical soundness of the numerical results, for example, to be certain that inadequate resolution or even programming errors are not causing spurious results. The same applies to LES; additionally, formulating SGS models requires at least as detailed an understanding of turbulence physics as Reynolds-averaged models.

The first important point we must consider in DNS and LES is that of the smallest scales of turbulence. Our primary focus in devising Reynolds-averaged closure approximations has been on the dynamics of the largest eddies, which account for most of the transport of properties in a turbulent flow. Our use of dimensional analysis, in which molecular viscosity has been ignored, guarantees that the closure approximations involve length scales typical of the energy-bearing eddies whose Reynolds number — however defined — is much larger than unity except close to a solid surface, i.e. in the viscous sublayer, $y^+ < 3$, say. This is the reason that viscous-damping functions are often needed close to a solid boundary where the dissipating eddies dominate, and even the energy-bearing eddies have Reynolds numbers of order unity. DNS is supposed to resolve the whole range of eddy sizes, while in LES we try to resolve all the important (larger) eddies so that the SGS model for the small eddies does not have a critical influence on the overall results. In both cases we need to know the typical scales of the smallest eddies.

As shown in Subsection 1.3.3, the smallest scales of turbulence are the **Kolmogorov scales** of length, time and velocity, viz.,

$$\eta \equiv \left(\nu^3/\epsilon\right)^{1/4}, \quad \tau \equiv (\nu/\epsilon)^{1/2}, \quad \upsilon \equiv (\nu\epsilon)^{1/4} \qquad (8.1)$$

where $\nu$ is kinematic viscosity and $\epsilon$ is dissipation rate. Note that the Reynolds number $\upsilon\eta/\nu$ is equal to unity, which is plausible in view of the basic definition of Reynolds number as a ratio of inertial forces to viscous forces. Necessarily, inertial and viscous effects just balance in the smallest eddies (this is merely an order-of-magnitude argument, and $\upsilon\eta/\nu$ comes out as exactly unity simply because the above definitions contain no numerical factors). To relate the Kolmogorov length scale to the length scale we have been dealing with in standard turbulence models, consider the following. By hypothesis, we have been using the length scale appropriate to the energy-bearing eddies, $\ell$. This length scale is often chosen as the **integral length scale** in statistical turbulence theory, and is related to $\epsilon$ by Equation (4.10), so that

$$\frac{\eta}{\ell} \sim Re_T^{-3/4} \tag{8.2}$$

where $Re_T = k^{1/2}\ell/\nu$ is the usual turbulence Reynolds number. Since values of $Re_T$ in excess of $10^4$ are typical of fully-developed turbulent boundary layers and $\ell \sim 0.1\delta$ where $\delta$ is boundary-layer thickness, the Kolmogorov length scale, $\eta$, outside the viscous wall region is less than one ten-thousandth times the thickness of the boundary layer.

DNS and LES studies also make use of another length scale from the statistical theory of turbulence, the **Taylor microscale**, $\lambda$ [c.f., Tennekes and Lumley (1983) or Hinze (1975)]. The basic definition is

$$\lambda^2 = \frac{\overline{u'^2}}{\overline{(\partial u'/\partial x)^2}} \tag{8.3}$$

For locally isotropic turbulence (i.e. turbulence in which the small scales are statistically isotropic even if the large ones are not, which is usually the case at high Reynolds numbers), the exact expression for dissipation rate, $\epsilon$, leads to

$$\epsilon = 15\nu \overline{\left(\frac{\partial u'}{\partial x}\right)^2} \equiv 15\nu \frac{\overline{u'^2}}{\lambda^2} \tag{8.4}$$

Other definitions of $\lambda$ can be constructed by using different velocity components and gradients in the basic definition, but in locally-isotropic turbulence they are simply related. Using Equation (4.10), and assuming $k \sim \overline{u'^2}$, we conclude that

$$\frac{\lambda}{\ell} \sim Re_T^{-1/2} \quad \text{or} \quad \lambda \sim (\ell\eta^2)^{1/3} \tag{8.5}$$

Thus, in general we can say that for high-Reynolds-number turbulence there is a distinct separation of these scales, i.e.,

$$\eta \ll \lambda \ll \ell \tag{8.6}$$

Now the basic definition shows that $\lambda$ is a composite quantity, depending on properties of the large-scale eddies as well as the small ones. Unlike $\ell$ and $\eta$, it cannot be identified with any meaningful range of eddy sizes. However, results of numerical simulations are often characterized in terms of the **microscale Reynolds number**, $Re_\lambda$, defined by

$$Re_\lambda = k^{1/2}\lambda/\nu \tag{8.7}$$

Substitution for $\lambda$ from Equation (8.4) leads to

$$Re_\lambda \sim (k^{1/2}L_\epsilon/\nu)^{1/2} \tag{8.8}$$

where $L_\epsilon \equiv k^{3/2}/\epsilon$ is the "dissipation length scale", actually the typical length scale of the stress-bearing motion used implicitly in all two-equation models. Now, $L_\epsilon$ is of the same order as $\ell$ so it follows from Equation (8.8) that also

$$Re_\lambda \sim Re_T^{-1/2} \tag{8.9}$$

Thus although $\lambda$ is not a very meaningful length scale, $Re_\lambda$ is an alternative to the Reynolds number of the energy-containing eddies. Finally, the **eddy turnover time**, $\tau_{turnover}$, is simply the ratio of the macroscales for length, $\ell$ or $L_\epsilon$, and velocity, $k^{1/2}$, and is given by

$$\tau_{turnover} \sim \ell/k^{1/2} \sim L_\epsilon/k^{1/2} \tag{8.10}$$

The eddy turnover time is a measure of the time it takes an eddy to interact with its surroundings. As can be seen from the definition of $L_\epsilon$ it is also the reciprocal of the specific dissipation rate, $\omega \sim \epsilon/k$.

A second important consideration is the spectral representation of turbulence properties (see Subsection 1.3.4), which replaces the vague idea of "eddy size." If $\kappa$ denotes wavenumber, defined as $2\pi/$wavelength, and $E(\kappa)d\kappa$ is the turbulence kinetic energy contained between wavenumbers $\kappa$ and $\kappa + d\kappa$, we can write

$$k \equiv \frac{1}{2}\overline{u_i'u_i'} = \int_0^\infty E(\kappa)\,d\kappa \tag{8.11}$$

Recall that $k$ is half the trace of the autocorrelation tensor, $\mathcal{R}_{ij}$, defined in Equation (2.43), at zero time delay. Correspondingly, the **energy spectral density** or **energy spectrum function**, $E(\kappa)$, is related to the Fourier transform of half the trace of $\mathcal{R}_{ij}$. In general, we regard a spectral representation as a decomposition into wavenumbers ($\kappa$). In loose discussions of "eddy size," we regard the reciprocal of $\kappa$ as the eddy size associated with $\kappa$ – small $\kappa$ equals large wavelength equals large eddies, and conversely. Of course turbulence is not a superposition of simple waves; any definition of an "eddy" based on observed flow patterns will actually cover a range of wavenumbers, and is still vague. However the definition of spectral density and the associated analysis are precise. The present discussion is simplified: see Tennekes and Lumley (1983) for a detailed discussion of energy spectra.

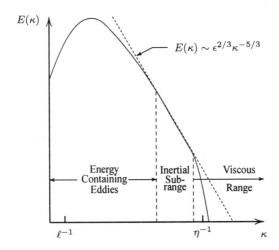

Figure 8.1: *Energy spectrum for a turbulent flow — log-log scales.*

Again using dimensional analysis, Section 1.3.4 shows that, for wavenumbers small enough that viscosity does not affect the motion, but large enough that the overall dimensions of the flow such as boundary-layer thickness do not matter,

$$E(\kappa) = C_K \epsilon^{2/3} \kappa^{-5/3}, \qquad \frac{1}{\ell} \ll \kappa \ll \frac{1}{\eta} \qquad (8.12)$$

where $C_K$ is the Kolmogorov constant. This is the famous Kolmogorov $-5/3$ law that characterizes the **inertial subrange**. Figure 8.1 shows a typical energy spectrum for a turbulent flow. With these preliminary remarks in hand, we are now in a position to discuss DNS and LES in the next two sections.

## 8.2 Direct Numerical Simulation

A direct numerical simulation, or DNS for short, means a complete three-dimensional and time-dependent solution of the Navier-Stokes and continuity equations. The value of such simulations is obvious: they are, in principle, numerically-accurate solutions of exact equations of motion and − in principle − the proper solution to the turbulence problem. From a practical standpoint, statistics computed from DNS results can be used to test proposed closure approximations in engineering models. At the most fundamental level, DNS can be used to obtain understanding of turbulence structure and processes that can be of value in developing turbulence-control methods (e.g., drag reduction) or prediction methods. DNS can also be viewed as an additional source of experimental data, taken with unobtrusive measuring techniques. This is especially useful for obtaining information about essentially-unmeasurable properties like pressure fluctuations.

All of these comments assume the DNS is free of significant numerical, and other, forms of error. This is a nontrivial consideration, and the primary concerns in DNS are related to numerical accuracy, specification of boundary and initial conditions, and making optimum use of available computer resources. In this section, we discuss these issues only briefly. For more detail at an introductory level, with extensive references to recent research work, see the review article by Moin and Mahesh (1998). As a final reminder, remember that even the numerical solution of the exact equations of motion requires detailed understanding of the physics of turbulence if the solutions are to be economical and accurate.

Estimating the number of grid points and timesteps needed to perform an accurate DNS reveals the complexity of the problem from a computational point of view. As an example, consider incompressible turbulent flow in a channel of height $H$. The computational domain must be of sufficient extent to accommodate the largest turbulence scales. In channel flow, eddies are elongated in the direction parallel to the channel walls, and their length $\Lambda$ is known to be about $2H$. Also, in principle, the grid must be fine enough to resolve the smallest eddies whose size is of the order of the Kolmogorov length scale, $\eta$. Assuming that at least 4 grid points in each direction are needed to resolve an eddy (since we need adequate resolution of derivatives), we estimate that the total number of grid points for uniform spacing, $N_{uniform}$, is

$$N_{uniform} \approx \left[ 4\frac{\Lambda}{\eta} \right]^3 = \left[ 8H \left( \frac{\epsilon}{\nu^3} \right)^{1/4} \right]^3 \tag{8.13}$$

Now, in channel flow, the average dissipation is $\epsilon \approx 2u_\tau^2 U_m/H$ where $U_m$ is the average velocity across the channel, and $U_m/u_\tau \approx 20$. Substituting these estimates into Equation (8.13), we arrive at

$$N_{uniform} \approx (110 Re_\tau)^{9/4}, \quad Re_\tau = \frac{u_\tau H/2}{\nu} \tag{8.14}$$

In practice, it is wasteful to use uniformly-spaced grid points since there are regions where $\epsilon$ is small and the Kolmogorov length scale is much larger than it is near the surface where $\epsilon$ is largest. By using stretched grids to concentrate points where the smallest eddies reside, experience [Moser and Moin (1984), Kim, Moin and Moser (1987)] shows that the factor of 110 in Equation (8.14) can be replaced by about 3. Thus, the actual number of grid points typically used in a DNS of channel flow, $N_{DNS}$, is

$$N_{DNS} \approx (3 Re_\tau)^{9/4} \tag{8.15}$$

Similarly, the timestep in the computation, $\Delta t$, should be of the same order as the Kolmogorov time scale, $\tau = (\nu/\epsilon)^{1/2}$. Based on results of the computations done by Kim, Moin and Moser (1987), the timestep must be

$$\Delta t \approx \frac{.003}{\sqrt{Re_\tau}} \frac{H}{u_\tau} \qquad (8.16)$$

To appreciate how prohibitive these constraints are, consider the channel-flow experiments done by Laufer (1951) at Reynolds numbers of 12,300, 30,800 and 61,600 and the experiment of Comte-Bellot (1963) at a Reynolds number of 230,000. Table 8.1 lists the number of grid points and timesteps required to perform a DNS, assuming the time required to reach a statistically-steady state is $100H/U_m \approx 5H/u_\tau$. Clearly, computer memory limitations make all but the lowest Reynolds numbers considered by Laufer impractical with the computers of the late 1990's. The development of massively-parallel machines over the last few years has reduced execution times, but storage is still a problem, both during the computation and for later archiving of "fields" of raw data at selected timesteps.

Table 8.1: *Grid Point/Timestep Requirements for Channel-Flow DNS and LES*

| $Re_H$ | $Re_\tau$ | $N_{DNS}$ | DNS Timesteps | $N_{LES}$ |
|--------|-----------|-----------|---------------|-----------|
| 12,300 | 360 | $6.7 \cdot 10^6$ | 32,000 | $6.1 \cdot 10^5$ |
| 30,800 | 800 | $4.0 \cdot 10^7$ | 47,000 | $3.0 \cdot 10^6$ |
| 61,600 | 1,450 | $1.5 \cdot 10^8$ | 63,000 | $1.0 \cdot 10^7$ |
| 230,000 | 4,650 | $2.1 \cdot 10^9$ | 114,000 | $1.0 \cdot 10^8$ |

The computations of Kim, Moin and Moser (1987) provide an example of the computer resources required for DNS of the geometrically simple case of channel flow. To demonstrate grid convergence of their methods, they compute channel flow with $Re_\tau = 180$, corresponding to $Re_H \approx 6,000$ using grids with $2 \cdot 10^6$ and $4 \cdot 10^6$ points. For the finer grid, the CPU time on a Cray X/MP supercomputer (about one fourth the time on a present-day personal computer: see problems section) was 40 seconds per timestep. The calculation was run for a total time $5H/u_\tau$, and required 250 CPU hours.

Both second-order accurate and fourth-order accurate numerical algorithms have been used in DNS research to advance the solution in time. There are two primary concerns regarding numerical treatment of the spatial directions. The first is achieving accurate representations of derivatives, especially at the smallest scales (or, equivalently, the highest wavenumbers). Spectral methods — Fourier series in the spatial directions — can be used to insure accurate computation of derivatives. Finite-difference methods usually underestimate derivatives of a given velocity field, leading to inaccuracies in the smallest (dissipating) scales. The dissipation as such is set by the rate of energy transfer from the larger eddies, so the underestimated derivatives are compensated by an excess in spectral density at the highest wavenumbers to achieve the right value for the dissipation as expressed by the right-hand side of Equation (4.6). This is usually just called "numerical dissipation," but is in no sense an addition to the dissipation rate set by the energy transfer. Thus, the first concern in demonstrating grid convergence of a DNS is to verify that the energy spectrum, $E(\kappa)$, displays a rapid decay, often referred to as the **rolloff**, near the Kolmogorov length scale, $\eta$. The second concern is to avoid a phenomenon known as **aliasing**. This occurs when nonlinear interactions among the resolved wavenumbers produce waves with wavenumbers greater than $\kappa_{max}$, which can be misinterpreted numerically. If special precautions are not taken, this can result in a spurious transfer of energy to small wavenumbers [Ferziger (1976)].

While spectral methods are more accurate for computing derivatives at the smallest scales, they are difficult to use with arbitrarily nonuniform grids. Because of the wish to extend DNS – and LES – to more realistic geometries, bringing the need for more complicated grids, there has been a general swing towards finite-difference methods, but a higher order of accuracy is needed than for spectral methods. "Unstructured" grids, now well established in conventional CFD [e.g., Venkatakrishnan (1996)], are being introduced into DNS and LES for complicated geometries, but they carry a further penalty in storage and CPU time.

In their grid-convergence study, Kim, Moin and Moser (1987) show that their energy spectra display the characteristic rolloff approaching wavenumber $\kappa \approx 1/\eta$ where $\eta$ is the Kolmogorov length scale. This corresponds to a wavelength of $2\pi\eta \approx 6\eta$, roughly the top of the dissipating range.

The primary difficulty with boundary conditions in any Navier-Stokes calculation, DNS, LES or Reynolds-averaged, is at open boundaries. Because of the elliptic nature of the problem, the flow at such boundaries depends on the unknown flow outside the computational domain. In LES and DNS this problem is circumvented by imposing periodic boundary conditions for directions in which the flow is statistically homogeneous (e.g., the streamwise and spanwise directions in channel flow). Most simulations done to date have been homogeneous or periodic in at least one spatial direction, which has the additional advantage that statistics can be obtained by averaging over the homogeneous direction as well as over time, thus reducing the time sample needed to get converged statistics.

Flows that grow in the streamwise direction in a nearly self-similar manner (e.g., equilibrium boundary layers) can be reduced to approximate homogeneity by a coordinate transformation [Spalart (1986), Spalart (1988), Spalart (1989)]. Alternatives are to use the results of a previous simulation to give the incoming flow at the upstream boundary [Le, Moin and Kim (1997)] or use a suitably-rescaled version of the outgoing flow at the downstream boundary [the "fringe" method of Spalart; see Bertolotti et al. (1992)]. Finally one can add a synthetic random fluctuating velocity field to a prescribed mean-velocity field. After a few eddy-turnover times, the correct statistics evolve, but this may correspond to an unacceptably large downstream distance. In nonperiodic flows the downstream boundary condition is usually taken as zero streamwise gradient of all variables. This is acceptable if the statistics of the real flow are changing slowly in the streamwise direction, because this implies negligible upstream influence – usually equivalent to validity of the boundary-layer approximation, which leads to parabolic equations.

Solid boundaries, where the no-slip velocity boundary condition applies, pose no special problems for DNS.

DNS results illustrate one of the curious features of turbulence and other chaotic systems. Suppose we generate a solution from a given set of initial conditions, and then repeat the computation with a very small perturbation in the initial conditions. We find that, after a few eddy-turnover times, the second solution, or **realization**, is very different from the first. However, in terms of all statistical measures, the two flows are identical! This is the classical problem of **predictability** discussed, for example, by Sandham and Kleiser (1992): see also Section 8.4. It is a real phenomenon and has nothing to do with numerical error. Also, it occurs in everyday life, although usually with finite initial perturbations. As a simple example, two strangers in a crowd, initially side by side, tend to drift apart – that is, a small difference in initial position tends to grow indefinitely, and the standard deviation of the difference, averaged over many trials, certainly grows. The public recognizes this in an empirical way: if one stranger steps on another's foot *twice*, the steppee is likely to suspect the stepper of doing it on purpose. Thus, while somewhat disconcerting to the mathematician, this phenomenon should come as no great surprise to the engineer.

DNS matured rapidly during the 1980's and continues to develop as more and more powerful computers appear, although Reynolds numbers are still well below those found in most branches of engineering. As an example, DNS data are currently available for the following flows, and the list of applications continues to grow.

- Curved channel flow

- Channel flow, with and without heat transfer — values of $Re_H$ as high as 13,750 have been achieved

- Two-dimensional boundary layers in various pressure gradients — values of $Re_\theta$ as high as 1,410 have been achieved

- Two-dimensional separating and reattaching flows

    (a) shallow separation bubble on a flat surface

    (b) flow over a backward-facing step

- Boundary layers on curved surfaces

- Two-dimensional time- and spatially-developing mixing layers

- Two-dimensional buoyant flows

- Two-dimensional homogeneously-strained flows

- Two-dimensional, constant-density homogeneous turbulent flows that include chemical reactions

- Three-dimensional homogeneous turbulent flows with miscellaneous extra strain rates

- Three-dimensional Ekman layer

- Three-dimensional flow over a swept wing

- Incompressible and compressible jets

- Acoustic fields generated by several turbulent flows

- Compressible homogeneous turbulent flows with mean shear

- Compressible homogeneous turbulent flows with bulk compression in one, two or three directions

- Transitional and turbulent compressible boundary layers and mixing layers

## 8.3  Large Eddy Simulation

A Large Eddy Simulation, or LES for short, is a computation in which the large eddies are computed and the smallest, **subgrid-scale (SGS)** eddies are modeled. The underlying premise is that the largest eddies are directly affected by the boundary conditions, carry most of the Reynolds stress, and must be computed. The small-scale turbulence is weaker, contributing less to the Reynolds stresses, and is therefore less critical. Also, it is more nearly isotropic and has nearly-universal characteristics; it is thus more amenable to modeling. Recent reviews include Ferziger (1996), Lesieur and Metais (1996) [who concentrate on work at the University of Grenoble] and Rodi (1997, 1998) on LES for flows with large regions of separation.

Because LES involves modeling the smallest eddies, the smallest finite-difference cells can be much larger than the Kolmogorov length, and much larger timesteps can be taken than are possible in a DNS. Hence, for a given computing cost, it is possible to achieve much higher Reynolds numbers with LES than with DNS, or conversely to obtain a solution at a given Reynolds number more cheaply. See Table 8.1 for a comparison of estimated DNS and LES grid point requirements in a simple flow. An actual example, for a more complex flow, is comparison of calculations of the flow over a backward-facing step by DNS [Le, Moin and Kim (1997)] and by LES [Akselvoll and Moin (1993)], both at the low Reynolds number of 5100 based on step height. The LES took 3 percent of the number of points taken by the DNS and the computer time taken was 2 percent of that for the DNS: agreement with experiment was equally good.

Aside from the issue of the need to resolve the smallest eddies, the comments regarding DNS numerics, boundary and initial conditions in the previous section hold for LES as well. The primary issue in accuracy remains that of computing derivatives for the smallest scales (highest wavenumbers) resolved. The ultimate test of grid convergence is again the requirement that excessive energy must not accumulate in the smallest scales. The primary requirement is to get the dissipation rate right; details of the dissipating eddies are unimportant in LES. [DNS nominally requires accurate simulation of the dissipating eddies, and the achievement of this in the classical channel simulation of Kim, Moin and Moser (1987) is verified by the accuracy of the dissipation-rate budget evaluated by Mansour, Kim and Moin (1988). However, marginally-resolved DNS often includes some numerical dissipation.] If spectral or pseudo-spectral methods are used, the same boundary-condition difficulties hold in both DNS and LES.

A major difficulty in "Large" Eddy Simulation is that near a solid surface all eddies are small — to the extent that the stress-bearing and dissipation ranges of eddy size overlap. If one requires LES to resolve most of the stress-bearing range, the grid spacing, and timestep, required by LES gradually fall towards that needed for full DNS as the surface is approached. This is, of course, a serious limitation on Reynolds number for LES, and later we will discuss the ways in which it can be avoided.

## 8.3.1   Filtering

To understand the primary difference between DNS and LES, we must introduce the concept of **filtering**. Note first that the values of flow properties at discrete points in a numerical simulation represent averaged values. To see this explicitly, consider the central-difference approximation for the first derivative of a continuous variable, $u(x)$, in a grid with points spaced a distance $h$ apart. We can write this as follows.

$$\frac{u(x+h) - u(x-h)}{2h} = \frac{d}{dx}\left[\frac{1}{2h}\int_{x-h}^{x+h} u(\xi)\, d\xi\right] \qquad (8.17)$$

This shows that the central-difference approximation can be thought of as an operator that **filters out scales smaller than the mesh size**. Furthermore, the approximation yields the derivative of an averaged value of $u(x)$.

There are many kinds of filters that can be used. The simplest type of filter is the **volume-average box filter** implemented by Deardorff (1970), one of the earliest LES researchers. The filter is:

$$\bar{u}_i(\mathbf{x}, t) = \frac{1}{\Delta^3} \int_{x-\frac{1}{2}\Delta x}^{x+\frac{1}{2}\Delta x} \int_{y-\frac{1}{2}\Delta y}^{y+\frac{1}{2}\Delta y} \int_{z-\frac{1}{2}\Delta z}^{z+\frac{1}{2}\Delta z} u_i(\boldsymbol{\xi}, t)\, d\xi\, d\eta\, d\zeta \qquad (8.18)$$

The quantity $\bar{u}_i$ denotes the **resolvable-scale filtered velocity**. The **subgrid-scale (SGS)** velocity, $u_i'$, and the **filter width**, $\Delta$, are given by

$$u_i' = u_i - \bar{u}_i \quad \text{and} \quad \Delta = (\Delta x \Delta y \Delta z)^{1/3} \qquad (8.19)$$

Leonard (1974) defines a generalized filter as a convolution integral, viz.,

$$\bar{u}_i(\mathbf{x}, t) = \iiint G(\mathbf{x} - \boldsymbol{\xi}; \Delta)\, u_i(\boldsymbol{\xi}, t)\, d^3\boldsymbol{\xi} \qquad (8.20)$$

The **filter function**, $G$, is normalized by requiring that

$$\iiint G(\mathbf{x} - \boldsymbol{\xi}; \Delta)\, d^3\boldsymbol{\xi} = 1 \qquad (8.21)$$

In terms of the filter function, the volume-average box filter as defined in Equation (8.18) is:

$$G(\mathbf{x} - \boldsymbol{\xi}; \Delta) = \begin{cases} 1/\Delta^3, & |x_i - \xi_i| < \Delta x_i/2 \\ 0, & \text{otherwise} \end{cases} \qquad (8.22)$$

The Fourier transform of Equation (8.20) is $\overline{\mathcal{U}}_i(\kappa, t) = \mathcal{G}(\kappa)\mathcal{U}_i(\kappa, t)$ where $\mathcal{U}_i$ and $\mathcal{G}$ represent the Fourier transforms of $u_i$ and $G$. Fourier spectral methods implicitly filter with

$$\mathcal{G}(\kappa; \Delta) = 0 \quad \text{for} \quad |\kappa| > \kappa_{max} = 2\pi/\Delta \qquad (8.23)$$

As an example, Orszag et al. [see Ferziger (1976)] use the **Fourier cutoff filter**, i.e.,

$$G(\mathbf{x} - \boldsymbol{\xi}; \Delta) = \frac{1}{\Delta^3} \prod_{i=1}^{3} \frac{\sin{(x_i - \xi_i)/\Delta}}{(x_i - \xi_i)/\Delta} \qquad (8.24)$$

The **Gaussian filter** [Ferziger (1976)] is popular in LES research, and is defined by

$$G(\mathbf{x}\text{-}\boldsymbol{\xi}; \Delta) = \left(\frac{6}{\pi\Delta^2}\right)^{3/2} \exp\left(-6\frac{|\mathbf{x} - \boldsymbol{\xi}|^2}{\Delta^2}\right) \qquad (8.25)$$

Many other filters have been proposed and used, some of which are neither isotropic nor homogeneous. In all cases however, the filter introduces a scale $\Delta$ that represents the smallest turbulence scale allowed by the filter.

The filter provides a formal definition of the averaging process and separates the **resolvable scales** from the **subgrid scales**. We use filtering to derive the **resolvable-scale equations**. For incompressible flow, the continuity and Navier-Stokes equations assume the following form.

$$\frac{\partial \overline{u}_i}{\partial x_i} = 0 \tag{8.26}$$

$$\frac{\partial \overline{u}_i}{\partial t} + \frac{\partial}{\partial x_j}\left(\overline{u_i u_j}\right) = -\frac{1}{\rho}\frac{\partial \overline{p}}{\partial x_i} + \nu\frac{\partial^2 \overline{u}_i}{\partial x_k \partial x_k} \tag{8.27}$$

Now, the convective flux is given by

$$\overline{u_i u_j} = \overline{u}_i \overline{u}_j + L_{ij} + C_{ij} + R_{ij} \tag{8.28}$$

where

$$\left.\begin{array}{l} L_{ij} = \overline{\overline{u}_i \overline{u}_j} - \overline{u}_i \overline{u}_j \\[2mm] C_{ij} = \overline{\overline{u}_i u_j'} + \overline{\overline{u}_j u_i'} \\[2mm] R_{ij} = \overline{u_i' u_j'} \end{array}\right\} \tag{8.29}$$

Note that, with the exception of the Fourier cutoff filter [Equation (8.24)], filtering differs from standard averaging in one important respect:

$$\overline{\overline{u}}_i \neq \overline{u}_i \tag{8.30}$$

i.e., a second averaging yields a different result from the first averaging. The tensors $L_{ij}$, $C_{ij}$ and $R_{ij}$ are known as the **Leonard stress, cross-term stress** and the **SGS Reynolds stress**, respectively.

Leonard (1974) shows that the Leonard-stress term removes significant energy from the resolvable scales. It can be computed directly and need not be modeled. This is sometimes inconvenient, however, depending on the numerical method used. Leonard also demonstrates that since $\overline{u}_i$ is a smooth function, $L_{ij}$ can be computed in terms of its Taylor-series expansion, the first term of which is

$$L_{ij} \approx \frac{\gamma_\ell}{2}\nabla^2\left(\overline{u}_i \overline{u}_j\right), \qquad \gamma_\ell = \iiint |\boldsymbol{\xi}|^2 G(\boldsymbol{\xi})\, d^3\boldsymbol{\xi} \tag{8.31}$$

Clark et al. (1979) verify that this representation is very accurate, at low Reynolds number, by comparing with DNS results. However, as shown by Shaanan, Ferziger and Reynolds (1975), the Leonard stresses are of the same order as the truncation error when a finite-difference scheme of second-order accuracy is used, and they are thus implicitly represented.

The cross-term stress tensor, $C_{ij}$, also drains significant energy from the resolvable scales. An expansion similar to Equation (8.31) can be made for

$C_{ij}$. However, most current efforts model the sum of $C_{ij}$ and $R_{ij}$. Clearly, the accuracy of a LES depends critically upon the model used for these terms.

We can now rearrange Equation (8.27) into a more conventional form, i.e.,

$$\frac{\partial \overline{u}_i}{\partial t} + \frac{\partial}{\partial x_j}\left(\overline{u}_i \overline{u}_j\right) = -\frac{1}{\rho}\frac{\partial P}{\partial x_i} + \frac{\partial}{\partial x_j}\left[\nu\frac{\partial \overline{u}_i}{\partial x_j} + \tau_{ij}\right] \qquad (8.32)$$

where

$$\left.\begin{array}{l} \tau_{ij} = -\left(Q_{ij} - \frac{1}{3}Q_{kk}\delta_{ij}\right) \\[2mm] P = \overline{p} + \frac{1}{3}\rho Q_{kk}\delta_{ij} \\[2mm] Q_{ij} = R_{ij} + C_{ij} \end{array}\right\} \qquad (8.33)$$

At this point, **the fundamental problem of Large Eddy Simulation** is evident. Specifically, we must establish a satisfactory model for the SGS stresses as represented by the tensor $Q_{ij}$. To emphasize the importance of achieving an accurate SGS stress model, consider the following. In simulating the decay of homogeneous isotropic turbulence with $16^3 = 4,096$ and $32^3 = 32,768$ grid points, Ferziger (1976) reports that the SGS turbulence energy is 29% and 20%, respectively, of the total. Thus, the subgrid scales constitute a significant portion of the turbulence spectrum. The various attempts at developing a satisfactory SGS stress model during the past four decades resemble the research efforts on engineering models discussed in Chapters 3 – 6. That is, models have been postulated that range from a simple gradient-diffusion model [Smagorinsky (1963)], to a one-equation model [Lilly (1966)], to the analog of a second-order closure model [Deardorff (1973)]. Nonlinear stress-strain rate relationships have even been postulated [Bardina, Ferziger and Reynolds (1983)]. Only the analog of the two-equation model appears to have been overlooked, most likely because the filter width serves as a readily-available length scale.

## 8.3.2   Subgrid-Scale (SGS) Modeling

Smagorinsky (1963) was the first to postulate a model for the SGS stresses. The model assumes the SGS stresses follow a gradient-diffusion process, similar to molecular motion. Consequently, $\tau_{ij}$ is given by

$$\tau_{ij} = 2\nu_T S_{ij}, \qquad S_{ij} = \frac{1}{2}\left(\frac{\partial \overline{u}_i}{\partial x_j} + \frac{\partial \overline{u}_j}{\partial x_i}\right) \qquad (8.34)$$

where $S_{ij}$ is called the "resolved strain rate," $\nu_T$ is the **Smagorinsky eddy viscosity**

$$\nu_T = (C_s \Delta)^2 \sqrt{S_{ij} S_{ij}} \qquad (8.35)$$

and $C_s$ is the Smagorinsky coefficient. Note that Equation (8.35) is akin to a mixing-length formula with mixing length $C_s \Delta$. Obviously the grid scale $\Delta$, or

$(\Delta_1\Delta_2\Delta_3)^{1/3}$ if the steps in the three coordinate directions are different, is an overall scale of the SGS motion, but assuming it to be a unique one is clearly crude. If $\Delta$ were in the inertial subrange of eddy size in which Equation (8.12) holds, and sufficiently larger than the Kolmogorov viscous length scale $\eta$ that the viscous-dependent part of the SGS motion was a small fraction of the whole, then no other length scale would be relevant and the Smagorinsky constant would be universal. This is rarely the case.

For all of the reasons discussed in Chapter 3, the physical assumption behind the mixing-length formula, that eddies behave like molecules, is simply not true. Nevertheless, just as the mixing-length model can be calibrated for a given class of flows, so can the Smagorinsky coefficient, $C_S$. Its value varies from flow to flow, and from place to place within a flow. In the early days of LES, the basic Smagorinsky subgrid-scale model was widely used, $C_S$ being adjusted to get the best results for each flow [see e.g. Rogallo and Moin (1984) who quote a range $0.10 < C_S < 0.24$]. In the critical near-wall region, law-of-the-wall arguments — valid in well-behaved flows — suggest that $C_S$ should be a function of $\Delta/y$ and an increasingly strong function of $u_\tau y/\nu$ as the latter decreases. However there seems to be no record of attempts to calibrate this function: virtually all users of the basic model keep $C_S$ constant throughout the flow.

There are two key reasons why the basic Smagorinsky model enjoys some degree of success. First, the model yields sufficient diffusion and dissipation to stabilize the numerical computations. Second, low-order statistics of the larger eddies are usually insensitive to the SGS motions.

In an attempt to incorporate some representation of the dynamics of the subgrid scales, Lilly (1966) postulates that

$$\nu_T = C_L \Delta\, q \qquad (8.36)$$

where $q^2$ is the SGS kinetic energy, and $C_L$ is a closure coefficient. The subgrid-scale stress anisotropy now depends on the sign of the resolved strain rate, rather than on its magnitude as in the Smagorinsky formula. An equation for $q^2$ can be derived from a moment of the Navier-Stokes equation, which involves several terms that must be modeled. This model is very similar to Prandtl's one-equation model (Section 4.2), both in spirit and in results obtained. As pointed out by Schumann (1975) who used the model in his LES research, it is difficult to conclude that any significant improvement over the Smagorinsky model can be obtained with such a model. However we will meet equations for SGS kinetic energy later.

Germano et al. (1991) [see also Ghosal et al. (1995), Yang and Ferziger (1993) and Carati and Eijnden (1997) for later developments], proposed what is known as a **Dynamic SGS Model**. Their formulation begins with the Smagorinsky eddy-viscosity approximation. However, rather than fixing the value of $C_S$ a priori, they permit it to be computed as the LES proceeds. This is accomplished by using two filters, the usual LES filter at $\kappa = \kappa_{max}$ and the "test filter" which examines the resolved fluctuations between some lower wavenumber, usually $\kappa_{max}/2$, and $\kappa_{max}$ itself. It is then assumed that the subgrid stresses can be represented by rescaling the resolved stresses in the test-filter band: usually, this is done by evaluating the Smagorinsky coefficient, $C_S$, from the resolved fluctuations in the test-filter band, and then using the same coefficient to evaluate the SGS stresses at the same point in space on the next timestep (say). This "bootstrap" procedure could be rigorously justified on the same grounds as for the Smagorinsky formula, above: the test-filter band would have to lie in the inertial subrange and $\kappa_{max}$ would have to be well below the viscous region.

Dynamic models undoubtedly work, surprisingly well, in cases where rigorous justification is not valid. Jimemez (1995) points out that the essential feature of an SGS model is to dissipate the kinetic energy cascaded down to it. He also points out that the dynamic-model concept could be used with intrinsically more realistic models than Smagorinsky's. However it is clear that, whatever SGS model is used, the *test-filter* concept implies that the structure of the SGS turbulence is similar to that in the test-filter band, which will not be the case when the local turbulence Reynolds number, $Re_T$, is small, as near a solid surface. Unfortunately, this is the most critical region for an SGS model: unless the LES is to collapse into DNS, the SGS model must carry much of the Reynolds stresses.

It is a symptom of the inadequacy of the Smagorinsky mixing-length formula that dynamic-model values of $C_S$ evaluated from the calculated motion in the test-filter band fluctuate wildly in space and time — "dynamic" is all too apt a name for this model when used with the Smagorinsky formulation. A particular difficulty arising from this is that if the implied eddy viscosity is negative, kinetic energy can be transferred from the SGS motion to the resolved scales. This is in principle the real-life phenomenon of **backscatter** (reverse cascading of energy from smaller to larger eddies): the statistical-average energy transfer at high wavenumber is always towards the dissipating range, but this is not true instantaneously. However, negative eddy viscosity usually leads to instability of the calculation because there is nothing in the Smagorinsky formula to limit the depletion of SGS kinetic energy. A simple fix is to average $C_S$ over a direction of homogeneity of the flow. As an alternative, Ghosal et al. (1995) and others have modeled an equation for SGS kinetic energy and used it to cut off the SGS eddy viscosity when the SGS energy falls to zero.

A subgrid-scale model with some general similarities to the dynamic model has been suggested by Domaradzki and Saiki (1997). The resolved motion is interpolated on a length scale of half the grid size, and the phases of the resulting subgrid modes are adjusted to correspond to the phases of the subgrid modes that would be generated (in a DNS) by nonlinear interactions of the resolved modes in the LES. This "bootstrap" procedure seems to be rigorously justifiable and initial results are promising.

### 8.3.3 "Off the Wall" Boundary Conditions

If LES is to be applied to wall flows at indefinitely high Reynolds numbers, the *viscous sublayer* or *viscous wall region*, $u_\tau y/\nu < 30$, say, (Figure 1.5) must be excluded from the main computation. Moreover, the distance of the first LES grid point from the solid surface, at $y = y_2$, say, must be independent of Reynolds number — e.g., set at some suitable fraction of the shear-layer thickness, $\delta$ — so that the total number of grid points is independent of Reynolds number. (We use $y_2$ to denote the first LES point from the wall for conformity with the rest of the book, irrespective of the arrangements $y < y_2$.) If the first grid point is set at a given value of $u_\tau y/\nu$ and if, as usual, the $y$ step increases proportional to $y$, the number of grid points required in the $y$ direction increases proportional to $\ell n \delta^+$ where $\delta^+ = u_\tau \delta/\nu$ is the sublayer-scaled flow width (see problems section). This implies that the computer work required (number of grid points in the three-dimensional domain divided by timestep) will vary as $(\ell n \delta^+)^4$ approximately. Thus, a factor of 10 increase in Reynolds number means a factor of 30 increase in computer work.

Allowing the SGS model to bear more and more of the Reynolds stresses as the wall is approached [Speziale (1997b)] does not remove the Reynolds-number dependence of the grid-point count. This is so even if the SGS model limits to a reliable *Reynolds-Averaged Navier Stokes* (RANS) model for a coarse mesh. The mesh must still be graded, so the number of points needed in the $y$ direction is still proportional to $\ell n \delta^+$, with a different proportionality constant, even for full RANS calculations. For example, in Spalart's (1988) boundary-layer DNS, $u_\tau y_2/\nu$ is between 0.2 and 0.3, while most low-Reynolds-number RANS models require $u_\tau y_2/\nu < 1$. So, there is not usually an order-of-magnitude difference between the number of points in the $y$ direction for simulations and for RANS calculations (RANS models can of course use much larger steps in $x$ and $z$).

The current fashion in RANS modeling is integration to the wall rather than the use of off-the-wall boundary conditions (*wall functions* in RANS modeling terminology). For the geometrical-progression grid with **grading ratio**, $k_g = 1.14$ and $y_2^+ \not> 1$ recommended for the **EDDYBL** test case in Appendix D, Section D.5, about 50 points are needed out to $y = \delta$ at momentum-thickness Reynolds number 1410 ($\delta^+ = 650$). The number increases by 17-18 points for every factor of 10 increase in $\delta^+$. Spalart's DNS at this Reynolds number used 62 points out to $y = \delta$.

Note also that the above discussion is phrased, qualitatively at least, in "law-of-the-wall" language. However, if LES is to be a significant improvement on RANS models, it must deal with strongly-nonequilibrium flows, notably separated flows, in which the simple law of the wall is not valid. Indeed the status of the law of the wall is uncertain even in moderately three-dimensional boundary layers.

The earliest LES of a "laboratory" flow, by Deardorff (1970), used the logarithmic law of the wall for the mean velocity as an instantaneous boundary condition. This crude approach was followed by other early workers. However, experiments by Robinson (1986) showed that the instantaneous friction velocity is not a good scale for the instantaneous velocity in the logarithmic region.

If LES is to be applied to high-Reynolds-number engineering flows, not only must the "wall" boundary condition be applied at a distance $y_2$ from the surface which is a (not-too-small) fraction of $\delta$, but any calculation for the region $0 < y < y_2$ should be Reynolds-number independent. Quantitative use of law-of-the-wall concepts is not acceptable in the long term. For most purposes, details of the flow in $0 < y < y_2$ will not be important as long as the surface-flux rates (of momentum, heat and possibly mass) can be related to those at $y = y_2$.

Several groups are currently working on "off the wall" boundary conditions: see for example Piomelli, Yu and Adrian (1996), Cabot (1997) and Baggett (1997). On the one hand, used with care, the instantaneous log law works satisfactorily for simple flows. On the other hand, Cabot found it to be unsatisfactory for separated flows such as the backward-facing step (where the law of the wall does not apply in real life).

## 8.3.4  Applications of LES

Estimates for the numbers of grid points, and computing times, needed for LES calculations of, say, the flow over a complete aircraft [e.g., Spalart et al. (1997)] are far outside the capabilities of current computers, even if one assumes that a satisfactory "off the wall" boundary condition can be found. However, current Reynolds-averaged models perform acceptably well in two-dimensional or mildly three-dimensional boundary layers not too close to separation – that is, in most of the turbulent flow over an aircraft. Something better is needed in critical areas, such as wing-body junctions, tip vortices and separated flows. If LES is to be used in such areas, patched in to a Reynolds-averaged calculation for the rest of the flow, some means is needed for providing time-dependent boundary conditions at the upstream end, and the sides, of the LES computational domain. This may involve enlarging the LES domain so that it considerably overlaps the region of reliable Reynolds-averaged prediction, imposing rough-and-ready boundary conditions at the edges of this domain, and then rescaling the LES in some way so that its statistics match those of the Reynolds-averaged model on the boundary of reliability of the latter. Spalart et al. (1997) suggest **Detached Eddy Simulation** or DES, combining Reynolds-averaged models in the boundary layers and coarse-mesh LES after a massive separation.

Other engineering applications are less demanding than aircraft in terms of Reynolds number, but may be more demanding in terms of complex flow patterns. LES is likely to be applied first to internal flows.

Compressible flow and heat transfer present no special difficulties to LES except for the presence of more equations to solve. Extension of subgrid-scale models to variable-density flows is straightforward and true compressibility effects in the weak small-scale motion are likely to be negligible.

Combustion is notoriously difficult to model at the Reynolds-averaged level, and fine-mesh DNS with the large number of species equations required for a realistic combustion model is currently out of the question for engineering use. Therefore, there is some interest in LES for combustion. Clearly the subgrid-scale model has to reproduce the statistics of fine-scale mixing of reactants, leading to the essentially molecular diffusion that finally brings the reactants together, and this seems a very severe requirement. Recent work is reviewed by Veynante and Poinsot (1997).

Two-phase flows [see Crowe et al. (1996) for a review] are central to many processes in technology and nature, from combustion of droplets or particles to cloud physics and bubble flows in fluidized-bed reactors. Several DNS studies have been reported. LES, with the need to add subgrid-scale motion of particles or bubbles to the larger-scale computed trajectories, is a longer-term prospect. Wang and Squires (1996) report good agreement between LES and DNS.

In conclusion, LES holds promise as a future design tool, especially as computers continue to increase in speed and memory. Intense efforts are currently focused on devising a satisfactory SGS stress model, which is the primary deficiency of the method at this time. Even if LES is too expensive for modern design efforts, results of LES research can certainly be used to help improve engineering models of turbulence. The future of LES research appears very bright.

# 8.4 Chaos

Our final topic is chaos, a mathematical theory that has attracted considerable attention in recent years. At the present time, no quantitative predictions for properties such as the reattachment length behind a backward-facing step or even the skin friction on a flat plate have been made. Hence, its relevance to turbulence modeling thus far has not been as a competing predictive tool. Rather, the theory's value is in developing qualitative understanding of turbulent-flow phenomena.

Chaos abounds with colorful terminology including **fractals, folded towel diffeomorphisms, smooth noodle maps, homeomorphisms, Hopf bifurcation** and the all important **strange attractor**. Chaos theory stretches our imagination to think of noninteger dimensional space, and abounds with marvelous geometrical patterns with which the name Mandelbrot is intimately connected.

In the context of turbulence, the primary focus in chaos is upon **nonlinear dynamical systems**, i.e., a coupled system of nonlinear ordinary differential equations. Mathematicians have discovered that certain dynamical systems with a very small number of equations (degrees of freedom) possess extremely complicated (chaotic) solutions. Very simple models have been created that qualitatively reproduce observed physical behavior for nontrivial problems. For example, consider an initially motionless fluid between two horizontal heat-conducting plates in a gravitational field. Now suppose the lower plate is heated slightly. For small temperature difference, viscous forces are sufficient to suppress any mass motion. As the temperature is increased, a threshold is reached where fluid motion begins. A series of steady convective rolls forms, becoming more and more complicated as the temperature difference increases, and the flow ultimately becomes time-dependent and then nonperiodic/chaotic/turbulent. This is the **Rayleigh-Bénard instability**.

One of the famous successes of chaos theory is in qualitatively simulating the Rayleigh-Bénard flow with the following three coupled **ordinary** differential equations

$$\left.\begin{aligned} \frac{dX}{dt} &= (Y - X)/Pr_L \\ \frac{dY}{dt} &= -XZ + rX - Y \\ \frac{dZ}{dt} &= XY - bZ \end{aligned}\right\} \tag{8.37}$$

The quantity $Pr_L$ is the Prandtl number, $b$ and $r$ are constants, and $X$, $Y$ and $Z$ are related to the streamfunction and temperature. The precise details of the model are given by Bergé, Pomeau and Vidal (1984), and are not important for the present discussion. What is important is that this innocent-looking set of equations yields a qualitative analog to the convection problem, including the geometry of the convection rolls and a solution that resembles turbulent flow.

The central feature of these equations is that they describe what is known as a strange attractor. This particular attractor was the first to be discovered and is more specifically referred to as the **Lorenz attractor**. For the general case, in some suitably defined **phase space** in which each point characterizes the velocity field within a three-dimensional volume ($X$, $Y$ and $Z$ for the Lorenz attractor), the dynamical system sweeps out a curve that we call the attractor. The concept of a phase space is an extension of classical phase-plane analysis of ordinary differential equations [c.f. Bender and Orszag (1978)]. In phase-plane analysis, for example, linear equations have critical points such as the focus, the node and the saddle point. For a dynamical system, if the flow is steady, the "curve" is a single point, because the velocity is independent of time. If the flow is periodic in time the curve is closed and we have the familiar limit cycle. The interesting case in chaos is the unsteady, aperiodic case in which the curve asymptotically approaches the strange attractor. If the dynamical system is dissipative, as the Lorenz equations are, the solution trajectories always converge toward an attractor. Additionally, a slight change in the initial conditions for $X$, $Y$ and $Z$ causes large changes in the solution.

Chaos theory puts great emphasis on the strange attractor, and one of the primary goals of chaos research is to find a set of equations that correspond to the **turbulence attractor**. A dynamical regime is chaotic if two key conditions are satisfied:

1. Its power spectrum contains a continuous part, i.e., a broad band, regardless of the possible presence of peaks.

2. The autocorrelation function goes to zero in finite time.

Of course, both of these conditions are characteristic of turbulence. The latter condition means there is ultimately a loss of memory of the signal with respect to itself. This feature of chaos accounts for the strange attractor's **sensitive dependence on initial conditions**. That is, on a strange attractor, two neighboring trajectories always diverge, regardless of their initial proximity, so that the trajectory actually followed by the system is very sensitive to initial conditions. In chaos studies, this is known as the **butterfly effect** — the notion that a butterfly flapping its wings in Beijing today can change storm systems in New York next month. It goes by the more formal name of **predictability** and was mentioned in Section 8.2 in discussion of the sensitivity of DNS and LES to initial conditions. The **predictability horizon** is the time beyond which predictions become inaccurate, however precise the calculations. Ruelle (1994), in a useful review of possible applications of chaos theory, points out that the motion of the planets in our Solar system is chaotic, with a **predictability time** of about 5 million years — only about 20,000 times the orbital period of Pluto.

While all of these observations indicate there may be promise in using chaos theory to tackle the turbulence problem, there are some sobering realities that must be faced. The broad spectrum of wavelengths in the turbulence spectrum, ranging from the Kolmogorov length scale to the dimension of the flow, is far greater than that of the dynamical systems that have been studied. Hence, as deduced by Keefe (1990) from analysis of DNS data, the dimension of the turbulence attractor (in essence, the number of equations needed to describe the attractor) must be several hundreds even at Reynolds numbers barely large enough for turbulence to exist. It seems essentially unlikely that a low-dimensional dynamical system can emulate turbulence to engineering standards of accuracy.

The layman-oriented book by Gleick (1988) provides an excellent introduction to this fascinating theory in general. See also the abovementioned short review by Ruelle (1994). As a more focused reference, Deissler (1989) presents a review of chaos studies in fluid mechanics.

## 8.5   Further Reading

Fluid dynamics is sometimes called a "mature science," but the capabilities of CFD are expanding rapidly as computer power increases, and the subject will have advanced considerably before this edition of *Turbulence Modeling for CFD* is replaced. Many bibliographies such as Inspec (general physical science) and EI/Compendex (engineering) are available on the Internet, generally via site licenses to institutions. A selected bibliography, with abstracts, of turbulence and related subjects is freely available on the World Wide Web at

**http://www-tsd.stanford.edu/tsd/resp_b.html**

This bibliography goes back to 1980, with some earlier references, and is updated monthly. The reader who wishes to remain up to date should use all these resources.

# Problems

**8.1** To help gain an appreciation for how much computing power is needed for a DNS, consider the following. Table 8.1 lists the number of grid points and timesteps required to perform a DNS for channel flow as a function of Reynolds number. Evaluate the amount of CPU time on a 200 MHz Pentium-based microcomputer needed to do *a single multiplication* at each grid point and timestep for the four Reynolds numbers listed in the table. Assume the microcomputer is capable of 24 megaflops, where a flop is one floating-point operation (multiply, divide, add or subtract) per second. Express your CPU-time answers in days.

**8.2** A DNS of channel flow with $Re_\tau = 180$ using $4 \cdot 10^6$ grid points requires 250 hours of CPU time on a Cray X/MP. The computation was run for a total time, $T_{max} = 5H/u_\tau$. You can assume a Cray X/MP operates at 100 megaflops, where a flop is one floating-point operation (multiply, divide, add or subtract) per second.

(a) Estimate the number of timesteps taken in the computation.

(b) Ignoring time spent reading and writing to disk, estimate the number of floating-point operations per grid point, per iteration.

**8.3** To help gain an appreciation for how much computer memory is needed for a DNS and an LES, consider the following. Table 8.1 lists the number of DNS and LES grid points for channel flow as a function of Reynolds number. There are three velocity components and, on a 32-bit computer, each requires 4 bytes of memory. Compute the amount of memory needed to hold all of the velocity components in memory for the four Reynolds numbers listed in the table. Express your answers in megabytes, noting that there are $1024^2 = 1,048,576$ bytes in a megabyte.

**8.4** This problem focuses on comparative grid requirements for LES and RANS.

(a) Assume the first grid point above a solid surface, $y = y_2$, is located at the outer edge of the viscous wall region, $u_\tau y_2/\nu = 30$. Also, assume that a simple expanding grid with $y_{n+1} = k_g y_n$ (see Subsection 7.2.1) is used in $y_2 < y < \delta$. Verify that the number of grid points in the $y$ direction is $1 + \ell n[u_\tau \delta/(30\nu)]/\ell n k_g$. Start by showing that $y_n = k_g^{n-2} y_2$.

(b) Deduce that if $k_g = 1.14$, as in the sample run of **EDDYBL** in Appendix D, Section D.5, a factor of 10 increase in $u_\tau \delta/\nu$ requires about 18 more profile points.

(c) Near the stern of a ship 300 m long, traveling at 10 m/sec (corresponding to a Reynolds number based on length of $10^9$), $u_\tau \delta/\nu \approx 150,000$. Show that if $k_g = 1.14$ and $y_2^+ = 30$ (wall function for RANS or *off-the-wall* boundary condition for LES), about 66 points are needed in the $y$ direction. Also show that if $y_2^+ = 1$ (integration to the wall), about 92 points are needed.

# Appendix A

# Cartesian Tensor Analysis

The central point of view of tensor analysis is to provide a systematic way for transforming quantities such as vectors and matrices from one coordinate system to another. Tensor analysis is a very powerful tool for making such transformations, although the analysis generally is very involved. For our purposes, working with Cartesian coordinates is sufficient so that we only need to focus on issues of notation, nomenclature and some special tensors. This appendix presents elements of Cartesian tensor analysis.

We begin by addressing the question of notation. In Cartesian tensor analysis we make extensive use of subscripts. For consistency with general tensor-analysis nomenclature we use the terms subscript and index interchangeably. The components of an $n$-dimensional vector $\mathbf{x}$ are denoted as $x_1, x_2, \ldots, x_n$. For example, in three-dimensional space, we rewrite the coordinate vector $\mathbf{x} = (x, y, z)$ as $\mathbf{x} = (x_1, x_2, x_3)$. Now consider an equation describing a plane in three-dimensional space, viz.,

$$a_1 x_1 + a_2 x_2 + a_3 x_3 = c \tag{A.1}$$

where $a_i$ and $c$ are constants. This equation can be written as

$$\sum_{i=1}^{3} a_i x_i = c \tag{A.2}$$

In tensor analysis, we introduce the Einstein summation convention and rewrite Equation (A.2) in the shorthand form

$$a_i x_i = c \tag{A.3}$$

The Einstein summation convention is as follows:

**Repetition of an index in a term denotes summation with respect to that index over its range.**

The **range** of an index $i$ is the set of $n$ integer values 1 to $n$. An index that is summed over is called a **dummy index**; one that is not summed is called a **free index**. Since a dummy index simply indicates summation, it is immaterial what symbol is used. Thus, $a_i x_i$ may be replaced by $a_j x_j$, which is obvious if we simply note that

$$\sum_{i=1}^{3} a_i x_i = \sum_{j=1}^{3} a_j x_j \tag{A.4}$$

As an example of an equation with a free index, consider a unit normal vector **n** in three-dimensional space. If the unit normals in the $x_1$, $x_2$ and $x_3$ directions are $\mathbf{i}_1$, $\mathbf{i}_2$ and $\mathbf{i}_3$, then the direction cosines $\alpha_1$, $\alpha_2$ and $\alpha_3$ for the vector **n** are

$$\alpha_k = \mathbf{n} \cdot \mathbf{i}_k \tag{A.5}$$

There is no implied summation in Equation (A.5). Rather, it is a shorthand for the three equations defining the direction cosines. Because the length of a unit vector is one, we can take the dot product of $(\alpha_1, \alpha_2, \alpha_3)$ with itself and say that

$$\alpha_i \alpha_i = 1 \tag{A.6}$$

As another example, consider the total differential of a function of three variables, $p(x_1, x_2, x_3)$. We have

$$dp = \frac{\partial p}{\partial x_1} dx_1 + \frac{\partial p}{\partial x_2} dx_2 + \frac{\partial p}{\partial x_3} dx_3 \tag{A.7}$$

In tensor notation, this is replaced by

$$dp = \frac{\partial p}{\partial x_i} dx_i \tag{A.8}$$

Equation (A.8) can be thought of as the dot product of the gradient of $p$, namely $\nabla p$, and the differential vector $d\mathbf{x} = (dx_1, dx_2, dx_3)$. Thus, we can also say that the $i$ component of $\nabla p$, which we denote as $(\nabla p)_i$, is given by

$$(\nabla p)_i = \frac{\partial p}{\partial x_i} = p_{,i} \tag{A.9}$$

where a comma followed by an index is tensor notation for differentiation with respect to $x_i$. Similarly, the divergence of a vector **u** is given by

$$\nabla \cdot \mathbf{u} = \frac{\partial u_i}{\partial x_i} = u_{i,i} \tag{A.10}$$

where we again denote differentiation with respect to $x_i$ by ", $i$".

Thus far, we have dealt with scalars and vectors. The question naturally arises about how we might handle a matrix. The answer is we denote a matrix by using two subscripts, or indices. The first index corresponds to row number while the second corresponds to column number. For example, consider the $3 \times 3$ matrix $[A]$ defined by

$$[A] = \begin{bmatrix} A_{11} & A_{12} & A_{13} \\ A_{21} & A_{22} & A_{23} \\ A_{31} & A_{32} & A_{33} \end{bmatrix} \tag{A.11}$$

In tensor notation, we represent the matrix $[A]$ as $A_{ij}$. If we post-multiply an $m \times n$ matrix $B_{ij}$ by an $n \times 1$ column vector $x_j$, their product is an $m \times 1$ column vector $y_i$. Using the summation convention, we write

$$y_i = B_{ij}x_j \tag{A.12}$$

Equation (A.12) contains both a free index ($i$) and a dummy index ($j$). The product of a square matrix $A_{ij}$ and its inverse is the unit matrix, i.e.,

$$[A][A]^{-1} = \begin{bmatrix} 1 & 0 & 0 \\ 0 & 1 & 0 \\ 0 & 0 & 1 \end{bmatrix} \tag{A.13}$$

Equation (A.13) is rewritten in tensor notation as follows:

$$A_{ik}(A^{-1})_{kj} = \delta_{ij} \tag{A.14}$$

where $\delta_{ij}$ is the Kronecker delta defined by

$$\delta_{ij} = \begin{cases} 1, & i = j \\ 0, & i \neq j \end{cases} \tag{A.15}$$

We can use the Kronecker delta to rewrite Equation (A.6) as

$$\alpha_i \delta_{ij} \alpha_j = 1 \tag{A.16}$$

This corresponds to pre-multiplying the $3 \times 3$ matrix $\delta_{ij}$ by the row vector $(\alpha_1, \alpha_2, \alpha_3)$ and then post-multiplying their product by the column vector $(\alpha_1, \alpha_2, \alpha_3)^T$, where superscript $T$ denotes transpose.

The determinant of a $3 \times 3$ matrix $A_{ij}$ is

$$\begin{vmatrix} A_{11} & A_{12} & A_{13} \\ A_{21} & A_{22} & A_{23} \\ A_{31} & A_{32} & A_{33} \end{vmatrix} = \begin{matrix} A_{11}A_{22}A_{33} + A_{21}A_{32}A_{13} + A_{31}A_{12}A_{23} \\ -A_{11}A_{32}A_{23} - A_{12}A_{21}A_{33} - A_{13}A_{22}A_{31} \end{matrix} \tag{A.17}$$

Tensor analysis provides a shorthand for this operation as well. Specifically, we replace Equation (A.17) by

$$det(A_{ij}) = |A_{ij}| = \epsilon_{rst} A_{r1} A_{s2} A_{t3} \tag{A.18}$$

where $\epsilon_{rst}$ is the **permutation tensor** defined by

$$\left. \begin{array}{l} \epsilon_{123} = \epsilon_{231} = \epsilon_{312} = 1 \\ \epsilon_{213} = \epsilon_{321} = \epsilon_{132} = -1 \\ \epsilon_{111} = \epsilon_{222} = \epsilon_{333} = \epsilon_{112} = \epsilon_{113} = \epsilon_{221} = \epsilon_{223} = \epsilon_{331} = \epsilon_{332} = 0 \end{array} \right\} \tag{A.19}$$

In other words, $\epsilon_{ijk}$ vanishes whenever the values of any two indices are the same; $\epsilon_{ijk} = 1$ when the indices are a permutation of 1, 2, 3; and $\epsilon_{ijk} = -1$ otherwise.

As can be easily verified, the cross product of two vectors **a** and **b** can be expressed as follows.

$$(\mathbf{a} \times \mathbf{b})_i = \epsilon_{ijk} a_j b_k \tag{A.20}$$

In particular, the curl of a vector **u** is

$$(\nabla \times \mathbf{u})_i = \epsilon_{ijk} \frac{\partial u_k}{\partial x_j} = \epsilon_{ijk} u_{k,j} \tag{A.21}$$

The Kronecker delta and permutation tensor are very important quantities that appear throughout this book. They are related by the $\epsilon$-$\delta$ identity, which is the following.

$$\epsilon_{ijk} \epsilon_{ist} = \delta_{js} \delta_{kt} - \delta_{jt} \delta_{ks} \tag{A.22}$$

All that remains to complete our brief introduction to tensor analysis is to define a tensor. Tensors are classified in terms of their rank. To determine the rank of a tensor, we simply count the number of indices.

The lowest rank tensor is rank zero which corresponds to a scalar, i.e., a quantity that has magnitude only. Thermodynamic properties such as pressure and density are scalar quantities. Vectors such as velocity, vorticity and pressure gradient are tensors of rank one. They have both magnitude and direction. Matrices are rank two tensors. The stress tensor is a good example for illustrating physical interpretation of a second-rank tensor. It defines a force per unit area that has a magnitude and two associated directions, the direction of the force and the direction of the normal to the plane on which the force acts. For a normal stress, these two directions are the same; for a shear stress, they are (by convention) normal to each other.

As we move to tensors of rank three and beyond, the physical interpretation becomes more difficult to ascertain. This is rarely an issue of great concern since virtually all physically relevant tensors are of rank 2 or less. The permutation tensor is of rank 3, for example, and is simply defined by Equation (A.19).

A tensor $a_{ij}$ is **symmetric** if $a_{ij} = a_{ji}$. Many important tensors in mathematical physics are symmetric, e.g., stress, strain and strain-rate tensors, moment of inertia tensor, virtual-mass tensor. A tensor is **skew symmetric** if $a_{ij} = -a_{ji}$. The rotation tensor, $\Omega_{ij} = \frac{1}{2}(u_{i,j} - u_{j,i})$ is skew symmetric.

As a final comment, in performing tensor-analysis operations with tensors that are not differential operators, we rarely have to worry about preserving the order of terms as we did in Equation (A.16). There is no confusion in writing $\delta_{ij}\alpha_i\alpha_j$ in place of $\alpha_i\delta_{ij}\alpha_j$. This is only an issue when the indicated summations actually have to be done. However, care should be exercised when differentiation occurs. As an example, $\nabla \cdot \mathbf{u} = \partial u_i/\partial x_i$ is a scalar number while $\mathbf{u} \cdot \nabla = u_i\partial/\partial x_i$ is a scalar differential operator.

# Problems

**A.1** Use the $\epsilon$-$\delta$ identity to verify the well known vector identity

$$\mathbf{A} \times (\mathbf{B} \times \mathbf{C}) = (\mathbf{A} \cdot \mathbf{C})\mathbf{B} - (\mathbf{A} \cdot \mathbf{B})\mathbf{C}$$

**A.2** Show that, when $i, j, k$ range over 1, 2, 3

   (a) $\delta_{ij}\delta_{ji} = 3$

   (b) $\epsilon_{ijk}\epsilon_{jki} = 6$

   (c) $\epsilon_{ijk}A_j A_k = 0$

   (d) $\delta_{ij}\delta_{jk} = \delta_{ik}$

**A.3** Verify that $2S_{ij,j} = \nabla^2 u_i$ for incompressible flow, where $S_{ij}$ is the strain-rate tensor, i.e., $S_{ij} = \frac{1}{2}(u_{i,j} + u_{j,i})$.

**A.4** Show that the scalar product $S_{ij}\Omega_{ji}$ vanishes identically if $S_{ij}$ is a symmetric tensor and $\Omega_{ij}$ is skew symmetric.

**A.5** If $u_j$ is a vector, show that the tensor $\omega_{ik} = \epsilon_{ijk}u_j$ is skew symmetric.

**A.6** Show that if $A_{jk}$ is a skew-symmetric tensor, the unique solution of the equation $\omega_i = \frac{1}{2}\epsilon_{ijk}A_{jk}$ is $A_{mn} = \epsilon_{mni}\omega_i$.

**A.7** The incompressible Navier-Stokes equation in a coordinate system rotating with constant angular velocity $\Omega$ and with position vector $\mathbf{x} = x_k \mathbf{i}_k$ is

$$\frac{\partial \mathbf{u}}{\partial t} + \mathbf{u} \cdot \nabla \mathbf{u} + 2\Omega \times \mathbf{u} = -\nabla\left(\frac{p}{\rho}\right) - \Omega \times \Omega \times \mathbf{x} + \nu\nabla^2 \mathbf{u}$$

   (a) Rewrite this equation in tensor notation.

   (b) Using tensor analysis, show that for $\Omega = \Omega\mathbf{k}$ ($\mathbf{k}$ is a unit vector aligned with $\Omega$), the centrifugal force per unit mass is given by

$$-\Omega \times \Omega \times \mathbf{x} = \nabla(\frac{1}{2}\Omega^2 x_k x_k) - [\mathbf{k} \cdot \nabla(\frac{1}{2}\Omega^2 x_k x_k)]\mathbf{k}$$

**A.8** Using tensor analysis, prove the vector identity

$$\mathbf{u} \cdot \nabla \mathbf{u} = \nabla(\frac{1}{2}\mathbf{u} \cdot \mathbf{u}) - \mathbf{u} \times (\nabla \times \mathbf{u})$$

# Appendix B

# Rudiments of Perturbation Methods

When we work with perturbation methods, we are constantly dealing with the concept of **order of magnitude**. There are three conventional **order symbols** that provide a mathematical measure of the order of magnitude of a given quantity, viz., **Big O, Little o**, and $\sim$. They are defined as follows.

**Big O:**

   $f(\delta) = O[g(\delta)]$ as $\delta \to \delta_o$ if a neighborhood of $\delta_o$ exists and a constant $M$ exists such that $|f| \leq M|g|$, i.e., $f(\delta)/g(\delta)$ is bounded as $\delta \to \delta_o$.

**Little o:**

   $f(\delta) = o[g(\delta)]$ as $\delta \to \delta_o$ if, given any $\epsilon > 0$, there exists a neighborhood of $\delta_o$ such that $|f| \leq \epsilon|g|$, i.e., $f(\delta)/g(\delta) \to 0$ as $\delta \to \delta_o$.

$\sim$:

   $f(\delta) \sim g(\delta)$ as $\delta \to \delta_o$ if $f(\delta)/g(\delta) \to 1$ as $\delta \to \delta_o$.

For example, the Taylor series for the exponential function is

$$e^{-x} = 1 - x + \frac{1}{2}x^2 - \frac{1}{6}x^3 + \cdots \tag{B.1}$$

where "$\cdots$" is conventional shorthand for the rest of the Taylor series, i.e.,

$$\cdots = \sum_{n=4}^{\infty} \frac{(-1)^n x^n}{n!} \tag{B.2}$$

In terms of the ordering symbols, we can replace "$\cdots$" as follows.

$$e^{-x} = 1 - x + \frac{1}{2}x^2 - \frac{1}{6}x^3 + O(x^4) = 1 - x + \frac{1}{2}x^2 - \frac{1}{6}x^3 + o(x^3) \quad \text{(B.3)}$$

We define an **asymptotic sequence of functions** as a sequence $\phi_n(\delta)$ for $n = 1, 2, 3, \ldots$, satisfying the condition

$$\phi_{n+1}(\delta) = o[\phi_n(\delta)] \quad \text{as} \quad \delta \to \delta_o \quad \text{(B.4)}$$

Examples of asymptotic sequences are:

$$\left.\begin{array}{rclr}
\phi_n(\delta) & = & 1, (\delta - \delta_o), (\delta - \delta_o)^2, (\delta - \delta_o)^3, \ldots & \delta \to \delta_o \\[4pt]
\phi_n(\delta) & = & 1, \delta^{1/2}, \delta, \delta^{3/2}, \ldots & \delta \to 0 \\[4pt]
\phi_n(\delta) & = & 1, \delta, \delta^2 \ell n \delta, \delta^2, \ldots & \delta \to 0 \\[4pt]
\phi_n(x) & = & x^{-1}, x^{-2}, x^{-3}, x^{-4}, \ldots & x \to \infty
\end{array}\right\} \quad \text{(B.5)}$$

We say that $g(\delta)$ is **transcendentally small** if $g(\delta)$ is $o[\phi_n(\delta)]$ for all $n$. For example,

$$e^{-1/\delta} = o(\delta^n) \quad \text{for all } n \quad \text{(B.6)}$$

**An asymptotic expansion** is the sum of the first $N$ terms in an asymptotic sequence. It is the asymptotic expansion of a function $F(\delta)$ as $\delta \to \delta_o$ provided

$$F(\delta) = \sum_{n=1}^{N} a_n \phi_n(\delta) + o[\phi_N(\delta)] \quad \text{(B.7)}$$

The following are a few useful asymptotic expansions generated from simple Taylor-series expansions, all of which are convergent as $\delta \to 0$.

$$\left.\begin{array}{rcl}
(1+\delta)^n & \sim & 1 + n\delta + \frac{n(n-1)}{2}\delta^2 + O(\delta^3) \\[6pt]
\ell n(1+\delta) & \sim & \delta - \frac{1}{2}\delta^2 + \frac{1}{3}\delta^3 + O(\delta^4) \\[6pt]
(1-\delta)^{-1} & \sim & 1 + \delta + \delta^2 + O(\delta^3) \\[6pt]
e^\delta & \sim & 1 + \delta + \frac{1}{2}\delta^2 + O(\delta^3) \\[6pt]
\cos\delta & \sim & 1 - \frac{1}{2}\delta^2 + \frac{1}{24}\delta^4 + O(\delta^6) \\[6pt]
\sin\delta & \sim & \delta - \frac{1}{6}\delta^3 + \frac{1}{120}\delta^5 + O(\delta^7) \\[6pt]
\tan\delta & \sim & \delta + \frac{1}{3}\delta^3 + \frac{2}{15}\delta^5 + O(\delta^7)
\end{array}\right\} \quad \text{(B.8)}$$

Not all asymptotic expansions are developed as a Taylor series, nor are they necessarily convergent. For example, consider the complementary error function, erfc($x$), i.e.,

$$\text{erfc}(x) = \frac{2}{\sqrt{\pi}} \int_x^\infty e^{-t^2} \, dt \quad \text{(B.9)}$$

We can generate an asymptotic expansion using a succession of integration-by-parts operations. (To start the process, for example, multiply and divide the integrand by $t$ so that $t \exp(-t^2)$ becomes integrable in closed form.) The expansion is:

$$\text{erfc}(x) \quad \sim \quad \frac{2}{\sqrt{\pi}} e^{-x^2} \sum_{n=0}^{\infty} (-1)^n \frac{(1)(3)\dots(2n-1)}{2^{n+1} x^{2n+1}} \quad \text{as} \quad x \to \infty$$

$$\sim \quad \frac{2}{\sqrt{\pi}} e^{-x^2} \left\{ \frac{x^{-1}}{2} - \frac{x^{-3}}{4} + O(x^{-5}) \right\} \tag{B.10}$$

A simple ratio test shows that this series is divergent for all values of $x$. However, if we define the remainder after the first $N$ terms of the series as $R_N(x)$, there are two limits we can consider, viz.,

$$\lim_{x \to \infty} |R_N(x)|_{Fixed\ N} = 0 \quad \text{and} \quad \lim_{N \to \infty} |R_N(x)|_{Fixed\ x} = \infty \tag{B.11}$$

Thus this divergent series gives a good approximation to $\text{erfc}(x)$ provided we don't keep too many terms! This is often the case for an asymptotic series.

Part of our task in developing a perturbation solution is to determine the appropriate asymptotic sequence. It is usually obvious, but not always. Also, more than one set of $\phi_n(\delta)$ may be suitable, i.e., we are not guaranteed uniqueness in perturbation solutions. These problems, although annoying from a theoretical viewpoint, by no means diminish the utility of perturbation methods. Usually, we have physical intuition to help guide us in developing our solution. This type of mathematical approach is, after all, standard operating procedure for the engineer. We are, in essence, using the methods Prandtl and von Kármán used before perturbation analysis was given a name.

A **singular-perturbation problem** is one in which no single asymptotic expansion is uniformly valid throughout the field of interest. For example, while $\delta/x^{1/2} = O(\delta)$ as $\delta \to 0$, the singularity as $x \to 0$ means this expression is not uniformly valid. Similarly, $\delta \ell n x = O(\delta)$ as $\delta \to 0$ and is not uniformly valid as $x \to 0$ and as $x \to \infty$. The two most common situations that lead to a singular-perturbation problem are:

(a) the coefficient of the highest derivative in a differential equation is very small;

(b) difficulties arise in behavior near boundaries.

Case (b) typically arises in analyzing the turbulent boundary layer where logarithmic behavior of the solution occurs close to a solid boundary. The following second-order ordinary differential equation illustrates Case (a).

$$\delta \frac{d^2 F}{ds^2} + \frac{dF}{ds} + F = 0, \qquad 0 \le s \le 1 \tag{B.12}$$

We want to solve this equation subject to the following boundary conditions.

$$F(0) = 0 \quad \text{and} \quad F(1) = 1 \tag{B.13}$$

We also assume that $\delta$ is very small compared to 1, i.e.,

$$\delta \ll 1 \tag{B.14}$$

This equation is a simplified analog of the Navier-Stokes equation. The second-derivative term has a small coefficient just as the second-derivative term in the Navier-Stokes equation, in nondimensional form, has the reciprocal of the Reynolds number as its coefficient. An immediate consequence is that only one boundary condition can be satisfied if we set $\delta = 0$. This is similar to setting viscosity to zero in the Navier-Stokes equation, which yields Euler's equation, and the attendant consequence that only the normal-velocity surface boundary condition can be satisfied. That is, we cannot enforce the no-slip boundary condition for Euler-equation solutions.

The exact solution to this equation is

$$F(s; \delta) = \frac{e^{\alpha(1-s)} - e^{\alpha - \beta s/\delta}}{1 - e^{\alpha - \beta/\delta}} \tag{B.15}$$

where

$$\alpha = \frac{1 - \sqrt{1 - 4\delta}}{2\delta} \quad \text{and} \quad \beta = \frac{1 - \sqrt{1 - 4\delta}}{2} \tag{B.16}$$

which clearly satisfies both boundary conditions. If we set $\delta = 0$ in Equation (B.12), we have the following first-order equation:

$$\frac{dF}{ds} + F = 0 \tag{B.17}$$

and the solution, $F(s; 0)$, is

$$F(s; 0) = e^{1-s} \tag{B.18}$$

where we use the boundary condition at $s = 1$. However, the solution fails to satisfy the boundary condition at $s = 0$ because $F(0; 0) = e = 2.71828 \cdots$. Figure B.1 illustrates the solution to our simplified equation for several values of $\delta$.

As shown, the smaller the value of $\delta$, the more closely $F(s; 0)$ represents the solution throughout the region $0 < s \leq 1$. Only in the immediate vicinity of $s = 0$ is the solution inaccurate. The thin layer where $F(s; 0)$ departs from the exact solution is called a boundary layer, in direct analogy to its fluid-mechanical equivalent.

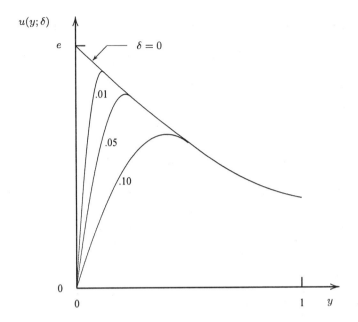

Figure B.1: *Solutions to the model equation for several values of $\delta$.*

To solve this problem using perturbation methods, we seek a solution that consists of two separate asymptotic expansions, one known as the **outer expansion** and the other as the **inner expansion**. For the outer expansion, we assume a solution of the form

$$F_{outer}(s;\delta) \sim \sum_{n=0}^{N} F_n(s)\phi_n(\delta) \tag{B.19}$$

where the asymptotic sequence functions, $\phi_n(\delta)$, will be determined as part of the solution. Substituting Equation (B.19) into Equation (B.12) yields the following.

$$\sum_{n=0}^{N} \left\{ \frac{d^2 F_n}{ds^2}\delta\phi_n(\delta) + \frac{dF_n}{ds}\phi_n(\delta) + F_n\phi_n(\delta) \right\} = 0 \tag{B.20}$$

Clearly, if we select

$$\phi_n(\delta) = \delta^n \tag{B.21}$$

we, in effect, have a power-series expansion. Equating like powers of $\delta$, the **leading-order** ($n = 0$) problem is Equation (B.17), while the second-derivative term makes its first appearance in the **first-order** ($n = 1$) problem. Thus, our perturbation solution yields the following series of problems for the **outer expansion**.

$$
\left.
\begin{aligned}
\frac{dF_0}{ds} + F_0 &= 0 \\
\frac{dF_1}{ds} + F_1 &= -\frac{d^2 F_0}{ds^2} \\
\frac{dF_2}{ds} + F_2 &= -\frac{d^2 F_1}{ds^2} \\
&\vdots
\end{aligned}
\right\}
\qquad (B.22)
$$

Provided we solve the equations in sequence starting at the lowest order ($n = 0$) equation, the right-hand side of each equation is known from the preceding solution and serves simply to make each equation for $n \geq 1$ nonhomogeneous. Consequently, to all orders, the equation for $F_n(s)$ is of first order. Hence, no matter how many terms we include in our expansion, we can satisfy only one of the two boundary conditions. As in the introductory remarks, we elect to satisfy $F(1) = 1$. In terms of our expansion [Equations (B.19) and (B.21)], the boundary conditions for the $F_n$ are

$$
F_0(1) = 1 \quad \text{and} \quad F_n(1) = 0 \quad \text{for} \quad n \geq 1 \qquad (B.23)
$$

The solution to Equations (B.22) subject to the boundary conditions specified in Equation (B.23) is as follows.

$$
\left.
\begin{aligned}
F_0(s) &= e^{1-s} \\
F_1(s) &= (1 - s)e^{1-s} \\
&\vdots
\end{aligned}
\right\}
\qquad (B.24)
$$

Hence, our **outer expansion** assumes the following form.

$$
F_{outer}(s; \delta) \sim e^{1-s} \left[ 1 + (1 - s)\delta + O(\delta^2) \right] \qquad (B.25)
$$

In general, for singular-perturbation problems, we have no guarantee that continuing to an infinite number of terms in the outer expansion yields a solution that satisfies both boundary conditions. That is, our expansion may or may not be convergent. Hence, we try a different approach to resolve the region near $s = 0$. We now generate an **inner expansion** in which we **stretch** the $s$ coordinate. That is, we define a new independent variable $\sigma$ as follows.

$$
\sigma = \frac{s}{\mu(\delta)} \qquad (B.26)
$$

We assume an inner expansion in terms of a new set of asymptotic-sequence functions, $\psi_n(\delta)$, i.e.,

$$F_{inner}(\sigma; \delta) \sim \sum_{n=0}^{N} f_n(\sigma)\psi_n(\delta) \qquad \text{(B.27)}$$

To best illustrate how we determine the appropriate stretching function, $\mu(\delta)$, consider the leading-order terms in the original differential equation, viz.,

$$\frac{d^2 f_0}{d\sigma^2}\left(\frac{\delta\psi_0}{\mu^2}\right) + \frac{df_0}{d\sigma}\left(\frac{\psi_0}{\mu}\right) + f_0\psi_0 = O\left(\frac{\delta\psi_1}{\mu^2}, \frac{\psi_1}{\mu}, \psi_1\right) \qquad \text{(B.28)}$$

First of all, we must consider the three possibilities for the order of magnitude of $\mu(\delta)$, viz., $\mu \gg 1$, $\mu \sim 1$ and $\mu \ll 1$. If $\mu \gg 1$, inspection of Equation (B.28) shows that $f_0 = 0$ which is not a useful solution. If $\mu \sim 1$, we have the outer expansion. Thus, we conclude that $\mu \ll 1$.

We are now faced with three additional possibilities: $\delta\psi_0/\mu^2 \gg \psi_0/\mu$; $\delta\psi_0/\mu^2 \sim \psi_0/\mu$; and $\delta\psi_0/\mu^2 \ll \psi_0/\mu$. Using the boundary condition at $s = 0$, assuming $\delta\psi_0/\mu^2 \gg \psi_0/\mu$ yields $f_0 = A\sigma$ where $A$ is a constant of integration. While this solution might be useful, we have learned nothing about the stretching function, $\mu(\delta)$. At the other extreme, $\delta\psi_0/\mu^2 \ll \psi_0/\mu$, we obtain the trivial solution, $f_0 = 0$, which doesn't help us in our quest for a solution. The final possibility, $\delta\psi_0/\mu^2 \sim \psi_0/\mu$, is known as the **distinguished limit**, and this is the case we choose. Thus,

$$\mu(\delta) = \delta \qquad \text{(B.29)}$$

Again, the most appropriate choice for the $\psi_n(\delta)$ is

$$\psi_n(\delta) = \delta^n \qquad \text{(B.30)}$$

The following sequence of equations and boundary conditions define the **inner expansion**.

$$\left.\begin{aligned}
\frac{d^2 f_0}{d\sigma^2} + \frac{df_0}{d\sigma} &= 0 \\
\frac{d^2 f_1}{d\sigma^2} + \frac{df_1}{d\sigma} &= -f_0 \\
\frac{d^2 f_2}{d\sigma^2} + \frac{df_2}{d\sigma} &= -f_1 \\
\vdots
\end{aligned}\right\} \qquad \text{(B.31)}$$

$$f_n(0) = 0 \quad \text{for all } n \geq 0 \qquad \text{(B.32)}$$

Solving the leading, or **zeroth**, order problem ($n = 0$) and the **first** order problem ($n = 1$), we find

$$\left.\begin{aligned}
f_0(\sigma) &= A_0(1 - e^{-\sigma}) \\
f_1(\sigma) &= (A_1 - A_0\sigma) - (A_1 + A_0\sigma)e^{-\sigma} \\
&\vdots
\end{aligned}\right\} \qquad (B.33)$$

where $A_0$ and $A_1$ are constants of integration. These integration constants arise because each of Equations (B.31) is of second order and we have used only one boundary condition.

To complete the solution, we perform an operation known as **matching**. To motivate the matching procedure, note that on the one hand, the boundary $s = 1$ is located at $\sigma = 1/\delta \to \infty$ as $\delta \to 0$. Hence, we need a boundary condition for $F_{inner}(\sigma; \delta)$ valid as $\sigma \to \infty$. On the other hand, the independent variable in the outer expansion is related to $\sigma$ by $s = \delta\sigma$. Thus, for any finite value of $\sigma$, the inner expansion lies very close to $s = 0$. We **match** these two asymptotic expansions by requiring that

$$\lim_{\sigma \to \infty} F_{inner}(\sigma; \delta) = \lim_{s \to 0} F_{outer}(s; \delta) \qquad (B.34)$$

The general notion is that on the scale of the outer expansion, the inner expansion is valid in an infinitesimally thin layer. Similarly, on the scale of the inner expansion, the outer expansion is valid for a region infinitely distant from $s = 0$. For the problem at hand,

$$\lim_{\sigma \to \infty} f_0(\sigma) = A_0 \quad \text{and} \quad \lim_{s \to 0} F_0(s) = e \qquad (B.35)$$

Thus, we conclude that

$$A_0 = e \qquad (B.36)$$

Equivalently, we can visualize the existence of an **overlap region** between the inner and outer solutions. In the overlap region, we stretch the $s$ coordinate according to

$$s^* = \frac{s}{\nu(\delta)}, \qquad \delta \ll \nu(\delta) \ll 1 \qquad (B.37)$$

In terms of this intermediate variable, for any finite value of $s^*$,

$$s \to 0 \quad \text{and} \quad \sigma \to \infty \quad \text{as} \quad \nu(\delta) \to 0 \qquad (B.38)$$

Using this method, we can match to as high an order as we wish. For example, matching to $n^{th}$ order, we perform the following limit operation.

$$\lim_{\delta \to 0} \left[ \frac{F_{inner} - F_{outer}}{\delta^n} \right] = 0 \qquad (B.39)$$

For the problem at hand, the independent variables $s$ and $\sigma$ become

$$s = \nu(\delta)s^* \quad \text{and} \quad \sigma = \frac{\nu(\delta)s^*}{\delta} \tag{B.40}$$

Hence, replacing $e^{-\nu(\delta)s^*}$ by its Taylor-series expansion, we find

$$F_{outer} \sim e\left\{1 - \nu(\delta)s^* + \delta + O[\delta\nu(\delta)]\right\} \tag{B.41}$$

Similarly, noting that $e^{-\nu(\delta)s^*/\delta}$ is transcendentally small as $\delta \to 0$, we have

$$F_{inner} \sim A_0 - A_0\nu(\delta)s^* + A_1\delta + O(\delta^2) \tag{B.42}$$

Thus, holding $s^*$ constant,

$$\lim_{\delta \to 0} \left[\frac{F_{inner} - F_{outer}}{\delta}\right] \sim \frac{(A_0 - e)(1 - \nu(\delta)s^*) + (A_1 - e)\delta + o(\delta)}{\delta} \tag{B.43}$$

Clearly, **matching to zeroth and first orders** can be achieved only if

$$A_0 = A_1 = e \tag{B.44}$$

In summary, the **inner and outer expansions** are given by

$$\left.\begin{aligned}
F_{outer}(s;\delta) &\sim e^{1-s}\left[1 + (1 - s)\delta + O(\delta^2)\right] \\
F_{inner}(\sigma;\delta) &\sim e\left\{(1 - e^{-\sigma}) + [(1 - \sigma) - (1 + \sigma)e^{-\sigma}]\delta + O(\delta^2)\right\} \\
\sigma &= s/\delta
\end{aligned}\right\} \tag{B.45}$$

Finally, we can generate a single expansion, known as a **composite expansion**, that can be used throughout the region $0 \leq s \leq 1$. Recall that in the matching operations above, we envisioned an **overlap region**. In constructing a composite expansion, we note that the inner expansion is valid in the inner region, the outer expansion is valid in the outer region, and both are valid in the overlap region. Hence, we define

$$F_{composite} \equiv F_{inner} + F_{outer} - F_{cp} \tag{B.46}$$

where $F_{cp}$ is the **common part**, i.e., the part of the expansions that cancel in the matching process. Again, for the case at hand, comparison of Equations (B.41) and (B.42) with $A_0$ and $A_1$ given by Equation (B.44) shows that

$$F_{cp} \sim e\left[1 + (1 - \sigma)\delta + O(\delta^2)\right] \tag{B.47}$$

where we use the fact that $\nu(\delta)s^* = \delta\sigma$. Hence, the **composite expansion** is

$$F_{composite} \sim \left[e^{1-s} - e^{1-s/\delta}\right] + \left[(1 - s)e^{1-s} - (1 + s/\delta)e^{1-s/\delta}\right]\delta + O(\delta^2) \tag{B.48}$$

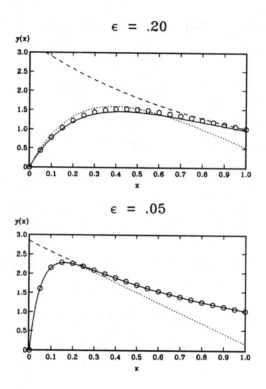

Figure B.2: *Comparison of asymptotic expansions and the exact solution for the sample boundary-value problem:* o *exact;* − − − *outer expansion;* · · · · · · *inner expansion;* —— *composite expansion.*

What we have done is combine two non-uniformly valid expansions to achieve a **uniformly-valid approximation** to the exact solution. Retaining just the zeroth-order term of the composite expansion yields an approximation to the exact solution that is accurate to within 7% for $\epsilon$ as large as 0.2! This is actually a bit fortuitous however, since the leading term in Equation (B.48) and the exact solution differ by a transcendentally small term. Figure B.2 compares the two-term inner, outer and composite expansions with the exact solution for $\epsilon = .05$ and $\epsilon = .20$.

For the obvious reason, perturbation analysis is often referred to as the theory of **matched asymptotic expansions**. The discussion here, although sufficient for our needs, is brief and covers only the bare essentials of the theory. For additional information, see the books by Van Dyke (1975), Bender and Orszag (1978), Kevorkian and Cole (1981), Nayfeh (1981) or Wilcox (1995a) on this powerful mathematical theory.

# Problems

**B.1** Consider the polynomial

$$x^3 - x^2 + \delta = 0$$

(a) For nonzero $\delta < 4/27$ this equation has three real and unequal roots. Why is this a singular-perturbation problem in the limit $\delta \to 0$?

(b) Use perturbation methods to solve for the first two terms in the expansions for the roots.

**B.2** Determine the first two terms in the asymptotic expansion for the roots of the following polynomial valid as $\delta \to 0$.

$$\delta x^3 + x + 2 + \delta = 0$$

**B.3** Consider the following nonlinear, first-order initial-value problem.

$$\frac{dy}{dt} + y + \delta y^2 = 0, \qquad y(0) = 1$$

Determine the exact solution and classify this problem as regular or singular in the limit $\delta \to 0$. Do the classification first for $t > 0$ and then for $t < 0$.

**B.4** The following is an example of a perturbation problem that is singular because of nonuniformity near a boundary. Consider the following first-order equation in the limit $\epsilon \to 0$.

$$x^3 \frac{dy}{dx} = \epsilon y^2, \qquad y(1) = 1$$

The solution is known to be finite on the closed interval $0 \le x \le 1$.

(a) Solve for the first two terms in the outer expansion and show that the solution has a singularity as $x \to 0$.

(b) Show that there is a boundary layer near $x = 0$ whose thickness is of order $\epsilon^{1/2}$.

(c) Solve for the first two terms of the inner expansion. Note that the algebra simplifies if you do the zeroth-order matching before attempting to solve for the next term in the expansion.

**B.5** Generate the first two terms in a perturbation solution for the following initial-value problem valid as $\delta \to 0$.

$$\frac{dy}{dx} = y^2 + 2y, \qquad y(0) = -2 + 5\delta$$

**B.6** Generate the first two terms of the inner and outer expansions for the following boundary-value problem. Also, construct a composite expansion.

$$\delta \frac{d^2 y}{dx^2} + \frac{dy}{dx} - xy = 0, \qquad \delta \ll 1$$

$$y(0) = 0 \quad \text{and} \quad y(1) = e^{1/2}$$

**B.7** Generate the first two terms of the inner and outer expansions for the following boundary-value problem. Also, construct a composite expansion.

$$\delta \frac{d^2y}{dx^2} + \frac{dy}{dx} = \frac{1}{2}x^2, \qquad \delta \ll 1$$

$$y(0) = 1 \quad \text{and} \quad y(1) = 1/6$$

**B.8** This problem demonstrates that the **overlap region** is not a layer in the same sense as the boundary layer. Rather, its thickness depends upon how many terms we retain in the matching process. Suppose we have solved a boundary-layer problem and the first three terms of the inner and outer expansions valid as $\epsilon \to 0$ are:

$$y_{outer}(x;\epsilon) \sim 1 + \epsilon e^{-x^2} + \epsilon^2 e^{-2x^2} + O(\epsilon^3)$$

$$y_{inner}(x;\epsilon) \sim A(1 - e^{-\xi}) + \epsilon B(1 - e^{-\xi}) + \epsilon^2 C(1 - \xi^2) + O(\epsilon^3)$$

where

$$\xi \equiv \frac{x}{\epsilon^{1/2}}$$

Determine the coefficients $A$, $B$ and $C$. Explain why the thickness of the overlap region, $\nu(\epsilon)$, must lie in the range

$$\epsilon^{1/2} \ll \nu(\epsilon) \ll \epsilon^{1/4}$$

as opposed to the normally assumed range $\epsilon^{1/2} \ll \nu(\epsilon) \ll 1$.

# Appendix C

# Companion Software

## C.1 Overview

The software described in this appendix solves for: (a) free-shear-flow farfield behavior; (b) pipe and channel flow; and, (c) detailed sublayer and defect-layer behavior. In all cases, accurate algorithms are used that guarantee grid-independent solutions on any computer from an IBM PC to a Cray Y/MP. These programs serve two purposes. First, they solve basic **building-block** flow problems and can thus be helpful in developing or modifying a turbulence model. Second, these programs provide a definitive separation of turbulence-model error and numerical error.

As computers have increased in power, there has been a tendency away from analytical methods such as similarity solutions and singular-perturbation methods. The mathematics of these procedures can be tedious, and it can be a lot easier to use a parabolic marching program for the types of flows described above. Eventually, marching far enough in space, self-similarity is achieved with such a program. However, complete avoidance of analytical methods can lead to an ignorance of important flow details such as singularities and important asymptotic behavior that can be masked by numerical error. More alarmingly, improper treatment of such flow detail can be the source of numerical error.

While the view presented here may appear to be a bit overcautious, it is justified by the difficulties so often encountered in solving turbulence-transport equations. Exact solutions are virtually nonexistent. Experimental measurements cannot objectively be used to test for numerical accuracy. Consequently, it is difficult to assess the accuracy of a new turbulent-flow program. The programs described in this appendix generate very accurate solutions for a variety of simple turbulent flows and turbulence models, and can be used to assess numerical accuracy of more complicated programs.

419

The accompanying diskette includes source code for all of the programs, as well as several auxiliary routines referenced by the programs. Plotting programs are included that can be used on IBM PC and compatible computers with support for a variety of video displays and hardcopy devices. Only the executable versions of the plotting programs are included as they are based on proprietary plotting software. Section C.6 lists the files included on the distribution diskette. Appendix E presents details of the hardware supported by the plotting programs.

## C.1.1   Program Structure

All of the programs use time-marching methods to solve the nonlinear two-point boundary-value problems attending use of the similarity-solution method for simple turbulent flows. The solution algorithm used is based on implicit Crank-Nicolson differencing. To render straightforward and easy to modify programs, each equation of a given turbulence model is solved independently using a standard tridiagonal-matrix inversion algorithm.

In the interest of portability, the programs have been written so that they run on IBM PC and compatible microcomputers, SUN Workstations, VAX (VMS based) computers, the Silicon Graphics IRIS and Cray super computers. The programs all use an **include** file named **cpuid** that defines a single parameter called *icpu*. This parameter is passed to a subroutine named **NAMSYS** that returns system-dependent and compiler-specific parameters. The file **cpuid** contains the following statements.

```
c-----------------------------------------------------------------------
c    S E L E C T   A P P R O P R I A T E   C O M P I L E R / C P U
c-----------------------------------------------------------------------
c    icpu =  0...SVS Fortran (680x0/80x86)
c         =  1...Lahey Fortran/Microsoft Fortran (80x86)
c         =  2...VAX/VMS
c         =  3...SUN Fortran (68020/SPARC)
c         =  4...Cray/Unicos
c         =  5...Silicon Graphics Iris
c-----------------------------------------------------------------------
      icpu=1
```

Set *icpu* to the value appropriate for your system. If your computer and/or Fortran compiler is not listed, you will have to modify subroutine **NAMSYS**. See Subsection C.5.2 for a detailed explanation of what is required as well as a listing of subroutine **NAMSYS**.

The same basic structure has been used for all of the programs. A standardized set of subroutine, input and output file, variable and common block names has been used throughout. For example, if the program name is **PROG-NAME**, the input data file is **progname.dat** and the main output disk file is **progname.prt**.

The main program coordinates all computations and program logic by calling a collection of subroutines. In all programs, the computational sequence is as follows:

1. Call **NAMIN** to coordinate reading the input data file. Note that **NAMIN** calls **NAMSYS** to set all system-dependent and compiler-specific parameters.

2. Call **GRID** to set up the finite-difference grid.

3. Call **START** to set closure coefficients and initial conditions.

4. Enter the main computation loop and repeat the following steps until convergence is achieved.

   - Call **GETETA** to compute $\eta = \int \mu_T \, d\xi$ for programs that use the Rubel-Melnik (1984) transformation.
   - Call **CALCS** to compute eddy viscosity, vertical velocity, etc.
   - Call **TMESTP** to compute the timestep.
   - Call **EDDY** to advance the solution in time.

5. When the solution has either converged or the maximum allowable number of timesteps has been reached, call **EDIT** to either write program output to a disk file or directly to the printer. For the free-shear-layer programs, call **GROW** to compute spreading rate.

6. Write a disk file that can be used for making a plot of program output.

## C.1.2   Program Input

The programs use a standardized method, reminiscent of the non-Ansi-Standard NAMELIST scheme, to provide input to the programs. As noted above, if the program name is **PROGNAME**, input is provided in a disk file named **progname.dat** (sample input for each program is provided on the distribution diskette). The format for integer quantities is (1x,a12,i4) while the format for floating-point quantities is (1x,a12,e13.6). The (1x,a12) permits entering the variable name and an equal sign. Typical input thus appears as follows.

```
               iunit1    =    2
               iunit2    =    7
               model     =    0
               etin      =    1.000000e-07
               wtin      =    4.000000e-01
               |         |  | |          |
Column number:    2        12 15 17      26
```

### C.1.3  Program Output

The output from program **PROGNAME** consists of a disk file that can be used
to plot computed results and, depending upon user preference, printed output that
is directed to either the system printer, or to a disk file named **progname.prt**.
The name of the plotting-data disk file depends upon the turbulence model used
as follows.

| | |
|---|---|
| **bbarth.dat** | Baldwin-Barth model [Baldwin and Barth (1990)] |
| **blomax.dat** | Baldwin-Lomax model [Baldwin and Lomax (1978)] |
| **csmith.dat** | Cebeci-Smith model [Smith and Cebeci (1967)] |
| **kepsln.dat** | $k$-$\epsilon$ model [Launder and Sharma (1974)] |
| **komega.dat** | $k$-$\omega$ model |
| **mixlen.dat** | Mixing-length model with $\ell_{mix} = \alpha\delta(x)$ |
| **newmod.dat** | User-Defined model |
| **somega.dat** | Stress-$\omega$ model |
| **spalart.dat** | Spalart-Allmaras model [Spalart and Allmaras (1992)] |

All of the programs make provision for a user-defined model so that the
supplied plotting utilities can be used for customized versions of the various
programs. Usually, only a subset of the models listed above is supported by
any one program. Also, the contents of the plotting-data file are a bit different
for each program. See the appropriate section to determine which models are
supported by the program of interest and to determine the format of the plotting-
data disk file.

The first input parameter for all of the programs is an integer variable named
*iunit1*. Setting *iunit1* = 6 will cause the printed output to go to the system line
printer for IBM PC implementations. Any other integer will send the printed
output to a disk file whose name is **progname.prt**.

Printed output consists of three segments. First, all input data are printed.
Next, the maximum error and other flow properties such as spreading rate for
free shear flows are printed; this information is also shown on the video display
as the run proceeds. Finally, profiles of computed mean-flow and turbulence
properties are printed in a self-explanatory manner. The precise format of the
printed information differs slightly from one program to the next.

## C.2  Free Shear Flows

There are three free-shear-flow programs on the distribution diskette, viz., Programs **WAKE**, **MIXER** and **JET**. These programs solve the self-similar form of
the turbulent-flow equations that are asymptotically approached far downstream.
Section 4.5 of the main text presents the equations of motion in physical variables

and in similarity form. An additional transformation devised by Rubel and Melnik (1984) has been used that greatly improves numerical accuracy. Specifically, we introduce a new independent variable, $\xi$, defined in terms of the similarity variable, $\eta$, by

$$d\xi = \frac{d\eta}{N(\eta)} \quad \text{or} \quad \frac{d}{d\xi} = N(\eta)\frac{d}{d\eta} \tag{C.1}$$

where $N(\eta)$ is the dimensionless eddy viscosity appearing in the similarity solution. In terms of this variable, the equations for the $k$-$\omega$ and $k$-$\epsilon$ models, for example, assume the following form:

**Mean Momentum:**

$$\mathcal{V}\frac{d\mathcal{U}}{d\xi} - \frac{1}{\eta^j}\frac{d}{d\xi}\left[\eta^j\frac{d\mathcal{U}}{d\xi}\right] = S_u N\mathcal{U} \tag{C.2}$$

**Turbulence Kinetic Energy:**

$$\mathcal{V}\frac{dK}{d\xi} - \frac{1}{\eta^j}\frac{d}{d\xi}\left[\sigma^*\eta^j\frac{dK}{d\xi}\right] = S_k NK + \left(\frac{d\mathcal{U}}{d\xi}\right)^2 - \beta^* K^2 \tag{C.3}$$

**$k$-$\omega$ Model:**

$$\mathcal{V}\frac{dW}{d\xi} - \frac{1}{\eta^j}\frac{d}{d\xi}\left[\sigma\eta^j\frac{dW}{d\xi}\right] = S_w NW + \alpha\frac{W}{K}\left(\frac{d\mathcal{U}}{d\xi}\right)^2 - \beta KW \tag{C.4}$$

$$N = \frac{K}{W} \tag{C.5}$$

**$k$-$\epsilon$ Model:**

$$\mathcal{V}\frac{dE}{d\xi} - \frac{1}{\eta^j}\frac{d}{d\xi}\left[\frac{\eta^j}{\sigma_\epsilon}\frac{dE}{d\xi}\right] = S_e NE + C_{\epsilon1}\frac{E}{K}\left(\frac{d\mathcal{U}}{d\xi}\right)^2 - C_\mu C_{\epsilon2} KE \tag{C.6}$$

$$N = C_\mu\frac{K^2}{E} \tag{C.7}$$

**Mixing-Length Model:**

$$N = \alpha\sqrt{\left|\frac{d\mathcal{U}}{d\xi}\right|} \tag{C.8}$$

where $\mathcal{U}$, $\mathcal{V}$, $K$, $W$ and $E$ are the transformed velocity components, turbulence energy, specific dissipation rate and dissipation rate, respectively. See Section 4.5

of the main text for additional details on notation and other features of the similarity solution.

This transformation greatly improves numerical accuracy primarily because it removes numerical difficulties that are associated with the presence of sharp turbulent/nonturbulent interfaces. The edge of the shear layer that occurs at a finite value of $\eta$ is moved to infinity in terms of the transformed independent variable $\xi$. Inspection of converged solutions shows a well-behaved asymptotic approach to freestream conditions, a feature rarely observed when the equations are solved without the transformation. Consequently, a much tighter convergence criterion can be satisfied. Additionally, there is weaker coupling amongst the turbulence-model equations which also improves the convergence rate.

The only drawback to this transformation is the need to determine an appropriate maximum value of $\xi$. Using too large or too small a value can slow convergence and even cause the solution to blow up. All of the programs automatically compute the value of $\xi_{max}$ that is suitable for the turbulence models implemented. If you add a new turbulence model, it may be necessary to empirically determine a suitable value for $\xi_{max}$.

Boundary conditions for these equations must be satisfied at $\xi = 0$ and as $\xi \to \infty$, so that we must solve a two-point boundary-value problem. This is conveniently done by adding unsteady terms to the left-hand sides of Equations (C.2), (C.3), (C.4) and (C.6). We then make an initial guess and let the solution evolve in time. The solution to the desired two-point boundary-value problem is obtained when temporal variations vanish. Thus, for example, we replace the mean-momentum equation by the following.

$$\frac{\partial \mathcal{U}}{\partial t} + \mathcal{V}\frac{\partial \mathcal{U}}{\partial \xi} - \frac{1}{\eta^j}\frac{\partial}{\partial \xi}\left[\eta^j \frac{\partial \mathcal{U}}{\partial \xi}\right] = S_u N \mathcal{U} \tag{C.9}$$

The resulting time-dependent system of equations is solved using implicit Crank-Nicolson differencing that is second-order accurate in both $t$ and $\xi$. Using 101 mesh points, all of the free-shear-flow programs typically require computing times of less than 5 seconds on a 133 MHz Pentium-based microcomputer.

## C.2.1   Program WAKE: Far Wake

Program **WAKE** computes two-dimensional flow in the far wake of an object in an incompressible stream.

**Input-parameter description:**
Program **WAKE** reads the following input parameters in the order listed below from disk file **wake.dat**. Integer quantities must be formatted according to (1x,a12,i4) while floating-point quantities must be formatted as (1x,a12,e13.6). See Subsection C.1.2 for a sample input-data file.

*iunit1*    Output-file unit number
                = 6 Printed output sent to printer
                $\neq$ 6 Printed output saved in disk file **wake.prt**
*iunit2*   Plotting-data disk file unit number
*model*   Turbulence-model identification flag
                0  $k$-$\omega$ model
                1  Stress-$\omega$ model
                2  $k$-$\epsilon$ model (see *kerng* also)
                4  Spalart-Allmaras model
                9  Mixing-length model with $\ell_{mix} = \alpha\delta(x)$
                99  User-defined model
*kerng*   $k$-$\epsilon$ model identification flag
                = 0  Standard $k$-$\epsilon$ model
                $\neq$ 0  RNG $k$-$\epsilon$ model
*etin*    Freestream value of transformed turbulence kinetic energy
*wtin*    Freestream value of transformed specific dissipation rate
*jmax*   Number of grid points
*maxn*   Maximum number of timesteps
*nedit*   Profile-print modulus; profiles are printed every *nedit* steps
*nfreq*   Short-print modulus; maximum error, shear-layer growth rate, etc. are
             printed every *nfreq* steps

**Output description:**
   VIDEO OUTPUT includes the timestep number, maximum error and spreading rate every *nfreq* timesteps. PRINTED OUTPUT is sent to the line printer throughout the run if *iunit1* = 6. DISK-FILE OUTPUT is saved in the appropriate plotting-data disk file (see Subsection C.1.3 for file names). The file is **unformatted** and is created using the following statements:

```
jaxi=10
write(iunit2) jmax,jaxi,0.,0.,0.,'PLOTF'
write(iunit2) (eta(j),uoum(j),j=1,jmax)
```

where *eta(j)* is $\eta_j$ and *uoum(j)* is $\mathcal{U}(\eta_j)$. The parameter *jaxi* is used by plotting program **PLOTF**. Additionally, for any value of *iunit1* other than 6, printed output is saved in disk file **wake.prt** at the conclusion of the run.

**Comments:**
   Program **WAKE** runs most efficiently for *jmax* $\geq$ 101. A smaller number of grid points tends to slow convergence.

## C.2.2   Program MIXER: Mixing Layer

Program **MIXER** computes two-dimensional flow in the mixing layer between
two streams of differing velocity, including effects of compressibility.

**Input-parameter description:**

   Program **MIXER** reads the following input parameters in the order listed
below from disk file **mixer.dat**. Integer quantities must be formatted according
to (1x,a12,i4) while floating-point quantities must be formatted as (1x,a12,e13.6).
See Subsection C.1.2 for a sample input-data file.

*iunit1*   Output-file unit number
        = 6 Printed output sent to printer
        $\neq$ 6 Printed output saved in disk file **mixer.prt**
*iunit2*   Plotting-data disk file unit number
*model*   Turbulence-model identification flag
        0 $k$-$\omega$ model
        1 Stress-$\omega$ model
        2 $k$-$\epsilon$ model (see *kerng* also)
        4 Spalart-Allmaras model
        9 Mixing-length model with $\ell_{mix} = \alpha\delta(x)$
        99 User-defined model
*kerng*   $k$-$\epsilon$ model identification flag
        = 0 Standard $k$-$\epsilon$ model
        $\neq$ 0 RNG $k$-$\epsilon$ model
*etin*   Freestream value of transformed turbulence kinetic energy
*gam*   Specific-heat ratio, $\gamma$
*prt*   Turbulent Prandtl number, $Pr_T$
*rho2*   Density ratio, $\rho_2/\rho_1$ where subscript 2 corresponds to the slow stream;
    required only for incompressible flow
*u2ou1*   Velocity ratio, $U_2/U_1$ where subscript 2 corresponds to the slow
    stream; required only for incompressible flow
*wtin*   Freestream value of transformed specific dissipation rate
*xis*   Compressibility-modification closure coefficient $\xi^*$
*xma1*   Mach number of the fast stream, $M_1$
*xma2*   Mach number of the slow stream, $M_2$
*xmt0*   Compressibility-modification closure coefficient $M_{to}$
*imach*   Compressibility-modification model flag
        0 Sarkar's model
        1 Wilcox's model
        2 Zeman's model
*jmax*   Number of grid points
*jzero*   Index of grid point at the dividing streamline ($\eta = 0$)

| | |
|---|---|
| *maxn* | Maximum number of timesteps |
| *nedit* | Profile-print modulus; profiles are printed every *nedit* steps |
| *nfreq* | Short-print modulus; maximum error, shear-layer growth rate, etc. are printed every *nfreq* steps |
| *nthick* | Mixing-layer thickness definition flag for short print |

        -1  Bogdanoff's vorticity thickness

         0  Birch's energy thickness

         1  Roshko's pitot thickness

         2  Sullin's momentum thickness

**Output description:**

VIDEO OUTPUT includes the timestep number, maximum error and spreading rate every *nfreq* timesteps. PRINTED OUTPUT is sent to the line printer throughout the run if *iunit1* = 6. DISK-FILE OUTPUT is saved in the appropriate plotting-data disk file (see Subsection C.1.3 for file names). The file is **unformatted** and is created using the following statements:

```
jaxi=99
write(iunit2) jmax,jaxi,0.,0.,0.,'PLOTF'
write(iunit2) (yox(j),u(j),j=1,jmax)
```

where *yox(j)* is $\eta_j$ and *u(j)* is $\mathcal{U}(\eta_j)$. The parameter *jaxi* is used by plotting program **PLOTF**. Additionally, for any value of *iunit1* other than 6, printed output is saved in disk file **mixer.prt** at the conclusion of the run.

**Comments:**

- Program **MIXER** runs most efficiently for *jmax* $\geq$ 101. A smaller number of grid points tends to slow convergence.

- Program **MIXER** runs most efficiently for Mach numbers up to 5. It will run for larger values, although smaller timesteps are needed. The timestep can be reduced by changing the value of *cflmax* in the main program. The parameter *cflmax* is the maximum value of the Courant-Friedrichs-Lewy number; the timestep is computed as the product of a number less than or equal to 1 (for the first 200 timesteps), *cflmax*, and the maximum timestep required for stability of an explicit scheme.

## C.2.3   Program JET: Plane, Round and Radial Jet

Program **JET** computes far-field flow for plane, round and radial jets issuing
into a quiescent incompressible fluid.

**Input-parameter description:**

Program **JET** reads the following input parameters in the order listed below from disk file **jet.dat**. Integer quantities must be formatted according to
(1x,a12,i4) while floating-point quantities must be formatted as (1x,a12,e13.6).
See Subsection C.1.2 for a sample input-data file.

*iunit1*   Output-file unit number
  = 6 Printed output sent to printer
  $\neq$ 6 Printed output saved in disk file **jet.prt**

*iunit2*   Plotting-data disk file unit number

*model*   Turbulence-model identification flag
  0 $k$-$\omega$ model
  1 Stress-$\omega$ model
  2 $k$-$\epsilon$ model (see *kerng* also)
  4 Spalart-Allmaras model
  9 Mixing-length model with $\ell_{mix} = \alpha\delta(x)$
  99 User-defined model

*kerng*   $k$-$\epsilon$ model identification flag
  = 0 Standard $k$-$\epsilon$ model
  $\neq$ 0 RNG $k$-$\epsilon$ model

*etin*   Freestream value of transformed turbulence kinetic energy

*wtin*   Freestream value of transformed specific dissipation rate

*ipope*   Round-jet modification flag (for the $k$-$\epsilon$ model only)
  0 Omit Pope's round-jet correction
  1 Use Pope's round-jet correction

*jaxi*   Geometry flag
  -1 Radial jet
  0 Plane jet
  1 Round jet

*jmax*   Number of grid points

*maxn*   Maximum number of timesteps

*nedit*   Profile-print modulus; profiles are printed every *nedit* steps

*nfreq*   Short-print modulus; maximum error, shear-layer growth rate, etc. are
  printed every *nfreq* steps

**Output description:**

VIDEO OUTPUT includes the timestep number, maximum error and spreading rate every *nfreq* timesteps. PRINTED OUTPUT is sent to the line printer

throughout the run if *iunit1* = 6. DISK-FILE OUTPUT is saved in the appropriate plotting-data disk file (see Subsection C.1.3 for file names). The file is **unformatted** and is created using the following statements:

```
write(iunit2) jmax,jaxi,0.,0.,0.,'PLOTF'
write(iunit2) (eta(j),uoum(j),j=1,jmax)
```

where *eta(j)* is $\eta_j$ and *uoum(j)* is $\mathcal{U}(\eta_j)$. The parameter *jaxi* is used by plotting program **PLOTF**. Additionally, for any value of *iunit1* other than 6, printed output is saved in disk file **jet.prt** at the conclusion of the run.

**Comments:**

Program **JET** runs most efficiently for *jmax* $\geq$ 101. A smaller number of grid points tends to slow convergence.

## C.2.4   Program PLOTF: Plotting Utility

Program **PLOTF** creates video and hardcopy plots of free-shear-flow velocity profiles computed with programs **WAKE**, **MIXER** and **JET** on IBM PC's and compatibles. The program automatically detects the turbulence model used and the type of shear flow for which computations have been done.

**Input-parameter description:**

Program **PLOTF** reads the following nine input parameters in the order listed below from disk file **plotf.dat**. Integer quantities must be formatted according to (7x,i6) while floating-point quantities must be formatted as (7x,f6.2). This is similar to the format used for Programs **WAKE**, **MIXER** and **JET**.

| | |
|---|---|
| *mon* | Monitor type (see Appendix E) |
| *ifore* | Foreground color (see Appendix E) |
| *iback* | Background color (see Appendix E) |
| *nprin* | Printer type (see Appendix E) |
| *mode* | Graphics-mode flag for printers; number of pens for plotters (see Appendix E) |
| *ksize* | Plot scaling factor. Using 100 yields a 5″ by 5″ hardcopy plot. Smaller values yield a hardcopy plot reduced by *ksize* per cent. For example, *ksize* = 50 produces a 2.5″ by 2.5″ plot. |
| *symsiz* | Size of experimental data symbols, in inches |
| *height* | Physical height, in inches, of the video display |
| *width* | Physical width, in inches, of the video display |

Next, Program **PLOTF** reads a single, free formatted, line to indicate where hardcopy print is directed and where required font files are located. This line comes immediately after the specified value for *width* and defines the following five additional parameters.

| | |
|---|---|
| *devid* | Device name of type character*4; valid devices are LPT1, LPT2, LPT3, COM1, COM2, COM3, COM4 |
| *nbaud* | Baud rate for a serial port; valid baud rates are 110, 150, 300, 600, 1200, 2400, 4800, 9600 |
| *parity* | Parity of type character*3 or character*4 for a serial port; valid parity settings are 'even', 'odd' and 'none' |
| *nstop* | Number of stop bits for a serial port; either 1 or 2 |
| *lword* | Word length for a serial port; either 7 or 8 |

**Output description:**

A 5″ by 5″ video plot (see Figure C.1) is created centered on the screen. When the plot is complete, the following message appears:

### Hardcopy output (y/n)?

Enter a *y* or a *Y* to create a hardcopy plot. Pressing any other key terminates the run without creating a hardcopy plot.

**Comments:**

- The following is a sample input data file for a machine with a standard VGA monitor and an HP DeskJet connected to serial port COM1:.

```
mon    =    18        (Standard VGA monitor)
ifore =    15         (Bright-white foreground)
iback =     1         (Blue background)
nprin =     2         (HP DeskJet)
mode  =     3         (300 dots per inch resolution)
ksize =   100         (Full size plot)
symsiz=  .080         (.08" experimental data symbols)
height= 7.500         (7.5" high video display)
width = 9.250         (9.25" wide video display)
'com1' , 9600 , 'none' , 1 , 8
```

The last line indicates the printer is connected to serial port COM1: and the port is set at 9600 baud, no parity, 1 stop bit and 8 data bits.

- If disk file **plotf.dat** is not available, Program **PLOTF** uses the following set of default values:

*mon* = 21, *ifore* = 15, *iback* = 1, *nprin* = 5, *mode* = 4, *ksize* = 100, *symsiz* = .08, *height* = 7.5, *width* = 9.25, *devid* = 'LPT1'

Note that *nbaud*, *parity*, *nstop* and *lword* are not used for parallel ports.

- As many as 6 different curves, each with a unique line style, can be drawn on a single plot.

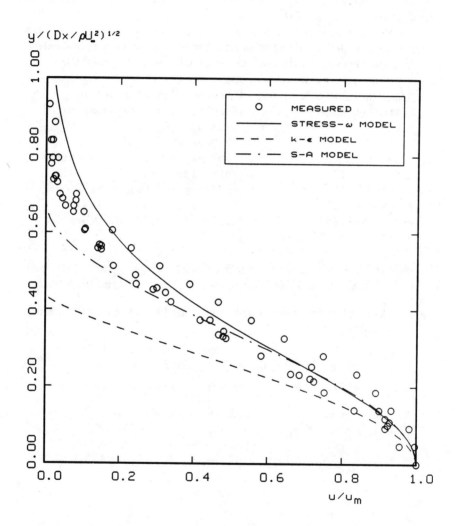

Figure C.1: *Sample plot created by Program PLOTF.*

# C.3 Channel and Pipe Flow

Program **PIPE** can be used to compute incompressible channel flow or pipe flow with several turbulence models. Subsections 3.5.1 and 4.8.1 of the main text describe the channel- and pipe-flow equations. No additional transformations are introduced in Program **PIPE**.

As with the free-shear-flow programs, we add unsteady terms to the various turbulence-model equations to facilitate solution of the two-point boundary-value problem. However, the momentum equation is solved at each timestep by trapezoidal-rule integration. For example, in the case of a two-equation model, we advance the turbulence parameters in time. Then, after updating the eddy viscosity, we determine the velocity by integration of the following equation.

$$\frac{dU^+}{dy^+} = \frac{1 - y^+/R^+}{1 + \mu_T^+} \qquad (C.10)$$

All notation in Equation (C.10) is identical to that used in Subsections 3.5.1 and 4.8.1 of the main text.

The only other subtle feature of the program is the way the specific dissipation rate, $\omega$, in the $k$-$\omega$ model and the Stress-$\omega$ model is computed close to the solid boundary. To eliminate numerical error associated with computing the singular behavior of $\omega$ for perfectly-smooth and slightly-rough surfaces, the exact asymptotic behavior of $\omega$ is prescribed close to the surface (see Subsection 7.2.1). That is, we use the fact that, for $y^+ < 2.5$, $\omega^+ = \nu\omega/u_\tau^2$ is given by:

$$\omega^+ \to \frac{6}{\beta_o(y^+)^2} \qquad \text{as} \quad y \to 0 \quad \text{(smooth wall)} \qquad (C.11)$$

$$\omega^+ \to \frac{\omega_w^+}{\left[1 + \sqrt{\frac{\beta_o\omega_w^+}{6}}\, y^+\right]^2} \qquad \text{as} \quad y \to 0 \quad \text{(rough wall)} \qquad (C.12)$$

The exact analytical behavior of $\omega$ is imposed for a prescribed number of mesh points, *jskip*, next to the surface. Using 201 mesh points, Program **PIPE** typically requires computing times of less than 5 seconds on a 133 MHz Pentium-based microcomputer.

## C.3.1   Program PIPE: Channel and Pipe Flow

Program **PIPE** computes incompressible, fully-developed flow in either a two-dimensional channel or a pipe of circular cross section.

**Input-parameter description:**
   Program **PIPE** reads the following input parameters from disk file **pipe.dat** in the order listed below.  Integer quantities must be formatted according to (1x,a12,i4) while floating-point quantities must be formatted as (1x,a12,e13.6). See Subsection C.1.2 for a sample input-data file.

*iunit1*   Output-file unit number
   = 6 Printed output sent to printer
   $\neq$ 6 Printed output saved in disk file **pipe.prt**
*iunit2*   Plotting-data disk file unit number
*model*   Turbulence-model identification flag
      0 $k$-$\omega$ model
      1 Stress-$\omega$ model
      3 Baldwin-Barth model
      4 Spalart-Allmaras model
      5 Cebeci-Smith model
      6 Baldwin-Lomax model
      7 Johnson-King model
      99 User-defined model
*omegw*   Surface value of dimensionless specific dissipation rate, $\omega_w^+$
*retau*   $Re_\tau = R^+ = u_\tau R/\nu$, dimensionless channel half height/pipe radius
*yone*   Value of $y^+$ at the first grid point above the wall
*iruff*   Surface-roughness flag
      0 Rough surface
      1 Smooth surface
*jaxi*   Geometry flag
      0 Channel flow
      1 Pipe flow
*jmax*   Number of grid points
*jskip*   Grid-point number below which the exact asymptotic solution for specific dissipation rate is used
*maxn*   Maximum number of timesteps
*nedit*   Profile-print modulus; profiles are printed every *nedit* steps
*nfreq*   Short-print modulus; maximum error, Reynolds number, skin friction, etc. are printed every *nfreq* steps
*nvisc*   Viscous-modification flag
      = 0 No $k$-$\omega$/Stress-$\omega$ model viscous modifications
      $\neq$ 0 Use $k$-$\omega$/Stress-$\omega$ model viscous modifications

**Output description:**

VIDEO OUTPUT includes timestep number, maximum error, Reynolds number and skin friction every *nfreq* timesteps. PRINTED OUTPUT is sent to the line printer throughout the run if *iunit1* = 6. DISK-FILE OUTPUT is saved in the appropriate plotting-data disk file (see Subsection C.1.3 for file names). The file is **unformatted** and is created using the following statements.

```
      write(iunit2) jmax,jaxi,retau,reh,cf,'PLOTP'
      if(model.ge.3.and.model.le.7) then
         write(iunit2) (yoh(j),uoum(j),yplus(j),uplus(j),tau(j),
     *                 j=1,jmax)
      else
         do 60 j=2,jmax
            diss(j)=-betas*fbetas(j)*bbeta(j)*et(j)*wt(j)
            prod(j)=tau(j)*dudy(j)
60       continue
         if(nvisc.ne.0) diss(1)=2.*diss(2)-diss(3)
         prod(1)=0.
         write(iunit2) (yoh(j),uoum(j),yplus(j),uplus(j),tau(j),
     *                 et(j),diss(j),prod(j),j=1,jmax)
      endif
```

The quantity *reh* is Reynolds number based on the average velocity and channel height/pipe diameter, while *cf* is skin friction based on average velocity. Also, *yoh(j)* is $y/R$, *uoum(j)* is $U(y)/U(0)$, *yplus(j)* is $y^+$, *uplus(j)* is $U^+$, *upvp(j)* is $\tau_{xy}/u_\tau^2$, *et(j)* is $k/u_\tau^2$, *diss(j)* is $\beta^* f_{\beta^*} \nu \omega k / u_\tau^4$, and *prod(j)* is $\nu \tau_{xy}(dU/dy)/u_\tau^4$. Additionally, for any value of *iunit1* other than 6, printed output is saved in disk file **pipe.prt** at the conclusion of the run.

## C.3.2    Program PLOTP: Plotting Utility

Program **PLOTP** creates video and hardcopy plots of channel- or pipe-flow properties computed with program **PIPE** on IBM PC's and compatibles. The program automatically detects the turbulence model used and the type flow for which computations have been done.

### Input-parameter description:

Program **PLOTP** reads the following seven input parameters in the order listed below from disk file **plotp.dat**. Integer quantities must be formatted according to (7x,i6) while floating-point quantities must be formatted as (7x,f6.2). This is similar to the format used for Program **PIPE**.

| | |
|---|---|
| *mon* | Monitor type (see Appendix E) |
| *ifore* | Foreground color (see Appendix E) |
| *iback* | Background color (see Appendix E) |
| *nprin* | Printer type (see Appendix E) |
| *mode* | Graphics-mode flag for printers; number of pens for plotters (see Appendix E) |
| *ksize* | Plot scaling factor. Using 100 yields a full-size hardcopy plot. Smaller values yield a hardcopy plot reduced by *ksize* per cent. For example, *ksize* = 50 produces a half-size plot. |
| *symsiz* | Size of experimental data symbols, in inches |

Next, Program **PLOTP** reads a single, free formatted, line to indicate where hardcopy print is directed and where required font files are located. This line comes immediately after the specified value for *symsiz* and defines the following five additional parameters.

| | |
|---|---|
| *devid* | Device name of type character∗4; valid devices are LPT1, LPT2, LPT3, COM1, COM2, COM3, COM4 |
| *nbaud* | Baud rate for a serial port; valid baud rates are 110, 150, 300, 600, 1200, 2400, 4800, 9600 |
| *parity* | Parity of type character∗3 or character∗4 for a serial port; valid parity settings are 'even', 'odd' and 'none' |
| *nstop* | Number of stop bits for a serial port; either 1 or 2 |
| *lword* | Word length for a serial port; either 7 or 8 |

In addition to disk file **plotp.dat**, an optional disk file including skin-friction data can be included. This information includes results from a series of runs and must be prepared by the user. The name of the optional disk file depends upon the turbulence model used, viz.:

| **cfbb.dat** | Baldwin-Barth model |
|---|---|
| **cfbl.dat** | Baldwin-Lomax model |
| **cfcs.dat** | Cebeci-Smith model |
| **cfjk.dat** | Johnson-King model |
| **cfkw.dat** | $k\text{-}\omega$ model |
| **cfsa.dat** | Spalart-Allmaras model |
| **cfsw.dat** | Stress-$\omega$ model |
| **cfus.dat** | User-supplied model |

The first line of the disk file must contain the number of input data pairs with format (i4). This line is followed by Reynolds number/skin friction data pairs with format (2e13.6). For example, a series of 10 runs with the Baldwin-Lomax model for channel flow yields the following results.

```
10
1.021000e 03 1.919000e-02
2.625000e 03 1.161000e-02
5.510000e 03 8.538000e-03
1.020000e 04 6.922000e-03
1.413000e 04 6.252000e-03
2.718000e 04 5.308000e-03
4.071000e 04 4.828000e-03
6.429000e 04 4.355000e-03
8.835000e 04 4.100000e-03
1.003000e 05 3.990000e-03
```

For channel flow, confine Reynolds number to the range $10^3 \leq Re_H \leq 10^5$. For pipe flow, Reynolds number should be in the range $10^3 \leq Re_D \leq 10^6$.

**Output description:**

A video plot with six graphs (see Figure C.2) is created on the screen. When the plot is complete, the following message appears:

**Hardcopy output (y/n)?**

Enter a $y$ or a $Y$ to create a hardcopy plot. Pressing any other key terminates the run without creating a hardcopy plot.

**Comments:**

- The following is a sample input data file for a machine with a standard VGA monitor and an HP DeskJet connected to serial port COM1:.

```
mon    =    18              (Standard VGA monitor)
ifore =    15              (Bright-white foreground)
iback =     1              (Blue background)
nprin =     2              (HP DeskJet)
mode  =     3              (300 dots per inch resolution)
ksize =   100              (Full size plot)
symsiz=  .080              (.08" experimental data symbols)
'com1' , 9600 , 'none' , 1 , 8
```

The last line indicates the printer is connected to serial port COM1: and the port is set at 9600 baud, no parity, 1 stop bit and 8 data bits.

- If disk file **plotp.dat** is not available, Program **PLOTP** uses the following set of default values:

  $mon = 21$, $ifore = 15$, $iback = 1$, $nprin = 5$, $mode = 4$, $ksize = 100$, $symsiz = .08$, $devid = $ 'LPT1'

  Note that $nbaud$, $parity$, $nstop$ and $lword$ are not used for parallel ports.

- The following data are used for comparison with computed results:

  | | |
  |---|---|
  | Channel Flow — $Re_\tau < 287$, | Mansour et al. (1988) $[Re_\tau = 180]$ |
  | Channel Flow — $Re_\tau \geq 287$, | Mansour et al. (1988) $[Re_\tau = 395]$ |
  | Pipe Flow — All $Re_\tau$, | Laufer (1952) $[Re_\tau = 1058]$ |

- As many as 6 different curves, each with a unique line style, can be drawn on a single plot.

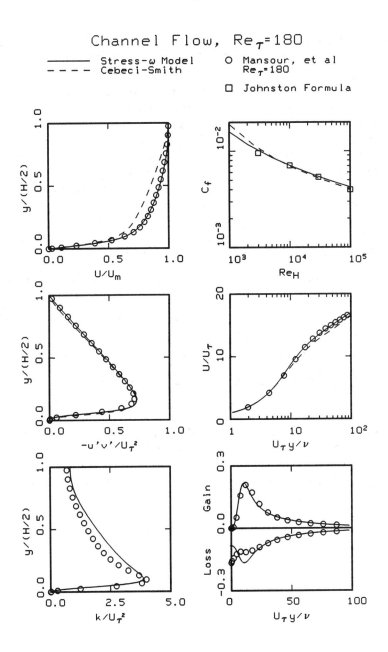

Figure C.2: *Sample plot created by Program PLOTP.*

## C.4   Boundary-Layer Perturbation Analysis

Programs **SUBLAY** and **DEFECT** can be used to compute turbulence-model predicted flow properties in the incompressible viscous sublayer and defect layer, respectively. Section 4.6 of the main text describes the sublayer and defect-layer equations. No additional transformations are introduced in Program **SUBLAY**. Program **DEFECT** uses the Rubel-Melnik (1984) transformation.

As with the free-shear-flow and pipe-flow programs, we add unsteady terms to the turbulence-model equations to facilitate solution of the two-point boundary-value problems appropriate for the sublayer and the defect layer. In Program **SUBLAY**, the momentum equation is solved at each timestep by trapezoidal-rule integration. For example, in the case of a two-equation model, we advance the turbulence parameters in time. Then, after updating the eddy viscosity, we determine the velocity by integration of the following equation.

$$\frac{dU^+}{dy^+} = \frac{1}{1 + \mu_T^+} \tag{C.13}$$

All notation in Equation (C.13) is identical to that used in Section 4.6 of the main text.

The only other subtle feature of Program **SUBLAY** is the way the specific dissipation rate, $\omega$, in the $k$-$\omega$ model and the Stress-$\omega$ model is computed close to the solid boundary. To eliminate numerical error associated with computing the singular behavior of $\omega$ for perfectly-smooth and slightly-rough surfaces, the exact asymptotic behavior of $\omega$ is prescribed close to the surface (see Subsection 7.2.1). That is, we use the fact that, for $y^+ < 2.5$, $\omega^+ = \nu\omega/u_\tau^2$ is given by:

$$\omega^+ \to \frac{6}{\beta_o(y^+)^2} \qquad \text{as} \quad y \to 0 \qquad \text{(smooth wall)} \tag{C.14}$$

$$\omega^+ \to \frac{\omega_w^+}{\left[1 + \sqrt{\frac{\beta_o\omega_w^+}{6}}\, y^+\right]^2} \qquad \text{as} \quad y \to 0 \qquad \text{(rough wall)} \tag{C.15}$$

The exact analytical behavior of $\omega$ is imposed for a prescribed number of mesh points, *jskip*, next to the surface.

In terms of the transformation devised by Rubel and Melnik, the defect-layer equations are as follows. Note that to avoid numerical difficulties, Program **DEFECT** uses a small nonzero value for $K_0(\xi)$ when $\xi \to \infty$. The nonvanishing boundary conditions quoted for $W_0(\xi)$ and $E_0(\xi)$ as $\xi \to \infty$ are the only choices consistent with the similarity solution.

**All Models:**

$$\eta = \int_{-\infty}^{\xi} N_0(\xi') \, d\xi' \tag{C.16}$$

$$\frac{d^2 U_1}{d\xi^2} + (1 + 2\beta_T)\eta \frac{dU_1}{d\xi} + 2\beta_T N_0 U_1 = 0 \tag{C.17}$$

$$\sigma^* \frac{d^2 K_0}{d\xi^2} + 2\beta_T K_0 + (1 + 2\beta_T)\eta \frac{dK_0}{d\xi} + \sqrt{\beta_o^*} \left[ \left( \frac{dU_1}{d\xi} \right)^2 - N_0 E_0 \right] = 0 \tag{C.18}$$

**$k$-$\omega$ Model:**

$$\sigma \frac{d^2 W_0}{d\xi^2} + (1 + 2\beta_T)\eta \frac{dW_0}{d\xi} + (1 + 4\beta_T)N_0 W_0$$
$$+ \frac{\sqrt{\beta_o^*}}{K_0} \left[ \alpha \left( \frac{dU_1}{d\xi} \right)^2 - \frac{\beta_o}{\beta_o^*} K_0^2 \right] W_0 = 0 \tag{C.19}$$

$$W_0(\xi) \to \frac{(1 + 4\beta_T)\sqrt{\beta_o^*}}{\beta_o} \quad \text{as} \quad \xi \to \infty \tag{C.20}$$

$$N_0 = K_0/W_0 \quad \text{and} \quad N_0 E_0 = f_{\beta^*} K_0^2 \tag{C.21}$$

**$k$-$\epsilon$ Model:**

$$\sigma_\epsilon^{-1} \frac{d^2 E_0}{d\xi^2} + (1 + 2\beta_T)\eta \frac{dE_0}{d\xi} + (1 + 6\beta_T)N_0 E_0$$
$$+ \frac{\sqrt{C_\mu}}{K_0} \left[ C_{\epsilon 1} \left( \frac{dU_1}{d\xi} \right)^2 - C_{\epsilon 2} K_0^2 \right] E_0 = 0 \tag{C.22}$$

$$E_0(\xi)/K_0(\xi) \to \frac{(1 + 2\beta_T)}{C_{\epsilon 2}\sqrt{C_\mu}} \quad \text{as} \quad \xi \to \infty \tag{C.23}$$

$$N_0 = K_0^2/E_0 \tag{C.24}$$

## C.4.1  Program SUBLAY: Viscous Sublayer

Program **SUBLAY** computes incompressible viscous sublayer flow, including surface roughness and surface mass transfer.

**Input-parameter description:**

Program **SUBLAY** reads the following input parameters from disk file **sublay.dat** in the order listed below. Integer quantities must be formatted according to (1x,a12,i4) while floating-point quantities must be formatted as (1x,a12,e13.6). See Subsection C.1.2 for a sample input-data file.

*iunit1*   Output-file unit number

  $= 6$ Printed output sent to printer

  $\neq 6$ Printed output saved in disk file **sublay.prt**

*iunit2*   Plotting-data disk file unit number

*model*   Turbulence-model identification flag

  0 $k$-$\omega$ model

  1 Stress-$\omega$ model

  99 User-defined model

*omegw*   Surface value of dimensionless specific dissipation rate, $\omega_w^+$

*vwplus*   Dimensionless vertical velocity at the surface, $v_w/u_\tau$.

*ymax*   Maximum value of $y^+$

*yone*   Value of $y^+$ at the first grid point above the wall

*iruff*   Surface-roughness flag

  0 Rough surface

  1 Smooth surface

*jmax*   Number of grid points

*jskip*   Grid-point number below which the exact asymptotic solution for specific dissipation rate is used

*maxn*   Maximum number of timesteps

*nedit*   Profile-print modulus; profiles are printed every *nedit* steps

*nfreq*   Short-print modulus; maximum error and $C = u^+ - \frac{1}{\kappa}\ell n y^+$ at the specified value of $y^+ = ymax$ are printed every *nfreq* steps

*nvisc*   Viscous-modification flag

  $= 0$ No $k$-$\omega$/Stress-$\omega$ model viscous modifications

  $\neq 0$ Use $k$-$\omega$/Stress-$\omega$ model viscous modifications

**Output description:**

VIDEO OUTPUT includes the timestep number, maximum error, and the constant in the law of the wall, $C = u^+ - \frac{1}{\kappa}\ell n y^+$, at $y^+ = ymax$ every *nfreq* timesteps. PRINTED OUTPUT is sent to the line printer throughout the run if *iunit1* = 6. DISK-FILE OUTPUT is saved in the appropriate plotting-data disk file (see Subsection C.1.3 for file names). The file is **unformatted** and is created using the following statements.

```
      do 50 j=2,jmax
        wtm(j)=-betas*fbetas(j)*bbeta(j)*et(j)*wt(j)
        if(model.ne.1) tau(j)=eps(j)*dudy(j)
        etm(j)=tau(j)*dudy(j)
50    continue
      if(nvisc.ne.0) wtm(1)=2.*wtm(2)-wtm(3)
      etm(1)=0.
      write(iunit2) jmax,0.,0.,0.,0.,'PLOTS'
      write(iunit2) (yplus(j),uplus(j),wtm(j),etm(j),j=1,jmax)
```

The quantity *yplus(j)* is $y^+$, *uplus(j)* is $U^+$, *wtm(j)* is $\beta^* f_{\beta^*} \nu \omega k / u_\tau^4$, and *etm(j)* is $\nu \tau_{xy} (dU/dy) / u_\tau^4$. Additionally, for any value of *iunit1* other than 6, printed output is saved in disk file **sublay.prt** at the conclusion of the run.

## C.4.2    Program DEFECT: Defect Layer

Program **DEFECT** computes properties of the incompressible defect-layer including effects of pressure gradient.

**Input-parameter description:**

Program **DEFECT** reads the following input parameters from disk file **defect.dat** in the order listed below. Integer quantities must be formatted according to (1x,a12,i4) while floating-point quantities must be formatted as (1x,a12,e13.6). See Subsection C.1.2 for a sample input-data file.

| | |
|---|---|
| *iunit1* | Output-file unit number |
| | = 6 Printed output sent to printer |
| | $\neq$ 6 Printed output saved in disk file **defect.prt** |
| *iunit2* | Plotting-data disk file unit number |
| *model* | Turbulence-model identification flag |
| | 0 $k$-$\omega$ model |
| | 2 $k$-$\epsilon$ model (see *kerng* also) |
| | 3 Baldwin-Barth model |
| | 4 Spalart-Allmaras model |
| | 99 User-defined model |
| *kerng* | $k$-$\epsilon$ model identification flag |
| | = 0 Standard $k$-$\epsilon$ model |
| | $\neq$ 0 RNG $k$-$\epsilon$ model |
| *betat* | Pressure-gradient parameter, $\beta_T$ |
| *jmax* | Number of grid points |
| *maxn* | Maximum number of timesteps |
| *nedit* | Profile-print modulus; profiles are printed every *nedit* steps |
| *nfreq* | Short-print modulus; maximum error, wake strength, etc. are printed every *nfreq* steps |

**Output description:**

VIDEO OUTPUT includes timestep number, maximum error, wake strength, etc. every *nfreq* timesteps. PRINTED OUTPUT is sent to the line printer throughout the run if *iunit1* = 6. DISK-FILE OUTPUT is saved in the appropriate plotting-data disk file (see Subsection C.1.3 for file names). The file is **unformatted** and is created using the following statements.

```
write(iunit2) jmax,betat,0.,0.,0.,'PLOTD'
write(iunit2) (eta(j),u(j),j=1,jmax)
```

where *betat* is $\beta_T$, *eta(j)* is $\eta_j$ and *u(j)* is $[U_e - U(\eta_j)]/u_\tau$. Additionally, for any value of *iunit1* other than 6, printed output is saved in disk file **defect.prt** at the conclusion of the run.

## C.4.3   Program PLOTS: Sublayer Plotting Utility

Program **PLOTS** creates video and hardcopy plots of viscous sublayer properties computed with program **SUBLAY** on IBM PC's and compatibles. The program automatically detects the turbulence model used.

**Input-parameter description:**

Program **PLOTS** reads the following seven input parameters from disk file **plots.dat** in the order listed below. Integer quantities must be formatted according to (7x,i6) while floating-point quantities must be formatted as (7x,f6.2). This is similar to the format used for Program **SUBLAY**.

| | |
|---|---|
| *mon* | Monitor type (see Appendix E) |
| *ifore* | Foreground color (see Appendix E) |
| *iback* | Background color (see Appendix E) |
| *nprin* | Printer type (see Appendix E) |
| *mode* | Graphics-mode flag for printers; number of pens for plotters (see Appendix E) |
| *ksize* | Plot scaling factor. Using 100 yields a full-size hardcopy plot. Smaller values yield a hardcopy plot reduced by *ksize* per cent. For example, *ksize* = 50 produces a half-size plot. |
| *symsiz* | Size of experimental data symbols, in inches |

Next, Program **PLOTS** reads a single, free formatted, line to indicate where hardcopy print is directed and where required font files are located. This line comes immediately after the specified value for *symsiz* and defines the following five additional parameters.

| | |
|---|---|
| *devid* | Device name of type character∗4; valid devices are LPT1, LPT2, LPT3, COM1, COM2, COM3, COM4 |
| *nbaud* | Baud rate for a serial port; valid baud rates are 110, 150, 300, 600, 1200, 2400, 4800, 9600 |
| *parity* | Parity of type character∗3 or character∗4 for a serial port; valid parity settings are 'even', 'odd' and 'none' |
| *nstop* | Number of stop bits for a serial port; either 1 or 2 |
| *lword* | Word length for a serial port; either 7 or 8 |

**Output description:**

A video plot with two graphs (see Figure C.3) is created on the screen. When the plot is complete, the following message appears:

**Hardcopy output (y/n)?**

Enter a *y* or a *Y* to create a hardcopy plot. Pressing any other key terminates the run without creating a hardcopy plot.

**Comments:**

- The following is a sample input data file for a machine with a standard VGA monitor and an HP DeskJet connected to serial port COM1:.

```
mon   =    18              (Standard VGA monitor)
ifore =    15              (Bright-white foreground)
iback =     1              (Blue background)
nprin =     2              (HP DeskJet)
mode  =     3              (300 dots per inch resolution)
ksize =   100              (Full size plot)
symsiz=  .080              (.08" experimental data symbols)
'com1' , 9600 , 'none' , 1 , 8
```

The last line indicates the printer is connected to serial port COM1: and the port is set at 9600 baud, no parity, 1 stop bit and 8 data bits.

- If disk file **plots.dat** is not available, Program **PLOTS** uses the following set of default values:

  *mon* = 21, *ifore* = 15, *iback* = 1, *nprin* = 5, *mode* = 4, *ksize* = 100, *symsiz* = .08, *devid* = 'LPT1'

  Note that *nbaud*, *parity*, *nstop* and *lword* are not used for parallel ports.

- As many as 3 different curves, each with a unique line style, can be drawn on a single plot.

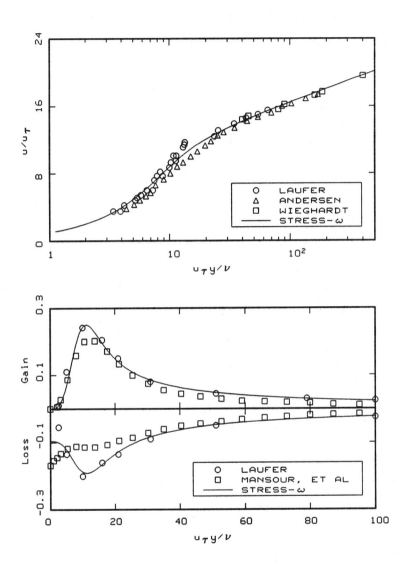

Figure C.3: *Sample plot created by Program PLOTS.*

## C.4.4   Program PLOTD: Defect-Layer Plotting Utility

Program **PLOTD** creates video and hardcopy plots of the defect-layer velocity
profile computed with program **DEFECT** on IBM PC's and compatibles. The
program automatically detects the turbulence model used.

**Input-parameter description:**
   Program **PLOTD** reads the following nine input parameters from disk file
**plotd.dat** in the order listed below. Integer quantities must be formatted according-
ing to (7x,i6) while floating-point quantities must be formatted as (7x,f6.2). This
is similar to the format used for Program **DEFECT**.

| | |
|---|---|
| *mon* | Monitor type (see Appendix E) |
| *ifore* | Foreground color (see Appendix E) |
| *iback* | Background color (see Appendix E) |
| *nprin* | Printer type (see Appendix E) |
| *mode* | Graphics-mode flag for printers; number of pens for plotters (see Appendix E) |
| *ksize* | Plot scaling factor. Using 100 yields a 5″ by 5″ hardcopy plot. Smaller values yield a hardcopy plot reduced by *ksize* per cent. For example, *ksize* = 50 produces a 2.5″ by 2.5″ plot. |
| *symsiz* | Size of experimental data symbols, in inches |
| *height* | Physical height, in inches, of the video display |
| *width* | Physical width, in inches, of the video display |

   Next, Program **PLOTD** reads a single, free formatted, line to indicate where
hardcopy print is directed and where required font files are located. This line
comes immediately after the specified value for *width* and defines the following
five additional parameters.

| | |
|---|---|
| *devid* | Device name of type character∗4; valid devices are LPT1, LPT2, LPT3, COM1, COM2, COM3, COM4 |
| *nbaud* | Baud rate for a serial port; valid baud rates are 110, 150, 300, 600, 1200, 2400, 4800, 9600 |
| *parity* | Parity of type character∗3 or character∗4 for a serial port; valid parity settings are 'even', 'odd' and 'none' |
| *nstop* | Number of stop bits for a serial port; either 1 or 2 |
| *lword* | Word length for a serial port; either 7 or 8 |

**Output description:**
   A 5″ by 5″ video plot (see Figure C.4) is created centered on the screen.
When the plot is complete, the following message appears:

## Hardcopy output (y/n)?

Enter a *y* or a *Y* to create a hardcopy plot. Pressing any other key terminates the run without creating a hardcopy plot.

## Comments:

- The following is a sample input data file for a machine with a standard VGA monitor and an HP DeskJet connected to serial port COM1:.

```
mon    =     18        (Standard VGA monitor)
ifore =     15         (Bright-white foreground)
iback =      1         (Blue background)
nprin =      2         (HP DeskJet)
mode  =      3         (300 dots per inch resolution)
ksize =    100         (Full size plot)
symsiz=   .080         (.08" experimental data symbols)
height= 7.500          (7.5" high video display)
width = 9.250          (9.25" wide video display)
'com1' , 9600 , 'none' , 1 , 8
```

The last line indicates the printer is connected to serial port COM1: and the port is set at 9600 baud, no parity, 1 stop bit and 8 data bits.

- If disk file **plotd.dat** is not available, Program **PLOTD** uses the following set of default values:

  *mon* = 21, *ifore* = 15, *iback* = 1, *nprin* = 5, *mode* = 4, *ksize* = 100, *symsiz* = .08, *height* = 7.5, *width* = 9.25, *devid* = 'LPT1'

  Note that *nbaud*, *parity*, *nstop* and *lword* are not used for parallel ports.

- The following experimental data from Coles and Hirst (1969) are used:

$$\beta_T < 0, \quad \text{Flow 2800: Herring-Norbury } (\beta_T = -0.35)$$
$$0 \le \beta_T \le 2, \quad \text{Flow 1400: Wieghardt } (\beta_T = 0.0)$$
$$2 < \beta_T < 7, \quad \text{Flow 2600: Bradshaw } (\beta_T = 4.0)$$
$$\beta_T \ge 7, \quad \text{Flow 2300: Clauser } (\beta_T = 8.7)$$

- As many as 5 different curves, each with a unique line style, can be drawn on a single plot.

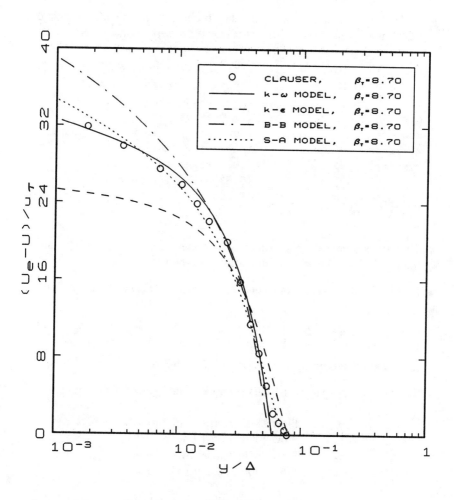

Figure C.4: *Sample plot created by Program PLOTD.*

# C.5 Miscellaneous Routines

This section includes several utility routines called by the various programs described in the preceding sections. They implement several standard mathematical procedures such as the Runge-Kutta predictor-corrector method for integrating a system of ordinary differential equations, the Newton iteration method for solving transcendental equations and Thomas' algorithm for solving a tridiagonal matrix system. These routines are of general usefulness in computational fluid mechanics as well as for the programs that are the main topic of this appendix. We assume the user is familiar with these algorithms and thus include only instructions on use of the subroutines. Users unfamiliar with the techniques should refer to texts such as Abramowitz and Stegun (1965), Hildebrand (1976), Chapra and Canale (1985) and Press, Flannery, Teukolsky and Vetterling (1987).

There is one routine, **NAMSYS**, that is called by all of the programs. The purpose of this routine is to make the programs portable to a variety of computers. The routine sets several system-dependent variables used in opening files, assigning video and hardcopy units, etc. The standardization of Fortran-77 and Fortran-90, including VAX extensions, by most compiler writers makes it possible to confine virtually all system-dependent parameters to this single subroutine.

## C.5.1   Function ERF: Error Function

Function **ERF** computes the error function $\mathrm{erf}(x)$ defined by

$$\mathrm{erf}(x) = \frac{2}{\sqrt{\pi}} \int_0^x e^{-t^2}\, dt$$

**Usage:**  $<$real variable$> = \mathrm{erf}(x)$

**Input-parameter description:**

$x$        Error function argument, $x$

**Output-parameter description:**

*erf*      Computed value, $\mathrm{erf}(x)$

**Comments:**

- A polynomial approximation is used. When $x < 0$, the function uses the fact that

$$\mathrm{erf}(-x) = -\mathrm{erf}(x)$$

## C.5.2 Subroutine NAMSYS: Fortran Portability

Subroutine **NAMSYS** returns several system-dependent and compiler-specific parameters to aid in portability of Fortran programs.

**Usage:** call namsys(icpu,iin,iv,msdos,newfil,pform)

### Input-parameter description:

*icpu*      CPU identification flag
- 0 SVS Fortran (680x0, 80x86)
- 1 Lahey/Microsoft Fortran (8088, 80x86)
- 2 VAX/VMS
- 3 SUN Fortran (SUN Workstation, Definicon SPARC)
- 4 Cray Fortran/Unicos (Cray X/MP, Y/MP)
- 5 Silicon Graphics Iris

### Output-parameter description:

*iin*      Input data file logical unit number; set to unit 15 for all CPU's

*iv*      Standard console unit number; set to 5 for all CPU's

*msdos*    Open-printer flag
- 0 Printer opened as *'prn'*
- 1 Don't open *'prn'*

*newfil*    Character*7 string used in opening new files
- *'new'* if compiler writes over an existing file
- *'unknown'* for Ansi-77 standard operation

*pform*    Character*9 string used as format type for printer output that is redirected to a disk file
- *'printer'* for SVS Fortran
- *'print'* for SUN Fortran
- *'formatted'* for all others

### Comments:

- This routine is currently configured for the CPU's and Fortran compilers listed in the Input-parameter description. Other CPU's and compilers can be included by adding the appropriate statements to the routine. The following page includes an abbreviated listing of **NAMSYS**.

```fortran
      subroutine namsys(icpu,iin,iv,msdos,newfil,pform)
      character newfil*7,pform*9
      iin=15
      iv=5
c-----------------------------------
c  SVS Fortran (680x0 and 80x86)
      if(icpu.eq.0) then
        msdos=0
        pform='printer'
        newfil='new'
c-----------------------------------
c  Lahey/Microsoft Fortran (80x86)
      elseif(icpu.eq.1) then
        msdos=0
        pform='formatted'
        newfil='unknown'
c-----------------------------------
c             VAX/VMS
      elseif(icpu.eq.2) then
        msdos=1
        pform='formatted'
        newfil='new'
c-----------------------------------
c  SUN Fortran...SUN Workstation
c      and Definicon SPARC
      elseif(icpu.eq.3) then
        msdos=0
        pform='print'
        newfil='unknown'
c-----------------------------------
c  Cray Fortran...Unicos
      elseif(icpu.eq.4) then
        msdos=1
        pform='formatted'
        newfil='unknown'
c-----------------------------------
c     Silicon Graphics Iris
      elseif(icpu.eq.5) then
        msdos=0
        pform='formatted'
        newfil='unknown'
c-----------------------------------
c     Error...say so and quit
      else
        write(*,*) 'icpu = ',icpu,' is not supported!!!'
        stop
      endif
      return
      end
```

## C.5.3 Subroutine RKGS: Runge-Kutta Integration

Subroutine **RKGS** solves a system of first-order ordinary differential equations defined in an external subroutine with given initial values. The system of equations is of the form

$$\frac{dy_i}{dx} = f_i(x, y_j) \quad \text{for} \quad x_0 \leq x \leq x_1$$

**Usage:** call rkgs(prmt,y,yp,ndim,ihlf,fct,outp,aux)

**Input-parameter description:**

*fct*     The name of an external subroutine used to compute the right hand side vector, $\mathbf{f} = yp$. The argument list to this subroutine must be *(x,y,yp)*, and the subroutine must leave the values of $x$ and $y$ unchanged.

*ndim*    Number of equations in the system

*prmt*    Input and output array with dimension $\geq 5$, that specifies interval and accuracy parameters and that serves for communication between output subroutine *outp* (furnished by the user) and subroutine **RKGS**. With the exception of *prmt(5)*, the components are not destroyed by subroutine **RKGS** and they are:

    *prmt(1)* Lower bound of the interval, $x_0$

    *prmt(2)* Upper bound of the interval, $x_1$

    *prmt(3)* $\Delta x_0$, initial increment of $x$

    *prmt(4)* Upper error bound. If the absolute error is greater than *prmt(4)*, the increment is halved. If the increment is less than $\Delta x_0$ and the absolute error is less than *prmt(4)*/50, the increment is doubled. If desired, the user can change *prmt(4)* in output subroutine *outp*.

    *prmt(5)* Termination parameter. Subroutine **RKGS** initially sets *prmt(5)* = 0. In order to terminate subroutine **RKGS** at any output point, change *prmt(5)* to a nonzero value in subroutine *outp*.

*y*       Input vector of initial values, $y_0$.

*yp*      Input vector of error weights; the sum of its components must be equal to 1.

**Output-parameter description:**

*aux*     An *(8 x ndim)* auxiliary storage array

*ihlf*    Number of bisections of the initial increment. If *ihlf* exceeds 10, subroutine **RKGS** returns *ihlf* = 11. Additionally, if $\Delta x_0 = 0$, **RKGS** returns with *ihlf* = 12 while if $\Delta x_0$ and $(x_1 - x_0)$ differ in sign, **RKGS** returns *ihlf* = 13.

*outp*    The name of an external subroutine used for program output. The form of its argument list must be *(x,y,yp,ihlf,ndim,prmt)*. None of these parameters (except, if necessary, *prmt(4)*, *prmt(5)*, ...) should be changed by subroutine *outp*. If *prmt(5)* is changed to a nonzero value, subroutine **RKGS** is terminated.

*prmt*    Input and output array with dimension $\geq 5$, that specifies interval and accuracy parameters and that serves for communication between output subroutine *outp* (furnished by the user) and subroutine **RKGS**.

> *prmt(6)* Although not required by subroutine **RKGS**, additional parameters can be included in array *prmt* provided its dimension is declared to be $> 5$. Such parameters may be useful for passing values to the routine calling **RKGS** which are obtained by special manipulations of output data in subroutine *outp*.

*y*       Output vector of computed values, **y**, at intermediate points.

*yp*      Output vector of derivatives, corresponding to function values *y* at point *x*.

**Comments:**

- Computation is done using the fourth-order accurate Runge-Kutta method with Gill's modification. Accuracy is tested by comparing the results of the procedure with single and double increments of the independent variable, $\Delta x$. Subroutine **RKGS** automatically adjusts the increment during the computation by halving or doubling $\Delta x$. The procedure terminates and returns to the calling routine, if any of the following conditions occur.

  1. More than 10 bisections of the initial increment are necessary to achieve satisfactory accuracy (error flag $ihlf = 11$);

  2. Either the initial increment $\Delta x_0 = 0$ or has the wrong sign (error flags $ihlf = 12$ or $ihlf = 13$);

  3. The end of the integration interval, $x = x_1$, reached;

  4. Subroutine *outp* has changed *prmt(5)* to a nonzero value.

- The calling routine must declare the two user-supplied subroutines named *outp(x,y,yp,ihlf,ndim,prmt)* and *fct(x,y,yp)* as external by including the statements **EXTERNAL FCT** and **EXTERNAL OUTP**.

## C.5.4 Subroutine RTNI: Newton's Iterations

Subroutine **RTNI** solves a general equation of the form $f(x) = 0$ using Newton's iteration method. The function $f(x)$ is specified by the user in a SUBROUTINE subprogram.

**Usage:** call rtni(x,f,fp,fct,xst,eps,iend,ier)

### Input-parameter description:

| | |
|---|---|
| *eps* | Upper bound on the error in $x$ |
| *fct* | Name of the external subroutine used. It computes $f = f(x)$ and $fp = df/dx$ for a given value of $x$. Its argument list must be *(x,f,fp)*. |
| *iend* | Maximum number of iterations allowed |
| *xst* | Initial guess of the root $x_{st}$ |

### Output-parameter description:

| | |
|---|---|
| *f* | Computed value of $f(x)$ at root $x$ |
| *fp* | Computed value of $df/dx$ at root $x$ |
| *ier* | Error flag |
| |    0 No error |
| |    1 No convergence after *iend* iterations |
| |    2 $df/dx = 0$ encountered |
| *x* | Computed root of $f(x) = 0$ |

### Comments:

- Solution of the equation $f(x) = 0$ is obtained using Newton's iteration method, which starts at the initial guess $x_{st}$ of the root $x$. Convergence is quadratic if the value of $df/dx$ at root $x$ is not equal to zero. One iteration step requires one evaluation of $f(x)$ and one evaluation of $df/dx$.

- The subroutine returns with the error flag *ier* = 2 if, at any iteration step, $df/dx$ vanishes.

- The calling routine must declare the subroutine *fct(x,f,fp)* as external by including an **EXTERNAL FCT** statement.

## C.5.5    Subroutine TRI: Tridiagonal Matrix Inversion

Subroutine **TRI** solves the tridiagonal matrix equation

$$a_i x_{i-1} + b_i x_i + c_i x_{i+1} = d_i \quad for \quad I_l + 1 \le i \le I_u - 1$$

subject to either Dirichlet or Neumann boundary conditions.

**Usage:**    call tri(a,b,c,d,il,iu,ibcl,dl,ibcu,du)

**Input-parameter description:**

$a$      Array of matrix elements left of the diagonal, $a_i$; destroyed in the computation.

$b$      Array of matrix diagonal elements, $b_i$; destroyed in the computation.

$c$      Array of matrix elements right of the diagonal, $c_i$.

$d$      Input right-hand-side vector, $d_i$. This vector is replaced by the solution vector, $x_i$.

$dl$      $d_{I_l}$, lower boundary-condition value

$du$      $d_{I_u}$, upper boundary-condition value

$ibcl$      Lower boundary-condition flag
         0   Dirichlet, $x_{I_l} = d_{I_l}$
         1   Neumann, $x_{I_l+1} - x_{I_l} = d_{I_l}$

$ibcu$      Upper boundary-condition flag
         0   Dirichlet, $x_{I_u} = d_{I_u}$
         1   Neumann, $x_{I_u} - x_{I_u-1} = d_{I_u}$

$il$      $I_l$, lower bound on $i$

$iu$      $I_u$, upper bound on $i$; also length of $a_i$, $b_i$, $c_i$ and $d_i$.

**Output-parameter description:**

$d$      Solution vector, $x_i$. This vector replaces input vector $d_i$.

**Comments:**

- The solution is obtained using Thomas' algorithm. The input arrays $a_i$, $b_i$, $c_i$ and $d_i$ need only be specified for indices in the range

$$I_l + 1 \le i \le I_u - 1$$

# C.6  Diskette Contents

The following files are written from the distribution disk to your hard disk in a master directory, **\dcwinc**, and several subdirectories as indicated. The master directory also includes a file named **readme.txt** that gives up-to-date information.

IBM PC Executable Programs (\**dcwinc**):

| | |
|---|---|
| **defect.exe** | Program **DEFECT** |
| **jet.exe** | Program **JET** |
| **mixer.exe** | Program **MIXER** |
| **pipe.exe** | Program **PIPE** |
| **plotd.exe** | Plotting program for Program **DEFECT** |
| **plotf.exe** | Plotting program for **JET, MIXER** and **WAKE** |
| **plotp.exe** | Plotting program for Program **PIPE** |
| **plots.exe** | Plotting program for Program **SUBLAY** |
| **sublay.exe** | Program **SUBLAY** |
| **wake.exe** | Program **WAKE** |

Flowfield Program Source (\**dcwinc\appendc**):

| | |
|---|---|
| **cpuid** | Include file specifying CPU type |
| **defect.for** | Source code for Program **DEFECT** |
| **jet.for** | Source code for Program **JET** |
| **mixer.for** | Source code for Program **MIXER** |
| **pipe.for** | Source code for Program **PIPE** |
| **sublay.for** | Source code for Program **SUBLAY** |
| **wake.for** | Source code for Program **WAKE** |

Miscellaneous Routine Source (\**dcwinc\appendc**):

| | |
|---|---|
| **erf.for** | Source code for Function **ERF** |
| **namsys.for** | Source code for Subroutine **NAMSYS** |
| **rkgs.for** | Source code for Subroutine **RKGS** |
| **rtni.for** | Source code for Subroutine **RTNI** |
| **tri.for** | Source code for Subroutine **TRI** |

Input Data (\**dcwinc\data**):

| | |
|---|---|
| **defect.dat** | Input data for Program **DEFECT** |
| **jet.dat** | Input data for Program **JET** |
| **mixer.dat** | Input data for Program **MIXER** |
| **pipe.dat** | Input data for Program **PIPE** |
| **plotd.dat** | Input data for Program **PLOTD** |
| **plotf.dat** | Input data for Program **PLOTF** |
| **plotp.dat** | Input data for Program **PLOTP** |
| **plots.dat** | Input data for Program **PLOTS** |
| **sublay.dat** | Input data for Program **SUBLAY** |
| **wake.dat** | Input data for Program **WAKE** |

# Appendix D

# Program EDDYBL

## D.1  Overview

This appendix is the user's guide for Program **EDDYBL**, a two-dimensional and axisymmetric, compressible boundary-layer program for laminar, transitional and turbulent boundary layers that is included on the distribution diskette. An overview of the program's operation is given, along with instructions for installing the program and its menu-driven input-data preparation utility, **SETEBL**, on your computer. Two bench-mark runs are described that can be used to make sure the program is operating properly. The software includes a plotting utility for both video and hardcopy plots on IBM PC and compatible computers.

### D.1.1  Acknowledgments

Program **EDDYBL** is a compressible, two-dimensional and axisymmetric program suitable for computing properties of laminar, transitional and turbulent boundary layers. The program embodies a wide variety of turbulence models ranging from mixing-length oriented algebraic models to a complete stress-transport model. This program has evolved over the past two and a half decades and can thus be termed a mature software package. Many U. S. Government Agencies have contributed to development of the program that is based on a computer code originally developed by Price and Harris (1972).

Additionally, important improvements have been made to this software package as a result of feedback from users, most notably from the outstanding fluid mechanics students at USC and UCLA. The author owes special thanks to Dr. G. Brereton of the University of Michigan whose personal efforts resulted in the addition of the option to use either USCS or SI units, and to J. Morrison of AS&M for adapting the software to a SUN Workstation.

461

## D.1.2   Required Hardware and Software

Versions of the program are available for the following computers.

- Cray X-MP and Y-MP
- VAX 11 and 8600
- SUN Workstations
- Silicon Graphics Iris

- 80386/80486/Pentium Microcomputers
- Definicon 68020/68030 Coprocessor Boards
- Definicon SPARC Coprocessor Boards
- IBM PC/XT/AT and Compatibles

The program requires at least 320 kilobytes of memory. To achieve sensible computing times, IBM PC/XT/AT and compatibles must have an 8087 or 80287 math coprocessor, and must use Microsoft Fortran Version 5.0 or higher. Intel 80386 based machines must have either an 80387 or Weitek math coprocessor.

# D.2   Getting Started Quickly

Because **EDDYBL** and its input-data preparation utility, **SETEBL**, run on many different computers, installation of the software is a little different for each machine. The main difference occurs in the commands needed to compile and link the programs. To install the software on an IBM PC or compatible, execute the batch file **setup.bat**, which places executable programs, source code and input-data files in a directory \**dcwinc** and several subdirectories (see Section D.12 for additional details). The diskette label explains how to run **setup.bat**. To install the software on a computer other than an IBM PC or compatible, you must first extract the source files from the ".zip" files included on the distribution diskette. Then, skip ahead to Sections D.3 and D.4. Regardless of the computer you are using, once you have executable programs, complete the installation as follows.

1. Read the contents of the file **readme.txt** in subdirectory \**dcwinc** or directly from the distribution diskette. This file will tell you of any program revisions. Section D.12 indicates the location of all pertinent program files copied from the diskette. Now, copy the following files to your working directory:

   | | | |
   |---|---|---|
   | eddybl.exe | blocrv.dat | ploteb.exe |
   | instl.exe | heater.dat | ploteb.dat |
   | setebl.exe | presur.dat | exper.dat |

   Omit the files **ploteb.exe**, **ploteb.dat** and **exper.dat** if you are using a computer other than an IBM PC or compatible microcomputer.

2. Run Program **INSTL** and answer all questions posed by the program. This program generates a file named **grafic.dat** that should be saved in your working directory.

3. If your computer is an IBM PC or compatible, install the **ansi.sys** driver supplied with your MS-DOS or Windows-95 operating system by adding the following command to your **config.sys** file:

**device=ansi.sys**

Make sure the file **ansi.sys** is available in your path. If you have not previously had this command in your **config.sys** file, you must now re-boot your computer to install the **ansi.sys** driver.

A simple bench-mark case is built into the software package to allow you to quickly determine that everything is operating properly, and to see how easy it is to use Program **EDDYBL**. Because the input-data preparation utility, **SETEBL**, is menu driven, you will find that very little explanation of the program's operation is needed. After successfully completing the following bench-mark run, the first time user should nevertheless do the example of Section D.5 to be sure the software is properly installed and to learn some of the more subtle features of Program **SETEBL**.

1. The first step is to run **SETEBL**. If you have not installed **SETEBL**, you will be notified with a brief message after which the program will immediately terminate. If this happens, refer to Section D.3 and perform the installation procedure.

2. Assuming the program is properly installed, you will see a message informing you that file **eddybl.dat** does not exist. The message asks if you want to create a new file named **eddybl.dat**. For this sample session, you should answer yes by typing the letter *Y* or *y* followed by pressing the *ENTER* key.

3. Having performed Step 2, you are presented with the main menu (Figure D.1) on which ten options are listed. **SETEBL** has default values for all input parameters that correspond to an incompressible (Mach .096) flat-plate boundary layer. When you eventually exit **SETEBL**, these data will be written into an Ascii data file named **eddybl.dat**. For this bench-mark case, if you selected USCS units as the default when you ran Program **INSTL**, you have no need to change any data. However, this case must be done using USCS units. If you selected SI units as the default when you ran **INSTL**, you must change to USCS units. Type either a *U* or a *u* (for Units - note that the *U* is in reverse video on your display) and press the *ENTER* key to make the change. The menu will change to indicate which units are in effect. Before exiting, you must generate two other data files, viz., **table.dat** and **input.dat** that are needed in order to run **EDDYBL**.

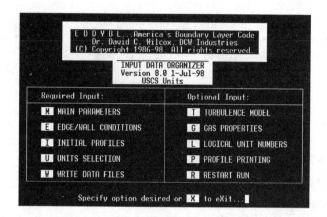

Figure D.1: *Opening menu of Program SETEBL.*

4. To generate these files, select the **Write Data Files** option. To do this, type a *W* or a *w* followed by pressing the *ENTER* key. After a short wait, you will be notified that the binary data file **table.dat** has been successfully written. You are now presented with the following query in reverse video.

   **Save the profiles in Ascii form?  (X=eXit, Y=Yes, ENTER=No)...**

   If you desire a copy of the initial profiles to be saved in a disk file named **setebl.prt** for inspection at a later time, respond with a *Y, ENTER* sequence; otherwise press the *ENTER* key. After you have responded to this query, a second query will appear, viz.,

   **Display the profiles on the video?  (X=eXit, Y=Yes, ENTER=No)...**

   If you want to see the profiles on your video display, respond accordingly. Otherwise, press *ENTER*. After you have responded, your screen clears again and a message appears indicating initial profiles are being generated. If you elected to display profiles on your video display, they will now be displayed, a screen at a time. Press *ENTER* to advance to the next screen. Regardless of the options you have chosen, the precise values of the integral parameters for your computed initial profiles are displayed. Finally, a message appears indicating the binary data file **input.dat** has been successfully written and that you must press *ENTER* in order to continue.

5. After you press *ENTER*, control returns to the main menu. At this point you have prepared all input-data files for the bench-mark run. Exit by typing *X, ENTER*.

6. All that remains now is to run Program **EDDYBL**. Program output will be directed to a disk file named **eddybl.prt**. The file **eddybl.prt** supplied on the distribution diskette contains the printout for the bench-mark run on a personal computer. Your results should agree to within several decimal places with those in the sample printout. For reference, Table D.1 summarizes approximate computing times required for several computers.

7. If you are using an IBM PC or compatible computer, you can generate a video and hardcopy plot of the computational results by running Program **PLOTEB**. Before executing this program, be sure to modify the input-data file, **ploteb.dat**, as required for your system. Section D.8 describes all input parameters in the file. If you are running under Windows 3.1, be sure to use a full-screen DOS window when you run PLOTEB.

Table D.1: *Computing Time for the Bench-Mark Case*

| Computer | CPU (MHz) | FPU (MHz) | CPU Time(sec) |
|----------|-----------|-----------|---------------|
| Cray 2 | - | - | 1 |
| PC-Pentium | 80586 (133) | 80587 (133) | 2 |
| PC-486/DX4 | 80486 (100) | 80487 (100) | 3 |
| CDC-7600 | - | - | 3 |
| PC-486/DX2 | 80486 ( 66) | 80487 ( 66) | 5 |
| PC-486/Weitek | 80486 ( 25) | mW4167 ( 25) | 7 |
| PC-486/DX | 80486 ( 33) | 80487 ( 33) | 9 |
| PC-386/Weitek | 80386 ( 33) | mW3167 ( 33) | 12 |
| VAX 8600 | - | - | 17 |
| SPARC | 7C601 ( 20) | 8847 ( 20) | 21 |
| PC-386/DX | 80386 ( 33) | 80387 ( 33) | 26 |
| Tandy 4000 | 80386 ( 16) | mW1167 ( 16) | 30 |
| DSI-785+ | 68020 ( 20) | 68882 ( 20) | 42 |
| VAX 11/785 | - | - | 63 |
| PC-286/DX | 80286 ( 12) | 80287 ( 8) | 270 |

## D.3 Installing SETEBL

To use the supplied executable version of **SETEBL** on an IBM PC or compatible microcomputer, including the default values specified for all input-data parameters, simply copy the executable file to your working directory. Otherwise, if you wish to change some of the default values, or if you are using a computer other than an IBM PC, the first step required to install Program **SETEBL** is to compile and link the program. The main program is the file named **setebl.for**, and the various subroutines are listed in Section D.12. All routines reference three **include** files, **chars**, **comeb1** and **comeb2**. Section D.10 summarizes the commands required to compile and link Program **SETEBL**.

> **The first step required to install SETEBL for your computer is to either copy the executable file to your working directory or to compile and link Program SETEBL.**

In order to use Program **SETEBL**, you must first install it for your particular console. The program makes extensive use of reverse video, direct cursor positioning, and some graphics characters. Since no uniform standard exists for such console characteristics, the appropriate sequences used by your console must be defined for Program **SETEBL**.

## D.3.1  Boot-Console Installation

In order to install **SETEBL** on your main (or boot) console, you must generate a binary data file named **grafic.dat** that contains all of the information needed by **SETEBL**. The source code for a program that generates **grafic.dat** customized for your console has been supplied as part of this software package. The program is called **INSTL**, and the source is contained in **instl.for**. If you customize Program **INSTL** or if you are using a computer other than an IBM PC, you must first compile and link Program **INSTL**. Then:

> **The second step required to install SETEBL for your console is to run Program INSTL.**

When you run Program **INSTL**, you will be given the option of specifying whether you want the default units to be USCS or SI. Make the choice best suited to your needs. You will also have to specify the type of computer you have and, in some cases, the type of console.

When you have successfully run Program **INSTL**, the required binary data file **grafic.dat** will be created and **INSTL** will print a message to that effect. Whenever you wish to run **SETEBL**, simply make sure **grafic.dat** is present in your directory. If it is not present, **SETEBL** displays a message informing you that you are attempting to run an uninstalled version of **SETEBL** and promptly terminates. If you are running Program **SETEBL** on an IBM PC based system, you must also install the **ansi.sys** driver. Thus,

> The third step required to install **SETEBL** on an **IBM PC** based
> system is to install the ansi.sys driver by adding the following
> command to your config.sys file.

**device=ansi.sys**

If you have not previously had this command in your **config.sys** file, it will not
take effect until you re-boot your computer.

### D.3.2   Remote-Terminal Installation

For a remote terminal whose characteristics are different from those of your boot
console, you can create another **grafic.dat** by making appropriate changes to
Program **INSTL**. The program is heavily commented, and customization should
be straightforward.

## D.4   Installing EDDYBL

To use the supplied executable version of **EDDYBL** on an IBM PC or compatible microcomputer, simply copy the executable file to your working directory.
Otherwise, if you wish to make program changes, or if you are using a computer
other than an IBM PC, the first step required to install Program **EDDYBL** is
to compile and link the program. The main program is the file **eddybl.for** that
also makes use of the include files **common** and **cpuid**. Be sure to link with
the **/e** option for the Microsoft Fortran version or the **-pack** option with SVS
Fortran-386/Phar Lap to reduce the size of the executable file.

> The only step required to install **EDDYBL** for your computer is
> to either copy the executable file to your working directory or
> to compile and link Program **EDDYBL**.

## D.5   Running a General Case

This section explores, in detail, all of the salient features of the input-data preparation utility, **SETEBL**. You will be guided through the various menus and, in
the process, you will set up a constant-pressure boundary-layer computation for
a Mach 1 freestream. For the case you will do, freestream conditions are as
follows.

| | | |
|---|---|---|
| Total pressure, $p_{t_\infty}$ | = | 482.7 lb/ft$^2$ (23112 N/m$^2$) |
| Total temperature, $T_{t_\infty}$ | = | 468° R (260 K) |
| Mach number, $M_\infty$ | = | 1 |

The surface will be slightly cooled so that surface temperature is 95% of the adiabatic-wall temperature.

Your goal is to initiate the computation at a plate-length Reynolds number, $Re_x$, of one million and determine the point where the momentum-thickness Reynolds number, $Re_\theta$, is 8000. You might want to do this, for example, in order to provide upstream profiles for a Navier-Stokes computation. You know from a correlation of experimental data that when $Re_x = 1.0 \cdot 10^6$, the boundary layer has the following integral properties:

| | | |
|---|---|---|
| Skin friction, $c_f$ | = | .0038 |
| Shape factor, $H$ | = | 1.80 |
| Boundary-layer thickness, $\delta$ | = | 11.9 $\theta$ |
| Reynolds number, $Re_\theta$ | = | 1500 |

Finally, the surface is perfectly smooth, there is no surface mass transfer, and the Spalart-Allmaras model will be used.

## D.5.1  Preliminary Operations

To perform this exercise, delete any existing **eddybl.dat** data file that might be in your directory. Although this is not generally necessary, for the purposes of this section it will be easier if you begin with the default values.

As with the bench-mark case of Section D.2, the very first step is to run Program **SETEBL**. If you have not installed the program, you will be notified with a brief message after which the program will immediately terminate. If this happens, go back to Section D.3 and perform the installation procedure.

Assuming the program is properly installed, you will see a message informing you that file **eddybl.dat** does not exist. You will be asked if you want to create a new file named **eddybl.dat**. For this sample session, you should answer yes by typing the letter $Y$ followed by pressing the *ENTER* key.

## D.5.2  Units Selection

This case can be done in either USCS or SI units. Examine the main menu to determine which units are in effect. If you wish to change units, type a $U$ followed by pressing *ENTER*. The menu will reflect the change in units immediately. If you change your mind and wish to go back to the original units, repeat the $U$, *ENTER* sequence. In the following sections, values are quoted in USCS units followed by corresponding SI values in parentheses.

## D.5.3   Main Parameters

At this point, you will be presented with the main menu on which ten options
are listed. Begin by entering the **Main Parameters** sub-menu. To enter this
sub-menu, type an *M* followed by pressing the *ENTER* key.

Yet another sub-menu will now appear that gives you the choice of entering
input data for either **Freestream Conditions** or **Body Parameters**. There is a
third option that allows you to **eXit**. The latter option permits you to return to
the previous menu. You will eventually do so, but first you will do some actual
data preparation.

**Freestream Parameters.** Type an *F* followed by pressing the *ENTER* key to
descend to the **Freestream Conditions** menu. You will now see a display that
includes seven of the primary quantities that specify freestream flow conditions,
including freestream total pressure, total temperature, Mach number, shock-wave
angle, and some turbulence parameters. The bottom row provides instructions
on how to proceed. Press the *ENTER* key several times, for example, and you
will see the arrow move from one input variable name to the next. When you
reach the last variable, pressing the *ENTER* key again will cause the arrow to
move to the uppermost variable. You may make as many passes through the list
of variables as you wish.

This particular menu includes a help option to further explain the meaning of
the more obscure input quantities. To display the **Help** menu, type an *H* followed
by pressing the *ENTER* key. After reading this **Help** menu, pressing the *ENTER*
key returns you to the **Freestream Conditions** data-entry menu.

Having returned from the Help menu, you will now exercise the **Change**
option. First, position the arrow in front of Mach number. You accomplish this
by pressing *ENTER* twice. Now, type the letter *C* (for Change) followed by
pressing *ENTER*. The bottom line of the menu will now change. You are told to
specify the new value, and that the FORMAT must be the standard FORTRAN
floating-point format E13.6. The default value assigned to the Mach number
is .096, corresponding to essentially incompressible flow conditions. Change
the Mach number to one by typing 1. (the exponent E+00 is unnecessary but
the decimal point is mandatory — this is normal FORTRAN I/O). As with all
commands to **SETEBL**, nothing will happen until you press the *ENTER* key.
Before you do however, watch the line near the bottom of your display entitled
**Static Conditions.** Keeping your eyes on the static conditions line, press the
*ENTER* key. If you have done this step correctly, the new static conditions should
appear in place of the old. Also, if you look at the value assigned to the Mach
number you will find it has been changed to one.

At this point, you can change any of the seven input quantities. In addition to Mach number, you must change total pressure and temperature. Press

the *ENTER* key five times in order to position the arrow in front of PT1, the total pressure. Using the change procedure, i.e., type a *C* followed by *ENTER*, insert the desired total pressure of 482.7 lb/ft² (23112. N/m²). You may enter 4.827e+02 (2.3112e+04) or 4.827E2 (2.3112E4), etc. if you wish. Note that your keyboard's normal destructive backspace key can be used to correct typing errors. When your desired new total pressure is correctly entered, press *ENTER* and the change will be made. Verify that the new value for PT1 shown on the display is 4.827000E+02 (2.311200E+04). If you made any mistakes, repeat the change operation until you get it right.

Now press the *ENTER* key to position the arrow in front of TT1, the total temperature. Using the change procedure, change the value of TT1 to 468. (260.). Don't forget the decimal point or else your total temperature will be .000468 (.000260). Verify that the new value for TT1 shown on the display is 4.680000E+02 (2.600000E+02).

If you have changed Mach number, total pressure and total temperature correctly, the value listed below for static pressure will be very close to 255 lb/ft² (12209 N/m²) and the unit Reynolds number is approximately $1.24 \cdot 10^6$ ft⁻¹ ($4.07 \cdot 10^6$ m⁻¹). Verify that the static conditions you have entered match these two values. If they do not, find and correct any errors you have made before continuing.

Jot the values of static pressure and freestream unit Reynolds number on a slip of paper for reference later. In general, knowing these values often helps expedite preparation of your input data. You can always return to this menu to find their values, of course. Later on, we will see an example of using both parameters to determine input quantities on other sub-menus.

You have now finished this sub-menu. In order to exit, simply type an *X* followed by pressing the *ENTER* key. Note that, with the exception of Help menus for which only *ENTER* is needed, you return to the previous menu by the *X*, *ENTER* sequence. Also, if you are ever in doubt about what to do, look at the last line of the display for instructions.

**Body Parameters, Etc.** Now you are back to the **Main Parameters** sub-menu that provides the options of altering freestream conditions, body parameters, etc. Descend to the **Body Parameters** sub-menu by typing the letter *B* followed by pressing *ENTER*. You will be presented with a menu similar to the **Freestream Conditions** sub-menu. As before, press *ENTER* several times to move the arrow. Scan the input variable definitions and default values. Examine the Help menu. In other words, begin discovering that you already know most of what is needed in order to operate **SETEBL**!

There are only two input quantities you need to change, viz., ISHORT and SSTOP. Because you are looking for the point where momentum-thickness Reynolds number is 8000, you have no need for the long printout that gives far more detail than you are interested in. Consequently, you should position the arrow next to ISHORT and change its value to 0. This is done by typing *C* and a carriage return; no value need be entered. As an experiment, you might want to try repeating this sequence. If you do, the value of ISHORT will change back to 1. Be sure you have changed ISHORT to 0 after you finish experimenting.

Turning now to SSTOP, use the *ENTER* key to position the arrow next to SSTOP. This is the maximum value of plate length to which you will permit computation to proceed. Imagine that you are certain the momentum-thickness Reynolds number will reach 8000 at a plate-length Reynolds number somewhere between three and five million. Hence, you might want to terminate your run when Reynolds number reaches five million. Referring to the unit Reynolds number of $1.24 \cdot 10^6$ ft$^{-1}$ ($4.07 \cdot 10^6$ m$^{-1}$) that you jotted down earlier, a quick computation shows that a plate-length Reynolds number of five million occurs when plate length is 4.03 ft (1.23 m). Thus, change SSTOP to 4.03 (1.23).

There are no further changes you need to make at this time on this menu, so you should now exit by typing *X* followed by pressing the *ENTER* key. At this point, you are done with the **Main Parameters** sub-menu. In order to return to the main menu, type another *X* followed by pressing the *ENTER* key. Remember, nothing happens in **SETEBL** until you press the *ENTER* key.

## D.5.4  Taking a Lunch Break

Before continuing setting up a new run, you are going to simulate a lunch break. Imagine that it's time to break for lunch and the systems people upstairs are notorious for causing your mainframe computer to crash during the lunch hour. Any file you leave open will be lost as a result of a crash. In order to protect your work from such a disaster, simply exercise the exit option by typing yet another *X* followed by pressing the *ENTER* key.

Inspection of your directory will show that a new file named **eddybl.dat** has been created. Verify that the file exists at this time. If it does not, go back to Subsection D.5.1 and omit the mistake you made that caused you to reach this point unsuccessfully.

Now imagine you have returned from lunch, and your microcomputer (which never crashes during lunch because there are no system people to cause it to) is ready to continue serving your data-processing needs. At this point, run Program **SETEBL** again. Because the data file **eddybl.dat** exists, the program will go directly to the main menu.

## D.5.5  Edge/Wall Conditions

From the main menu, you should now proceed to the **Edge/Wall Conditions** sub-menu by typing an *E* followed by pressing *ENTER*. This sub-menu contains five options, viz., Pressure Distribution, Heat Transfer, Mass Transfer, Body Geometry and eXit. Type a *P* followed by *ENTER*. The **Pressure Distribution** sub-menu explains that you must prepare a file **presur.dat** that defines the pressure distribution. You must prepare the file with an editor such as MS-DOS's EDIT, Windows' WordPad, DEC's EDT, UNIX's vi, etc. All you can change in this menu is the number of points you plan on using. You will not change NUMBER because your run will have constant pressure. Thus, you need to specify pressure at two values of plate length. Note that this menu describes in detail the contents and format of **presur.dat**. Exit this menu with the usual *X*, *ENTER* sequence.

Now go to the **Heat Transfer** sub-menu. You are presented with a description of data file **heater.dat** that must be created with your own editor. Note that the adiabatic-wall temperature is given for your information and the value listed should be 459.4° R (255.2 K). The one parameter you can change on this menu is KODWAL which determines whether you plan on specifying surface heat flux or surface temperature. Type a *C* followed by *ENTER* to change KODWAL. Note that the display now indicates temperature is prescribed at the surface. Jot down the adiabatic-wall temperature for later reference. Exit this sub-menu by typing an *X* followed by *ENTER*.

Now go to the **Mass Transfer** sub-menu. This sub-menu describes a file, **blocrv.dat**, that must be prepared externally. You can alter the one parameter NFLAG. The default value is 0, which means **blocrv.dat** is not required to prepare your edge and surface conditions. You have no need to change its value for this application. Note that your display indicates the file **blocrv.dat** will not be required. Exit this sub-menu.

Having exited the **Mass Transfer** sub-menu, you have now made your way back to the **Edge/Wall Conditions** sub-menu. Proceed to the **Body Geometry** sub-menu. No, you didn't make a typing error. This is the same menu you just completed. It has been included for planned future enhancements to **SETEBL**. Exit back to the **Edge/Wall Conditions** sub-menu with an *X*, *ENTER* sequence.

The final option is to eXit. Do so by typing another *X*, *ENTER* sequence. You are now back at the main menu. You cannot continue until you have prepared input-data files **presur.dat** and **heater.dat** (**blocrv.dat** is not needed because NFLAG is 0). Hence, it is time to exit **SETEBL** and save your work.

## D.5.6   Preparing Edge/Wall Condition Data Files

The easiest way to prepare data files **presur.dat** and **heater.dat** (and **blocrv.dat** as well) is to use an editor such as EDIT, WordPad, EDT, vi, etc. to modify existing files from a previous run. That is why you left the files from the benchmark run in your directory. You can delete **blocrv.dat** now if you wish as it won't be needed for this run.

If you have followed all of the instructions correctly, you have the static pressure of 255 lb/ft$^2$ (12209 N/m$^2$) and the adiabatic-wall temperature of 459.4° R (255.2 K) jotted down somewhere. Using your favorite editor, change **presur.dat** to one of the following, depending on the units you have chosen:

```
USCS Units:        0.000000E 00   2.550000E 02
                   1.000000E 01   2.550000E 02
                   0.000000E 00   0.000000E 00


SI Units:          0.000000E 00   1.220900E 04
                   1.000000E 01   1.220900E 04
                   0.000000E 00   0.000000E 00
```

As explained in the **Pressure Distribution** sub-menu, the first two lines of this file are arc-length/pressure pairs presented in format (2E14.6). The final line is the pressure gradient at the beginning and end of the interval given in (2E14.6) format also. You have specified pressure at a plate length of zero and ten feet (meters). This interval must at least cover the planned integration range. The value of the pressure is the static pressure you jotted down earlier.

Turning now to surface temperature, note that 95% of the adiabatic-wall temperature is approximately 436° R (242 K), which is the value you should use. Use your editor to modify **heater.dat** as required, noting that the values of arc length at which you specify wall conditions must match the values used for the pressure distribution. As explained on the Heat Transfer sub-menu, this file must consist of one of the following sets of four lines, depending upon which set of units (SI or USCS) you have selected:

```
USCS Units:        0.000000E 00   4.360000E 02   0.000000E 00
                   1.000000E 01   4.360000E 02   0.000000E 00
                   0.000000E 00   0.000000E 00
                   0.000000E 00   0.000000E 00


SI Units:          0.000000E 00   2.420000E 02   0.000000E 00
                   1.000000E 01   2.420000E 02   0.000000E 00
                   0.000000E 00   0.000000E 00
                   0.000000E 00   0.000000E 00
```

The format of the first two lines is (3E14.6), while the last two lines have format (2E14.6). The first column for the first two lines is arc length, the second is wall temperature, and the third is surface heat flux. Note that since you have chosen to specify surface temperature rather than heat flux, any value can be entered for the heat flux — it won't be used in the computation. Similarly, if you choose to specify surface heat flux, the value assigned to surface temperature is arbitrary. The third line gives surface temperature slope (in the streamwise direction) at the beginning and end of the interval, while the last line gives surface heat flux slope. Of course, you are not limited to constant properties in the most general case. You may prescribe as many as 50 different values for edge pressure, surface temperature, surface heat flux, etc., which should be sufficient for most boundary-layer computations. **Make sure the arc-length values match those used for the pressure distribution.**

At this point, you have prepared all of the freestream conditions, body parameters, and (from an external editor) the two data files **presur.dat** and **heater.dat**. Before reentering Program **SETEBL**, examine your directory. In addition to the input data file **eddybl.dat**, you should find another file named **eddybl.bak**. The former file is your most recent version of **eddybl.dat**. The latter file is the version you created just before taking your lunch break. Program **SETEBL** always saves your previous work in **eddybl.bak** to provide you with a little extra protection. You no longer need **eddybl.bak**, so delete it if you wish.

## D.5.7   Generating Edge/Wall Conditions

Run **SETEBL** again. When the main menu appears, use a *W, ENTER* sequence to execute the Write Data Files option. A message will appear briefly indicating that edge conditions are being generated. When all computations are complete, a message appears telling you that a data file named **table.dat** has been successfully written. If you receive any other message, there are probably errors in the files you created with your editor in Subsection D.5.6, and you must correct them before you can continue setting up your run. What you are doing in this step is executing a subroutine in **SETEBL** that accomplishes two ends. First, you are generating data file **table.dat** in binary form that is used by **EDDYBL**. Second, you are computing several parameters appearing in data file **eddybl.dat** that are needed in preparing the initial profiles that are required by **EDDYBL**.

## D.5.8   Initial Profiles

In addition to the message that **table.dat** has been successfully written, you also receive the message

**Save the profiles in Ascii form?  (X=eXit, Y=Yes, ENTER=No)...**

Since you have not yet prepared the data needed to generate initial profiles, type an *X, ENTER* sequence. Upon returning to the main menu, you are now ready to go to the Initial Profiles sub-menu. Type an *I* followed by pressing the *ENTER* key. The sub-menu that appears has three options, viz., **Integral Parameters**, **Grid Parameters** and **eXit**.

**Integral Parameters.** Go to the **Integral Parameters** sub-menu first by entering another *I, ENTER* sequence. Press *ENTER* once to position the arrow next to skin friction. Change the value to 0.0038 in the usual manner. Press *ENTER* again to move the arrow in front of boundary-layer thickness, $\delta$. For the conditions specified above, a quick calculation shows that $\delta$ for your unit Reynolds number is .0144 ft (.004389 m). Change the value of DELTA to 0.0144 (.004389). Now move the arrow to shape factor and change its value to 1.8. Finally, move the arrow one more time to momentum-thickness Reynolds number and change its value to 1500., being careful to remember the decimal point. Inspect your work for possible errors. When you have made all entries correctly, exit this sub-menu.

**Grid Parameters.** Now exercise the *G* option to enter the **Grid Parameters** sub-menu. The first quantity you should change is the initial streamwise stepsize, DS. For this constant pressure case, you can use a stepsize as big as triple the boundary-layer thickness. Hence, change the value of DS to 0.04 ft (0.0122 m). In order to start the computation at a plate-length Reynolds number of one million, the initial plate length (arc length) must be 0.806 ft (0.246 m), a fact you can deduce by using the freestream unit number of $1.24 \cdot 10^6$ ft$^{-1}$ ($4.07 \cdot 10^6$ m$^{-1}$) you jotted down earlier. Hence, change SI to 0.806 (0.246). The next parameter is XK, the geometric-progression ratio. A coarser grid can be used for this case than the default grid. Change the value to 1.14, which corresponds to grid increments increasing in a geometric progression at a 14% rate. Finally, change the number of grid points normal to the surface, IEDGE, to 51. Again, inspect your work for possible errors. When your entries are error free, exit this sub-menu. Having returned to the **Initial Profiles** sub-menu, exercise the exit option with the usual *X, ENTER* sequence to return to the main menu.

**Generating Initial Profiles.** As in Subsection D.5.7, exercise the **Write Data Files** option by entering a *W, ENTER* sequence. This will regenerate **table.dat** and you are again presented with the following message.

Save the profiles in Ascii form? (X=eXit, Y=Yes, ENTER=No)...

If you desire a copy of the profiles, in Ascii form, to be sent to a disk file named **setebl.prt** that can be printed and/or examined with an editor after exiting Program **SETEBL**, respond with a *Y, ENTER* sequence; otherwise simply press the *ENTER* key. After you have responded to this query, a second query will appear as follows.

Display the profiles on the video? (X=eXit, Y=Yes, ENTER=No)...

If you want to see the profiles on your video display, respond with a *Y, ENTER* sequence. Otherwise, press *ENTER*. After you have responded, your screen clears again and a message appears indicating initial profiles are being generated. If you elected to display profiles on your video display, they will now be displayed, a screen at a time. Press *ENTER* to view the next screen. Regardless of the options you have chosen, the precise values of the integral parameters for your computed initial profiles are displayed. Finally, a message appears indicating the binary data file **input.dat** has been successfully written and that you must press *ENTER* in order to continue.

Notice that the value of the conventional sublayer coordinate, $y^+$, for the point nearest the surface is printed and its value is 0.175. Subroutine START will alert you if this value ever exceeds unity as Program **EDDYBL** requires the value of $y^+$ nearest the surface to be less than 1 in order to remain numerically stable. If this ever happens, you must either increase XK or IEDGE. When you press *ENTER*, control returns to the main menu.

You will receive a warning if you use the $k$-$\epsilon$ model and the value of $y^+$ for the point nearest the surface is less than 0.1. Values smaller than 0.1 slow the convergence rate for the $k$-$\epsilon$ model, and may even cause your run to crash.

## D.5.9  Selecting a Turbulence Model

In order to select the Spalart-Allmaras model, go to the **Turbulence Model** submenu. Type a *T* followed by *ENTER*, and you will find that the fourth quantity listed is a flag called MODEL. Press *ENTER* three times to position the arrow in front of MODEL. Change its value to 4 by typing *C* followed by entering a 4 and pressing *ENTER*. Note that the highlighted bar below the menu now indicates you are using the Spalart-Allmaras model for the computation. Exit back to the main menu.

For general reference, there are 15 turbulence models implemented in Program **EDDYBL**, and the two input parameters MODEL and NVISC are used to make the selection. The choices are listed in Table D.2.

## D.5.10  Logical Unit Numbers and Plotting Files

Your final input-data changes will cause printed output to go to your line printer rather than to disk file **eddybl.prt**. You will also verify that two disk files named **profil.dat** and **wall.dat** will be created that can be used as starting conditions for another program or as input to a plotting program. Go to the **Logical Unit Numbers** sub-menu by typing an *L, ENTER* sequence. The first parameter is IUNIT1 which, by default, is disk file **eddybl.prt**. For Lahey, Microsoft or SVS Fortran versions, change its value to 6. For all other versions, use your normal operating system procedure to direct the contents of **eddybl.prt** to a line printer. Verify that the value for the parameter IUPLOT is some value other than 0. If it is 0, change its value to 10 (or any other convenient value excluding unit 15 and any previously assigned unit number).

Table D.2: *Turbulence Models Included in EDDYBL*

| MODEL | NVISC | Turbulence Model |
|-------|-------|------------------|
| -1 | - | None (Laminar Flow) |
| 0 | 0 | $k$-$\omega$, viscous corrections excluded |
| 0 | 1 | $k$-$\omega$, viscous corrections included |
| 1 | 0 | Stress-$\omega$, viscous corrections excluded |
| 1 | 1 | Stress-$\omega$, viscous corrections included |
| 2 | 0 | $k$-$\epsilon$, Jones-Launder |
| 2 | 1 | $k$-$\epsilon$, Launder-Sharma |
| 2 | 2 | $k$-$\epsilon$, Lam-Bremhorst |
| 2 | 3 | $k$-$\epsilon$, Chien |
| 2 | 4 | $k$-$\epsilon$, Yang-Shih |
| 2 | 5 | $k$-$\epsilon$, Fan-Lakshminarayana-Barnett |
| 3 | - | $\nu_T$, Baldwin-Barth |
| 4 | - | $\nu_T$, Spalart-Allmaras |
| 5 | - | Algebraic, Cebeci-Smith |
| 6 | - | Algebraic, Baldwin-Lomax |
| 7 | - | Half-Equation, Johnson-King |

Disk file **wall.dat** is written as an unformatted file, each record of which can be read by another FORTRAN program according to the following program fragment.

```
      i=1
10    read(iunit) s(i),res(i),cfe(i),rethet(i),
     *            h(i),che(i),anue(i),pe(i),tw(i)
      if(s(i).ne.-999.) then
        i=i+1
        go to 10
      endif
```

The various quantities saved in disk file **wall.dat** are:

| Quantity | Description | Dimensions |
|----------|-------------|------------|
| s | $s$, arc length along surface | ft (m) |
| res | $Re_s$, Reynolds number based on $s$ | None |
| cfe | $c_{fe} = 2\tau_w/\rho_e u_e^2$, skin friction | None |
| rethet | $Re_\theta$, Reynolds number based on $\theta$ | None |
| h | $H$, shape factor | None |
| che | $\dot{h}/\rho_e u_e c_p$, Stanton number | None |
| anue | $Pr\,\dot{sh}/\mu_e c_p$, Nusselt number | None |
| pe | $\bar{p}$, pressure | lb/ft$^2$ (N/m$^2$) |
| tw | $T_w$, wall temperature | °R (K) |

The first line of the file **profil.dat** generated by Program **EDDYBL** contains the streamwise step number, M, and the number of mesh points normal to the surface, IEDGE. The format for this line is (2I6). The remainder of the file consists of IEDGE lines of data, format (12E11.4), containing the following boundary-layer profile data, with quantities written on each line in the order listed. Note that for the $k$-$\epsilon$ model, the specific dissipation rate, $\omega$, is defined by

$$\omega = \frac{\epsilon}{C_\mu k} \tag{D.1}$$

Also, the tensor $T_{ij}$ is specific to the Stress-$\omega$ model and is given by (see Subsection D.11.2):

$$T_{ij} = \tau_{ij} + \frac{2}{3} k \delta_{ij} \tag{D.2}$$

| Quantity | Description | Dimensions |
|---|---|---|
| $y$ | Distance normal to surface | ft (m) |
| $\tilde{u}$ | Horizontal velocity | ft/sec (m/sec) |
| $\tilde{T}$ | Temperature | °R (K) |
| $\bar{\rho}$ | Density | slug/ft$^3$ (kg/m$^3$) |
| $k$ | Turbulence kinetic energy | ft$^2$/sec$^2$ (m$^2$/sec$^2$) |
| $\omega$ | Specific dissipation rate | sec$^{-1}$ (sec$^{-1}$) |
| $T_{zz}$ | Specific Reynolds zz-normal stress deviator | ft$^2$/sec$^2$ (m$^2$/sec$^2$) |
| $T_{xx}$ | Specific Reynolds xx-normal stress deviator | ft$^2$/sec$^2$ (m$^2$/sec$^2$) |
| $T_{xy}$ | Specific Reynolds shear stress | ft$^2$/sec$^2$ (m$^2$/sec$^2$) |
| $T_{yy}$ | Specific Reynolds yy-normal stress deviator | ft$^2$/sec$^2$ (m$^2$/sec$^2$) |
| $y^+$ | Compressible sublayer-scaled distance | None |
| $U^+$ | Compressible sublayer-scaled velocity | None |

You can now exit back to the main menu. All of your input data are prepared and you are ready to run **EDDYBL**. Exit Program **SETEBL**.

## D.5.11  Running the Boundary-Layer Program

Run Program **EDDYBL**. Examination of program output reveals that your run didn't go far enough to determine the point where momentum-thickness Reynolds number reaches 8000. After 87 steps, the program stops at $s = 4.03$ ft (1.23 m), and $Re_\theta$ is only 7550. Linear extrapolation of your computed $Re_\theta$ indicates you needed to integrate to about $s = 4.27$ ft (1.30 m).

## D.5.12  Restart Run

You could go back to **SETEBL**, increase SSTOP and simply rerun **EDDYBL**. On your little sister's hand-me-down IBM PC/AT without an 80287, that's another 8 or 9 minutes. Since you might not really want to take another coffee break while your job runs, you might prefer a less time-consuming solution. Program **SETEBL** provides such a possibility through its **Restart** option.

Examine your directory and verify that Program **EDDYBL** has created a new file named **output.dat**. This file contains sufficient information to restart your program. Now, run Program **SETEBL** again. From the main menu, go to the Restart Run sub-menu by typing an *R* followed by pressing *ENTER*. This menu will permit you to change IEND1, the maximum streamwise step number, and SSTOP, the maximum value of arc (plate) length. Since your value for IEND1 is clearly large enough, you need only change SSTOP. With the usual procedure, change SSTOP from 4.03 ft (1.23 m) to 4.43 ft (1.35 m). Now type *X* followed by pressing *ENTER* in order to return to the main menu. Before returning, you will receive a message in reverse video as follows:

**"Do you wish to copy OUTPUT.DAT to INPUT.DAT? (Y/N)..."**

Respond Yes by typing *Y* followed by pressing *ENTER*. At this point, for all but the VAX version, **SETEBL** will inform you that it first copies **input.dat** to a new file named **input.bak**. For all versions, **SETEBL** then copies **output.dat** to **input.dat**. The final output of your original run becomes input for the restart run. Additionally, your original **input.dat** has effectively been renamed as **input.bak** (VAX/VMS creates its own backup file so this file is unnecessary). Upon completion of the copy operation, control returns to the main menu (for the VAX version, you are instructed to press *ENTER* to continue). Exit Program **SETEBL**.

At this point, data files **eddybl.dat** and **input.dat** have been modified as needed to continue your run from where you left off. The file **table.dat** requires no modification as SSTOP remains smaller than the top end of the interval for which you have defined edge and surface properties. Had we made SSTOP larger than 10, you would have to make appropriate changes to **presur.dat** and **heater.dat** to make sure edge and surface conditions are defined at least up to the new value of SSTOP. You would then have to regenerate **table.dat** via the Write Data Files option (Subsection D.5.7).

Now, run Program **EDDYBL** again. If you have made no errors, inspection of the printout combined with a little interpolation shows that $Re_\theta$ is 8000 at a plate length of approximately 4.31 ft (1.32 m).

## D.5.13   Gas Properties and Profile Printing

At this point, you have seen virtually all of Program **SETEBL**'s menus and options. There are two sub-menus we didn't use in this exercise, viz., **Gas Properties** and **Profile Printing**. Both menus are self explanatory and operate in the same manner as the menus you've already explored.

The **Gas Properties** menu allows you to modify thermodynamic properties such as specific-heat ratio, perfect-gas constant, and viscosity-law coefficients. The default values are set up for air with the Sutherland viscosity law. You can implement a **power-law** viscosity relationship by setting SU = 0.  Note that if you want a viscosity law of the form $\mu = \mu_o T^\omega$ then you must set input parameters VISCON = $\mu_o$ and VISPOW = $\omega + 1$.

The **Profile Printing** sub-menu permits you to print velocity, temperature, turbulent energy, etc. profiles at specified streamwise stations.  Program **ED-DYBL** always prints profiles at the final station.  Also, whenever **EDDYBL** prints profiles, disk file **output.dat** is automatically written.

## D.5.14   Selecting Laminar, Transitional or Turbulent Flow

Program **EDDYBL** can run in three different modes corresponding to (1) pure laminar flow, (2) transition from laminar to turbulent flow, and (3) pure turbulent flow.  The two test cases exercise **EDDYBL** in its pure turbulent mode in which integral parameters are specified and IBOUND is set to 1.

To run in transitional mode, simply select IBOUND = 0 in the **Initial Profiles/Integral Parameters** menu.  As a result, exact laminar velocity and temperature profiles will be generated in conjunction with approximate laminar profiles for the various turbulence-model parameters.  The transition point is determined automatically by the model equations and depends strongly upon the freestream values of $k$ and $\omega$ that are specified in the **Main Parameters/Freestream Conditions** menu in terms of ZIOTAE and ZIOTAL.  To obtain physically realistic transition Reynolds numbers you must include low-Reynolds-number corrections in the $k$-$\omega$ and Stress-$\omega$ models by setting NVISC = 1 in the **Turbulence Model** sub-menu.  Although the $k$-$\epsilon$ models are capable of predicting transition, extremely small streamwise steps are needed with **EDDYBL**, and stable computation is very difficult to achieve.  Even if you are not interested in transition, this mode is nevertheless useful as it provides an alternate method for generating turbulent starting profiles, e.g., by starting laminar and running up to a desired value of $Re_\theta$.

The $k$-$\omega$ and Stress-$\omega$ models are very robust and can be integrated through transition with and without low-Reynolds-number corrections.  By contrast, many of the other models require smaller streamwise steps than the $k$-$\omega$ and Stress-$\omega$ models, and, in general, cannot be integrated through transition unless extremely small steps are taken.  If integral properties are unknown and a solution with a model other than $k$-$\omega$ or Stress-$\omega$ is desired, the optimum procedure is to start laminar with the $k$-$\omega$ model and integrate through transition.  Then, select the desired model and use the **Restart** option to continue the run.

Finally, to run **EDDYBL** as a pure laminar boundary-layer program, the turbulence model can be suppressed by setting MODEL = $-1$ in the **Turbulence Model** sub-menu.  When this is done, turbulence-model computations are bypassed and no transition to turbulence occurs.

# D.6  Applicability and Limitations

Program **EDDYBL** applies to attached, compressible, two-dimensional and axisymmetric boundary layers. The program computes properties of turbulent boundary layers using the $k$-$\omega$ model or the Stress-$\omega$ model, including effects of surface roughness, surface mass transfer, surface curvature, and low-Reynolds-number corrections. The program also includes several other algebraic, one-equation and two-equation models.

Computations can be initiated either from turbulent starting profiles that are generated from specified integral properties, or from laminar profiles that are automatically generated.

The $k$-$\omega$ and Stress-$\omega$ models are very robust and can be integrated through transition from laminar to turbulent flow with and without low-Reynolds-number corrections. By contrast, the $k$-$\epsilon$ model requires smaller streamwise steps than the $k$-$\omega$ and Stress-$\omega$ models, and, in general, cannot be integrated through transition unless extremely small steps are taken.

If integral properties are unknown and a $k$-$\epsilon$ (or any other model included in the program) solution is desired, the optimum procedure is to start laminar with the $k$-$\omega$ model and integrate through transition. Then, select the desired $k$-$\epsilon$ model and use the **Restart** option to continue the run.

Finally, if numerical difficulties are encountered with the $k$-$\epsilon$ model, try changing input parameter *PSIEPS* in the Turbulence Model menu. Table 7.2 lists the default value of this parameter ($\psi_\epsilon$); its purpose is explained in Section 7.3.

# D.7  EDDYBL Output Parameters

Printed output from Program **EDDYBL** consists of dimensionless boundary-layer profiles, and integral parameters, some of which are dimensional. The various quantities are listed in the following two tables.

Dimensionless-profiles portion of **EDDYBL** output:

| Name | Symbol/Equation | Definition |
|------|-----------------|------------|
| i | $i$ | Mesh point number |
| y/delta | $y/\delta$ | Dimensionless normal distance |
| u/Ue | $\bar{u}/u_e$ | Dimensionless velocity |
| yplus | $y^+ = u_\tau y/\nu_w$ | Compressible sublayer-scaled distance |
| uplus | $U^+ = u^*/u_\tau$ | Compressible sublayer-scaled velocity |
| k/Ue**2 | $k/u_e^2$ | Dimensionless turbulence energy |
| omega | $\nu_e\omega/u_e^2$ | Dimensionless dissipation rate |
| eps/mu | $\mu_T/\mu$ | Dimensionless eddy viscosity |
| L/delta | $\sqrt{k/\beta_o^*}/(\omega\delta)$ | Dimensionless turbulence length scale |
| uv/tauw | $-\overline{\rho u'v'}/\tau_w$ | Dimensionless Reynolds shear stress |
| T/Te | $\bar{T}/T_e$ | Temperature ratio |

## Integral-parameter portion of **EDDYBL** output:

| Symbol | Meaning | USCS Units | SI Units |
|--------|---------|------------|----------|
| F | Force | pounds (lb) | Newtons (N) |
| L | Length | feet (ft) | meters (m) |
| M | Mass | slugs (sl) | kilograms (kg) |
| Q | Heat flux | Btu/second (Btu/sec) | Watts (W) |
| T | Time | seconds (sec) | seconds (sec) |
| $\Theta$ | Temperature | °Rankine | Kelvins |

| Name | Symbol/Equation | Definition | Dimensions |
|------|-----------------|------------|------------|
| beta | $\beta = (2\xi/u_e)\,du_e/d\xi$ | Pressure-gradient parameter | None |
| Cfe | $c_{fe} = 2\tau_w/\rho_e u_e^2$ | Skin friction based on $\bar{\rho}_e$ | None |
| Cfw | $c_{fw} = 2\tau_w/\rho_w u_e^2$ | Skin friction based on $\bar{\rho}_w$ | None |
| delta | $\delta$ | Boundary-layer thickness | L |
| delta* | $\delta^*$ | Displacement thickness | L |
| dPe/ds | $d(\bar{p}/\rho_\infty U_\infty^2)/d\bar{s}$ | Dimensionless pressure gradient | None |
| dTe/ds | $d(T_e/T_{ref})/d\bar{s}$ | Dimensionless temperature gradient | None |
| dUe/ds | $d(u_e/U_\infty)/d\bar{s}$ | Dimensionless velocity gradient | None |
| H | $H = \delta^*/\theta$ | Shape factor | None |
| hdot | $\dot{h} = q_w/(T_w - T_{aw})$ | Heat-transfer coefficient | $QL^{-2}\Theta^{-1}$ |
| Iedge | $N$ | Total number of mesh points in B.L. | None |
| Itro | | Number of iterations | None |
| kmax | $\sqrt{\beta_o^*(\bar{\rho}k)_{max}/\tau_w}$ | Maximum turbulence energy | None |
| M | $m$ | Streamwise step number | None |
| Me | $M_e$ | Edge Mach number | None |
| Mue | $\mu_e$ | Edge molecular viscosity | $ML^{-1}T^{-1}$ |
| Ne | $N_e$ | Mesh point number at B.L. edge | None |
| Negtiv | | Number of points where $k, \omega, \epsilon < 0$ | None |
| Nerror | | Number of points not converged | None |
| Nskip | | Number of points below $U^+$ = USTOP | None |
| Nste | $\dot{h}/\rho_e u_e c_p$ | Stanton number based on $\bar{\rho}_e$ | None |
| Nstw | $\dot{h}/\rho_w u_e c_p$ | Stanton number based on $\bar{\rho}_w$ | None |
| Nue | $Pr\,s\dot{h}/\mu_e c_p$ | Nusselt number based on $\mu_e$ | None |
| Nuw | $Pr\,s\dot{h}/\mu_w c_p$ | Nusselt number based on $\mu_w$ | None |
| Pe | $\bar{p}$ | Edge pressure | $FL^{-2}$ |
| qw | $q_w$ | Surface heat flux | $QL^{-2}$ |
| radius | $r_o$ | Body radius | L |
| Recov | $r$ | Recovery factor | None |
| Redel* | $Re_{\delta^*} = \rho_e u_e \delta^*/\mu_e$ | Reynolds number based on $\delta^*$ | None |
| Res | $Re_s = \rho_e u_e s/\mu_e$ | Reynolds number based on $s$ | None |
| Rethet | $Re_\theta = \rho_e u_e \theta/\mu_e$ | Reynolds number based on $\theta$ | None |
| Rhoe | $\rho_e$ | Edge density | $ML^{-3}$ |
| rho*vw | $\rho_w v_w$ | Surface mass flux | $ML^{-2}T^{-1}$ |
| s | $s$ | Arc length | L |
| tauw | $\tau_w$ | Surface shear stress | $FL^{-2}$ |
| Te | $T_e$ | Edge temperature | $\Theta$ |
| theta | $\theta$ | Momentum thickness | L |
| Ue | $u_e$ | Edge velocity | $LT^{-1}$ |
| utau | $u_\tau$ | Friction velocity | $LT^{-1}$ |
| xi | $\xi = \int_0^s \rho_e u_e \mu_e r_o^{2j}\,ds$ | Transformed streamwise coordinate | L |
| yplus | $y_2^+$ | Value of $y^+$ nearest the surface | None |
| z | $z$ | Axial distance | L |

# D.8 Program PLOTEB: Plotting Utility

Program **PLOTEB** creates video and hardcopy plots of skin friction, $c_f$, or Stanton number, $St$, versus arc length, $s$, and a $U^+$ versus $y^+$ velocity profile computed with Program **EDDYBL** on IBM PC's and compatibles.

**Input-parameter description:**

Program **PLOTEB** reads the following sixteen input parameters from disk file **ploteb.dat** in the order listed below. Integer quantities must be formatted according to (7x,i6) while floating-point quantities must be formatted as (7x,f6.2).

| | |
|---|---|
| *mon* | Monitor type (see Appendix E) |
| *ifore* | Foreground color (see Appendix E) |
| *iback* | Background color (see Appendix E) |
| *nprin* | Printer type (see Appendix E) |
| *mode* | Graphics-mode flag for printers; number of pens for plotters (see Appendix E) |
| *metric* | Input arc-length units flag |
| |     -1 Input arc length is in meters |
| |     0 Input arc length is in feet |
| |     1 Convert from feet to meters |
| *ideccf* | Number of decimal places for $c_f/St$ scale |
| *idecx* | Number of decimal places for arc-length scale |
| *idecup* | Number of decimal places for $U^+$ scale |
| *isymb* | Symbol type for experimental data points |
| |     0 Circle |
| |     1 Triangle |
| |     2 Square |
| |     3 Diamond |
| *jstart* | Number of $c_f/St$ points to skip over at beginning of computation; this is sometimes useful in order to skip over transient behavior at the beginning of a computation. The sign also determines what is plotted. |
| |     > 0 Plot $c_f$ versus $s$ |
| |     < 0 Plot $St$ versus $s$ |
| *kcyccf* | Increment between points to be plotted for $c_f/St$ versus $s$ |
| *kcycup* | Increment between points to be plotted for $U^+$ versus $y^+$ |
| *kfilt* | 0 to suppress data filtering; otherwise use filtering. The filtering algorithm generates a smoothed curve. |
| *ksize* | Plot scaling factor. Using 100 yields a full-size hardcopy plot. Smaller values yield a hardcopy plot reduced by *ksize* per cent. For example, selecting *ksize* = 50 produces a half-size plot. |
| *symsiz* | Size of experimental-data symbols, in inches |

Next, Program **PLOTEB** reads a single, free-formatted, line to indicate where hardcopy print is directed. This line comes immediately after the specified value for *symsiz* and defines the following five additional parameters.

*devid*   Device name of type character∗4; valid devices are LPT1, LPT2, LPT3, COM1, COM2, COM3, COM4

*nbaud*   Baud rate for a serial port; valid baud rates are 110, 150, 300, 600, 1200, 2400, 4800, 9600

*parity*  Parity of type character∗3 or character∗4 for a serial port; valid parity settings are 'even', 'odd' and 'none'

*nstop*   Number of stop bits for a serial port; either 1 or 2

*lword*   Word length for a serial port; either 7 or 8

In addition to disk file **ploteb.dat**, an optional file named **exper.dat** containing measured skin-friction and velocity-profile data can be included. The first line of the file must contain the number of input data pairs with format (i6). If no $c_f$ or $St$ data are available, place a zero on this line. If $c_f$ or $St$ data are available, this line is followed by $s$-$c_f$ (or $s$-$St$) data pairs with format (2e11.4).

Next, enter the data source; as many as twenty characters can be used. The final $c_f/St$ entry is the location of the box citing the data source. Enter a 1 for upper left, 2 for upper right, 3 for lower right, and 4 for lower left (see Figure D.2). This option is provided to help avoid having the curves pass through the citations box. The format is (7x,i6). A similar sequence of input parameters follows for velocity-profile data. The order of the data pairs is $y^+$ first and $U^+$ last. For example, the bench-mark case is an incompressible flat-plate boundary layer. Experimental data for this flow are given by Coles and Hirst (1969). The sample **exper.dat** included on the distribution diskette is as follows.

```
        19
1.5978e 00 3.4500e-03
2.0899e 00 3.3700e-03
2.5820e 00 3.1700e-03
3.0741e 00 3.1700e-03
3.5663e 00 3.0800e-03
4.0584e 00 3.0100e-03
4.7146e 00 2.9300e-03
5.5348e 00 2.8400e-03
6.5190e 00 2.7800e-03
7.5033e 00 2.6900e-03
8.4875e 00 2.6600e-03
9.4718e 00 2.6000e-03
1.0456e 01 2.6000e-03
1.1440e 01 2.5600e-03
1.2425e 01 2.5300e-03
1.3409e 01 2.4700e-03
1.4393e 01 2.4700e-03
1.5377e 01 2.4600e-03
```

```
1.6362e 01 2.4300e-03
WIEGHARDT
iposcf=     2
     14
3.8100e 01 1.4240e 01
7.6100e 01 1.5900e 01
1.5230e 02 1.7390e 01
3.8070e 02 1.9520e 01
7.6150e 02 2.1300e 01
1.1422e 03 2.2500e 01
1.5229e 03 2.3330e 01
1.9037e 03 2.4220e 01
2.2844e 03 2.4940e 01
3.0459e 03 2.6290e 01
3.8074e 03 2.7410e 01
4.5688e 03 2.8270e 01
5.3303e 03 2.8590e 01
6.0918e 03 2.8700e 01
WIEGHARDT
iposup=     1
```

**Program Output:** A video plot with two graphs (see Figure D.2) is created on the screen. When the plot is complete, the following message appears:

### Hardcopy output (y/n)?

Enter a *y* or a *Y* to create a hardcopy plot. Pressing any other key terminates the run without creating a hardcopy plot.

**Comments:**

- The following is a sample input-data file, **ploteb.dat**, for a typical Personal Computer with a standard VGA monitor and an HP LaserJet printer connected to serial port COM1:.

```
mon    =    18        (Standard VGA monitor)
ifore  =    15        (Bright-white foreground)
iback  =    1         (Blue background)
nprin  =    5         (HP LaserJet Series III or IV)
mode   =    3         (300 dots per inch resolution)
metric=     1         (Convert feet to meters)
ideccf=     1         (One decimal place on Cf scale)
idecs  =    1         (One decimal place on s scale)
idecup=    -1         (Integers on u+ scale)
isymb  =    0         (Circles for experimental data)
jstart=     2         (Start plot at the second point)
kcyccf=     1         (Plot every Cf point)
```

```
kcycup=     1                  (Plot every u+ point)
kfilt =     1                  (Use filtering)
ksize =   100                  (Full size plot)
symsiz=   .080                 (.08" experimental data symbols)
'com1' , 9600 , 'none' , 1 , 8
```

The last line indicates the printer is connected to serial port COM1: and the port is set at 9600 baud, no parity, 1 stop bit and 8 data bits.

If disk file **ploteb.dat** is not available, Program **PLOTEB** uses the following set of default values:

*mon* = 21, *ifore* = 15, *iback* = 1, *nprin* = 5, *mode* = 4, *symsiz* = .08, *ksize* = 100, *devid* = 'LPT1'

Note that *nbaud*, *parity*, *nstop* and *lword* are not used for parallel ports.

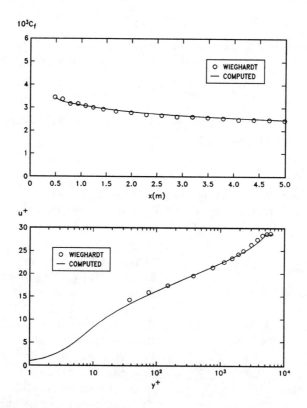

Figure D.2: *Sample plot created by Program PLOTEB.*

# D.9 Adapting to Other Compilers/Systems

If you change computers or compilers, the appropriate modifications may already be included in the source code provided. If your Fortran compiler is an ANSI-77 Standard compiler and supports most of the standard VAX extensions, only three categories of changes are needed.

1. You must determine the correct syntax used by your Fortran compiler for the **include** command. Then, note that the source code provided uses the VAX syntax, which is more-or-less standard in most modern Fortran compilers. Make the appropriate change throughout the source code for your compiler. Examples of VAX and other syntax are:

| Fortran Compiler | Include Syntax |
|---|---|
| VAX, SVS, Lahey, Microsoft | `include 'filename'` |
| Cray (UNICOS), SUN | `include 'filename'` |
| Microsoft (older versions) | `$include: 'filename'` |
| SVS (older versions) | `$include filename` |
| Cray (COS) | `*CALL FILENAME` |
| | ↑ |
| | Column 1 |

2. Change the value of *icpu* defined in the include file named **cpuid**. The values currently assigned are:

| | |
|---|---|
| *icpu* = 0 | SVS Fortran (680x0, 80x86) |
| *icpu* = 1 | Lahey/Microsoft Fortran (8088, 80x86) |
| *icpu* = 2 | VAX/VMS |
| *icpu* = 3 | SUN Fortran (68020, SPARC) |
| *icpu* = 4 | Cray (UNICOS) |
| *icpu* = 5 | Silicon Graphics Iris |

3. The only other compiler-specific syntax differences are located in a subroutine called **NAMSYS** that appears in **eddybl.for**. This subroutine opens disk files depending upon the value of *icpu*. Make any changes required for your system.

4. Search the **EDDYBL** and **SETEBL** source code for occurrences of *icpu* to see if the correct action is taken for your compiler and/or operating system. Make any changes required for your system.

5. Modify Program **INSTL** as required for your video display and/or compiler specific requirements.

# D.10   Compile and Link Commands

This section describes the commands required to compile and link Programs
**SETEBL**, **INSTL** and **EDDYBL** for the various Fortran compilers supported.
Be sure that you have selected the appropriate value for *icpu* in the include file
**cpuid**.

### ICPU = 0:  SVS Fortran-386 ... Phar Lap and $C^3$

**Special Comments:**  For the Phar Lap version, add the **+w1167** option to compile
for a Weitek math coprocessor.  Linker options for either **fastlink** or **386link**
can be specified in an environment variable by including the following in your
**autoexec.bat** file ...

> 80387 version
> set 386link=-l libf28 libp28 -pack -maxr ffffh -s 40000
> Weitek version
> set 386link=-l libf28w libp28w -pack -maxr ffffh -s 40000

**Compile and Link:**
> svs instl.for
> svs eddybl.for
> svs setebl.for edge.for grafic.for initil.for ioebl.for main0.for misc.for

### ICPU = 0:  SVS Fortran-020

**Special Comments:**  Use **pload** in place of **load** for Definicon PM-020 and
PM-030 boards.

**Compile and Link:**
> load fc instl -lk
> load fc eddybl -lk
> load fc setebl edge grafic initil ioebl main0 misc -lk

### ICPU = 1:  Lahey Fortran ... LF90

**Special Comments:** None.

**Compile and Link:**
> lf90 instl
> lf90 eddybl
> lf90 setebl edge grafic initil ioebl main0 misc

## ICPU = 1: Lahey Fortran ... F77L-EM/32
**Special Comments:** None.

**Compile and Link:**
    f77l3 instl
    386link instl
    f77l3 eddybl
    386link eddybl
    f77l3 setebl
    f77l3 edge
    f77l3 grafic
    f77l3 initil
    f77l3 ioebl
    f77l3 main0
    f77l3 misc
    386link setebl edge grafic initil ioebl main0 misc

## ICPU = 1: Lahey Fortran ... F77L
**Special Comments:** None.

**Compile and Link:**
    f77l instl
    optlink instl;
    f77l eddybl
    optlink eddybl;
    f77l setebl
    f77l edge
    f77l grafic
    f77l initil
    f77l ioebl
    f77l main0
    f77l misc
    optlink setebl+edge+grafic+initil+ioebl+main0+misc;

## ICPU = 1:  Microsoft Fortran

**Special Comments:** Using the /e option reduces executable file size. You must split eddybl.for into 2 files as this compiler cannot handle large files. The commands below assume you split it into files named eddybl.for and eddybl2.for.

**Compile and Link:**
    fl instl.for
    fl /c eddybl.for eddybl2.for
    link eddybl eddybl2,,nul, /e;
    fl /c setebl.for edge.for grafic.for initil.for ioebl.for main0.for misc.for
    link setebl edge grafic initil ioebl main0 misc,,nul, /e;

## ICPU = 2:  VAX Fortran

**Special Comments:** None.

**Compile and Link:**
    for instl
    link instl
    for eddybl
    link eddybl
    for setebl
    for edge
    for grafic
    for initil
    for ioebl
    for main0
    for misc
    link setebl,edge,grafic,initil,ioebl,main0,misc

## ICPU = 3:  SUN Fortran ... SUN/OS or MS-DOS/SP-1

**Special Comments:** Using the -O3 option yields maximum optimization.

**Compile and Link:**
    f77 instl.f -O3 -o instl
    f77 eddybl.f -O3 -o eddybl
    f77 setebl.f edge.f grafic.f initil.f ioebl.f main0.f misc.f -O3 -o setebl

## ICPU = 4: Cray Fortran ... UNICOS
**Special Comments:** None.

**Compile and Link:**
    cf77 -o instl instl.f
    cf77 -o eddybl eddybl.f
    cf77 -o setebl setebl.f edge.f grafic.f initil.f ioebl.f main0.f misc.f

## ICPU = 5: Silicon Graphics Iris
**Special Comments:** None.

**Compile and Link:**
    f77 -o instl instl.f
    f77 -o eddybl eddybl.f
    f77 -o setebl setebl.f edge.f grafic.f initil.f ioebl.f main0.f misc.f

# D.11  Additional Technical Information

The program uses the conventional Levy-Lees variables [see Hayes and Probstein (1959)] and much of the program notation follows that of Harris and Blanchard (1982). The numerical procedure is the Blottner (1974) variable-grid method augmented with an algorithm devised by Wilcox (1981b) to permit large streamwise steps. Section 7.3 of the main text provides an in-depth discussion of the algorithm. This section first presents the governing equations for mean-flow properties and all turbulence-model equations implemented in the program. Then, the transformed, nondimensional form of the equations is presented for the $k$-$\omega$, Stress-$\omega$ and $k$-$\epsilon$ models.

## D.11.1  Mean-Flow Equations

The equations governing conservation of mass, momentum and mean energy for all models are the same. For compressible two-dimensional ($j = 0$) and axisymmetric ($j = 1$) boundary layers, the program uses body-oriented coordinates $(s, n)$, where $s$ is arc length and $n$ is distance normal to the surface. The equations are as follows.

$$\frac{\partial}{\partial s}\left(\bar{\rho}\tilde{u}\right) + \frac{1}{r^j}\frac{\partial}{\partial n}\left(r^j\bar{\rho}\tilde{v}\right) = 0 \tag{D.3}$$

$$\bar{\rho}\tilde{u}\frac{\partial\tilde{u}}{\partial s} + \bar{\rho}\tilde{v}\frac{\partial\tilde{u}}{\partial n} = -\frac{dP}{ds} + \frac{1}{r^j}\frac{\partial}{\partial n}\left[r^j\left(\mu\frac{\partial\tilde{u}}{\partial n} + \bar{\rho}\tau\right)\right] \tag{D.4}$$

$$\bar{\rho}\tilde{u}\frac{\partial \tilde{h}}{\partial s} + \bar{\rho}\tilde{v}\frac{\partial \tilde{h}}{\partial n} = \tilde{u}\frac{dP}{ds} + \mu\left(\frac{\partial \tilde{u}}{\partial n}\right)^2 + \bar{\rho}\epsilon + \frac{1}{r^j}\frac{\partial}{\partial n}\left[r^j\left(\frac{\mu}{Pr_L} + \frac{\mu_T}{Pr_T}\right)\frac{\partial \tilde{h}}{\partial n}\right]$$
(D.5)

The perfect-gas law is used as the equation of state and the fluid is assumed calorically perfect so that

$$P = \bar{\rho}R\tilde{T} \quad \text{and} \quad \tilde{h} = c_p\tilde{T} \tag{D.6}$$

In Equations (D.3) through (D.6): $\tilde{u}$ and $\tilde{v}$ are streamwise and normal mass-averaged velocity components; $\bar{\rho}$, $P$ and $\tilde{h}$ are fluid density, pressure and enthalpy; $\mu$ and $\mu_T$ are molecular and eddy viscosity; $\tau$ is specific Reynolds shear stress; $\epsilon$ is turbulence dissipation rate; $Pr_L$ and $Pr_T$ are laminar and turbulent Prandtl numbers; $\tilde{T}$ is mass-averaged temperature; $R$ is the perfect-gas constant; and $c_p$ is specific heat at constant pressure.

## D.11.2   $k$-$\omega$ and Stress-$\omega$ Model Equations

For both the $k$-$\omega$ and Stress-$\omega$ models the dissipation, $\epsilon$, is given by

$$\epsilon = \beta^*\omega k \tag{D.7}$$

where $k$ is turbulence kinetic energy and $\omega$ is specific dissipation rate. The equations for $k$ and $\omega$ applicable to compressible boundary layers are as follows.

$$\bar{\rho}\tilde{u}\frac{\partial k}{\partial s} + \bar{\rho}\tilde{v}\frac{\partial k}{\partial n} = \bar{\rho}\tau\frac{\partial \tilde{u}}{\partial n} - \beta^*\bar{\rho}\omega k + \frac{1}{r^j}\frac{\partial}{\partial n}\left[r^j\left(\mu + \sigma^*\mu_T\right)\frac{\partial k}{\partial n}\right] \tag{D.8}$$

$$\bar{\rho}\tilde{u}\frac{\partial \omega}{\partial s} + \bar{\rho}\tilde{v}\frac{\partial \omega}{\partial n} = \alpha\frac{\omega}{k}\bar{\rho}\tau\frac{\partial \tilde{u}}{\partial n} - \beta\bar{\rho}\omega^2 + \frac{1}{r^j}\frac{\partial}{\partial n}\left[r^j\left(\mu + \sigma\mu_T\right)\frac{\partial \omega}{\partial n}\right] \tag{D.9}$$

For the $k$-$\omega$ model, the Reynolds shear stress is given by

$$\bar{\rho}\tau = \alpha^*\mu_T\frac{\partial \tilde{u}}{\partial n} \tag{D.10}$$

For the Stress-$\omega$ model, the Reynolds stresses are computed from the following equations:

$$\bar{\rho}\tilde{u}\frac{\partial \tau}{\partial s} + \bar{\rho}\tilde{v}\frac{\partial \tau}{\partial n} = \left[(1-\hat{\alpha})\sigma_y - \hat{\beta}\sigma_x + \frac{2}{3}(1-\hat{\alpha}-\hat{\beta}+\frac{3}{4}\hat{\gamma})k\right]\bar{\rho}\frac{\partial \tilde{u}}{\partial n}$$
$$-C_1\beta^*\bar{\rho}\omega\tau + \frac{1}{r^j}\frac{\partial}{\partial n}\left[r^j\left(\mu + \sigma^*\mu_T\right)\frac{\partial \tau}{\partial n}\right] \tag{D.11}$$

$$\bar{\rho}\tilde{u}\frac{\partial\sigma_x}{\partial s} + \bar{\rho}\tilde{v}\frac{\partial\sigma_x}{\partial n} = \frac{2}{3}\left[2(1-\hat{\alpha})+\hat{\beta}\right]\bar{\rho}\tau\frac{\partial\tilde{u}}{\partial n} - C_1\beta^*\bar{\rho}\omega\sigma_x$$
$$+\frac{1}{r^j}\frac{\partial}{\partial n}\left[r^j\left(\mu+\sigma^*\mu_T\right)\frac{\partial\sigma_x}{\partial n}\right] \qquad (D.12)$$

$$\bar{\rho}\tilde{u}\frac{\partial\sigma_y}{\partial s} + \bar{\rho}\tilde{v}\frac{\partial\sigma_y}{\partial n} = -\frac{2}{3}\left[(1-\hat{\alpha})+2\hat{\beta}\right]\bar{\rho}\tau\frac{\partial\tilde{u}}{\partial n} - C_1\beta^*\bar{\rho}\omega\sigma_y$$
$$+\frac{1}{r^j}\frac{\partial}{\partial n}\left[r^j\left(\mu+\sigma^*\mu_T\right)\frac{\partial\sigma_y}{\partial n}\right] \qquad (D.13)$$

The quantities $\sigma_x$ and $\sigma_y$ are the stress deviator components given in terms of the normal Reynolds stress by

$$\sigma_x = \frac{\overline{\rho u'^2}}{\bar{\rho}} - \frac{2}{3}k \quad \text{and} \quad \sigma_y = \frac{\overline{\rho v'^2}}{\bar{\rho}} - \frac{2}{3}k \qquad (D.14)$$

The various closure coefficients, viz., $\alpha$, $\beta$, $\beta^*$, $\sigma$, $\sigma^*$, $C_1$, $\hat{\alpha}$, $\hat{\beta}$ and $\hat{\gamma}$ are given by the following. First, we define the fully turbulent (subscript $\infty$), incompressible (subscript $i$) values by

$$\alpha_\infty = \frac{13}{25}, \quad \alpha^*_\infty = 1, \quad \beta_i = \frac{9}{125}, \quad \beta^*_\infty = \frac{9}{100}, \quad \sigma = \frac{1}{2}, \quad \sigma^* = \frac{1}{2} \qquad (D.15)$$

$$\hat{\alpha}_\infty = \frac{8+C_2}{11}, \quad \hat{\beta}_\infty = \frac{8C_2-2}{11}, \quad \hat{\gamma}_\infty = \frac{60C_2-4}{55} \qquad (D.16)$$

$$C_1 = \frac{9}{5}, \qquad C_2 = \frac{13}{25} \qquad (D.17)$$

$$f_{\beta^*} = \begin{cases} 1, & \chi_k \le 0 \\ \dfrac{1+C_\chi\chi_k^2}{1+400\chi_k^2}, & \chi_k > 0 \end{cases}, \quad \chi_k \equiv \frac{1}{\omega^3}\frac{\partial k}{\partial y}\frac{\partial\omega}{\partial y} \qquad (D.18)$$

where

$$C_\chi = \begin{cases} 680, & k\text{-}\omega \text{ model} \\ 640, & \text{Stress-}\omega \text{ model} \end{cases} \qquad (D.19)$$

On the one hand, if low-Reynolds-number corrections are excluded from the $k$-$\omega$ and Stress-$\omega$ models, we simply use:

$$\alpha^* = \alpha^*_\infty, \quad \alpha = \alpha_\infty, \quad \beta_i^* = \beta^*_\infty, \quad \hat{\alpha} = \hat{\alpha}_\infty, \quad \hat{\beta} = \hat{\beta}_\infty, \quad \hat{\gamma} = \hat{\gamma}_\infty \qquad (D.20)$$

On the other hand, if low-Reynolds-number corrections are included in the $k$-$\omega$ model, we use the following:

$$\left.\begin{aligned}
\alpha^* &= \alpha^*_\infty \frac{\alpha^*_o + Re_T/R_k}{1 + Re_T/R_k} \\
\alpha &= \alpha_\infty \cdot \frac{\alpha_o + Re_T/R_\omega}{1 + Re_T/R_\omega} \cdot (\alpha^*)^{-1} \\
\beta^*_i &= \beta^*_\infty \cdot \frac{4/15 + (Re_T/R_\beta)^4}{1 + (Re_T/R_\beta)^4}
\end{aligned}\right\} \tag{D.21}$$

where

$$\alpha^*_o = \beta_i/3, \quad \alpha_o = \frac{1}{9}, \quad R_\beta = 8, \quad R_k = 6, \quad R_\omega = 2.95 \tag{D.22}$$

The quantity $Re_T$ is turbulence Reynolds number defined by

$$Re_T = \frac{k}{\omega\nu} \tag{D.23}$$

If low-Reynolds-number corrections are included in the Stress-$\omega$ model, we use the following:

$$\left.\begin{aligned}
\alpha^* &= \frac{\alpha^*_o + Re_T/R_k}{1 + Re_T/R_k} \\
\alpha &= \alpha_\infty \cdot \frac{\alpha_o + Re_T/R_\omega}{1 + Re_T/R_\omega} \cdot \frac{3 + Re_T/R_\omega}{3\alpha^*_o + Re_T/R_\omega} \\
\beta^* &= \beta^*_\infty \cdot \frac{4/15 + Re_T/R_\beta}{1 + Re_T/R_\beta} \\
\hat{\alpha} &= \frac{1 + \hat{\alpha}_\infty Re_T/R_k}{1 + Re_T/R_k} \\
\hat{\beta} &= \hat{\beta}_\infty \cdot \frac{Re_T/R_k}{1 + Re_T/R_k} \\
\hat{\gamma} &= \hat{\gamma}_\infty \cdot \frac{\hat{\gamma}_o + Re_T/R_k}{1 + Re_T/R_k} \\
C_1 &= \frac{9}{5} \cdot \frac{5/3 + Re_T/R_k}{1 + Re_T/R_k}
\end{aligned}\right\} \tag{D.24}$$

where

$$\alpha^*_o = \frac{\beta_i}{3}, \quad \alpha_o = \frac{21}{200}, \quad R_\beta = 12, \quad R_k = 12, \quad R_\omega = 6.20 \tag{D.25}$$

Finally, the compressible values of $\beta$ and $\beta^*$ are

$$\beta = \beta_i \left[ 1 - \frac{\beta_i^*}{\beta_i} \xi^* F(M_t) \right], \quad \beta^* = \beta_i^* \left[ 1 + \xi^* F(M_t) \right], \quad \xi^* = 3/2 \quad \text{(D.26)}$$

The compressibility function $F(M_t)$ is given by

$$F(M_t) = \left\{ \begin{array}{ll} 0, & M_t \leq M_{to} \\ M_t^2 - M_{to}^2, & M_t > M_{to} \end{array} \right. \quad \text{(D.27)}$$

where $M_t^2 \equiv 2k/a^2$, $a$ is the speed of sound, and $M_{to}$ is given by

$$M_{to} = 1/4 \quad \text{(D.28)}$$

## D.11.3 $k$-$\epsilon$ Model Equations

For the $k$-$\epsilon$ model the equations for $k$ and $\epsilon$ are:

$$\bar{\rho}\tilde{u}\frac{\partial k}{\partial s} + \bar{\rho}\tilde{v}\frac{\partial k}{\partial n} = \bar{\rho}\tau\frac{\partial \tilde{u}}{\partial n} - \bar{\rho}\epsilon + \frac{1}{r^j}\frac{\partial}{\partial n}\left[ r^j (\mu + \mu_T/\sigma_k)\frac{\partial k}{\partial n} \right] \quad \text{(D.29)}$$

$$\bar{\rho}\tilde{u}\frac{\partial \tilde{\epsilon}}{\partial s} + \bar{\rho}\tilde{v}\frac{\partial \tilde{\epsilon}}{\partial n} = f_1 C_{\epsilon 1}\frac{\tilde{\epsilon}}{k}\bar{\rho}\tau\frac{\partial \tilde{u}}{\partial n} - f_2 C_{\epsilon 2}\bar{\rho}\frac{\tilde{\epsilon}^2}{k} + \bar{\rho}E + \frac{1}{r^j}\frac{\partial}{\partial n}\left[ r^j (\mu + \mu_T/\sigma_\epsilon)\frac{\partial \tilde{\epsilon}}{\partial n} \right] \quad \text{(D.30)}$$

where

$$\epsilon = \tilde{\epsilon} + \epsilon_o \quad \text{(D.31)}$$

and the eddy viscosity is

$$\mu_T = C_\mu f_\mu \rho k^2 / \tilde{\epsilon} \quad \text{(D.32)}$$

Program **EDDYBL** includes six low-Reynolds-number versions of the $k$-$\epsilon$ model. The models differ in the form of the damping functions $f_\mu$, $f_1$, $f_2$, $\epsilon_o$, $E$, in the values of the closure coefficients, and in the surface boundary condition imposed on $\tilde{\epsilon}$. The damping functions depend upon one or more of the following three dimensionless parameters.

$$Re_T = \frac{k^2}{\tilde{\epsilon}\nu}, \quad R_y = \frac{k^{1/2}n}{\nu}, \quad y^+ = \frac{u_\tau n}{\nu} \quad \text{(D.33)}$$

The damping functions, closure coefficients and surface boundary condition on $\tilde{\epsilon}$ for the six models are as follows.

### Jones-Launder Model

$$\left. \begin{array}{l} f_\mu = e^{-2.5/(1+Re_T/50)} \\[2mm] f_1 = 1 \\[2mm] f_2 = 1 - 0.3 e^{-Re_T^2} \\[1mm] \epsilon_o = 2\nu \left( \dfrac{\partial \sqrt{k}}{\partial n} \right)^2 \\[3mm] E = 2\nu\nu_T \left( \dfrac{\partial^2 \tilde{u}}{\partial n^2} \right)^2 \\[2mm] C_{\epsilon 1} = 1.55, \quad C_{\epsilon 2} = 2.00, \quad C_\mu = 0.09, \quad \sigma_k = 1.0, \quad \sigma_\epsilon = 1.3 \\[2mm] \tilde{\epsilon} = 0 \quad \text{at} \quad n = 0 \end{array} \right\} \quad \text{(D.34)}$$

### Launder-Sharma Model

$$\left. \begin{array}{l} f_\mu = e^{-3.4/(1+Re_T/50)^2} \\[2mm] f_1 = 1 \\[1mm] f_2 = 1 - 0.3 e^{-Re_T^2} \\[1mm] \epsilon_o = 2\nu \left( \dfrac{\partial \sqrt{k}}{\partial n} \right)^2 \\[3mm] E = 2\nu\nu_T \left( \dfrac{\partial^2 \tilde{u}}{\partial n^2} \right)^2 \\[2mm] C_{\epsilon 1} = 1.44, \quad C_{\epsilon 2} = 1.92, \quad C_\mu = 0.09, \quad \sigma_k = 1.0, \quad \sigma_\epsilon = 1.3 \\[2mm] \tilde{\epsilon} = 0 \quad \text{at} \quad n = 0 \end{array} \right\} \quad \text{(D.35)}$$

### Lam-Bremhorst Model

$$\left. \begin{array}{l} f_\mu = \left( 1 - e^{-0.0165 R_y} \right)^2 \left( 1 + 20.5/Re_T \right) \\[2mm] f_1 = 1 + (0.05/f_\mu)^3 \\[2mm] f_2 = 1 - e^{-Re_T^2} \\[1mm] \epsilon_o = 0 \\[1mm] E = 0 \\[1mm] C_{\epsilon 1} = 1.44, \quad C_{\epsilon 2} = 1.92, \quad C_\mu = 0.09, \quad \sigma_k = 1.0, \quad \sigma_\epsilon = 1.3 \\[1mm] \tilde{\epsilon} = \nu \dfrac{\partial^2 k}{\partial n^2} \quad \text{at} \quad n = 0 \end{array} \right\} \quad \text{(D.36)}$$

## Chien Model

$$
\left.
\begin{aligned}
f_\mu &= 1 - e^{-0.0115y^+} \\
f_1 &= 1 \\
f_2 &= 1 - 0.22e^{-(Re_T/6)^2} \\
\epsilon_o &= 2\nu\frac{k}{n^2} \\
E &= -2\nu\frac{\tilde{\epsilon}}{n^2}e^{-y^+/2} \\
C_{\epsilon 1} &= 1.35, \quad C_{\epsilon 2} = 1.80, \quad C_\mu = 0.09, \quad \sigma_k = 1.0, \quad \sigma_\epsilon = 1.3 \\
\tilde{\epsilon} &= 0 \quad \text{at} \quad n = 0
\end{aligned}
\right\} \quad (\text{D.37})
$$

## Yang-Shih Model

$$
\left.
\begin{aligned}
f_\mu &= \frac{\left[1 - exp\left(-1.5\cdot 10^{-4}R_y - 5\cdot 10^{-7}R_y^3 - 10^{-10}R_y^5\right)\right]^{1/2}}{\left(1 + 1/\sqrt{Re_T}\right)} \\
f_1 &= \sqrt{Re_T}/\left(1 + \sqrt{Re_T}\right) \\
f_2 &= \sqrt{Re_T}/\left(1 + \sqrt{Re_T}\right) \\
\epsilon_o &= 0 \\
E &= \nu\nu_T\left(\frac{\partial^2\tilde{u}}{\partial n^2}\right)^2 \\
C_{\epsilon 1} &= 1.44, \quad C_{\epsilon 2} = 1.92, \quad C_\mu = 0.09, \quad \sigma_k = 1.0, \quad \sigma_\epsilon = 1.3 \\
\tilde{\epsilon} &= 2\nu\left(\frac{\partial\sqrt{k}}{\partial n}\right)^2 \quad \text{at} \quad n = 0
\end{aligned}
\right\} \quad (\text{D.38})
$$

## Fan-Lakshminarayana-Barnett Model

$$
\left.
\begin{aligned}
f_\mu &= 0.4\frac{f_w}{\sqrt{Re_T}} + \left(1 - 0.4\frac{f_w}{\sqrt{Re_T}}\right)\left(1 - e^{-R_y/42.63}\right)^3 \\
f_1 &= 1 \\
f_2 &= \left[1 - 0.22e^{-(Re_T/6)^2}\right]f_w^2 \\
f_w &= 1 - exp\left[-\frac{\sqrt{R_y}}{2.30} + \left(\frac{\sqrt{R_y}}{2.30} - \frac{R_y}{8.89}\right)\left(1 - e^{-R_y/20}\right)^3\right] \\
\epsilon_o &= 0 \\
E &= 0 \\
C_{\epsilon 1} &= 1.39, \quad C_{\epsilon 2} = 1.80, \quad C_\mu = 0.09, \quad \sigma_k = 1.0, \quad \sigma_\epsilon = 1.3 \\
\frac{\partial\tilde{\epsilon}}{\partial n} &= 0 \quad \text{at} \quad n = 0
\end{aligned}
\right\} \quad (\text{D.39})
$$

## D.11.4   Transformed Equations

The boundary-layer equations are singular at the leading edge of a body. As noted above, the program uses conventional Levy-Lees variables $(\xi, \eta)$ to remove this singularity. Body oriented physical coordinates $(s, n)$ are related to transformed coordinates $(\xi, \eta)$ according to

$$d\xi = \bar{\rho}_e \tilde{u}_e \mu_e r_o^{2j}\, ds \quad \text{and} \quad d\eta = \frac{\bar{\rho} \tilde{u}_e (r_o + n)^j\, dn}{\sqrt{2\xi}} \tag{D.40}$$

where $r_o$ is body radius and subscript $e$ denotes boundary-layer edge. Equivalently, we can write

$$\xi(s) = \int_0^s \bar{\rho}_e \tilde{u}_e \mu_e r_o^{2j}\, ds \quad \text{and} \quad \eta(s, n) = \frac{\bar{\rho}_e \tilde{u}_e r_o^j}{\sqrt{2\xi}} \int_0^n \left( \frac{\bar{\rho}}{\bar{\rho}_e} \right) t^j\, dn \tag{D.41}$$

where $t$ is the transverse curvature defined by

$$t = \frac{r}{r_o} \tag{D.42}$$

The relations between derivatives in the physical $(s, n)$ and transformed $(\xi, \eta)$ coordinate system are as follows:

$$\left( \frac{\partial}{\partial s} \right)_n = \bar{\rho}_e \tilde{u}_e \mu_e r_o^{2j} \left( \frac{\partial}{\partial \xi} \right)_\eta + \left( \frac{\partial \eta}{\partial s} \right)_n \left( \frac{\partial}{\partial \eta} \right)_\xi \tag{D.43}$$

$$\left( \frac{\partial}{\partial n} \right)_s = \frac{\bar{\rho}_e \tilde{u}_e r_o^j t^j}{\sqrt{2\xi}} \left( \frac{\bar{\rho}}{\bar{\rho}_e} \right) \left( \frac{\partial}{\partial \eta} \right)_\xi \tag{D.44}$$

The dependent variables are also transformed according to:

$$\left. \begin{aligned} & F(\xi, \eta) = \frac{\tilde{u}}{\tilde{u}_e}, \quad \Theta(\xi, \eta) = \frac{\tilde{T} - \tilde{T}_e}{\tilde{T}_e} \\[2mm] & V(\xi, \eta) = \frac{2\xi}{\bar{\rho}_e \tilde{u}_e \mu_e r_o^{2j}} \left[ F \left( \frac{\partial \eta}{\partial s} \right) + \frac{\bar{\rho} \tilde{v} r_o^j t^j}{\sqrt{2\xi}} \right] \\[2mm] & K(\xi, \eta) = \frac{k}{\tilde{u}_e^2} \quad \hat{W}(\xi, \eta) = \frac{2\xi \omega}{\tilde{u}_e^2}, \quad \hat{\mathcal{E}}(\xi, \eta) = \frac{2\xi \tilde{\epsilon}}{\tilde{u}_e^4} \end{aligned} \right\} \tag{D.45}$$

The transformed equations for the $k$-$\omega$ and $k$-$\epsilon$ models can then be expressed as follows.

$$2\xi \frac{\partial F}{\partial \xi} + \frac{\partial V}{\partial \eta} + F = 0 \tag{D.46}$$

$$2\bar{\xi}F\frac{\partial F}{\partial\bar{\xi}} + V\frac{\partial F}{\partial\eta} - \frac{\partial}{\partial\eta}\left[t^{2j}L\left(1+\bar{\mu}_T\right)\frac{\partial F}{\partial\eta}\right] + \bar{\beta}\left(F^2 - \Theta - 1\right) = 0 \quad \text{(D.47)}$$

$$2\bar{\xi}F\frac{\partial\Theta}{\partial\bar{\xi}} + V\frac{\partial\Theta}{\partial\eta} - \frac{\partial}{\partial\eta}\left[t^{2j}L\left(\frac{1}{Pr_L} + \frac{\bar{\mu}_T}{Pr_T}\right)\frac{\partial\Theta}{\partial\eta}\right]$$
$$-\bar{\alpha}t^{2j}L\left(\frac{\partial F}{\partial\eta}\right)^2 - \frac{\bar{\alpha}}{\hat{\rho}_e\hat{\mu}_e\hat{r}_o^{2j}}\left(\mathcal{E} + \mathcal{E}_o\right) = 0 \quad \text{(D.48)}$$

**$k$-$\omega$ Model:**

$$2\bar{\xi}F\frac{\partial K}{\partial\bar{\xi}} + V\frac{\partial K}{\partial\eta} - \frac{\partial}{\partial\eta}\left[t^{2j}L\left(1+\sigma^*\bar{\mu}_T\right)\frac{\partial K}{\partial\eta}\right]$$
$$+2\bar{\beta}FK - t^{2j}L\bar{\mu}_T\left(\frac{\partial F}{\partial\eta}\right)^2 + \frac{\beta^*}{\hat{\rho}_e\hat{\mu}_e\hat{r}_o^{2j}}WK = 0 \quad \text{(D.49)}$$

$$2\bar{\xi}F\frac{\partial W}{\partial\bar{\xi}} + V\frac{\partial W}{\partial\eta} - \frac{\partial}{\partial\eta}\left[t^{2j}L\left(1+\sigma\bar{\mu}_T\right)\frac{\partial W}{\partial\eta}\right]$$
$$+2(\bar{\beta}-1)FW - \alpha\frac{W}{K}t^{2j}L\bar{\mu}_T\left(\frac{\partial F}{\partial\eta}\right)^2$$
$$+\frac{\beta}{\hat{\rho}_e\hat{\mu}_e\hat{r}_o^{2j}}W^2 = 0 \quad \text{(D.50)}$$

$$\bar{\mu}_T = \frac{2\bar{\xi}\hat{\rho}_e}{\hat{\mu}_e\hat{\epsilon}^2}\frac{K}{L(1+\Theta)^2W}, \qquad \mathcal{E} = \beta^*KW, \qquad \mathcal{E}_o = 0 \quad \text{(D.51)}$$

**$k$-$\epsilon$ Model:**

$$2\bar{\xi}F\frac{\partial K}{\partial\bar{\xi}} + V\frac{\partial K}{\partial\eta} - \frac{\partial}{\partial\eta}\left[t^{2j}L\left(1+\frac{\bar{\mu}_T}{\sigma_k}\right)\frac{\partial K}{\partial\eta}\right]$$
$$+2\bar{\beta}FK - t^{2j}L\bar{\mu}_T\left(\frac{\partial F}{\partial\eta}\right)^2 + \frac{1}{\hat{\rho}_e\hat{\mu}_e\hat{r}_o^{2j}}\left(\mathcal{E} + \mathcal{E}_o\right) = 0 \quad \text{(D.52)}$$

$$2\bar{\xi}F\frac{\partial\mathcal{E}}{\partial\bar{\xi}} + V\frac{\partial\mathcal{E}}{\partial\eta} - \frac{\partial}{\partial\eta}\left[t^{2j}L\left(1+\frac{\bar{\mu}_T}{\sigma_\epsilon}\right)\frac{\partial\mathcal{E}}{\partial\eta}\right]$$
$$+2(2\bar{\beta}-1)F\mathcal{E} - C_{\epsilon1}f_1\frac{\mathcal{E}}{K}t^{2j}L\bar{\mu}_T\left(\frac{\partial F}{\partial\eta}\right)^2$$
$$+\frac{C_{\epsilon2}f_2}{\hat{\rho}_e\hat{\mu}_e\hat{r}_o^{2j}}\frac{\mathcal{E}^2}{K} - \Sigma = 0 \quad \text{(D.53)}$$

$$\bar{\mu}_T = C_\mu f_\mu \frac{2\bar{\xi}\hat{\rho}_e}{\hat{\mu}_e \hat{\epsilon}^2} \frac{K^2}{L(1+\Theta)^2 \mathcal{E}}, \qquad \mathcal{E}_o = \frac{A}{U_\infty^3} \frac{2\bar{\xi}}{\hat{u}_e^4} \epsilon_o, \qquad \Sigma = \frac{A^2}{U_\infty^4} \frac{(2\bar{\xi})^2}{\hat{\rho}_e \hat{u}_e^6 \hat{\mu}_e \hat{r}_o^{2j}} E$$

(D.54)

The quantities $\bar{\alpha}$, $\bar{\beta}$ and $L$ are defined by

$$\bar{\alpha} \equiv \frac{\tilde{u}_e^2}{c_p \tilde{T}_e}, \qquad \bar{\beta} \equiv \frac{2\xi}{\tilde{u}_e} \frac{d\tilde{u}_e}{d\xi}, \qquad L \equiv \frac{\bar{\rho}}{\bar{\rho}_e} \frac{\mu}{\mu_e}$$

(D.55)

and the following dimensionless quantities have been introduced:

$$\left.\begin{array}{ccc}
\bar{\xi} = \dfrac{\xi}{\rho_\infty U_\infty \mu_r A^{2j+1}}, & \hat{r}_o = \dfrac{r_o}{A}, & \hat{u}_e = \dfrac{\tilde{u}_e}{U_\infty} \\[3mm]
\hat{\rho}_e = \dfrac{\bar{\rho}_e}{\rho_\infty}, & \hat{T}_e = \dfrac{\tilde{T}_e}{T_r}, & \hat{\mu}_e = \dfrac{\mu_e}{\mu_r} \\[3mm]
\hat{\epsilon} = \sqrt{\dfrac{\mu_r}{\rho_\infty U_\infty A}}, & W = \dfrac{\hat{W}}{\rho_\infty \mu_r A^{2j}}, & \mathcal{E} = \dfrac{\hat{\mathcal{E}}}{\rho_\infty \mu_r A^{2j}}
\end{array}\right\}$$

(D.56)

Finally, note that subscript $\infty$ denotes freestream flow condition, $A$ is a reference length, $T_r$ is the reference temperature defined as

$$T_r = U_\infty^2 / c_p$$

(D.57)

and $\mu_r$ is the value of $\mu$ for $T = T_r$.

# D.12  Software Package Modules

The following files are written from the distribution disk to your hard disk in a master directory, \dcwinc, and several subdirectories as indicated. The master directory also includes a file named **readme.txt** that gives up-to-date information.

IBM PC Executable Programs (\dcwinc):

| | |
|---|---|
| **eddybl.exe** | Program **EDDYBL** |
| **instl.exe** | Program **INSTL** |
| **ploteb.exe** | Program **PLOTEB** |
| **setebl.exe** | Program **SETEBL** |

Boundary-Layer Program Source (\dcwinc\eddybl):

| | |
|---|---|
| **eddybl.for** | Source code for Program **EDDYBL** |
| **common** | Include file for Program **EDDYBL** |
| **cpuid** | Include file specifying CPU type |

Data-Preparation Utility Source (\dcwinc\setebl):

| | |
|---|---|
| **setebl.for** | Source code for the main program |
| **edge.for** | Source code for edge condition menus |
| **grafic.for** | Source code for reading graphics data |
| **initil.for** | Source code for initial profile menus |
| **instl.for** | Source code for Program **INSTL** |
| **ioebl.for** | Source code for I/O subroutines |
| **main0.for** | Source code for main input parameter menus |
| **misc.for** | Source code for miscellaneous menus |
| **chars** | Include file for Program **SETEBL** |
| **comeb1** | Include file for Program **SETEBL** |
| **comeb2** | Include file for Program **SETEBL** |

Bench-Mark Case Input/Output Data (\dcwinc\data):

| | |
|---|---|
| **blocrv.dat** | Mass-transfer, body-curvature data file |
| **eddybl.prt** | Output from bench-mark test case |
| **exper.dat** | Experimental data file for plotting program |
| **heater.dat** | Heat-transfer, surface-temperature data file |
| **ploteb.dat** | Primary plotting-program data file |
| **presur.dat** | Pressure-distribution data file |

# Appendix E

# Plotting Program Details

The various plotting programs described in Appendices C and D, viz., **PLOTEB**, **PLOTD**, **PLOTF**, **PLOTP** and **PLOTS**, run on IBM PC and compatible micro-computers with 640k of memory. They create both video and hardcopy plots for many of the standard display devices currently available. The information in this appendix pertains to all five of the plotting programs, the executable versions of which are provided on the distribution diskette accompanying this book. Read the file **readme.txt** in the root directory of the distribution diskette to determine the location of the executable and input-data files for the plotting programs.

## E.1   Plotting Colors

Input parameters *ifore* and *iback* set foreground and background plotting colors, respectively. The standard MS-DOS color coding scheme is used to set these colors. The MS-DOS color code is summarized below, where *icolor* is the color number and 'color' is the corresponding color.

| *icolor* | color | *icolor* | color |
|---|---|---|---|
| 0 | Black | 8 | Dark Gray |
| 1 | Blue | 9 | Light Blue |
| 2 | Green | 10 | Light Green |
| 3 | Cyan | 11 | Light Cyan |
| 4 | Red | 12 | Light Red |
| 5 | Magenta | 13 | Light Magenta |
| 6 | Brown or Yellow | 14 | Yellow or Light Yellow |
| 7 | White | 15 | Bright White |

# E.2   Video Devices

Input parameter *mon* selects monitor type and resolution; valid devices are listed below, including the number of columns, rows and colors. The VESA modes should work on most 80486- and Pentium-based computers manufactured after 1993. **NOTE:** To run the plotting programs from Windows 3.1, you must use a full-screen window. Windows 95 will switch to a full-screen window automatically.

| *mon* | Description | Columns | Rows | Colors |
|---|---|---|---|---|
| 1 | Hercules | 720 | 348 | 2 |
| 2 | AT&T 6300 | 640 | 400 | 2 |
| 3 | Wyse-700 | 1280 | 800 | 2 |
| 4 | CGA | 320 | 200 | 4 |
| 5 | CGA Monochrome | 320 | 200 | 2 |
| 6 | CGA | 640 | 200 | 2 |
| 7 | SVGA: Video-7 | 800 | 600 | 16 |
| 8 | SVGA: Paradise | 800 | 600 | 16 |
| 9 | SVGA: ATI | 800 | 600 | 16 |
| 10 | SVGA: Genoa/Orchid | 800 | 600 | 16 |
| 11 | SVGA: Trident | 800 | 600 | 16 |
| 12 | SVGA: Oak | 800 | 600 | 16 |
| 13 | EGA | 320 | 200 | 16 |
| 14 | EGA | 640 | 200 | 16 |
| 15 | EGA Monochrome | 640 | 350 | 16 |
| 16 | EGA | 640 | 350 | 16 |
| 17 | MCGA/VGA | 640 | 480 | 2 |
| 18 | VGA | 640 | 480 | 16 |
| 19 | VGA | 320 | 200 | 256 |
| 20 | VESA | 800 | 600 | 16 |
| 21 | VESA | 1024 | 768 | 16 |
| 22 | VESA | 1280 | 1024 | 16 |
| 23 | SVGA: Unknown Card | 800 | 600 | 16 |
| 24 | SVGA: Everex | 800 | 600 | 16 |
| 25 | SVGA: Cirrus Logic | 800 | 600 | 16 |
| 26 | SVGA: Chips & Tech | 800 | 600 | 16 |
| 27 | SVGA: Genoa 5200-10 | 1024 | 768 | 16 |
| 28 | VESA (Alternate) | 1024 | 768 | 16 |
| 29 | VESA (Alternate) | 1280 | 1024 | 16 |

# E.3   Hardcopy Devices

Two input parameters are required for hardcopy plots, *nprin* and *mode*. The parameter *nprin* specifies the printer or plotter type. For a printer, input parameter *mode* specifies the resolution. For a plotter *mode* is the number of pens. Valid devices are as follows. If your printer or plotter is not listed, note that most popular printers and plotters used with personal computers are compatible with the Hewlett Packard or Epson printer control languages. Check your owner's manual to determine which language has been used.

| *nprin* | Printer type |
|---|---|
| 0 | Hewlett-Packard LaserJet Series I |
| 1 | Hewlett-Packard LaserJet Series II |
| 2 | Hewlett-Packard DeskJet Black and White |
| 3 | Hewlett-Packard DeskJet 1-Cartridge Color |
| 4 | Hewlett-Packard DeskJet 2-Cartridge Color |
| 5 | Hewlett-Packard LaserJet Series III and IV |
| 8 | Epson with 8-pin printer head (older models) |
| 24 | Epson with 8-pin and 24-pin printer heads |
| 25 | Epson color dot-matrix printer |
| 80 | HPGL plotter without built-in circle drawing |
| 90 | HPGL plotter with built-in circle drawing |
| 115 | CGP-115 4-pen color plotter |

Resolution for each printer and mode is given below as horizontal by vertical numbers of dots per inch.

**nprin = 0 - 5 ... Hewlett-Packard**

| *mode* | dots/in | *mode* | dots/in |
|---|---|---|---|
| 0 | 75 x 75 | 3 | 300 x 300 |
| 1 | 100 x 100 | 4 | 600 x 600 |
| 2 | 150 x 150 | | |

**nprin = 8...Epson 8-pin**

| *mode* | dots/in | *mode* | dots/in |
|---|---|---|---|
| 0 | 60 x 72 | 4 | 80 x 72 |
| 1 | 120 x 72 | 5 | 72 x 72 |
| 3 | 240 x 72 | 6 | 90 x 72 |

**nprin = 24,25...Epson 24-pin**

| *mode* | dots/in | *mode* | dots/in |
|---|---|---|---|
| 0 | 60 x 60 | 32 | 60 x 180 |
| 1 | 120 x 60 | 33 | 120 x 180 |
| 3 | 240 x 60 | 38 | 90 x 180 |
| 4 | 80 x 60 | 39 | 180 x 180 |
| 6 | 90 x 60 | 40 | 360 x 180 |

**nprin = 80,90...HPGL Plotter**

| *mode* | number of pens | *mode* | number of pens |
|---|---|---|---|
| 1 | 1 | 6 | 6 |
| 2 | 2 | 8 | 8 |

# Bibliography

Abid, R. and Speziale, C. G. (1993), "Predicting Equilibrium States with Reynolds Stress Closures in Channel Flow and Homogeneous Shear Flow," *Physics of Fluids A*, Vol. 5, pp. 1776.

Abramowitz, M. and Stegun, I. A. (1965), *Handbook of Mathematical Functions*, Dover Publications, Inc., New York, NY.

Afzal, N. and Narasimha, R. (1976), "Axisymmetric Turbulent Boundary Layer Along a Circular Cylinder," *Journal of Fluid Mechanics*, Vol. 74, pp. 113-128.

Ahmed, S. R., Ramm, G. and Faltin, G. (1984), "Some Salient Features of the Time-Averaged Ground Vehicle Wake," SAE Paper 840300, Society of Automotive Engineers, Warrendale, PA.

Akselvoll, K. and Moin, P. (1993), "Large-Eddy Simulation of a Backward-Facing Step Flow," Engineering Turbulence Modelling and Experiments 2, (W. Rodi and F. Martelli, eds.), Elsevier, pp. 303-313.

Andersen, P. S., Kays, W. M. and Moffat, R. J. (1972), "The Turbulent Boundary Layer on a Porous Plate: An Experimental Study of the Fluid Mechanics for Adverse Free-Stream Pressure Gradients," Report No. HMT-15, Dept. Mech. Eng., Stanford University, CA.

Anderson, D. A., Tannehill, J. C. and Pletcher, R. H. (1984), *Computational Fluid Dynamics and Heat Transfer*, Hemisphere Publishing, Washington, DC.

Baggett, J. S. (1997), "Some Modeling Requirements for Wall Models in Large Eddy Simulation," NASA Ames/Stanford Center for Turbulence Research, *Annual Research Briefs*, pp. 123-134.

Baldwin, B. S. and Lomax, H. (1978), "Thin-Layer Approximation and Algebraic Model for Separated Turbulent Flows," AIAA Paper 78-257.

Baldwin, B. S. and Barth, T. J. (1990), "A One-Equation Turbulence Transport Model for High Reynolds Number Wall-Bounded Flows," NASA TM-102847 [see also AIAA Paper 91-0610 (1991)].

Bardina, J., Ferziger, J. H. and Reynolds, W. C. (1983), "Improved Turbulence Models Based on Large Eddy Simulation of Homogeneous, Incompressible, Turbulent Flows," Report No. TF-19, Dept. Mech. Eng., Stanford University, CA.

Barenblatt, G. I. (1979), *Similarity, Self-Similarity and Intermediate Asymptotics*, Consultants Bureau, NY.

Barenblatt, G. I. (1991), "Scaling Laws (Incomplete Self-Similarity with Respect to Reynolds Numbers) for the Developed Turbulent Flows in Pipes," *CR Acad. Sc. Paris*, Series II, Vol. 313, pp. 307-312.

Barenblatt, G. I. (1993), "Scaling Laws for Fully Developed Turbulent Shear Flows. Part 1: Basic Hypotheses and Analysis," *Journal of Fluid Mechanics*, Vol. 248, pp. 513-520.

Barenblatt, G. I., Chorin, A. J. and Prostokishin, V. M. (1997), "Scaling Laws for Fully Developed Flow in Pipes," *Applied Mechanics Reviews*, Vol. 50, No. 7, pp. 413-429.

Barnwell, R. W. (1992), "Nonadiabatic and Three-Dimensional Effects in Compressible Turbulent Boundary Layers," *AIAA Journal*, Vol. 30, No. 4, pp. 897-904.

Beam, R. M. and Warming, R. F. (1976), "An Implicit Finite-Difference Algorithm for Hyperbolic Systems in Conservation Law Form," *Journal of Computational Physics*, Vol. 22, pp. 87-110.

Bender, C. M. and Orszag, S. A. (1978), *Advanced Mathematical Methods for Scientists and Engineers*, McGraw-Hill, New York, NY.

Bergé, P., Pomeau, Y. and Vidal, C. (1984), *Order within Chaos: Towards a Deterministic Approach to Turbulence*, John Wiley & Sons, New York, NY.

Bertolotti, F. P., Herbert, T. and Spalart, P. R. (1992), "Linear and Nonlinear Stability of the Blasius Boundary Layer," *Journal of Fluid Mechanics*, Vol. 242, pp. 441-474.

Birch, S. F. and Eggers, J. M. (1972), "Free Turbulent Shear Flows," NASA SP-321, Vol. 1, pp. 11-40.

Blair, M. F. and Werle, M. J. (1981), "Combined Influence of Free-Stream Turbulence and Favorable Pressure Gradients on Boundary Layer Transition and Heat Transfer," United Technologies Report No. R81-914388-17.

Blair, M. F. (1983), "Influence of Free-Stream Turbulence on Boundary Layer Heat Transfer and Mean Profile Development, Part 1 — Experimental Data," *Transactions of the ASME, Journal of Heat Transfer*, Vol. 105, pp. 33-40.

Blottner, F. G. (1974), "Variable Grid Scheme Applied to Turbulent Boundary Layers," *Comput. Meth. Appl. Mech. & Eng.*, Vol. 4, No. 2, pp. 179-194.

Boussinesq, J. (1877), "Théorie de l'Écoulement Tourbillant," *Mem. Présentés par Divers Savants Acad. Sci. Inst. Fr.*, Vol. 23, pp. 46-50.

Bradbury, L. J. S. (1965), "The Structure of a Self-Preserving Turbulent Plane Jet," *Journal of Fluid Mechanics*, Vol. 23, pp. 31-64.

Bradshaw, P., Ferriss, D. H. and Atwell, N. P. (1967), "Calculation of Boundary Layer Development Using the Turbulent Energy Equation," *Journal of Fluid Mechanics*, Vol. 28, Pt. 3, pp. 593-616.

Bradshaw, P. (1969), "The Response of a Constant-Pressure Turbulent Boundary Layer to the Sudden Application of an Adverse Pressure Gradient," R. & M. Number 3575, British Aeronautical Research Council.

Bradshaw, P. (1972), "The Understanding and Prediction of Turbulent Flow," *The Aeronautical Journal*, Vol. 76, No. 739, pp. 403-418.

Bradshaw, P. (1973a), "Effects of Streamline Curvature on Turbulent Flow," AGARD-AG-169.

Bradshaw, P. (1973b), "The Strategy of Calculation Methods for Complex Turbulent Flows," Imperial College Aero. Report No. 73-05.

Bradshaw, P. and Perot, J. B. (1993), "A Note on Turbulent Energy Dissipation in the Viscous Wall Region," *Physics of Fluids A*, Vol. 5, p. 3305.

Bradshaw, P. (1994), "Turbulence: The Chief Outstanding Difficulty of Our Subject," *Experiments in Fluids*, Vol. 16, pp. 203-216.

Bradshaw, P. and Huang, G. P. (1995), "The Law of the Wall in Turbulent Flow," *Proc. R. Soc. A*, Vol. 451, pp. 165-188.

Briggs, D. A., Ferziger, J. H., Koseff, J. R. and Monismith, S. G. (1996), "Entrainment in a Shear-Free Turbulent Mixing Layer," *Journal of Fluid Mechanics*, Vol. 310, pp. 215-241.

Brown, G. L. and Roshko, A. (1974), "On Density Effects and Large Structure in Turbulent Mixing Layers," *Journal of Fluid Mechanics*, Vol. 64, p. 775.

Brown, J. D. (1986), "Two Component LDV Investigation of Shock Related Turbulent Boundary Layer Separation with Increasing Three Dimensionality," PhD Thesis, U. C. Berkeley, Berkeley, CA.

Brusniak, L. and Dolling, D. S. (1996), "Engineering Estimation of Fluctuating Loads in Shock Wave/Turbulent Boundary-Layer Interactions," *AIAA Journal*, Vol. 34, No. 12, pp. 2554-2561.

Bush, W. B. and Fendell, F. E. (1972), "Asymptotic Analysis of Turbulent Channel and Boundary-Layer Flow," *Journal of Fluid Mechanics*, Vol. 56, Pt. 4, pp. 657-681.

Cabot, W. (1997), "Wall Models in Large Eddy Simulation of Separated Flow," NASA Ames/Stanford Center for Turbulence Research, *Annual Research Briefs*, pp. 97-106.

Carati, D. and Eijnden, E. V. (1997), "On the Self-Similarity Assumption in Dynamic Models for Large Eddy Simulations," *Physics of Fluids*, Vol. 9, pp. 2165-2167.

Castro, I. P. and Bradshaw, P. (1976), "The Turbulence Structure of a Highly Curved Mixing Layer," *Journal of Fluid Mechanics*, Vol. 73, p. 265.

Cazalbou, J. B., Spalart, P. R. and Bradshaw, P. (1994), "On the Behavior of Two-Equation Models at the Edge of a Turbulent Region," *Physics of Fluids*, Vol. 6, No. 5, pp. 1797-1804.

Cebeci, T. and Smith, A. M. O. (1974), *Analysis of Turbulent Boundary Layers*, Ser. in Appl. Math. & Mech., Vol. XV, Academic Press, Orlando, FL.

Chambers, T. L. and Wilcox, D. C. (1977), "Critical Examination of Two-Equation Turbulence Closure Models for Boundary Layers," *AIAA Journal*, Vol. 15, No. 6, pp. 821-828.

Champagne, F. H., Harris, V. G. and Corrsin, S. (1970), "Experiments on Nearly Homogeneous Turbulent Shear Flow," *Journal of Fluid Mechanics*, Vol. 41, Pt. 1, pp. 81-139.

Champney, J. (1989), "Modeling of Turbulence for Compression Corner Flows and Internal Flows," AIAA Paper 89-2344, Monterey, CA.

Chapra, S. C. and Canale, R. P. (1985), *Numerical Methods for Engineers: With Personal Computer Applications*, McGraw-Hill, New York, NY.

Chien, K.-Y. (1982), "Predictions of Channel and Boundary-Layer Flows with a Low-Reynolds-Number Turbulence Model," *AIAA Journal*, Vol. 20, No. 1, pp. 33-38.

Choi, K. S. and Lumley, J. L. (1984), "Return to Isotropy of Homogeneous Turbulence Revisited," *Turbulence and Chaotic Phenomena in Fluids*, T. Tatsumi, ed., New York: North-Holland, NY, pp. 267-272.

Chou, P. Y. (1945), "On the Velocity Correlations and the Solution of the Equations of Turbulent Fluctuation," *Quart. Appl. Math.*, Vol. 3, p. 38.

Clark, J. A. (1968), "A Study of Incompressible Turbulent Boundary Layers in Channel Flows," *Journal of Basic Engineering*, Vol. 90, pp. 455.

Clark, R. A., Ferziger, J. H. and Reynolds, W. C. (1979), "Evaluation of Subgrid-Scale Models Using an Accurately Simulated Turbulent Flow," *Journal of Fluid Mechanics*, Vol. 91, pp. 1-16.

Clauser, F. H. (1956), "The Turbulent Boundary Layer", *Advances in Applied Mechanics*, Vol. IV, Academic Press, New York, NY, pp. 1-51.

Clemens, N. T. and Mungal, M. G. (1995), "Large-Scale Structure and Entrainment in the Supersonic Mixing Layer," *Journal of Fluid Mechanics*, Vol. 284, p. 171.

Coakley, T. J. (1983), "Turbulence Modeling Methods for the Compressible Navier-Stokes Equations," AIAA Paper 83-1693, Danvers, MA.

Coakley, T. J. and Huang, P. G. (1992), "Turbulence Modeling for High Speed Flows," AIAA Paper 92-0436, Reno, NV.

Coleman, G. N. and Mansour, N. N. (1991), "Simulation and Modeling of Homogeneous Compressible Turbulence under Isotropic Mean Compression," Eighth Symposium on Turbulent Shear Flows, Munich, Germany.

Coleman, G. N., Kim, J. and Moser, R. D. (1995), "A Numerical Study of Turbulent Supersonic Isothermal-Wall Channel Flow," *Journal of Fluid Mechanics*, Vol. 305, p. 159.

Coles, D. E. and Hirst, E. A. (1969), *Computation of Turbulent Boundary Layers-1968 AFOSR-IFP-Stanford Conference*, Vol. II, Thermosciences Division, Stanford University, CA.

Comte-Bellot, G. (1963), "Contribution a l'Étude de la Turbulence de Conduite," PhD Thesis, University of Grenoble, France.

Comte-Bellot, G. (1965), "Écoulement Turbulent entre Deux Parois Parallèles," Publ. Sci. Tech. Ministère de l'Air, No. 419.

Comte-Bellot, G. and Corrsin, S. (1971), "Simple Eulerian Time Correlation of Full- and Narrow-Band Velocity Signals in Grid Generated Isotropic Turbulence," *Journal of Fluid Mechanics*, Vol. 48, pp. 273-337.

Corrsin, S. and Kistler, A. L. (1954), "The Free-Stream Boundaries of Turbulent Flows," NACA TN 3133.

Courant, R. and Hilbert, D. (1966), *Methods of Mathematical Physics*, Vol. II, Interscience Publishers, John Wiley & Sons, New York, NY.

Courant, R., Friedrichs, K. and Lewy, H. (1967), "On the Partial Difference Equations of Mathematical Physics," *IBM Journal*, pp. 215-234.

Craft, T. J., Fu, S., Launder, B. E. and Tselepidakis, D. P. (1989), "Developments in Modeling the Turbulent Second-Moment Pressure Correlations," Report No. TFD/89/1, Mech. Eng. Dept., Manchester Institute of Science and Technology, England.

Craft, T. J. and Launder, B. E. (1992), "New Wall-Reflection Model Applied to the Turbulent Impinging Jet," *AIAA Journal*, Vol. 30, No. 12, pp. 2970-2972.

Crank, J. and Nicolson, P. (1947), "A Practical Method for Numerical Evaluation of Solutions of Partial Differential Equations of the Heat-Conduction Type," *Proceedings of the Cambridge Philosophical Society*, Vol. 43, No. 50, pp. 50-67.

Crow, S. C. (1968), "Viscoelastic Properties of Fine-Grained Incompressible Turbulence," *Journal of Fluid Mechanics*, Vol. 33, Pt. 1, pp. 1-20.

Crowe, C. T., Troutt, T. R. and Chung, J. N. (1996), "Numerical Models for Two-Phase Turbulent Flows," *Annual review of Fluid Mechanics*, Vol. 28, p. 11.

Daly, B. J. and Harlow, F. H. (1970), "Transport Equations in Turbulence," *Physics of Fluids*, Vol. 13, pp. 2634-2649.

Davidov, B. I. (1961), "On the Statistical Dynamics of an Incompressible Fluid," *Doklady Akademiya Nauk SSSR*, Vol. 136, p. 47.

Deardorff, J. W. (1970), "A Numerical Study of Three-Dimensional Turbulent Channel Flow at Large Reynolds Numbers," *Journal of Fluid Mechanics*, Vol. 41, Pt. 2, pp. 453-480.

Deardorff, J. W. (1973), "The Use of Subgrid Transport Equations in a Three-Dimensional Model of Atmospheric Turbulence," *ASME, Journal of Fluids Engineering*, Vol. 95, pp. 429-438.

Degani, D. and Schiff, L. B. (1986), "Computation of Turbulent Supersonic Flows Around Pointed Bodies Having Crossflow Separation," *Journal of Computational Physics*, Vol. 66, No. 1, pp. 173-196.

Deissler, R. G. (1989), "On the Nature of Navier-Stokes Turbulence," NASA TM-109183.

Demuren, A. O. (1991), "Calculation of Turbulence-Driven Secondary Motion in Ducts with Arbitrary Cross Section," *AIAA Journal*, Vol. 29, No. 4, pp. 531-537.

Dhawan, S. and Narasimha, R. (1958), "Some Properties of Boundary Layer Flow During the Transition from Laminar to Turbulent Motion," *Journal of Fluid Mechanics*, Vol. 3, pp. 418-436.

Domaradzki, J. A. and Saiki, E. M. (1997), "A Subgrid-Scale Model Based on the Estimation of Unresolved Scales of Turbulence," *Physics of Fluids*, Vol. 9, pp. 2148-2164.

Donaldson, C. duP. and Rosenbaum, H. (1968), "Calculation of the Turbulent Shear Flows Through Closure of the Reynolds Equations by Invariant Modeling," ARAP Report 127, Aeronautical Research Associates of Princeton, Princeton, NJ.

Donaldson, C. duP. (1972), "Construction of a Dynamic Model of the Production of Atmospheric Turbulence and the Dispersal of Atmospheric Pollutants," ARAP Report 175, Aeronautical Research Associates of Princeton, Princeton, NJ.

Driver, D. M. and Seegmiller, H. L. (1985), "Features of a Reattaching Turbulent Shear Layer in Divergent Channel Flow," *AIAA Journal*, Vol. 23, No. 1, pp. 163-171.

Driver, D. M. (1991), "Reynolds Shear Stress Measurements in a Separated Boundary Layer," AIAA Paper 91-1787, Honolulu, HI.

Dryden, H. L. (1959), *Aerodynamics and Jet Propulsion*, Vol. V, University Press, Princeton, NJ.

DuFort, E. C. and Frankel, S. P. (1953), "Stability Conditions in the Numerical Treatment of Parabolic Differential Equations," *Math. Tables and Other Aids to Computation*, Vol. 7, pp. 135-152.

Durbin, P. A. (1991), "Near Wall Turbulence Closure Modeling Without 'Damping Functions'," *Theoretical and Computational Fluid Dynamics*, Vol. 3, No. 1, pp. 1-13.

Durbin, P. A. (1995), "Separated Flow Computations with the $k$-$\epsilon$-$v^2$ Model," *AIAA Journal*, Vol. 33, No. 4, pp. 659-664.

Dutoya, D. and Michard, P. (1981), "A Program for Calculating Boundary Layers Along Compressor and Turbine Blades," *Numerical Methods in Heat Transfer*, edited by R. W. Lewis, Morgan and O. C. Zienkiewicz, John Wiley & Sons, New York, NY.

Eaton, J. K. and Johnston, J. P. (1980), "Turbulent Flow Reattachment: An Experimental Study of the Flow and Structure Behind a Backward-Facing Step," Report No. MD-39, Dept. Mech. Eng., Stanford University, CA.

Emmons, H. W. (1954), "Shear Flow Turbulence," Proceedings of the $2^{nd}$ U. S. Congress of Applied Mechanics, ASME.

Escudier, M. P. (1966), "The Distribution of Mixing-Length in Turbulent Flows Near Walls," Imperial College, Heat Transfer Section Report TWF/TN/12.

Fage, A. and Falkner, V. M. (1932), "Note on Experiments on the Temperature and Velocity in the Wake of a Heated Cylindrical Obstacle," *Proc. R. Soc., Lond.*, Vol. A135, pp. 702-705.

Fan, S., Lakshminarayana, B. and Barnett, M. (1993), "A Low-Reynolds Number $k$-$\epsilon$ Model for Unsteady Turbulent Boundary Layer Flows," *AIAA Journal*, Vol. 31, No. 10, pp. 1777-1784.

Favre, A. (1965), "Equations des Gaz Turbulents Compressibles," *Journal de Mecanique*, Vol. 4, No. 3, pp. 361-390.

Fendell, F. E. (1972), "Singular Perturbation and Turbulent Shear Flow Near Walls," *Journal of the Astronautical Sciences*, Vol. 20, No. 3, pp. 129-165.

Fernholz, H. H. and Finley, P. J. (1981), "A Further Compilation of Compressible Boundary Layer Data with a Survey of Turbulence Data," AGARDograph 263.

Ferziger, J. H. (1976), "Large Eddy Numerical Simulations of Turbulent Flows," AIAA Paper 76-347, San Diego, CA.

Ferziger, J. H. (1989), "Estimation and Reduction of Numerical Error," Forum on Methods of Estimating Uncertainty Limits in Fluid Flow Computations, ASME Winter Annual Meeting, San Francisco, CA.

Ferziger, J. H. (1996), "Recent Advances in Large-Eddy Simulation," Engineering Turbulence Modelling and Experiments 3, (W. Rodi and G. Begeles, eds.), Elsevier, p. 163.

Ferziger, J. H. and Perić, M. (1996), *Computational Methods for Fluid Dynamics*, Springer-Verlag, Berlin, Germany.

Fisher, D. F. and Dougherty, N. S. (1982), "Transition Measurements on a 10° Cone at Mach Numbers from 0.5 to 2.0," NASA TP-1971.

Fu, S., Launder, B. E. and Tselepidakis, D. P. (1987), "Accommodating the Effects of High Strain Rates in Modelling the Pressure-Strain Correlation," Report No. TFD/87/5, Mech. Eng. Dept., Manchester Institute of Science and Technology, England.

Fureby, C., Tabor, G., Weller, H. G. and Gosman, A. D. (1997), "A Comparative Study of Subgrid Scale Models in Homogeneous Isotropic Turbulence," *Physics of Fluids*, Vol. 9, pp. 1416-1424.

Gatski, T. B. and Speziale, C. G. (1992), "On Explicit Algebraic Stress Models for Complex Turbulent Flows," *Journal of Fluid Mechanics*, Vol. 254, pp. 59-78.

Gee, K., Cummings, R. M. and Schiff, L. B. (1992), "Turbulence Model Effects on Separated Flow About a Prolate Spheroid," *AIAA Journal*, Vol. 30, No. 3, pp. 655-664.

George, W. K. Knecht, P. and Castillo, L. (1992), "The Zero-Pressure Gradient Turbulent Boundary Layer Revisited," *Proceedings of the 13th Biennial Symposium on Turbulence*, University of Missouri–Rolla.

Germano, M., Piomelli, U., Moin, P. and Cabot, W. (1991), "A Dynamic Subgrid-Scale Eddy Viscosity Model," *Physics of Fluids A*, Vol. 3, p. 1760.

Gerolymos, G. A. (1990), "Implicit Multiple-Grid Solution of the Compressible Navier-Stokes Equations Using $k$-$\epsilon$ Turbulence Closure," *AIAA Journal*, Vol. 28, pp. 1707-1717.

Ghalichi, F., Deng, X., DeChamplain, A., Douville, Y. and Guidoin, R. (1998), "Low Reynolds Number Turbulence Modeling of Blood Flow in Arterial Stenoses," *Biorheology*, to appear.

Ghosal, S., Lund, T. S., Moin, P. and Akselvoll, K. (1995), "A Dynamic Localization Model for Large-Eddy Simulation of Turbulent Flows," *Journal of Fluid Mechanics*, Vol. 286, pp. 229-255.

Gibson, M. M. and Launder, B. E. (1978), "Ground Effects on Pressure Fluctuations in the Atmospheric Boundary Layer," *Journal of Fluid Mechanics*, Vol. 86, Pt. 3, pp. 491-511.

Gibson, M. M. and Younis, B. A. (1986), "Calculation of Swirling Jets with a Reynolds Stress Closure," *Physics of Fluids*, Vol. 29, pp. 38-48.

Gleick, J. (1988), *Chaos: Making a New Science*, Penguin Books, New York, NY.

Glushko, G. (1965), "Turbulent Boundary Layer on a Flat Plate in an Incompressible Fluid," *Izvestia Akademiya Nauk SSSR, Mekh.*, No. 4, p. 13.

Goddard, F. E. Jr. (1959), "Effect of Uniformly Distributed Roughness on Turbulent Skin-Friction Drag at Supersonic Speeds," *J. Aero/Space Sciences*, Vol. 26, No. 1, pp. 1-15, 24.

Godunov, S. K. (1959), "Finite Difference Method for Numerical Computation of Discontinuous Solutions of the Equations of Fluid Dynamics," *Matematicheskii Sbornik*, Vol. 47, No. 3, pp. 271-306.

Goldberg, U. C. (1991), "Derivation and Testing of a One-Equation Model Based on Two Time Scales," *AIAA Journal*, Vol. 29, No. 8, pp. 1337-1340.

Gulyaev, A. N., Kozlov, A. N., Ye, V. and Sekundov, A. N. (1993), "Universal Turbulence Model $\nu_t$ - 92," ECOLEN, Science Research Center, Preprint No. 3, Moscow.

Halleen, R. M. and Johnston, J. P. (1967), "The Influence of Rotation on Flow in a Long Rectangular Channel - An Experimental Study," Report No. MD-18, Dept. Mech. Eng., Stanford University, CA.

Han, T. (1989), "Computational Analysis of Three-Dimensional Turbulent Flow Around a Bluff Body in Ground Proximity," *AIAA Journal*, Vol. 27, No. 9, pp. 1213-1219.

Hanjalić, K. (1970), "Two-Dimensional Flow in an Axisymmetric Channel," PhD Thesis, University of London.

Hanjalić, K. and Launder, B. E. (1976), "Contribution Towards a Reynolds-Stress Closure for Low-Reynolds-Number Turbulence," *Journal of Fluid Mechanics*, Vol. 74, Pt. 4, pp. 593-610.

Hanjalić, K., Jakirlić, S. and Hadzić, I. (1997), "Expanding the Limits of 'Equilibrium' Second-Moment Turbulence Closures," *Fluid Dynamics Research*, Vol. 20, pp. 25-41.

Harlow, F. H. and Nakayama, P. I. (1968), "Transport of Turbulence Energy Decay Rate," Los Alamos Sci. Lab., University of California Report LA-3854.

Harris, V. G., Graham, J. A. H. and Corrsin, S. (1977), "Further Experiments in Nearly Homogeneous Turbulent Shear Flow," *Journal of Fluid Mechanics*, Vol. 81, p. 657.

Harris, J. E. and Blanchard, D. K. (1982), "Computer Program for Solving Laminar, Transitional, or Turbulent Compressible Boundary-Layer Equations for Two-Dimensional and Axisymmetric Flow," NASA TM-83207.

Hassid, S. and Poreh, M. (1978), "A Turbulent Energy Dissipation Model for Flows with Drag Reduction," *ASME, Journal of Fluids Engineering*, Vol. 100, pp. 107-112.

Haworth, D. C. and Pope, S. B. (1986), "A Generalized Langevin Model for Turbulent Flows," *Physics of Fluids*, Vol. 29, pp. 387-405.

Hayes, W. D. and Probstein, R. F. (1959), *Hypersonic Flow Theory*, Academic Press, New York, NY, p. 290.

Henkes, R. A. W. M. (1998a), "Scaling of Equilibrium Boundary Layers Under Adverse Pressure Gradient Using Turbulence Models," *AIAA Journal*, Vol. 36, No. 3, pp. 320-326.

Henkes, R. A. W. M. (1998b), "Comparison of Turbulence Models for Basic Boundary Layers Relevant to Aeronautics," *Applied Scientific Research*, to appear.

Heskestad, G. (1965), "Hot-Wire Measurements in a Plane Turbulent Jet," *Journal of Applied Mechanics*, Vol. 32, No. 4, pp. 721-734.

Higuchi, H. and Rubesin, M. W. (1978), "Behavior of a Turbulent Boundary Layer Subjected to Sudden Transverse Strain," AIAA Paper 78-201, Huntsville, AL.

Hildebrand, F. B. (1976), *Advanced Calculus for Applications*, Second Edition, Prentice-Hall, Englewood Cliffs, NJ.

Hinze, J. O. (1975), *Turbulence*, Second Ed., McGraw-Hill, New York, NY.

Hoffmann, G. H. (1975), "Improved Form of the Low-Reynolds-Number $k$-$\epsilon$ Turbulence Model," *Physics of Fluids*, Vol. 18, pp. 309-312.

Horstman, C. C. (1992), "Hypersonic Shock-Wave/Turbulent-Boundary-Layer Interaction Flows," *AIAA Journal*, Vol. 30, No. 6, pp. 1480-1481.

Huang, P. G., Bradshaw, P. and Coakley, T. J. (1992), "Assessment of Closure Coefficients for Compressible-Flow Turbulence Models," NASA TM-103882.

Huang, P. G. and Coakley, T. J. (1992), "An Implicit Navier-Stokes Code for Turbulent Flow Modeling," AIAA Paper 92-547, Reno, NV.

Huang, P. G. and Liou, W. W. (1994), "Numerical Calculations of Shock-Wave/Boundary-Layer Flow Interactions," NASA TM 106694.

Huang, P. G., Coleman, G. N. and Bradshaw, P. (1995), "Compressible Turbulent Channel Flows - DNS Results and Modelling," *Journal of Fluid Mechanics*, Vol. 305, p. 185.

Hung, C. M. (1976), "Development of Relaxation Turbulence Models," NASA CR-2783.

Ibbetson, A. and Tritton, D. J. (1975), "Experiments on Turbulence in Rotating Fluid," *Journal of Fluid Mechanics*, Vol. 68, Pt. 4, pp. 639-672.

Inger, G. (1986), "Incipient Separation and Similitude Properties of Swept Shock/Turbulent Boundary Layer Interactions," AIAA Paper 86-345, Reno, NV.

Jayaraman, R., Parikh, P. and Reynolds, W. C. (1982), "An Experimental Study of the Dynamics of an Unsteady Turbulent Boundary Layer," Report No. TF-18, Dept. Mech. Eng., Stanford University, CA.

Jeans, J. (1962), *An Introduction to the Kinetic Theory of Gases*, Cambridge University Press, London.

Jimenez, J. (1995), "On why dynamic subgrid-scale models work," NASA Ames/Stanford Center for Turbulence Research, *Annual Research Briefs*, pp. 25-34.

Johnson, D. A. and King, L. S. (1985), "A Mathematically Simple Turbulence Closure Model for Attached and Separated Turbulent Boundary Layers," *AIAA Journal*, Vol. 23, No. 11, pp. 1684-1692.

Johnson, D. A. (1987), "Transonic Separated Flow Predictions with an Eddy-Viscosity/Reynolds-Stress Closure Model," *AIAA Journal*, Vol. 25, No. 2, pp. 252-259.

Johnson, D. A. and Coakley, T. J. (1990), "Improvements to a Nonequilibrium Algebraic Turbulence Model," *AIAA Journal*, Vol. 28, No. 11, pp. 2000-2003.

Johnston, J. P., Halleen, R. M. and Lezius, D. K. (1972), "Effects of a Spanwise Rotation on the Structure of Two-Dimensional Fully-Developed Turbulent Channel Flow," *Journal of Fluid Mechanics*, Vol. 56, pp. 533-557.

Jones, W. P. and Launder, B. E. (1972), "The Prediction of Laminarization with a Two-Equation Model of Turbulence," *International Journal of Heat and Mass Transfer*, Vol. 15, pp. 301-314.

Jones, S. A. (1985), "A Study of Flow Downstream of a Constriction in a Cylindrical Tube at Low Reynolds Numbers with Emphasis on Frequency Correlations," PhD Thesis, University of California.

Jovic, S. and Driver, D. (1994), "Backward-Facing Step Measurements at Low Reynolds Number," NASA TM-108870.

Keefe, L. (1990), "Connecting Coherent Structures and Strange Attractors," in *Near-Wall Turbulence – 1988 Zaric Memorial Conference*, S. J. Kline and N. H. Afgan, eds., Hemisphere, Washington, DC.

Kevorkian, J. and Cole, J. D. (1981), *Perturbation Methods in Applied Mathematics*, Springer-Verlag, New York, NY.

Kim, J., Kline, S. J. and Johnston, J. P. (1980), "Investigation of a Reattaching Turbulent Shear Layer: Flow Over a Backward-Facing Step," *ASME, Journal of Fluids Engineering*, Vol. 102, pp. 302-308.

Kim, J., Moin, P. and Moser, R. (1987), "Turbulence Statistics in Fully Developed Channel Flow at Low Reynolds Number," *Journal of Fluid Mechanics*, Vol. 177, pp. 133-166.

Klebanoff, P. S. (1954), "Characteristics of Turbulence in a Boundary Layer with Zero Pressure Gradient," NACA TN 3178.

Klebanoff, P. S. (1955), "Characteristics of Turbulence in a Boundary Layer with Zero Pressure Gradient," NACA TN 1247.

Kline, S. J., Morkovin, M. V., Sovran, G. and Cockrell, D. J. (1969), *Computation of Turbulent Boundary Layers – 1968 AFOSR-IFP-Stanford Conference*, Vol. I, Thermosciences Division, Stanford University, CA.

Kline, S. J., Cantwell, B. J. and Lilley, G. M. (1981), *1980-81 AFOSR-HTTM-Stanford Conference on Complex Turbulent Flows*, Thermosciences Division, Stanford University, CA.

Knight, D. D., Horstman, C. C., Shapey, B and Bogdanoff, S. (1987), "Structure of Supersonic Flow Past a Sharp Fin," *AIAA Journal*, Vol. 25, No. 10, pp. 1331-1337.

Knight, D. D. (1993), "Numerical Simulation of 3-D Shock Wave Turbulent Boundary-Layer Interactions," AGARD/FDP Short Course on Shock Wave-Boundary Layer Interactions in Supersonic and Hypersonic Flows, von Kármán Institute for Fluid Dynamics, Brussels, Belgium, (May 24-28, 1993).

Knight, D. D. (1997), "Numerical Simulation of Compressible Turbulent Flows Using the Reynolds-Averaged Navier-Stokes Equations," AGARD Report R-819.

Kolmogorov, A. N. (1941), "Local Structure of Turbulence in Incompressible Viscous Fluid for Very Large Reynolds Number," *Doklady Akademiya Nauk SSSR*, Vol. 30, pp. 299-303.

Kolmogorov, A. N. (1942), "Equations of Turbulent Motion of an Incompressible Fluid," *Izvestia Academy of Sciences, USSR; Physics*, Vol. 6, Nos. 1 and 2, pp. 56-58.

Lai, Y. G., So, R. M. C., Anwer, M. and Hwang, B. C. (1991), "Calculations of a Curved-Pipe Flow Using Reynolds Stress Closure," *Journal of Mechanical Engineering Science*, Vol. 205, Part C, pp. 231-244.

Lakshminarayana, B. (1986), "Turbulence Modeling for Complex Shear Flows," *AIAA Journal*, Vol. 24, No. 12, pp. 1900-1917.

Lam, C. K. G. and Bremhorst, K. A. (1981), "Modified Form of $k$-$\epsilon$ Model for Predicting Wall Turbulence," *ASME, Journal of Fluids Engineering*, Vol. 103, pp. 456-460.

Landau, L. D. and Lifshitz, E. M. (1966), *Fluid Mechanics*, Addison-Wesley, Reading, MA.

Landahl, M. T. and Mollo-Christensen, E. (1992), *Turbulence and Random Processes in Fluid Mechanics*, Second Ed., Cambridge University Press, New York, NY.

Laufer, J. (1950), "Some Recent Measurements in a Two-Dimensional Turbulent Channel," *Journal of the Aeronautical Sciences*, Vol. 17, pp. 277-287.

Laufer, J. (1951), "Investigation of Turbulent Flow in a Two Dimensional Channel," NACA Rept. 1053.

Laufer, J. (1952), "The Structure of Turbulence in Fully Developed Pipe Flow," NACA Rept. 1174.

Launder, B. E. and Spalding, D. B. (1972), *Mathematical Models of Turbulence*, Academic Press, London.

Launder, B. E. and Sharma, B. I. (1974), "Application of the Energy Dissipation Model of Turbulence to the Calculation of Flow Near a Spinning Disc," *Letters in Heat and Mass Transfer*, Vol. 1, No. 2, pp. 131-138.

Launder, B. E., Reece, G. J. and Rodi, W. (1975), "Progress in the Development of a Reynolds-Stress Turbulence Closure," *Journal of Fluid Mechanics*, Vol. 68, Pt. 3, pp. 537-566.

Launder, B. E., Priddin, C. H. and Sharma, B. I. (1977), "The Calculation of Turbulent Boundary Layers on Spinning and Curved Surfaces," *ASME, Journal of Fluids Engineering*, Vol. 99, p. 231.

Launder, B. E. and Morse, A. (1979), "Numerical Prediction of Axisymmetric Free Shear Flows with a Second-Order Reynolds Stress Closure," *Turbulent Shear Flows I*, edited by F. Durst, B. E. Launder, F. W. Schmidt and J. Whitelaw, Springer-Verlag, Berlin.

Launder, B. E. (Ed.) (1992), *Fifth Biennial Colloquium on Computational Fluid Dynamics*, Manchester Institute of Science and Technology, England.

Launder, B. E. and Li, S.-P. (1994), "On the Elimination of Wall-Topography Parameters from Second-Moment Closure," *Physics of Fluids*, Vol. 6, pp. 999-1006.

Laurence, D. and Durbin, P. A. (1994), "Modeling Near Wall Effects in Second Moment Closures by Elliptic Relaxation," NASA Ames/Stanford Center for Turbulence Research, Summer Program 1994, p. 323.

Lax, P. D. and Wendroff, B. (1960), "Systems of Conservation Laws," *Communications on Pure and Applied Mathematics*, Vol. 13, pp. 217-237.

Le, H., Moin, P. and Kim, J. (1997), "Direct Numerical Simulation of Turbulent Flow over a Backward-Facing Step," *Journal of Fluid Mechanics*, Vol. 330, pp. 349-374.

Le Penven, L., Gence, J. N. and Comte-Bellot, G. (1984), "On the Approach to Isotropy of Homogeneous Turbulence — Effect of the Partition of Kinetic Energy Among the Velocity Components," *Frontiers in Fluid Mechanics*, S. H. Davis and J. L. Lumley, eds., Springer-Verlag, New York, NY, p. 1.

Lele, S. K. (1985), "A Consistency Condition for Reynolds Stress Closures," *Physics of Fluids*, Vol. 28, p. 64.

Leonard, A. (1974), "Energy Cascade in Large-Eddy Simulations of Turbulent Fluid Flows," *Advances in Geophysics*, Vol. 18A, pp. 237-248.

Lesieur, M. and Metais, O. (1996), "New Trends in Large-Eddy Simulations of Turbulence," *Annual Review of Fluid Mechanics*, Vol. 28, p. 45.

Libby, P. A. (1996), *Introduction to Turbulence*, Taylor and Francis, Bristol, PA.

Liepmann, H. W. and Laufer, J. (1947), "Investigations of Free Turbulent Mixing," NACA TN 1257.

Lilly, D. K. (1965), "On the Computational Stability of Numerical Solutions of Time-Dependent Non-Linear Geophysical Fluid Dynamics Problems," *Monthly Weather Review*, U. S. Weather Bureau, Vol. 93, No. 1, pp. 11-26.

Lilly, D. K. (1966), "On the Application of the Eddy Viscosity Concept in the Inertial Subrange of Turbulence," NCAR Manuscript 123.

Liou, W. W. and Huang, P. G. (1996), "Calculations of Oblique Shock Wave/Turbulent Boundary-Layer Interactions with New Two-Equation Turbulence Models," NASA CR-198445.

Lumley, J. L. (1970), "Toward a Turbulent Constitutive Equation," *Journal of Fluid Mechanics*, Vol. 41, pp. 413-434.

Lumley, J. L. (1972), "A Model for Computation of Stratified Turbulent Flows," Int. Symposium on Stratified Flow, Novisibirsk.

Lumley, J. L. (1978), "Computational Modeling of Turbulent Flows," *Adv. Appl. Mech.*, Vol. 18, pp. 123-176.

MacCormack, R. W. (1969), "The Effect of Viscosity in Hypervelocity Impact Cratering," AIAA Paper 69-354, Cincinnati, OH.

MacCormack, R. W. (1985), "Current Status of Numerical Solutions of the Navier-Stokes Equations," AIAA Paper 85-32, Reno, NV.

Maise, G. and McDonald, H. (1967), "Mixing Length and Kinematic Eddy Viscosity in a Compressible Boundary Layer," AIAA Paper 67-199, New York, NY.

Mansour, N. N., Kim, J. and Moin, P. (1988), "Reynolds Stress and Dissipation Rate Budgets in Turbulent Channel Flow," *Journal of Fluid Mechanics*, Vol. 194, pp. 15-44.

Marshall, T. A. and Dolling, D. S. (1992), "Computation of Turbulent, Separated, Unswept Compression Ramp Interactions," *AIAA Journal*, Vol. 30, No. 8, pp. 2056-2065.

Mellor, G. L. (1972), "The Large Reynolds Number Asymptotic Theory of Turbulent Boundary Layers," *International Journal of Engineering Science*, Vol. 10, p. 851.

Mellor, G. L. and Herring, H. J. (1973), "A Survey of Mean Turbulent Field Closure Models," *AIAA Journal*, Vol. 11, No. 5, pp. 590-599.

Menter, F. R. (1992a), "Influence of Freestream Values on $k$-$\omega$ Turbulence Model Predictions," *AIAA Journal*, Vol. 30, No. 6, pp. 1657-1659.

Menter, F. R. (1992b), "Performance of Popular Turbulence Models for Attached and Separated Adverse Pressure Gradient Flows," *AIAA Journal*, Vol. 30, No. 8, pp. 2066-2072.

Menter, F. R. (1992c), "Improved Two-Equation $k$-$\omega$ Turbulence Models for Aerodynamic Flows," NASA TM-103975.

Menter, F. R. (1994), "A Critical Evaluation of Promising Eddy-Viscosity Models," International Symposium on Turbulence, Heat and Mass Transfer, August 9-12, 1994, Lisbon, Spain.

Meroney, R. N. and Bradshaw, P. (1975), "Turbulent Boundary-Layer Growth Over a Longitudinally Curved Surface," *AIAA Journal*, Vol. 13, pp. 1448-1453.

Millikan, C. B. (1938), "A Critical Discussion of Turbulent Flow in Channels and Circular Tubes," *Proceedings of the 5th International Conference of Theoretical and Applied Mechanics*, pp. 386-392, MIT Press, Cambridge, MA.

Minkowycz, W. J., Sparrow, E. M., Schneider, G. E. and Pletcher, R. H. (1988), *Handbook of Numerical Heat Transfer*, Wiley, New York.

Moin, P. and Mahesh, K. (1998), "Direct Numerical Simulation – A Tool In Turbulence Research," *Annual Review of Fluid Mechanics*, Vol. 30, pp. 539-578.

Morkovin, M. V. (1962), "Effects of Compressibility on Turbulent Flow," *The Mechanics of Turbulence*, A. Favre, Ed., Gordon and Breach, p. 367.

Morris, P. J. (1984), "Modeling the Pressure Redistribution Terms," *Physics of Fluids*, Vol. 27, No. 7, pp. 1620-1623.

Moser, R. D. and Moin, P. (1984), "Direct Numerical Simulation of Curved Turbulent Channel Flow," NASA TM-85974.

Myong, H. K. and Kasagi, N. (1990), "A New Approach to the Improvement of $k$-$\epsilon$ Turbulence Model for Wall-Bounded Shear Flows," *JSME International Journal*, Vol. 33, pp. 63-72.

Narayanswami, N., Horstman, C. C. and Knight, D. D. (1993), "Computation of Crossing Shock/Turbulent Boundary Layer Interaction at Mach 8.3," *AIAA Journal*, Vol. 31, No. 8, pp. 1369-1376.

Nayfeh, A. H. (1981), *Introduction to Perturbation Techniques*, John Wiley and Sons, Inc., New York, NY.

Nee, V. W. and Kovasznay, L. S. G. (1968), "The Calculation of the Incompressible Turbulent Boundary Layer by a Simple Theory," *Physics of Fluids*, Vol. 12, p. 473.

Ng, K. H. and Spalding, D. B. (1972), "Some Applications of a Model of Turbulence to Boundary Layers Near Walls," *Physics of Fluids*, Vol. 15, No. 1, pp. 20-30.

Oh, Y. H. (1974), "Analysis of Two-Dimensional Free Turbulent Mixing," AIAA Paper 74-594, Palo Alto, CA.

Papamoschou, D. and Roshko, A. (1988), "The Compressible Turbulent Shear Layer - An Experimental Study," *Journal of Fluid Mechanics*, Vol. 197, pp. 453-477.

Parneix, S., Laurence, D. and Durbin, P. A. (1998), "A Procedure for Using DNS Databases," *Journal of Fluids Engineering*, Vol. 120, pp. 40-46.

Patel, V. C. (1968), "The Effects of Curvature on the Turbulent Boundary Layer," Aeronautical Research Council Reports and Memoranda No. 3599.

Patel, V. C., Rodi, W. and Scheuerer, G. (1985), "Turbulence Models for Near-Wall and Low Reynolds Number Flows: A Review," *AIAA Journal*, Vol. 23, No. 9, pp. 1308-1319.

Patel, V. C. and Sotiropoulos, F. (1997), "Longitudinal Curvature Effects in Turbulent Boundary Layers," *Progress in Aerospace Sciences*, Vol. 33, p. 1.

Peaceman, D. W. and Rachford, H. H., Jr. (1955), "The Numerical Solution of Parabolic and Elliptic Differential Equations," *J. Soc. Indust. Applied Mathematics*, Vol. 3, No. 1, pp. 28-41.

Peng, S.-H., Davidson, L. and Holmberg, S. (1997), "A Modified Low-Reynolds Number $k$-$\omega$ Model for Recirculating Flows," *Journal of Fluids Engineering*, Vol. 119, pp. 867-875.

Peyret, R. and Taylor, T. D. (1983), *Computational Methods for Fluid Flow*, Springer-Verlag, New York, NY.

Piomelli, U., Yu, Y. and Adrian, R. J. (1996), "Subgrid-Scale Energy Transfer and Near-Wall Turbulence Structure," *Physics of Fluids*, Vol. 8, pp. 215-224.

Pope, S. B. (1975), "A More General Effective Viscosity Hypothesis," *Journal of Fluid Mechanics*, Vol. 72, pp. 331-340.

Pope, S. B. (1978), "An Explanation of the Turbulent Round-Jet/Plane-Jet Anomaly," *AIAA Journal*, Vol. 16, No. 3, pp. 279-281.

Prandtl, L. (1925), "Über die ausgebildete Turbulenz," *ZAMM*, Vol. 5, pp. 136-139.

Prandtl, L. (1945), "Über ein neues Formelsystem für die ausgebildete Turbulenz," *Nacr. Akad. Wiss. Göttingen, Math-Phys. Kl.*, pp. 6-19.

Press, W. H., Flannery, B. P., Teukolsky, S. A. and Vetterling, W. T. (1987), *Numerical Recipes: The Art of Scientific Computing*, Cambridge University Press, Cambridge.

Price, J. M. and Harris, J. E. (1972), "Computer Program for Solving Compressible Nonsimilar-Boundary-Layer Equations for Laminar, Transitional or Turbulent Flows of a Perfect Gas," NASA TM X-2458.

Rastogi, A. K. and Rodi, W. (1978), "Calculation of General Three Dimensional Turbulent Boundary Layers," *AIAA Journal*, Vol. 16, No. 2, pp. 151-159.

Reda, D. C., Ketter, F. C. Jr. and Fan, C. (1974), "Compressible Turbulent Skin Friction on Rough and Rough/Wavy Walls in Adiabatic Flow," AIAA Paper 74-574, Palo Alto, CA.

Reynolds, O. (1874), "On the Extent and Action of the Heating Surface for Steam Boilers," *Proc. Manchester Lit. Phil. Soc.*, Vol. 14, pp. 7-12.

Reynolds, O. (1895), "On the Dynamical Theory of Incompressible Viscous Fluids and the Determination of the Criterion," *Philosophical Transactions of the Royal Society of London, Series A*, Vol. 186, p. 123.

Reynolds, W. C. (1970), "Computation of Turbulent Flows-State of the Art," Report No. MD-27, Dept. Mech. Eng., Stanford University, CA.

Reynolds, W. C. (1976), "Computation of Turbulent Flows," *Annual Review of Fluid Mechanics*, Vol. 8, pp. 183-208.

Reynolds, W. C. (1987), "Fundamentals of Turbulence for Turbulence Modeling and Simulation," In Lecture Notes for von Kármán Institute, AGARD Lecture Series No. 86, pp. 1-66, New York: NATO.

Ristorcelli, J. R. (1993), "A Representation for the Turbulent Mass Flux Contribution to Reynolds Stress and Two-Equation Closures for Compressible Turbulence," ICASE Report No. 93-88, Univ. Space Research Assoc., Hampton, VA.

Ristorcelli, J. R., Lumley, J. L. and Abid, R. (1995), "A Rapid-Pressure Correlation Representation Consistent with the Taylor-Proudman Theorem Materially Frame-Indifferent in the 2-D Limit," *Journal of Fluid Mechanics*, Vol. 292, pp. 111-152.

Roache, P. J. (1972), *Computational Fluid Dynamics*, Hermosa Publishers, Albuquerque, NM.

Roache, P. J. (1990), "Need for Control of Numerical Accuracy," *Journal of Spacecraft and Rockets*, Vol. 27, No. 2, pp. 98-102.

Roache, P. J. and Salari, K. (1990), "Weakly Compressible Navier-Stokes Solutions with an Implicit Approximate Factorization Code," AIAA Paper 90-235, Reno, NV.

Robinson, D. F., Harris, J. E. and Hassan, H. A. (1995), "Unified Turbulence Closure Model for Axisymmetric and Planar Free Shear Flows," *AIAA Journal*, Vol. 33, No. 12, pp. 2324-2331.

Robinson, S. K. (1986), "Instantaneous Velocity Profile Measurements in a Turbulent Boundary Layer," *Chem. Eng. Commun.*, Vol. 43, pp. 347 [see also AIAA Paper 82-0963 (1982)].

Rodi, W. and Spalding, D. B. (1970), "A Two-Parameter Model of Turbulence and its Application to Free Jets," *Wärme und Stoffübertragung*, Vol. 3, p. 85.

Rodi, W. (1975), "A New Method of Analyzing Hot-Wire Signals in Highly Turbulent Flows and Its Evaluation in Round Jets," Disa Information, No. 17.

Rodi, W. (1976), "A New Algebraic Relation for Calculating Reynolds Stresses," *ZAMM*, Vol. 56, p. 219.

Rodi, W. (1981), "Progress in Turbulence Modeling for Incompressible Flows," AIAA Paper 81-45, St. Louis, MO.

Rodi, W. and Scheuerer, G. (1986), "Scrutinizing the $k$-$\epsilon$ Turbulence Model Under Adverse Pressure Gradient Conditions," *ASME, Journal of Fluids Engineering*, Vol. 108, pp. 174-179.

Rodi, W. (1991), "Experience with Two-Layer Models Combining the $k$-$\epsilon$ Model with a One-Equation Model Near the Wall," AIAA Paper 91-216, Reno, NV.

Rodi, W. (1997), "Comparison of LES and RANS Calculations of the Flow Around Bluff Bodies," *Journal of Wind Engineering and Industrial Aerodynamics*, Vol. 69-71, pp. 55-75.

Rodi, W. (1998), "Large-Eddy Simulations of the Flow Past Bluff Bodies – State-of-the Art," *Japan Society of Mechanical Engineers International Journal*, Series B, Vol. 41, No. 2.

Roe, P. L. (1981), "Approximate Riemann Solvers, Parameter Vectors, and Difference Schemes," *Journal of Computational Physics*, Vol. 43, pp. 357-372.

Rogallo, R. S. and Moin, P. (1984), "Numerical Simulation of Turbulent Flows," *Annual Review of Fluid Mechanics*, Vol. 16, pp. 99-137.

Rotta, J. C. (1951), "Statistische Theorie nichthomogener Turbulenz," *Zeitschrift für Physik*, Vol. 129, pp. 547-572.

Rotta, J. C. (1962), "Turbulent Boundary Layers in Incompressible Flow," *Progress in Aerospace Sciences*, Vol. 2, p. 1.

Rotta, J. C. (1968), "Über eine Methode zur Berechnung turbulenter Scherströmungen," Aerodynamische Versuchanstalt Göttingen, Rep. 69 A 14.

Rubel, A. and Melnik, R. E. (1984), "Jet, Wake and Wall Jet Solutions Using a $k$-$\epsilon$ Turbulence Model," AIAA Paper 84-1523, Snowmass, CO.

Rubel, A. (1985), "On the Vortex Stretching Modification of the $k$-$\epsilon$ Turbulence Model: Radial Jets," *AIAA Journal*, Vol. 23, No. 7, pp. 1129-1130.

Rubesin, M. W. (1989), "Turbulence Modeling for Aerodynamic Flows," AIAA Paper 89-606, Reno, NV.

Rubesin, M. W. (1990), "Extra Compressibility Terms for Favre Averaged Two-Equation Models of Inhomogeneous Turbulent Flows," NASA CR-177556.

Ruelle, D. (1994), "Where Can One Hope to Profitably Apply the Ideas of Chaos?" *Physics Today*, Vol. 47, No. 7, pp. 24-30.

Saad, A. A. and Giddens, D. P. (1983), "Velocity Measurements in Steady Flow through Axisymmetric Stenoses at Moderate Reynolds Numbers," *Journal Biomech.*, Vol. 16, pp. 505-516.

Saffman, P. G. (1970), "A Model for Inhomogeneous Turbulent Flow," *Proc. R. Soc., Lond.*, Vol. A317, pp. 417-433.

Saffman, P. G. and Wilcox, D. C. (1974), "Turbulence-Model Predictions for Turbulent Boundary Layers," *AIAA Journal*, Vol. 12, No. 4, pp. 541-546.

Saffman, P. G. (1976), "Development of a Complete Model for the Calculation of Turbulent Shear Flows," April 1976 Symposium on Turbulence and Dynamical Systems, Duke Univ., Durham, NC.

Sai, V. A. and Lutfy, F. M. (1995), "Analysis of the Baldwin-Barth and Spalart-Allmaras One-Equation Turbulence Models," *AIAA Journal*, Vol. 33, No. 10, pp. 1971-1974.

Sandham, N. D. and Kleiser, L. (1992), "The Late Stages of Transition to Turbulence in Channel Flow," *Journal of Fluid Mechanics*, Vol. 245, p. 319.

Sarkar, S., Erlebacher, G., Hussaini, M. Y. and Kreiss, H. O. (1989), "The Analysis and Modeling of Dilatational Terms in Compressible Turbulence," NASA CR-181959.

Sarkar, S. and Speziale, C. G. (1990), "A Simple Nonlinear Model for the Return to Isotropy in Turbulence," *Physics of Fluids A*, Vol. 2, pp. 84-93.

Sarkar, S., Erlebacher, G. and Hussaini, M. Y. (1991), "Compressible and Homogeneous Shear — Simulation and Modeling," $8^{th}$ Symposium on Turbulent Shear Flows, Munich, Paper No. 21-2.

Sarkar, S. (1992), "The Pressure-Dilatation Correlation in Compressible Flows," *Physics of Fluids A*, Vol. 4, pp. 2674-2682.

Schlichting, H. (1979), *Boundary Layer Theory*, Seventh Ed., McGraw-Hill, New York, NY.

Schubauer, G. B. and Skramstad, H. K. (1948), "Laminar-Boundary-Layer Oscillations and Transition on a Flat Plate," NACA 909.

Schumann, U. (1975), "Subgrid Scale Model for Finite Difference Simulations of Turbulent Flows in Plane Channels and Annuli," *Journal of Computational Physics*, Vol. 18, pp. 376-404.

Schwarz, W. R. and Bradshaw, P. (1994), "Term-by-Term Tests of Stress-Transport Turbulence Models in a Three-Dimensional Boundary Layer," *Physics of Fluids*, Vol. 6, p. 986.

Sekundov, A. N. (1971), "Application of the Differential Equation for Turbulent Viscosity to the Analysis of Plane Nonself-Similar Flows," *Akademiya Nauk SSSR*, Izvestiia, Mekhanika Zhidkosti i Gaza, pp. 114-127 (in Russian).

Settles, G. S., Vas, I. E. and Bogdonoff, S. M. (1976), "Details of a Shock Separated Turbulent Boundary Layer at a Compression Corner," *AIAA Journal*, Vol. 14, No. 12, pp. 1709-1715.

Shaanan, S., Ferziger, J. H. and Reynolds, W. C. (1975), "Numerical Simulation of Turbulence in the Presence of Shear," Report No. TF-6, Dept. Mech. Eng., Stanford University, CA.

Shang, J. S. and Hankey, W. L. (1975), "Numerical Solution of the Navier Stokes Equations for Compression Ramp," AIAA Paper 75-4, Pasadena, CA.

Shih, T. H. and Lumley, J. L. (1985), "Modeling of Pressure Correlation Terms in Reynolds Stress and Scalar Flux Equations," Report No. FDA-85-3, Cornell University, Ithaca, NY.

Shih, T. H., Mansour, N. and Chen, J. Y. (1987), " Reynolds Stress Models of Homogeneous Turbulence," *Studying Turbulence Using Numerical Simulation Databases*, NASA Ames/Stanford CTR-S87, p. 191.

Shih, T. H. and Hsu, A. T. (1991), "An Improved $k$-$\epsilon$ Model for Near-Wall Turbulence," AIAA Paper 91-611, Reno, NV.

Shih, T. I.-P., Steinthorsson, E. and Chyu, W. J. (1993), "Implicit Treatment of Diffusion Terms in Lower-Upper Algorithms," *AIAA Journal*, Vol. 31, No. 4, pp. 788-791.

Simpson, R. L. and Wallace, D. B. (1975), "Laminarescent Turbulent Boundary Layers: Experiments on Sink Flows," Project SQUID, Report SMU-1-PU.

Smagorinsky, J. (1963), "General Circulation Experiments with the Primitive Equations. I. The Basic Experiment," *Mon. Weather Rev.*, Vol. 91, pp. 99-164.

Smith, A. M. O. and Cebeci, T. (1967), "Numerical Solution of the Turbulent Boundary-Layer Equations," Douglas Aircraft Division Report DAC 33735.

Smith, B. R. (1990), "The $k$-$k\ell$ Turbulence and Wall Layer Model for Compressible Flows," AIAA Paper 90-1483, Seattle, WA.

Smith, B. R. (1994), "A near Wall Model for the $k$-$\ell$ Two Equation Turbulence Model," AIAA Paper 94-2386, Colorado Springs, CO.

So, R. M. C. and Mellor, G. L. (1972), "An Experimental Investigation of Turbulent Boundary Layers Along Curved Surfaces," NASA CR-1940.

So, R. M. C. and Mellor, G. L. (1978), "Turbulent Boundary Layers with Large Streamline Curvature Effects," *ZAMP*, Vol. 29, pp. 54-74.

So, R. M. C., Lai, Y. G., Hwang, B. C. and Yoo, G. J. (1988), "Low Reynolds Number Modeling of Flows Over a Backward Facing Step," *ZAMP*, Vol. 39, pp. 13-27.

So, R. M. C., Lai, Y. G., Zhang, H. S. and Hwang, B. C. (1991), "Second-Order Near-Wall Turbulence Closures: A Review," *AIAA Journal*, Vol. 29, No. 11, pp. 1819-1835.

So, R. M. C. and Yuan, S. P. (1998), "Near-Wall Two-Equation and Reynolds-Stress Modeling of Backstep Flow," *International Journal of Engineering Science*, Vol. 36, No, 3, pp. 283-298.

Spalart, P. R. (1986), "Numerical Study of Sink-Flow Boundary Layers," *Journal of Fluid Mechanics*, Vol. 172, pp. 307-328.

Spalart, P. R. (1988), "Direct Simulation of a Turbulent Boundary Layer up to $Re_\theta = 1400$," *Journal of Fluid Mechanics*, Vol. 187, pp. 61-98.

Spalart, P. R. (1989), "Direct Numerical Study of Leading-Edge Contamination," AGARD CP 438.

Spalart, P. R. and Allmaras, S. R. (1992), "A One-Equation Turbulence Model for Aerodynamic Flows," AIAA Paper 92-439, Reno, NV [see also *La Recherche Aerospatiale* No. 1, p. 5 (1994)].

Spalart, P. R., Jou, W.-H., Strelets, M. and Allmaras, S. R. (1997), "Comments on the Feasibility of LES for Wings, and on a Hybrid RANS/LES Approach," *Advances in DNS/LES* (C. Liu and Z. Liu, eds. - Proceedings of 1st AFOSR Internat. Conf. on DNS and LES, Louisiana Tech.), Greyden Press, Columbus, OH.

Speziale, C. G. (1985), "Modeling the Pressure-Gradient-Velocity Correlation of Turbulence," *Physics of Fluids*, Vol. 28, pp. 69-71.

Speziale, C. G. (1987a), "Second-Order Closure Models for Rotating Turbulent Flows," *Q. Appl. Math.*, Vol. 45, pp. 721-733.

Speziale, C. G. (1987b), "On Nonlinear $k$-$\ell$ and $k$-$\epsilon$ Models of Turbulence," *Journal of Fluid Mechanics*, Vol. 178, pp. 459-475.

Speziale, C. G., Abid, R. and Anderson, E. C. (1990), "A Critical Evaluation of Two-Equation Models for Near Wall Turbulence," AIAA Paper 90-1481, Seattle, WA.

Speziale, C. G. (1991), "Analytical Methods for the Development of Reynolds Stress Closures in Turbulence," *Annual Review of Fluid Mechanics*, Vol. 23, pp. 107-157.

Speziale, C. G. Sarkar, S. and Gatski, T. B. (1991), "Modeling the Pressure-Strain Correlation of Turbulence," *Journal of Fluid Mechanics*, Vol. 227, pp. 245-272.

Speziale, C. G., Abid, R. and Durbin, P. A. (1994), "On the Realizability of Reynolds Stress Turbulence Closures," *Journal of Scientific Computing*, Vol. 9, pp. 369-403.

Speziale, C. G. and So, R. M. C. (1996), "Turbulence Modeling and Simulation," Technical Report No. AM-96-015, Boston University, Boston, MA.

Speziale, C. G. and Xu, X. H. (1996), "Towards the Development of Second-Order Closure Models for Non-Equilibrium Turbulent Flows," *Int. Journal of Heat and Fluid Flow*, Vol. 17, pp. 238-244.

Speziale, C. G. (1997a), "Comparison of Explicit and Traditional Algebraic Stress Models of Turbulence," *AIAA Journal*, Vol. 35, No. 9, pp. 1506-1509.

Speziale, C. G. (1997b), "Turbulence Modeling for Time-Dependent RANS and VLES - a Review," AIAA Paper 97-2051.

Steger, J. and Warming, R. F. (1979), "Flux Vector Splitting of the Inviscid Gasdynamics Equations with Application to Finite Difference Methods," NASA TM-78605.

Stewartson, K. (1981), "Some Recent Studies in Triple-Deck Theory," in *Numerical and Physical Aspects of Aerodynamic Flows*, T. Cebeci, ed., Springer-Verlag, New York, NY, p. 142.

Stratford, B. S. (1959), "An Experimental Flow with Zero Skin Friction Throughout its Region of Pressure Rise," *Journal of Fluid Mechanics*, Vol. 5, pp. 17-35.

Tanaka, T. and Tanaka, E. (1976), "Experimental Study of a Radial Turbulent Jet," *Bulletin of the JSME*, Vol. 19, pp. 792-799.

Tavoularis, S. and Corrsin, S. (1981), "Experiments in Nearly Homogeneous Turbulent Shear Flow with Uniform Mean Temperature Gradient. Part I," *Journal of Fluid Mechanics*, Vol. 104, pp. 311-347.

Tavoularis, S. and Karnik, U. (1989), "Further Experiments on the Evolution of Turbulent Stresses and Scales in Uniformly Sheared Turbulence," *Journal of Fluid Mechanics*, Vol. 204, p. 457.

Taylor, G. I. (1935), "Statistical Theory of Turbulence," *Proc. R. Soc., Lond.*, Vol. A151, p. 421.

Tennekes, H. and Lumley, J. L. (1983), *A First Course in Turbulence*, MIT Press, Cambridge, MA.

Thangam, S. and Speziale, C. G. (1992), "Turbulent Flow Past a Backward Facing Step: A Critical Evaluation of Two-Equation Models," *AIAA Journal*, Vol. 30, No. 5, pp. 1314-1320.

Thomann, H. (1968), "Effect of Streamwise Wall Curvature on Heat Transfer in a Turbulent Boundary Layer," *Journal of Fluid Mechanics*, Vol. 33, pp. 283-292.

Townsend, A. A. (1956), "The Uniform Distortion of Homogeneous Turbulence," *Q. J. Mech. Appl. Math.*, Vol. 7, p. 104.

Townsend, A. A. (1976), *The Structure of Turbulent Shear Flow*, Second Ed., Cambridge University Press, Cambridge, England.

Tucker, H. J. and Reynolds, A. J. (1968), "The Distortion of Turbulence by Irrotational Plane Strain," *Journal of Fluid Mechanics*, Vol. 32, Pt. 4, pp. 657-673.

Uberoi, M. S. (1956), "Effect of Wind Tunnel Contraction on Free Stream Turbulence," *Journal of the Aeronautical Sciences*, Vol. 23, pp. 754-764.

Van Driest, E. R. (1951), "Turbulent Boundary Layer in Compressible Fluids," *Journal of the Aeronautical Sciences*, Vol. 18, pp. 145-160, 216.

Van Driest, E. R. (1956), "On Turbulent Flow Near a Wall," *Journal of the Aeronautical Sciences*, Vol. 23, p. 1007.

Van Dyke, M. D. (1975), *Perturbation Methods in Fluid Mechanics*, Parabolic Press, Stanford, CA.

Van Leer, B. (1982), "Flux-Vector Splitting for the Euler Equations," ICASE Report 82-30, Univ. Space Research Assoc., Hampton, VA.

Vasiliev, V. I., Volkov, D. V., Zaitsev, S. A. and Lyubimov, D. A. (1997), "Numerical Simulation of Channel Flows by a One-Equation Turbulence Model," *Journal of Fluids Engineering*, Vol. 119, pp. 885-892.

Venkatakrishnan, V. (1996), "Perspective in Unstructured Grid Flow Solvers," *AIAA Journal*, Vol. 34, pp. 533-547.

Veynante, D. and Poinsot, T. (1997), "Large Eddy Simulation of Combustion Instabilities in Turbulent Premixed Burners," NASA Ames/Stanford Center for Turbulence Research, *Annual Research Briefs*, pp. 253-274.

Viegas, J. R. and Horstman, C. C. (1979), "Comparison of Multiequation Turbulence Models for Several Shock Boundary-Layer Interaction Flows," *AIAA Journal*, Vol. 17, No. 8, pp. 811-820.

Viegas, J. R., Rubesin, M. W. and Horstman, C. C. (1985), "On the Use of Wall Functions as Boundary Conditions for Two-Dimensional Separated Compressible Flows," AIAA Paper 85-180, Reno, NV.

Vollmers, H. and Rotta, J. C. (1977), "Similar Solutions of the Mean Velocity, Turbulent Energy and Length Scale Equation," *AIAA Journal*, Vol. 15, No. 5, pp. 714-720.

von Kármán, T. (1930), "Mechanische Ähnlichkeit und Turbulenz," *Proc. Int. Congr. Appl. Mech., 3rd, Stockholm*, Part 1, pp. 85-105.

von Kármán, T. (1934), "Some Aspects of the Turbulence Problem," *Proc. Int. Congr. Appl. Mech., 4th, Cambridge*, p. 54.

von Kármán, T. (1937), "Turbulence," Twenty-Fifth Wilbur Wright Memorial Lecture, *Journal of the Aeronautical Sciences*, Vol. 41, p. 1109.

Vreman, A. W., Sandham, N. D. and Luo, K. H. (1996), "Compressible Mixing Layer Growth Rate and Turbulence Characteristics," *Journal of Fluid Mechanics*, Vol. 320, p. 235.

Wang, Q. and Squires, K. D. (1996), "Large Eddy Simulation of Particle-Laden Turbulent Channel Flow," *Physics of Fluids*, Vol. 8, pp. 1207-1223.

Weinstock, J. (1981), "Theory of the Pressure-Strain Rate Correlation for Reynolds Stress Turbulence Closures," *Journal of Fluid Mechanics*, Vol. 105, pp. 369-396.

Weygandt, J. H. and Mehta, R. D. (1995), "Three-Dimensional Structure of Straight and Curved Plane Wakes," *Journal of Fluid Mechanics*, Vol. 282, p. 279.

Wigeland, R. A. and Nagib, H. M. (1978), "Grid-Generated Turbulence With and Without Rotation About the Streamwise Direction," Fluids and Heat Transfer Report R78-1, Illinois Institute of Technology, Chicago, IL.

Wilcox, D. C. and Alber, I. E. (1972), "A Turbulence Model for High Speed Flows," *Proc. of the 1972 Heat Trans. & Fluid Mech. Inst.*, Stanford Univ. Press, Stanford, CA, pp. 231-252.

Wilcox, D. C. (1974), "Numerical Study of Separated Turbulent Flows," AIAA Paper 74-584, Palo Alto, CA.

Wilcox, D. C. and Chambers, T. L. (1975), "Further Refinement of the Turbulence Model Transition Prediction Technique," DCW Industries Report DCW-R-03-02, La Cañada, CA.

Wilcox, D. C. and Traci, R. M. (1976), "A Complete Model of Turbulence," AIAA Paper 76-351, San Diego, CA.

Wilcox, D. C. and Chambers, T. L. (1977), "Streamline Curvature Effects on Turbulent Boundary Layers," *AIAA Journal*, Vol. 15, No. 4, pp. 574-580.

Wilcox, D. C. (1977), "A Model for Transitional Flows," AIAA Paper 77-126, Los Angeles, CA.

Wilcox, D. C. and Rubesin, M. W. (1980), "Progress in Turbulence Modeling for Complex Flow Fields Including Effects of Compressibility," NASA TP-1517.

Wilcox, D. C. (1981a), "Alternative to the $e^9$ Procedure for Predicting Boundary Layer Transition," *AIAA Journal*, Vol. 19, No. 1, pp. 56-64.

Wilcox, D. C. (1981b), "Algorithm for Rapid Integration of Turbulence Model Equations on Parabolic Regions," *AIAA Journal*, Vol. 19, No. 2, pp. 248-251.

Wilcox, D. C. (1988a), "Reassessment of the Scale Determining Equation for Advanced Turbulence Models," *AIAA Journal*, Vol. 26, No. 11, pp. 1299-1310.

Wilcox, D. C. (1988b), "Multiscale Model for Turbulent Flows," *AIAA Journal*, Vol. 26, No. 11, pp. 1311-1320.

Wilcox, D. C. (1989), "Wall Matching, A Rational Alternative to Wall Functions," AIAA Paper 89-611, Reno, NV.

Wilcox, D. C. (1990), "Supersonic Compression-Corner Applications of a Multiscale Model for Turbulent Flows," *AIAA Journal*, Vol. 28, No. 7, pp. 1194-1198.

Wilcox, D. C. (1991), "Progress in Hypersonic Turbulence Modeling," AIAA Paper 91-1785, Honolulu, HI.

Wilcox, D. C. (1992a), "The Remarkable Ability of Turbulence Model Equations to Describe Transition," Fifth Symposium on Numerical and Physical Aspects of Aerodynamic Flows, 13-15 January 1992, California State University, Long Beach, CA.

Wilcox, D. C. (1992b), "Dilatation-Dissipation Corrections for Advanced Turbulence Models," *AIAA Journal*, Vol. 30, No. 11, pp. 2639-2646.

Wilcox, D. C. (1993a), "A Two-Equation Turbulence Model for Wall-Bounded and Free-Shear Flows," AIAA Paper 93-2905, Orlando, FL.

Wilcox, D. C. (1993b), "Comparison of Two-Equation Turbulence Models for Boundary Layers with Pressure Gradient," *AIAA Journal*, Vol. 31, No. 8, pp. 1414-1421.

Wilcox, D. C. (1994), "Simulation of Transition with a Two-Equation Turbulence Model," *AIAA Journal*, Vol. 32, No. 2, pp. 247-255.

Wilcox, D. C. (1995a), *Perturbation Methods in the Computer Age*, DCW Industries, Inc., La Cañada, CA.

Wilcox, C. C. (1995b), "Back of the Envelope Analysis of Turbulence Models," *IUTAM Symposium on Asymptotic Methods for Turbulent Shear Flows at High Reynolds Numbers*, pp. 309-322, Kluwer Academic Publishers, Dordrecht, The Netherlands.

Witze, P. O. and Dwyer, H. A. (1976), "The Turbulent Radial Jet," *Journal of Fluid Mechanics*, Vol. 75, pp. 401-417.

Wolfshtein, M. (1967), "Convection Processes in Turbulent Impinging Jets," Imperial College, Heat Transfer Section Report SF/R/2.

Wygnanski, I. and Fiedler, H. E. (1969), "Some Measurements in the Self-Preserving Jet," *Journal of Fluid Mechanics*, Vol. 38, pp. 577-612.

Yakhot, V. and Orszag, S. A. (1986), "Renormalization Group Analysis of Turbulence: 1. Basic Theory," *Journal of Scientific Computing*, Vol. 1, pp. 3-51.

Yakhot, V., Orszag, S. A., Thangam, S., Gatski, T. B. and Speziale, C. G. (1992), "Development of Turbulence Models for Shear Flows by a Double Expansion Technique," *Physics of Fluids A*, Vol. 4, p. 1510-1520.

Yang, K.-S. and Ferziger, J. H. (1993), "Large-Eddy Simulation of Turbulent Obstacle Flow Using a Dynamic Subgrid-Scale Model," *AIAA Journal*, Vol. 31, pp. 1406-1413.

Yang, Z. and Shih, T.-H. (1993), "A New Time Scale Based $k$-$\epsilon$ Model for Near Wall Turbulence," *AIAA Journal*, Vol. 31, No. 7, pp. 1191-1198.

Yap, C. (1987), "Turbulent Heat and Momentum Transfer in Recirculating and Impinging Flows," PhD Thesis, Faculty of Technology, University of Manchester.

Zagarola, M. V. (1996), "Mean Flow Scaling in Turbulent Pipe Flow," PhD Thesis, Princeton University, Princeton, NJ.

Zagarola, M. V., Perry, A. E. and Smits, A. J. (1997), "Log Laws or Power Laws: The Scaling in the Overlap Region," *Physics of Fluids*, Vol. 9, No. 7, pp. 2094-2100.

Zeierman, S. and Wolfshtein, M. (1986), "Turbulent Time Scale for Turbulent Flow Calculations," *AIAA Journal*, Vol. 24, No. 10, pp. 1606-1610.

Zeman, O. (1990), "Dilatational Dissipation: The Concept and Application in Modeling Compressible Mixing Layers," *Physics of Fluids A*, Vol. 2, No. 2, pp. 178-188.

Zeman, O. (1991), "The Role of Pressure-Dilatation Correlation in Rapidly Compressed Turbulence and in Boundary Layers," NASA Ames, Stanford Center for Turbulence Research Annual Research Briefs, p. 105.

Zeman, O. (1993), "A New Model for Super/hypersonic Turbulent Boundary Layers," AIAA Paper 93-0897.

Zhang, H. S., So, R. M. C., Speziale, C. G. and Lai, Y. G. (1993), "Near-Wall Two-Equation Model for Compressible Turbulent Flows," *AIAA Journal*, Vol. 31, pp. 196-199

Zijlema, M., Segal, A. and Wesselinh, P. (1995), "Finite Volume Computation of 2D Incompressible Turbulent Flows in General Coordinates on Staggered Grids," *Int. J. Numer. Methods in Fluids*, Vol. 20, pp. 621-640.

# Index

## A

Algebraic models, 49-91
    Baldwin-Lomax, 23, 76-79, 81-84, 86,
        88-89, 91, 93, 95, 98, 101-102,
        108-109, 113, 115, 166, 179,
        182, 210, 254-255, 266-267,
        422, 434, 437, 476
    Cebeci-Smith, 23, 74-76, 79, 81-86,
        88, 90-93, 95-98, 101-102, 109,
        166, 179, 182, 254-255, 268,
        319, 422, 434, 437, 476
    Prandtl eddy viscosity, 57, 73, 78, 97
    Prandtl mixing length, 2, 21, 23-24,
        49, 53-68, 72, 107-108, 132,
        422-423, 425-426, 428
Algebraic Stress Model, 281-284, 304, 329
Aliasing, 384
Anisotropic:
    dissipation, 286
    eddy viscosity, 282
    liquid, 35
    turbulence, 40, 274, 298-299
Anisotropy, 279, 283, 329, 391
Anisotropy tensor, 286, 290, 299-301
ASM (*see Algebraic Stress Model*)
Asymptotic consistency, 103, 179, 185-193,
        197-200, 309, 313, 315, 322
Asymptotic expansion (*defined*), 408
Asymptotic sequence (*defined*), 408
Attractor:
    Lorenz, 396
    strange, 395-397
    turbulence, 397
Autocorrelation, 42-43, 126, 380, 397
Averaging:
    ensemble, 30-31
    Favre, 229-231
    phase, 33, 46, 320
    Reynolds, 30-34
    spatial, 30-31
    time, 30-31

Axisymmetric flow:
    compressible, 326, 328, 461, 481, 491
    incompressible, 56-57, 65, 79-84, 86,
        88-89, 135-136, 148-149, 179-
        182, 201-203, 210, 312-313, 461

## B

Backscatter, 392
Backward-facing step, 115-117, 210-212,
        281, 292, 323-325, 329, 385,
        387, 394-395
Blood flow, 212-215
Boundary conditions:
    freestream, 140-144, 167-168, 345-353
    "off-the-wall", 393-394
    surface:
        porous-wall, 177-179, 308
        rough-wall, 175-177, 308
        slightly-rough-wall, 177, 343-344
        smooth-wall, 168-171, 188-189,
            306, 343-344
    wall-function, 173-174, 261-262, 305-
        306, 371
Boundary-layer applications:
    compressible:
        flat-plate, 254-255, 318
        nonadiabatic, 258-259, 318
        rough-wall, 258-260
        separated, 260-267, 326-329
        variable-pressure, 255, 258
    incompressible:
        curved-wall, 275-278, 317-318
        flat-plate, 84-85, 93-94, 109-114,
            182-185, 189, 204-205
        mass transfer, 177-179
        rough-wall, 17, 175-177
        separated, 89-91, 95, 102, 115-116,
            210-215, 225
        spinning-cylinder, 319
        transitional, 206-209
        unsteady, 320-322

## T

# Z

Zero-equation models (*see Algebraic models*)

Made in the USA

# Other Publications by DCW Industries

**Basic Fluid Mechanics, Second Edition, D. C. Wilcox (2000):** This book is appropriate for a two-term, junior or senior level undergraduate series of courses, or as an introductory text for graduate students with minimal prior knowledge of fluid mechanics. The first part of the book provides sufficient material for an introductory course, focusing primarily on the control-volume approach. With a combination of dimensional analysis and the control-volume method, the text discusses pipe flow, open-channel flow, elements of turbomachine theory and one-dimensional compressible flows. The balance of the text can be presented in a subsequent course, focusing on the differential equations of fluid mechanics. Topics covered include a rigorous development of the Navier-Stokes equation, potential-flow, exact Navier-Stokes solutions, boundary layers, simple viscous compressible flows, centered expansions and oblique shocks. The book is accompanied by a CD with a variety of useful programs, fluid mechanics photos and movies.

**Study Guide for Basic Fluid Mechanics, Second Edition, C. P. Landry and D. C. Wilcox (1999):** A companion for **Basic Fluid Mechanics** that contains 129 examples worked in complete detail to help readers benefit from step-by-step solutions. All of the problems are explicitly worked start to finish with comments indicating why steps are taken to achieve solutions.

**The Low-Down on Entropy and Interpretive Thermodynamics, S. J. Kline (1999):** This is the final book written by Professor Stephen Kline of Stanford University. It is a delightful treatise on the subtleties of entropy and the second law of thermodynamics.

**Clichés of Liberalism, D. C. Wilcox (1999):** An often humorous and always insightful discussion and analysis of the trite phrases that pass for political discourse in today's America. The book consists of 10 essays focusing on political philosophy, economics and individual liberty.

**Perturbation Methods in the Computer Age, D. C. Wilcox (1995):** Advanced undergraduate or first-year graduate text on asymptotic and perturbation methods. Discusses asymptotic expansion of integrals, including Laplace's method, stationary phase and steepest descent. Introduces the general principles of singular-perturbation theory, including examples for both ODE's and PDE's. Covers multiple-scale analysis, including the method of averaging and the WKB method. Shows, through a collection of practical examples, how useful asymptotics can be when used in conjunction with computational methods.

**Visit our World Wide Web Home Page (http://dcwindustries.com) for complete details about our books, software products and special sales that we conduct from time to time.**